T0297455

Electrochemistry and Corrosion Science

Nestor Perez

Electrochemistry and Corrosion Science

Second Edition

Nestor Perez
University of Puerto Rico
Mayaguez, Puerto Rico

Additional material to this book can be downloaded from http://extras.springer.com

ISBN 978-3-319-24845-5 ISBN 978-3-319-24847-9 (eBook)
DOI 10.1007/978-3-319-24847-9

Library of Congress Control Number: 2016943078

Printed on acid-free paper

This Springer imprint is published by Springer Nature
The registered company is Springer International Publishing AG Switzerland

My wife Neida
My daughters Jennifer and Roxie
My son Christopher

Foreword for the Second Edition

I have used extensively the first edition of this volume, *Electrochemistry and Corrosion Science* by Dr. Perez, in the two courses that I regularly offer at both graduate and senior undergraduate levels. The course on tribology deals with friction, wear, and lubrication in mechanical systems where corrosion is a major problem, and industries spend millions of dollars to minimize corrosion. Myself being a mechanical engineer, Dr. Perez's equations on electrochemical reactions had helped me a lot to explain the physical significance of corrosion, particularly in case of the corrosion of machinery.

In the second edition, Dr. Perez has added new mathematical analyses and their practical examples that help not only as illustrations in a classroom lecture but also in the laboratories for academic research and industrial testing. While most of the textbooks on corrosion emphasize descriptive details, this volume has gone deep into the analytical aspects of chemistry as well as physics of corrosion and their application in industries. Hence, such an addition in the second edition is a "quantum leap" over the previous one. The language is very clear—simple but not simplistic—and it explains some of the very complex phenomena of environmental corrosion in a straightforward fashion.

I have enjoyed reading the final draft of the second edition and shall recommend it as a graduate level and as an advanced undergraduate level textbook in mechanical engineering, chemical engineering, and industrial engineering as well as in chemistry and physics curricula at the universities worldwide. The first edition of this book has already been translated in Mandarin, and I am sure the second edition will follow suit. Besides, it is a good reference guide for the engineering practitioners in industries. For the retired professionals who want to stay in touch as casual consultants, this edition is a worthwhile addition in their personal libraries.

Jay K. Banerjee, Ph.D., P.E., M.Ed.
Professor
Mechanical Engineering Department
University of Puerto Rico at Mayaguez (UPRM)
Mayaguez, Puerto Rico, USA
2016

Preface

This second edition of the book retains all the features of the previous edition while new ones are added. The main work in this edition includes refined text in each chapter, new sections on corrosion of steel-reinforced concrete and on cathodic protection of steel-reinforced bars embedded in concrete, and some new solved examples.

The purpose of this book is to introduce mathematical and engineering approximation schemes for describing the thermodynamics and kinetics of electrochemical systems, which are the essence of corrosion science. The text in each chapter is easy to follow, giving clear definitions and explanations of theoretical concepts and full detail of derivation of formulae. Mathematics is kept simple so that the student does not have an obstacle for understanding the physical meaning of electrochemical processes, as related to the complex subject of corrosion. Hence, understanding and learning the corrosion behavior and metal recovery can be achieved when the principles or theoretical background is succinctly described with the aid of pictures, figures, graphs, and schematic models, followed by derivation of equations to quantify relevant parameters. Eventually, the reader's learning process may be enhanced by deriving mathematical models from principles of physical events followed by concrete examples containing clear concepts and ideas.

Example problems are included to illustrate the ease of application of electrochemical concepts and mathematics for solving complex corrosion problems in an easy and succinct manner.

The book has been written to suit the needs of metallurgical and mechanical engineering senior/graduate students and professional engineers for understanding corrosion science and corrosion engineering. Some mechanical engineering students comply with their particular curriculum requirement without taking a corrosion course, which is essential in their professional careers.

Chapter 1 includes definitions of different corrosion mechanisms that are classified as general corrosion and localized corrosion. A full description and detailed scientific approach of each corrosion mechanism under the above classification is not included since books on this topic are available in the literature.

Chapter 2 is devoted to a brief review of the physics of electrostatics and to the principles of electrochemical systems since they are significantly useful as technological tools for producing or consuming electrical energy in ionic mass transport processes. Standard electric potential for pure metals is also included in this chapter.

Chapter 3 includes an overview of thermodynamics of phases into solution. The concepts of charged particles' (ions') Gibbs free energy change are succinctly described as they relate to electric potential (voltage) difference between the metal electrodes. Pourbaix diagrams are also included in order to present data in a graphical form.

Chapter 4 deals with nanoelectrochemistry and related microscopic techniques, such as scanning tunneling microscopy (STM) for capturing images of a substrate surface at a nanoscale. The concepts of an electrical double layer and quantum tunneling are included.

Chapter 5 is concentrated on a mixed activation polarization and concentration polarization theory. Entire polarization curves are analyzed in order to determine the change in potential of a metal immersed in an electrolyte during oxidation and reduction.

Chapter 6 is devoted to mass transport by diffusion and migration. Fick's laws of diffusion are included for stationary and moving electrode boundaries, and the mass transport is described using total molar flux as per Nernst-Plank equation.

Chapter 7 deals with the degree of corrosivity of electrolytes for dissolving metals and the ability of metals immersed in these electrolytes to passivate or protect from further dissolution. Thus, passivity due to a current flow by external or natural means is studied in this chapter. Images of atom clusters are also included.

Chapter 8 provides schemes for designing against corrosion through cathode protection and anodic protection. This chapter also includes some procedures for protecting large engineering structures, such as underground steel pipelines and reinforced concrete.

Chapter 9 is devoted to electrodeposition of metals. This topic includes principles of electrochemistry which are essential in recovering metals from acid solutions.

Chapter 10 deals with high-temperature corrosion, in which the thermodynamics and kinetics of metal oxidation are included. The Pilling-Bedworth ratio and Wagner's parabolic rate constant theories are defined as related to formation of metal oxide scales, which, in turn, are classified as protective or nonprotective.

The content in this book can be summarized as shown in the flowchart given below. A solution manual is available for educators or teachers upon the consent of the book publisher. Also, all images, pictures, or data taken from reliable sources are included in this book for educational purposes and academic support only. Additional material to this book can be downloaded from http://extras.springer.com.

Mayaguez, Puerto Rico
2016

Nestor Perez

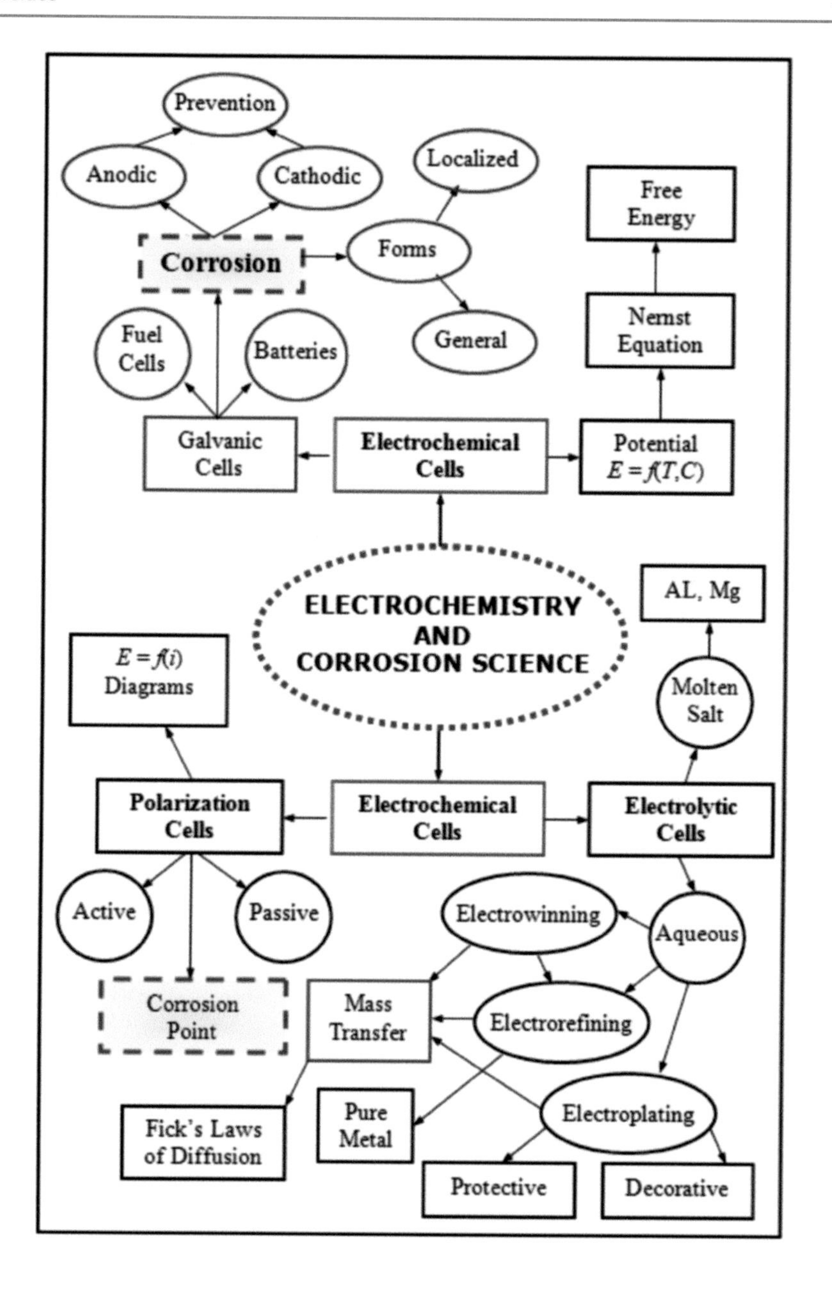
Prevention
Anodic
Cathodic
Localized
Free Energy
Corrosion
Forms
General
Nernst Equation
Fuel Cells
Batteries
Galvanic Cells
Electrochemical Cells
Potential $E = f(T,C)$
ELECTROCHEMISTRY AND CORROSION SCIENCE
AL, Mg
$E = f(i)$ Diagrams
Molten Salt
Polarization Cells
Electrochemical Cells
Electrolytic Cells
Active
Passive
Electrowinning
Aqueous
Corrosion Point
Mass Transfer
Electrorefining
Fick's Laws of Diffusion
Pure Metal
Electroplating
Protective
Decorative

Contents

1 Electrochemical Corrosion

Electrochemistry is a science that studies chemical reactions that involve electron (e^-) transfer at the interface between an electrical conductor (electrode) and an ionic conductor (electrolyte). An electrode is either a metal or semiconductor, and an electrolyte is a solution that conducts electricity due to ionic motion. Nowadays, electrochemistry can be treated as a science of interfaces. If an **electrochemical reaction** occurs on a metal surface, then this leads to metal deterioration or degradation, and the electrochemical process is called **corrosion**, which is represented by an **oxidation reaction**, $M \rightarrow M^{z+} + ze^-$. This oxidation reaction implies that M losses ze^- electrons to the immediate interface. As a result, M becomes a metallic ion (M^{z+}) known as a cation having a positive electrical charge.

In order to study the electrochemical behavior of a metal M, one needs an **electrochemical cell** with at least two electrodes connected through a wire to complete an electrical circuit. The scope of **electrochemical science** in this textbook includes the fundamentals of electrochemistry of metals (M) being exposed to a corrosive electrolyte.

Despite of the early achievements in electrochemistry, the scanning tunneling microscope (STM) and the atomic force microscopy (AFM) techniques provide means to reconstruct and record images of the atomic surface of a substrate (electrode) in an electrochemical cell. Due to lack of space, only the STM technique is described in a later chapter.

Having described the main topics included in this textbook, it is pertinent to introduce the forms of corrosion at a macroscale along with appropriate electrochemical reactions. Thus, **electrochemistry and corrosion science** come together in an organized manner in order to understand, in details, the mechanisms of metal oxidation and metal **reduction reaction** at low and high temperatures. Also, **corrosion engineering**, ionic mass, and electron transport mechanisms are also described.

N. Perez, *Electrochemistry and Corrosion Science*,
DOI 10.1007/978-3-319-24847-9_1

1.1 Forms of Corrosion

This introductory section includes basic definitions related to chemical and electrochemical reactions in the forward (f) and reserve (r) directions. The word **corrosion** stands for material or metal deterioration or surface damage in an aggressive environment. Metallic corrosion is an electrochemical oxidation process, in which the metal transfers electrons to the environment and undergoes a valence change from zero to a positive value z [1]. The environment may be a liquid, gas, or hybrid soil-liquid. These environments are called electrolytes since they have their own conductivity for ionic transfer.

An electrolyte is analogous to a conductive solution, which contains positively and negatively charged ions called cations and anions, respectively. An ion is an atom that has lost or gained one or more outer electron(s) and carries an electrical charge. These partial reactions are classified as anodic and cathodic reactions and are defined below for a metal M immersed in sulfuric acid (H_2SO_4) solution as an example. Hence, metal oxidation is referred to as metallic corrosion, during which the rate of oxidation and the rate of reduction are equal in terms of electron production and consumption [1]. Thus,

$$M \rightarrow M^{z+} + ze^- \qquad \text{(anodic} \equiv \text{oxidation)} \tag{1.1}$$

$$zH^+ + zSO_4^- + ze^- \rightarrow \frac{z}{2}H_2SO_4 \qquad \text{(cathodic} \equiv \text{reduction)} \tag{1.2}$$

$$M + zH^+ + zSO_4^- \rightarrow M^{z+} + \frac{z}{2}H_2SO_4 \quad \text{(overall} \equiv \text{redox)} \tag{1.3}$$

where M = metal $\quad M^{z+}$ = metal cation
H^+ = hydrogen cation $\quad SO_4^-$ = sulfate anion
Z = valence or oxidation state

Thus, **redox** (*red* = reduction and *ox* = oxidation) is the resultant reaction equation, Eq. (1.3), and represents the overall reaction at equilibrium where the anodic and cathodic reaction rates are equal. Observe that the anodic reaction is also referred to as an oxidation reaction since it has lost ze^- electrons, which have been gained by the cathodic reaction for producing sulfuric acid (H_2SO_4). Thus, a cathodic reaction is equivalent to a reduction reaction. The arrows in the above reactions indicate the reaction directions as written, and they represent irreversible processes. On the other hand, a reversible reaction is represented with an equal sign. Thus, the metal reaction can proceed to the right for oxidation or to the left for reduction as indicated by Eq. (1.4):

$$M = M^{z+} + ze^- \tag{1.4}$$

The most common corroded structures due to $M \rightarrow M^{z+} + ze^-$ are steel bridges, iron fences, steel pipes, boat fixtures, steel bars in concrete, bolts that secure a toilet to the floor, copper alloy household tubes, and so on. Plastics are often used to solve corrosion problems of metals; however, plastics do corrode also under certain conditions.

1.2 Classification of Corrosion

There is not a unique classification of the types of corrosion, but the following classification is adapted hereafter.

1.2.1 General Corrosion

This is the case when the exposed metal/alloy surface area is entirely corroded in an environment such as a liquid electrolyte (chemical solution, liquid metal), a gaseous electrolyte (air, CO_2, SO_2^-, etc.), or a hybrid electrolyte (solid and water, biological organisms, etc.). Some types of general corrosion and their description are given below [2]:

Atmospheric corrosion on steel tanks, steel containers, *Zn* parts, *Al* plates, etc.

Galvanic corrosion between dissimilar metals/alloys or microstructural phases (pearlitic steels, α-β copper alloys, α-β lead alloys).

High-temperature corrosion on carburized steels that forms a porous scale of several iron oxide phases.

Liquid-metal corrosion on stainless steel exposed to a sodium chloride ($NaCl$) environment.

Molten-salt corrosion on stainless steels due to molten fluorides (LiF, BeF_2, etc.).

Biological corrosion on steel, *Cu alloys*, *Zn alloys* in seawater.

Stray-current corrosion on a pipeline near a railroad.

1.2.2 Localized Corrosion

This term implies that specific parts of an exposed surface area corrode in a suitable electrolyte. This form of corrosion is more difficult to control than general corrosion. Localized corrosion can be classified as [3]:

Crevice corrosion which is associated with a stagnant electrolyte such as dirt, corrosion product, sand, etc. It occurs on a metal/alloy surface holes, underneath a gasket, on lap joints under bolts, and under rivet heads.

Filiform corrosion is basically a special type of crevice corrosion, which occurs under a protective film. It is common on food and beverage cans being exposed to the atmosphere.

Pitting corrosion is an extremely localized corrosion mechanism that causes destructive pits.

Oral corrosion occurs on dental alloys exposed to saliva.

Biological corrosion due to fouling organisms nonuniformly adhered on steel in marine environments.

Selective leaching corrosion is a metal removal process from the base alloy matrix, such as dezincification (*Zn* is removed) in *Cu-Zn* alloys and graphitization (*Fe* is removed) in cast irons. This type of corrosion is also known as dealloying of solid solution alloys.

1.3 Atmospheric Corrosion

This is a uniform and general attack, in which the entire metal surface area exposed to the corrosive environment is converted into its oxide form, provided that the metallic material has a uniform microstructure.

Aqueous corrosions of iron (Fe) in H_2SO_4 solution and of Zn in diluted H_2SO_4 solution are examples of uniform attack since Fe and Zn can dissolve (oxidize) at uniform rates according to the following anodic and cathodic reactions, respectively. Thus,

$$Fe \rightarrow Fe^{2+} + 2e^{-} \tag{1.5}$$

$$2H^{+} + 2e^{-} \rightarrow H_2 \uparrow \tag{1.6}$$

$$Fe + 2H^{+} \rightarrow Fe^{+2} + H_2 \uparrow \tag{1.7}$$

and

$$Zn \rightarrow Zn^{2+} + 2e^{-} \tag{1.8}$$

$$2H^{+} + 2e^{-} \rightarrow H_2 \uparrow \tag{1.9}$$

$$Zn + 2H^{+} \rightarrow Zn^{2+} + H_2 \uparrow \tag{1.10}$$

Here, $H_2 \uparrow$ is a hydrogen gas and SO_4^{-} are ions that do not participate in the reactions. The cathodic reaction is the common hydrogen evolution process. In fact, the aggressiveness of a solution to cause a metal to oxidize can be altered by additions of water, which is an amphoteric compound because it can act as an acid or base due to its dissociation as indicated below:

$$H_2O \rightarrow H^{+} + OH^{-} \tag{1.11}$$

Atmospheric corrosion of a steel structure is also a common example of uniform corrosion, which is manifested as a brown-color corrosion layer on the exposed steel surface. This layer is a ferric hydroxide compound known as **rust**. The formation of **brown rust** is as follows [1,4]:

$$(Fe \rightarrow Fe^{2+} + 2e^{-})(x2) \tag{1.12}$$

$$O_2 + 2H_2O + 4e^{-} \rightarrow 4OH^{-} \tag{1.13}$$

$$2Fe + O_2 + 2H_2O \rightarrow 2Fe^{2+} + 4OH^{-} \rightarrow 2Fe(OH)_2 \downarrow \tag{1.14}$$

$$2Fe(OH)_2 + \frac{1}{2}O_2 + H_2O \rightarrow 2Fe(OH)_3 \downarrow \equiv Fe_2O_3 \cdot 3H_2O \tag{1.15}$$

where $(x2)$ = multiplying factor for balancing the number of electrons
$2Fe(OH)_2$ = ferrous hydroxide (unstable compound)

$2Fe(OH)_3$ = ferric hydroxide (with Fe^{3+} cations)
$Fe_2O_3 \cdot 3H_2O$ = hydrated ferric hydroxide

In addition, *Zn* can uniformly corrode forming a white rust according to the following reactions [5–7]:

$$(Zn \rightarrow Zn^{2+} + 2e^{-})x2 \tag{1.16}$$

$$O_2 + 2H_2O + 4e^{-} \rightarrow 4OH^{-} \tag{1.17}$$

$$2Zn + O_2 + 2H_2O \rightarrow 2Zn^{2+} + 4OH^{-} \tag{1.18}$$

$$2Zn^{2+} + 4OH^{-} \rightarrow 2Zn(OH)_2 \tag{1.19}$$

$$2Zn(OH)_2 + CO_2 + O_2 + H_2O \rightarrow Zn_4CO_3 \cdot (OH)_6 \tag{1.20}$$

The compound $Zn_4CO_3 \cdot (OH)_6$ or $ZnCO_3 \cdot 3Zn(OH)_2$ is a zinc carbonate or white rust or wet-storage stain (porous). Atmospheric corrosion of aluminum is due to a passive oxide film formation instead of a porous layer. The gray-/black-color film may form as follows:

$$(Al \rightarrow Al^{3+} + 3e^{-})x2 \tag{1.21}$$

$$\frac{3}{2}O_2 + 3H_2O + 6e^{-} \rightarrow 6OH^{-} \tag{1.22}$$

$$2Al + \frac{3}{2}O_2 + 3H_2O \rightarrow 2Al^{3+} + 6OH^{-} \rightarrow Al_2O_3 \cdot 3H_2O \tag{1.23}$$

In general, the oxidation process can be deduced using a proper Pourbaix diagram, as schematically shown in Fig. 1.1. This diagram is a plot of electric potential of a metal as a function of *pH* of water at 25 °C [8].

This type of diagram indicates the possible electrochemical process on a metal surface if the potential and the pH of the electrochemical systems are known or estimated. In fact, corrosion rates cannot be determined from a Pourbaix diagrams. The diagram includes regions identified as corrosion where a metal oxidizes, passive region where a metal is protected by a stable oxide film being adhered on the metal surface, and immunity where corrosion or passivation is suppressed.

Furthermore, the **prevention of uniform corrosion** can be accomplished by selecting an adequate (1) material having a uniform microstructure; (2) coating or paint; (3) inhibitor(s) for retarding or suppressing corrosion, which are classified as adsorption-type hydrogen evolution poisons, scavengers, oxidizers, and vapor-phase inhibitors[1]; and (4) cathodic protection, which is an electrochemical process for suppressing corrosion in large steel structures. Figures 1.2 and 1.3 show atmospheric uniform corrosion on typical structures. Both the steel bridge structures and the pipeline were exposed to air by the ocean. Notice how the steel structures were subjected to chemical reactions, which proceeded uniformly over the exposed metal surface area.

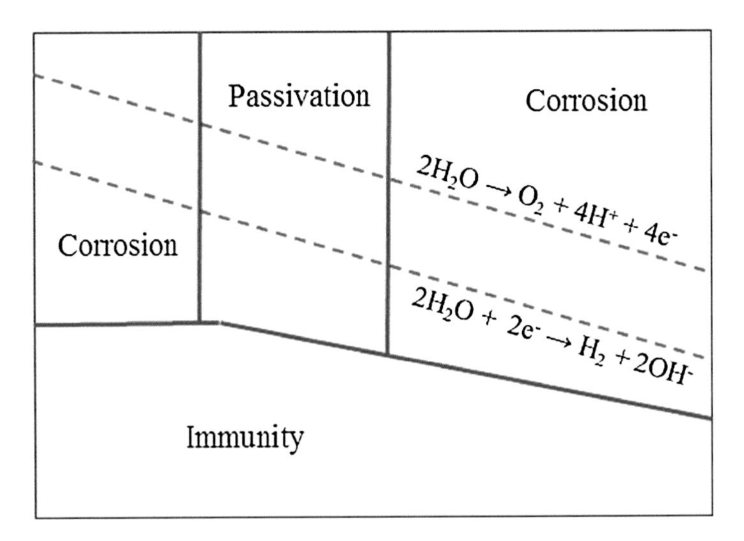

Fig. 1.1 Schematic Pourbaix diagram for a hypothetical metal

Fig. 1.2 Uniform corrosion (atmospheric attack) of a steel bridge

1.4 Galvanic Corrosion

Galvanic corrosion is either a chemical or an electrochemical corrosion. The latter is due to a potential difference between two different metals connected through a circuit for current flow to occur from the more active metal (more negative potential) to the more noble metal (more positive potential).

Galvanic coupling is a galvanic cell in which the anode is the less corrosion-resistant metal than the cathode. Figure 1.4 shows atmospheric galvanic corrosion

Fig. 1.3 Uniform corrosion (atmospheric attack) of a pipeline located on a concrete pier above the ocean water

Fig. 1.4 Galvanic corrosion of a (**a**) steel alloy chain and (**b**) carbon steel bolt-nut coupling

of a steel alloy chain and a bolt-nut holding a coated steel plate. Both the corroded bolt-nut and the steel box are the anodes having very small surface areas, while the coated steel plate and the steel post have very large cathode surface areas. The corrosion rate can be defined in terms of current density, such as $i = I/A$ where I is the current and A is the surface area. Therefore, the smaller A the larger i. This is an area effect on galvanic coupling. Thus, the driving force for corrosion or current flow is the potential (voltage) E between the anode and cathode. Subsequently, Ohm's law, $E = IR = iAR$, is applicable. Here, R is the galvanic cell resistance.

In addition, galvanic corrosion can be predicted by using the electromotive force (emf) or standard potential series for metal reduction listed in Table 2.1. These reactions are reversible. The standard metal potential is measured against the standard hydrogen electrode (SHE), which is a reference electrode having an arbitrary standard potential equal to zero. Details on types of reference electrodes are included in Chap. 2.

In selecting two metals or two alloys for a galvanic coupling, both metals should have similar potentials or be close to each other in the series in order to suppress

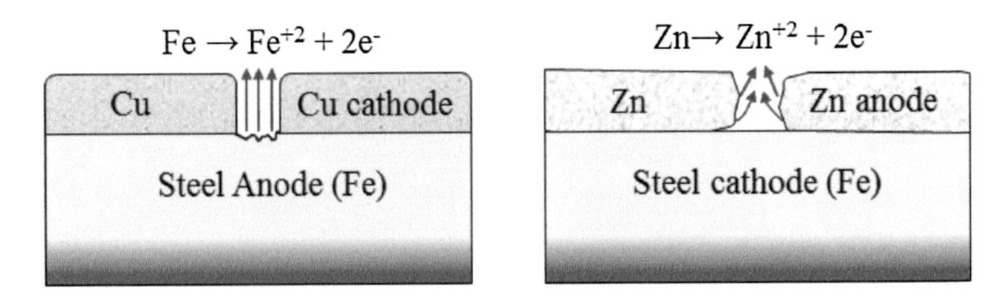

Fig. 1.5 Schematic galvanic couplings

galvanic corrosion. For example, $Fe - Cr$ or $Cu - Sn$ (bronze) couplings develop a very small potential differences since they are close to each other in their respective standard potential series. The given data in Table 2.1 is very appealing in designing against galvanic corrosion of pure metals. The closer the standard potentials of two metals, the weaker the galvanic effect; otherwise, the galvanic effect is enhanced.

Eventually, galvanic coupling can be used for cathodic protection purposes. In fact, in coupling two different metals, the metal with the lowest standard potential acts as the anode, and its standard potential sign is changed. Figure 1.5 shows two galvanic coupling cases, in which copper and zinc can be in the form of sheets or electroplated coatings [1]. Recall that iron (*Fe*) is the base metal for steel; therefore, *Fe* is to be protected against corrosion. Therefore, *Fe* is the anode for *Cu* and the cathode for *Zn* couplings. In the latter case, *Zn* becomes a sacrificial anode, which is the principle of coupling for galvanized steel sheets and pipes. On the other hand, if *Cu* coating breaks down, steel is then exposed to an electrolyte and becomes the anode, and therefore, it oxidizes.

Other types of galvanic coupling are batteries and fuel cells. Both are electrochemical power sources in which chemical energy is converted into electrochemical energy through controlled redox electrochemical reactions [9]. Subsequently, these electrochemical devices represent the beneficial application of galvanic corrosion. Among the reactions that occur in batteries, high hydrogen evolution is desirable. References [1–17] include details on several types of batteries and fuel cells. A detailed analysis of galvanic cells will be dealt with in Chap. 2.

1.4.1 Microstructural Effects

A mechanically deformed metal or alloy can experience galvanic corrosion due to differences in atomic plane distortion and a high dislocation density. In general, dislocations are line defects in crystals. Figure 1.6a shows a mechanically worked steel nail indicating localized anodes. The tip and the head of the nail act as stress cells for oxidation of iron to take place, provided that the nail is exposed to an aggressive environment. These two parts of the nail are examples of strain hardening, but are susceptible to corrode galvanically due to the localized crystal defects and the presence of mainly compressive residual stresses induced by the

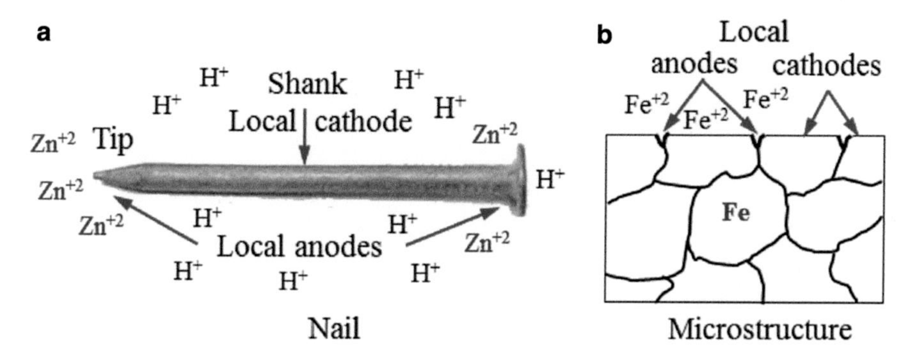

Fig. 1.6 Localized galvanic cells. (**a**) Stress cells and (**b**) microcells

mechanical deformation process. Furthermore, the nail shank acts as the cathode, and the tip/head shanks form galvanic cells in a corrosive environment.

In addition, improper heat treatment can cause nonuniform microstructure, and therefore, galvanic-phase corrosion is enhanced in corrosive media. In a crystalline metal, galvanic coupling can occur between grains and grain boundaries. Figure 1.6b shows a schematic microstructure of a metal subjected to corrosion along the grain boundaries, as in the case of a typical polished and etched microstructure. This type of corrosion can be referred to as grain-boundary corrosion because the grain boundaries act as anodes due to their atomic mismatch and possible segregation of impurities.

Galvanic corrosion can occur in a polycrystalline alloys, such as pearlitic steels, due to differences in microstructural phases. This leads to galvanic-phase coupling or galvanic microcells between ferrite (α-Fe) and cementite (Fe_3C) since each phase has different electrode potentials and atomic structure. Therefore, distinct localized anodic and cathodic microstructural areas develop due to microstructural inhomogeneities, which act as micro-electrochemical cells in the presence of a corrosive medium (electrolyte). This is an electrochemical action known as galvanic corrosion, which is mainly a metallic surface deterioration.

This form of corrosion is not always detrimental or fatal to metals. For instance, revealing the microstructure of pearlitic steels with a mild acid can be accomplished due to the formation of galvanic microcells. In this case, pearlite consists of ferrite and cementite, and when it is etched with a mild acid, which is the electrolyte, galvanic microcells between ferrite (*cathode*) and cementite (*anode*) are generated. Consequently, pearlite is revealed as dark cementite and white ferrite.

In addition, if a zinc (Zn) is immersed in hydrochloric (HCl) acid (reagent) at room temperature, then it spontaneously reacts in this strong corrosive environment. Figure 1.7 shows a galvanized steel nail that was immersed in such a solution.

Notice that the nail is covered with hydrogen bubbles. This is an example of hydrogen evolution that occurs in acid solutions. Thus, HCl acid solution acts as an oxidizer since chloride (Cl^-) ions are not involved in the reaction [1] and consequently, rate of corrosion of Zn is increased very rapidly. The electrochemical

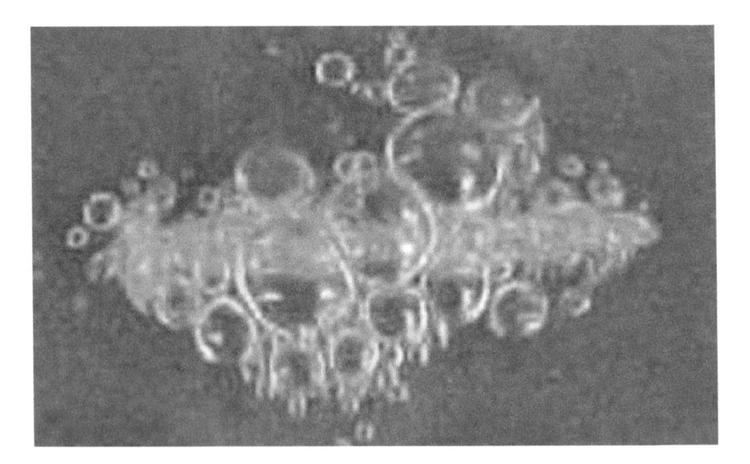

Fig. 1.7 A galvanized steel nail reacting in concentrated HCl acid solution

reactions for the case shown in Fig. 1.7 are similar to the reactions given by Eqs. (1.8) through (1.10). Thus, the galvanized steel nails are

$$Zn \longrightarrow Zn^{2+} + 2e^- \quad (1.24)$$

$$HCl \longrightarrow H^+ + Cl^- \quad (1.25)$$

$$2H^+ + 2e^- \longrightarrow H_2 \uparrow \quad (1.26)$$

$$Zn + HCl \longrightarrow Zn^{2+} + H_2 \uparrow + Cl^- \quad (1.27)$$

Furthermore, solid surfaces, such as automobile underbody parts, in contact with a mixture of mud, soil, and salt can deteriorate due to galvanic corrosion. In this case, the mixture is a stagnant electrolyte that causes the least galvanic action when compared with agitated electrolytes. In fact, agitation and temperature gradients can accelerate the galvanic action due to a higher current density, and consequently, galvanic corrosion is manifested as metal dissolution. If the corroding metal part is under the influence of a tensile stress, then it may become weak and may fail due to reduction in cross-sectional area.

In light of the above, in general, corrosion can be defined as the degradation or deterioration of materials by chemical and electrochemical processes. The former processes are used in this book for characterizing the corrosion science of metals and for implementing the fundamental concepts of electrochemical corrosion in order to protect metallic structures against corrosion. On the other hand, metal recovery from solution is also an interesting engineering topic that will be addressed in a later chapter.

In essence, electrochemistry and corrosion science are the main topics used in this book for characterizing metal oxidation known as corrosion, which, in turn, is undesirable in most engineering applications, and for reducing metal ions as an electrodeposition technique. Nevertheless, the fundamental concepts of electrochemical corrosion can be used to minimize or eliminate, to an extent, metal oxidation. Thus, protecting metals and alloys against the complexity of the corrosion phenomena dictates the use of the fundamental principles that govern corrosion in aqueous solutions and in other environments. This can be accomplished by implementing techniques known as cathodic protection, corrosion inhibition, coating, and so on.

1.4.1.1 Atomic Arrangement and Dislocations

Furthermore, the microstructure of crystalline solids is composed of grains, which are surrounded by grain boundaries. A single grain is composed of a regular and repeated array of atoms, which, in turn, form the atomic structure. The most common crystal lattices in engineering materials are the body-centered cubic (BCC), face-centered cubic (FCC), and hexagonal close packed (HCP). For instance, Fig. 1.8 shows the BCC crystal structure encountered in engineering metallic materials, such as chromium, iron, carbon steels, molybdenum, and the like.

On the other hand, brass and the 300-series stainless steels have an FCC crystal structure. Other crystal structures can be found in any book on physical metallurgy and material science. The close-packed spheres in Fig. 1.8 represent an atomic arrangement having unit cells, which repeat themselves forming the lattice crystal structure. Each atom is bounded to its neighbors, and each atom has its nucleus surrounded by electrons. The outer atoms forming the electrode surface exposed to a corrosive medium become electron deficient and are detached from the lattice and form part of the medium, such as an aqueous electrolyte, or react with atoms from the medium to form a surface corrosion product. The corrosion rate in terms of current density or penetration per time is the kinetic parameter that must be

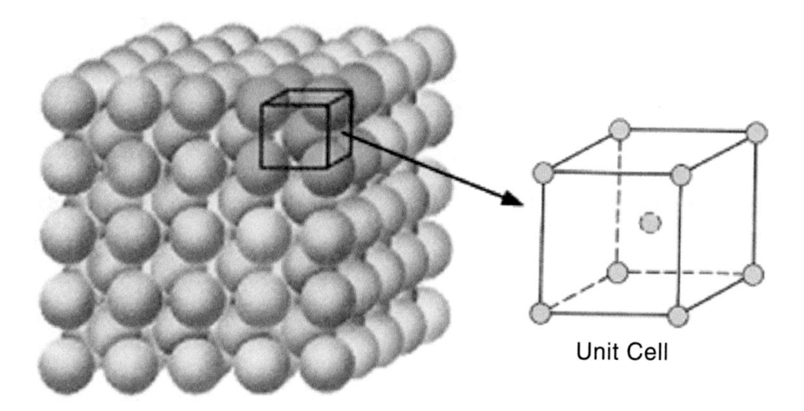

Fig. 1.8 Close-packed body-centered cubic structure [17]

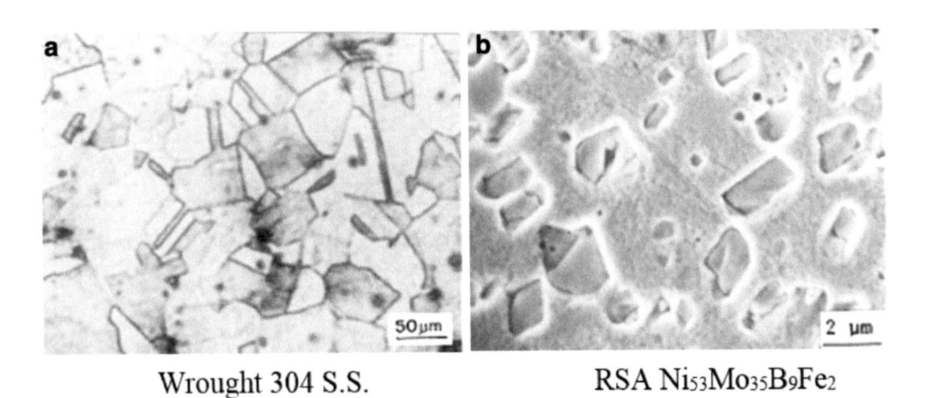

Fig. 1.9 Microstructure of annealed (**a**) AISI 304 at 200× and (**b**) rapidly solidified alloy (RSA) $Ni_{53}Mo_{35}B_9Fe_2$ at 5000× [18]

determined experimentally. Chapter 3 includes details on how to determined this parameter.

The grain boundaries in crystalline solids represent high-energy areas due to the atomic mismatch, and therefore, they are considered microstructural defects, which corrode more rapidly than the grain surfaces. Figure 1.9 illustrates two microstructures of AISI 304 stainless steel being annealed at 1000 °C for 0.5 h and 24 h at 1100 °C for a rapidly solidified and consolidated $Ni_{53}Mo_{35}B_9Fe_2$ alloy (RSA).

Figure 1.9a shows the grain boundaries as dark lines because of the severe chemical attack using an aqua regia etching solution (80 $\%HCl + 20\ \%HNO_3$). Figure 1.9b shows crystalline particles embedded in an *Ni-Mo* matrix after being etched with Marble's reagent. Denote that the RSA does not have visible grain boundaries, but it is clear that the severe chemical attack occurred along the matrix-particle interfaces due to localized galvanic cells [18]. These interfaces appear as bright areas due to optical effects. It suffices to say that one has to examine the surface microstructure and its topology for a suitable microstructural interpretation. Thus, the fundamental understanding of corrosion mechanism on crystalline and amorphous alloys is enhanced when the microstructural interpretation is satisfactorily accomplished. Corrosion mechanisms, for example, must be analyzed, preferably at a nanoscale, by separating the relative importance of microstructural effects on critical events such as stress corrosion cracking (*SCC*) initiation, subsequent growth, and consequent unstable fracture.

Another metallurgical aspect to consider is the dislocation network encounter in plastically deformed alloys. In general, dislocations are linear defect, which can act as high-energy lines, and consequently, they are susceptible to corrode as rapidly as grain boundaries in a corrosive medium. Figure 1.10 illustrates dislocation networks in an AISI 304 stainless steel and in RSA $Ni_{53}Mo_{35}B_9Fe_2$. The relevant pretreatment conditions can be found elsewhere [18].

With respect to Fig. 1.10b, there is a clear grain boundary shown as a dark horizontal line across the upper part of the TEM photomicrograph. The small white

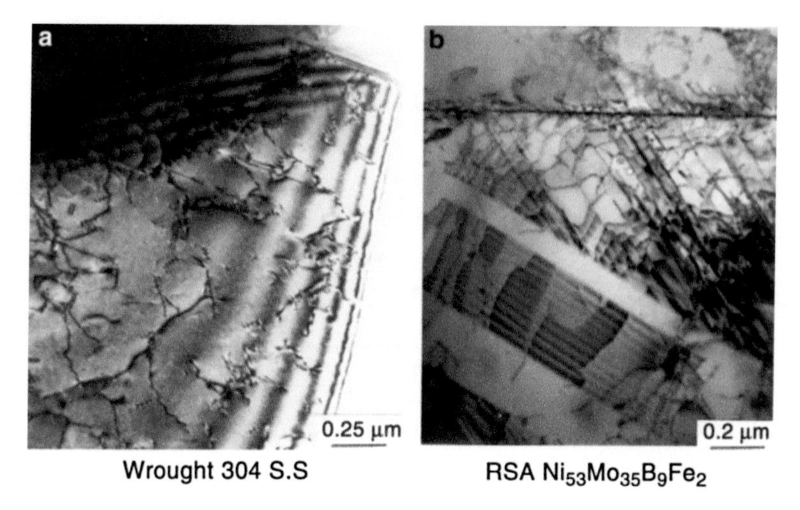

Fig. 1.10 Bright-field TEM photomicrographs showing dislocation networks in a (**a**) conventional and (**b**) rapidly solidified stainless steel (SS) alloy [18]

areas surrounded by dislocations are called sub-grains, which are crystal having an FCC structure for both alloys.

1.5 Pitting Corrosion

Pitting corrosion refers to a localized area in the form of a surface cavity or hole that may take different geometric shapes. During corrosion of certain materials, localized surface areas act as the anodes, while the surrounding materials become the cathodes. As a result, the localized anodes dissolve leaving cavities known as pits, which can act as stress risers in stress corrosion cracking (SCC) and fatigue testing. In principle, surface defects can be the source of stress concentration during quasi-static and dynamic testing, leading to cracking and subsequent premature failure in a particular environment.

This form of corrosion is extremely localized, and it manifests itself as holes on a metal surface. The initial formation of pits is difficult to detect due to the small size, but it requires a prolonged time for visual detection. Figure 1.11 shows a scanning electron microscope (SEM) photomicrograph of a 2195 Al-Li alloy (Weldalite 049 trade name) containing pits with an average diameter of approximately 4 μm. Also included in Fig. 1.11 is the model for pitting mechanism. Pitting corrosion may occur due to breakdown of a protective film (passive oxide film or organic coating). This form of corrosion can be found on aluminum and its alloys and automobile chromium-plated bumpers or body-coated (painted) parts due to film/coating breakdown at isolated surface sites. Pits vary in shape, but are very small surface holes due to the extremely localized anodic reaction sites.

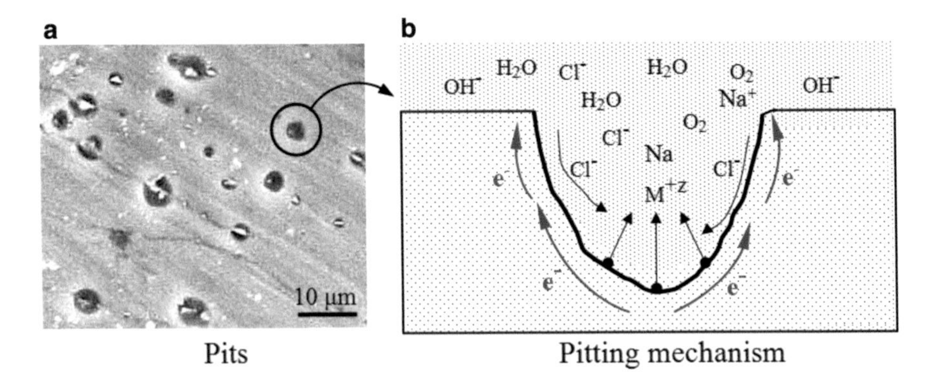

Fig. 1.11 (**a**) Localized corrosion on 2195 Al-Li alloy after a potentiodynamic polarization run in 3.5 % NaCl. The alloy was aged at 190 °C for 0.5 h. SEM photomicrograph taken by Laura Baca at a magnification of 2000×. (**b**) Model for pitting mechanism

The appearance of pits on a metal surface is not very appealing, but they can be harmless if perforation does not occur. The initiation of pits occurs at localized sites on a metal surface defects, which may be due to coating failure, mechanical discontinuities, or microstructural phase heterogeneities such as secondary phases. Besides the prolonged time needed for pit formation or pit growth, it is assumed that many anodic and cathodic reactions take place at localized sites. Both rates of anodic and cathodic reactions are slow; however, the reactions continue inward in the direction of gravity in most cases. This suggests that the bottom of pits are rich in metal M^{z+} ions due to the large number of anodic reactions.

In a water-based electrolyte containing chlorine Cl^- ions and oxygen molecules (O_2), the Cl^- ions migrate toward the bottom of the pits, and O_2 molecules react with water molecules on the metal surface [1]. Therefore, metal chloride $M^{z+}Cl^-$ and hydroxyl ions $(OH)^-$ are produced. This is an oxidation process known as metal dissolution. Prior to the formation of $M^{z+}Cl^-$, aqueous compound is produced; the initial governing reactions are as follows:

$$M \longrightarrow M^{z+} + ze^- \tag{1.1}$$

$$O_2 + 2H_2O + ze^- \longrightarrow 4(OH)^- \tag{1.28}$$

$$M^{z+} + Cl^- \longrightarrow M^{z+}Cl^- \tag{1.29}$$

Subsequently, $M^{z+}Cl^-$ is hydrolyzed by water molecules. Hence,

$$2M^{z+}Cl^- + O_2 + 2H_2O \longrightarrow 2M(OH)_2 + 2H^+Cl^- \tag{1.30}$$

where H^+Cl^- is the free hydrochloric acid that forms at the bottom of the pits increasing the acidity at these locations. This implies that the hydrogen ion concentration $\left[H^+\right]$ in $mol/liter$ is increased and the degree of acidity can be defined by

$$pH = -\log\left[H^{+}\right] \tag{1.31}$$

The metal hydroxide $2M(OH)_2$ compound is unstable, and therefore, it reacts with oxygen and water to form the final corrosion product. Hence,

$$2M(OH)_2 + \frac{1}{2}O_2 + H_2O \longrightarrow 2M(OH)_3 \tag{1.32}$$

Typical examples of specific formation of $M(OH)_3$-type corrosion product are given below:

$$M(OH)_3 = \begin{cases} Fe(OH)_3 \text{ for steel pits (rust)} \\ Al(OH)_3 \text{ for aluminum hydroxide pits} \\ Cr(OH)_3 \text{ for chromium hydroxide pits} \end{cases} \tag{1.33}$$

Should pitting occur, a heterogeneous mechanism for metal dissolution defines a localized attack that may involve metal penetration in thin structural sections. In the case of massive structural sections, pitting is usually of little significance. On the other hand, surface fatigue failures due to pitting mechanism are well documented in the literature. Hence, pitting develops with surface microcracks.

Furthermore, pitting depth (d) can be defined by the following empirical equation:

$$d = \lambda t^n \tag{1.34}$$

where λ, n = constants
t = time

1.6 Crevice Corrosion

Crevice corrosion is also a form of localized corrosion. Crevice cavities can form under surface deposits, gaskets, fastener heads, washers, and the like. A suitable explanation for this type of corrosion is that local anode surface areas undergo changes in local chemistry (depletion of oxygen, buildup of aggressive chloride ionic content, etc.) and therefore, irregular cavities or pitlike cavities form.

Relatively low-temperature electrochemical oxidation of a metal may occur as a sequence of localized anodic reactions according to Eq. (1.1) in a sheltered crevice surface containing a stagnant electrolyte (water, grease-sand mixture, or other insoluble substances). Figure 1.12 shows a schematic rivet-plate joint, in which the sheltered metal surface (joint inner surface) was in contact with a suitable electrolyte. The mechanism of crevice corrosion is electrochemical in nature, and

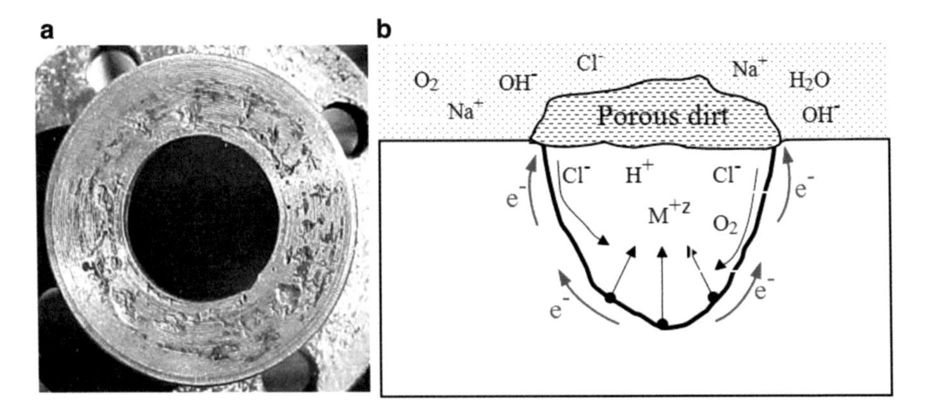

Fig. 1.12 Crevice corrosion [1]. (**a**) Stainless steel flange. (**b**) Crevice mechanism

it is also illustrated in Fig. 1.12. It requires a prolonged time to start the metal oxidation process, but it may be accelerated afterward [1].

Crevice corrosion is similar to pitting corrosion after its initiation stage in a stagnant electrolyte. This form of corrosion initiates due to changes in local chemistry such as depletion of oxygen in the crevice, increase in pH with increasing hydrogen concentration $[H^+]$, and increase of chlorine Cl^- ions. Oxygen depletion implies that cathodic reaction for oxygen reduction cannot be sustained within the crevice area and consequently, metal dissolution occurs. The problem of crevice corrosion can be eliminated or reduced using proper sealants and protective coatings

This is a case for illustrating the effects of nonmetallic materials (gasket, rubber, concrete, wood, plastic, and the like) in contact with a surface metal or alloy exposed to an electrolyte (stagnant water). For instance, as Fontana [1] pointed out, crevice attack can cut a stainless steel sheet by placing a stretched rubber band around it in seawater. Thus, metal dissolution occurs in the area of contact between the alloy and rubber band.

1.7 Spalling-Induced Corrosion

Spalling-induced corrosion can also be defined as a localized corrosion due to surface deterioration of a material surface area, exposing the base metal to a local environment. It can also occur in reinforced concrete due to cracking of the concrete cover.

Figure 1.13a shows a spalling-induced corrosion of a steel frame, which was initially protected by an organic coating (paint). Notice that spalling is a separation of the surface coating. This particular case is another atmospheric-related corrosion phenomenon. Spalling can also occur on metal oxides and refractory materials due to thermal cycling.

Spalling is a unique defect that represents local disruption of the original protective coating. Figure 1.13a illustrates a severe case of spalling since the

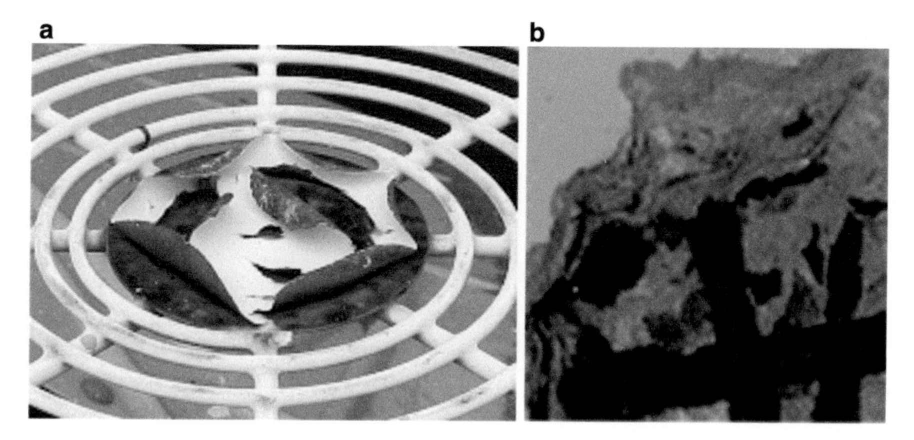

Fig. 1.13 Corrosion-induced spalling of an organic coating and reinforced concrete. (**a**) Paint spalling. (**b**) Concrete spalling

organic coating became detached over the central area of the steel structure. This type of defect normally takes a prolonged time to manifest its deleterious effects. Thus, corrosion-induced spalling may be attributable to the generation of molecular hydrogen, which is known as hydrogen evolution (H_2) beneath the organic coating. The pertinent reaction for hydrogen evolution is given by Eq. (1.9).

Furthermore, Fig. 1.13b depicts a severe case of spalling-induced corrosion of a house ceiling made of reinforced concrete. Denote the black color of the steel rebars which is an indication of a prolonged rust formation mechanism. Apparently, lack of proper maintenance of the house roof allowed, at least, chloride (Cl^-) ions to diffuse toward the steel rebars, and as a result, the structural steel corroded in a water-containing chloride ion solution. Nonetheless, spalling is also a common defect in concrete pavements that may become hazardous to roadway users. It occurs due to high compressive stresses in the concrete when cracks and joints are not properly closed or repaired. In fact, spalling tends to grow under repeated thermal stresses caused by traffic loadings, while concrete bridges may fail due to spalling and cracking.

1.8 Stress Corrosion Cracking

Structural parts subjected to a combination of a tensile stress and a corrosive environment may prematurely fail at a stress below the yield strength. This phenomenon is known as environmentally induced cracking (EIC), which is divided into the following categories: stress corrosion cracking (SCC), hydrogen-induced cracking (HIC), and corrosion-fatigue cracking (CFC). These three categories can develop under the influence of an applied potential related to polarization diagram. The former is extensively discussed in Chaps. 3 and 4. The EIC phenomenon has been studied for decades, but more research needs to be done in order to have a

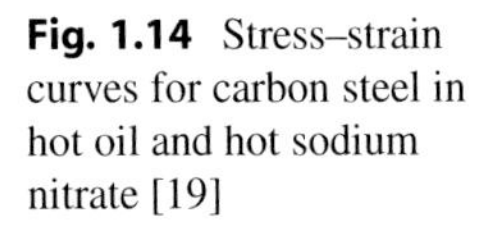

Fig. 1.14 Stress–strain curves for carbon steel in hot oil and hot sodium nitrate [19]

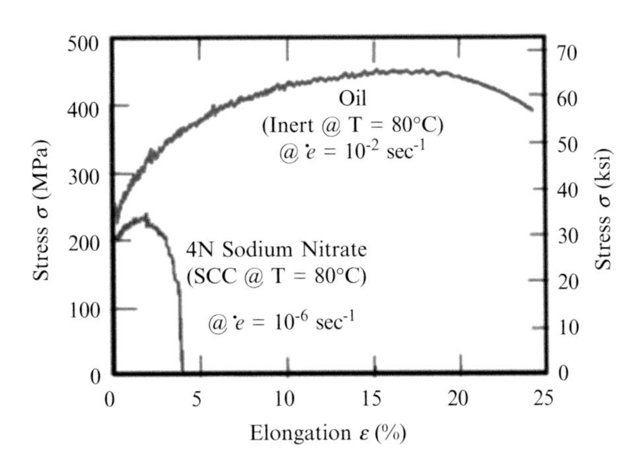

better understanding of corrosion. Since literature is abundant on this subject, it is convenient to include in this section a brief discussion on SCC and HIC.

It is known that SCC occurs under slow strain rate (SSR) and constant load (CL) only if the tensile strain rate or the applied potential is within a narrow and, yet, critical range; otherwise metals and alloys would appear to be immune to *SCC* either due to film repair at low strain rates or mechanical failure at high strain rates [19]. Figure 1.14 shows Parkins' classical stress–strain curves [20] for assessing SCC susceptibility of carbon steel in two different environments at relatively high temperature using the SSR technique. Obviously, the specimen exhibited SCC in hot sodium nitrate since the stress–strain curve shows approximately 4 % total elongation as compared to nearly 25 % elongation in the inert environment.

In order to assess SCC in details, there must exist a lower and an upper bound of strain rate at a constant applied potential and a potential range at constant strain rate for ductile materials. This is schematically illustrated in Fig. 1.15 after Kim and Wilde [21]. In this figure, the dashed rectangular shape indicated the range of strain rate and range of potential for activation of SCC. The critical SCC state is illustrated as the minimum ductility. Also indicated in Fig. 1.15 is the HIC continuous curve for brittle materials.

The detrimental effects of atomic hydrogen (H) diffusion on mechanical properties are schematically shown in Fig. 1.15. This particular mechanism is also known in the literature as hydrogen embrittlement (HE). If hydrogen atoms within the lattice defects, such as voids, react to form molecular hydrogen and these molecules form babbles, then the metallurgical damage is called blistering.

Experimental verification of Kim and Wilde's [21] SCC curve is shown in Fig. 1.16 for a rapidly solidified alloy (RSA) and ingot metallurgy (IM) AISI 304 stainless steels under tension testing using smooth round specimens in 0.10 N H_2SO_4 solution at room temperature [18].

These steels have the same chemical composition, but they were produced using different technologies. Their initial microstructural conditions were cold-rolled conditions. Ductility is characterized by a reduction in cross-sectional area

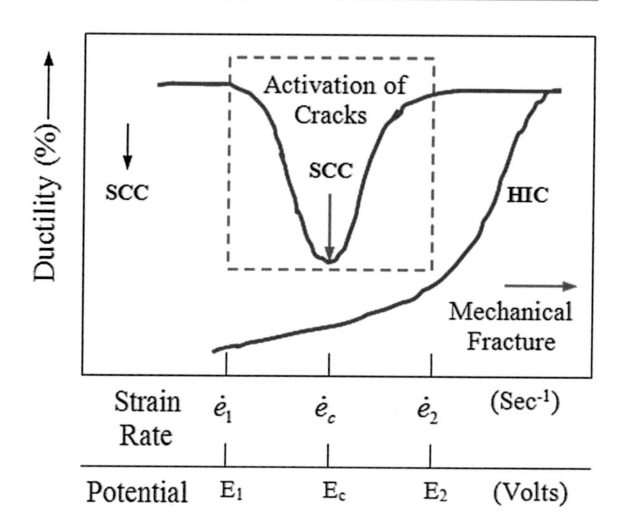

Fig. 1.15 Schematic effects of strain rate and potential on ductility using a slow strain-rate testing method [21]

(*%RA*). The IM 304 steel exhibited maximum SCC at zero potential, while the RSA 304 apparently was degraded due to HIC since RA decreases continuously as the potential decreases.

Figure 1.16 depicts SCC and polarization experimental data indicating the SCC mechanisms shown in Fig. 1.15. Hence, the effect of anodic polarization on stress corrosion cracking of RSA and IM 304 is clear. Therefore, the IM 304 alloy is susceptible to SCC, while the RSA 304 is to HIC. This difference is attributed to different solidification techniques used to produce these alloys.

Figure 1.17 shows typical secondary cracks on the specimen gage length [18]. Only half of the fractured specimen is shown since the other half exhibited similar cracking morphology.

These cracks are typically developed in ductile materials susceptible to *SCC*. Therefore, the combination of an applied stress and applied potential in a corrosive environment degrades the material mechanical properties, specifically due to the deterioration of the specimen surface at weak areas, such as microstructural and machining defects.

1.9 Nonmetallic Materials

Ceramics. These are brittle and corrosion-resistant compounds made out of metallic and nonmetallic elements. Some examples of ceramics are Al_2O_3 (alumina), SiC (silicon carbide), MgO (magnesia), Fe_3O_4 (magnetite), and ZrO (zirconia). Other ceramics are made of basic ceramics and are known as bricks, clay, concrete, porcelain, and the like. On the other hand, **refractories** are ceramics that withstand very high temperatures prior to melting, such as *NbC* (niobium carbide) @ 3615 °C and *MgO* @ 2852 °C. In addition, ceramics are immune to corrosion by almost all environments. Those which are not dissolve by chemical oxidation.

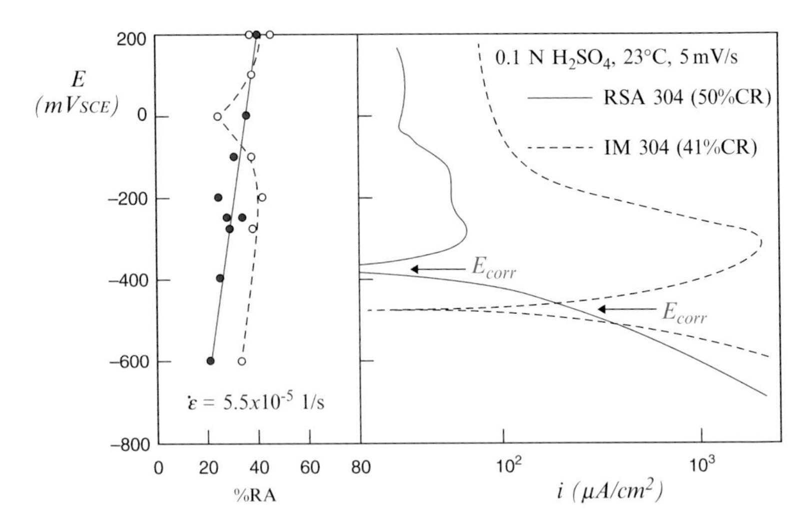

Fig. 1.16 Influence of slow strain rate and applied potential on the ductility of 41 % cold-rolled and annealed AISI 304 at 1000 °C for a 24-h heat treatment. The applied strain rate was $5.5 \times 10^{-5}\,s^{-1}$ [18]

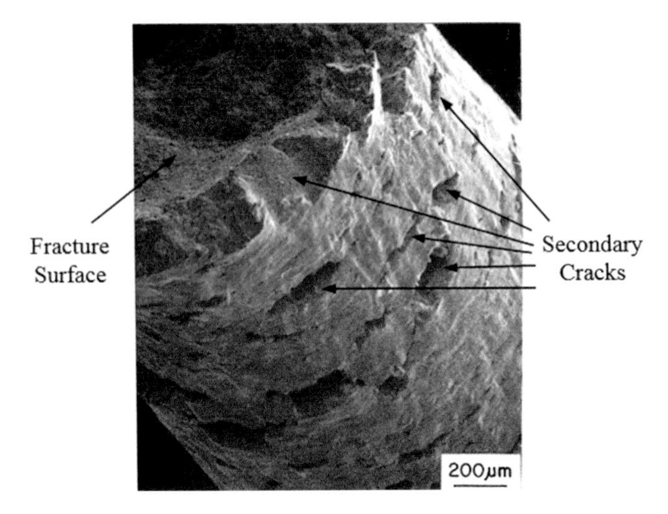

Fig. 1.17 SEM fractograph of annealed AISI 304 stainless steel at 1000 °C for a 24-h heat treatment. The specimen was tension tested in 0.10 N H_2SO_4 at 23 °C, $5.5 \times 10^{-5}\,s^{-1}$, and $E = 0\,mV$ [18]

Polymers that are nonmetallic materials are very common in today's society. A polymer is an organic compound, which means "poly" ≡ many and "meres" ≡ parts, and consists of repeated long-chain molecular structure bounded by covalent bonds [22, 23]. Natural polymers are known as proteins, silks, and deoxyribonucleic acid (DNA) among many others. On the other hand, synthetic polymers, such as nylon (polyamides), polyvinyl chloride (PVC), polyacrylonitrile (PAN), polyethylene, epoxy, and many more, are known as **plastics**, which are so important in nowadays society due to their vast and broad domestic and industrial applications. However, polymers are susceptible to degradation in natural and synthetic environments, such as high temperatures (thermal degradation), moisture, radiation, ultraviolet light, and mechanical agents. Degradation of polymers in natural environments is known as weathering due to the effect of ultraviolet radiation from the sunlight, moisture, and temperature. Some oxidation agents of polymers can be found elsewhere [24]. In addition, degradation or damage of polymers can be classified as (1) oxidation damage according to the oxidation reaction $R \longrightarrow R^{+} + e^{-}$ due to high-energy ionization radiation (radiolysis), such as electron beams, *γ-radiation*, and *x-rays*, and (2) swelling caused by moisture and oxygen [24]. Furthermore, the polymer R loses one electron leading to a degradation known a depolymerization.

Woods are organic in nature and corrosion resistant in water and diluted acids. Woods consist of cellulose fibers surrounded by lignin. The cellulose fibers are strong and, yet, flexible, whereas the lignin is stiffer. Some woods can dissolve in strong acids and diluted alkalies [1]. However, the wood texture and properties play an important role in the material selection scheme for making violins, guitar, pianos, furniture, and houses.

Concrete is mainly composed of cement, sand, gravel, and water. Initially, the cement reacts as a binder when wet with water, and the final product is a chemically inert liquidus composite mass, which is cast into a particular shape. As a result, the concrete dries out and becomes a hard and strong solid mass bonded together by cement and water having good compression strength and poor tension strength. Normally, concrete is reinforced with embedded steel bars since the steel bars carry all the tension loads and acts as corrosion and fire-protective layer. However, the reinforced concrete is a porous structural material that is susceptible to cracking and spalling when diffused chloride ions cause corrosion of the embedded steel bars.

1.10 Questions

1.1. During metallic corrosion there is loss of weight of the metal. Why metals undergo corrosion in a suitable environment?

1.2. Is copper corrosion an oxidation process?

1.3. It is known that corrosion on the surface of a metal is due to a direct reaction of atmospheric gases. Which gas is mainly responsible for the corrosion of most metallic iron and steel surfaces?

1.4. What are the two mechanisms for oxidation of iron? Write down the suitable electrochemical reactions.

1.5. Investigate the differences between dry and wet corrosion Write down at least three differences per type of corrosion.

1.6. What is bimetallic corrosion?

1.7. What form of corrosion will cause sand grains, dust particles, and an oxide scale on the surface of metals exposed to a corrosive medium?

1.8. Why stress corrosion is not uniform?

1.9. What form of corrosion develops at rivets and bolts?

1.10. Corrosion is a process of destruction of a metal surface by the surrounding environment. Environmental parameters like temperature, humidity, gases (CO_2, SO_2, NO_x), and pH play an important role in studying corrosion. What are the two factors that govern the corrosion process?

1.11. What are the main effects or consequences of corrosion of structures?

1.12. Metal coatings on base metals are used to prevent corrosion of the base metal from taking place or simply prolong the life span of the base metal. Why an anodic coating and a cathodic coating are used on base metal?

1.13. What method is used to prevent corrosion of iron by zinc coating?

1.14. The principle constituent of wood is cellulose, which is a polysaccharide. This is a polymer made of sugar molecules joined in long chains. Will these sugar molecules corrode a galvanized steel nail and weaken a wood beam in a hypothetical fishing vessel?

References

1. M.G. Fontana, *Corrosion Engineering* (McGraw-Hill Book Company, New York, 1986)
2. S.L. Pohlman, General corrosion, in *Corrosion*, vol. 13, 9th edn. (ASM International, Metals Park, 1987), p. 80
3. S.C. Dexter, Localized corrosion, in *Corrosion*, vol. 13 (ASM International, Materials Park, 1987), p. 104
4. K. Suda, S. Misra, K. Motohashi, Corrosion products of reinforcing bars embedded in concrete. Corros. Sci. **35**, 1543–1549 (1993)
5. A.R.L. Chivers, F.C. Porter, Zinc and zinc alloys, in *Corrosion: Metal/Environment Reactions*, vol. 4, 3rd edn., ed. by L.L. Shreir, R.A. Jarman, G.T. Burstein (Butterworth-Heinemann, Oxford, 1994), p. 172
6. C.H. Dale Nevison, Corrosion of zinc, in *Corrosion*, vol. 13, 9th edn. (ASM International, Materials Park, 1987), p. 756
7. C. Leygraf, T. Graedel, *Atmospheric Corrosion* (Wiley-Interscience/Wiley, New York, 2000), pp. 330–333
8. Wendy R. Cieslak (Chairman:), The ASM Committee on Corrosion of Electrochemical Power Sources, *Corrosion in Batteries and Fuel-Cell Power Sources*, ASM Handbook: Corrosion, 9th edition, Vol. 13, ASM International, Materials Park,OH, USA, 3297–3312 (1987)
9. W.R. Cieslak (Chairman), The ASM committee on corrosion of electrochemical power sources, in *Corrosion in Batteries and Fuel-Cell Power Sources*, 9th edn. ASM Handbook: Corrosion, vol. 13 (ASM International, Materials Park, 1987), pp. 3297–3312
10. P. Morriset, Zinc el Alliages, Zinc and zinc alloys, in *Corrosion: Metal-Environment Reactions*, vol. 20, 3rd edn., ed. by L.L. Shreir, R.A. Jarman, G.T. Burstein (Butterworth-Heinemann, Oxford, 1959), p. 15
11. J.P. Gabano, Lithium battery systems: an overview, in *Lithium Batteries*, ed. by J.P. Gabano (Academic, New York, 1983), pp. 1–12
12. P. Arora, R.E. White, M. Doyle, Capacity Fade Mechanisms and Side Reactions in Lithium-Ion Batteries. J. Electrochem. Soc. **145**(10), 3647–3667 (1998)
13. P. Arora, M. Doyle, R.E. White, Mathematical Modeling of the Lithium Deposition Overcharge Reaction in Lithium-Ion Batteries Using Carbon-Based Negative Electrodes. J. Electrochem. Soc. **146**(10), 3543–3553 (1999)

2 Electrochemistry

The goal in this chapter is to elucidate the fundamental characteristics and technological significance of electrochemical cells. A comprehensive review of the subject is excluded since the intention hereafter is to describe the principles of electrochemistry, which should provide the reader with relevant key definitions and concepts on metal reduction in electrolytic cells and metal oxidation in galvanic couplings. In general, electrochemistry deals with the chemical response of an electrode-electrolyte system, known as a half-cell, to an electrical stimulation. Subsequently, the electrochemical behavior of species (ions) can be assessed, including concentration, kinetics, and reaction mechanisms.

The most common electrochemical cells are classified as galvanic cells for producing energy or electrolytic cells for consuming energy. Of major interest hereafter is the function of an electrochemical cell when transferring electrons from metallic solid surfaces and allowing the movement of ions within its half-cells. This electron-transfer process requires that atoms in a solid electrode have the ability (1) to oxidize by losing an amount (z) of electrons (ze^-) and becoming electron-deficient atoms called cations (M^{z+}) or (2) to reduce those cations M^{z+} in a preeminent aqueous (aq) solution by gaining electrons (ze^-) at the electrode surface and become electron-efficient atoms. Hence, an electric field must exist among the electrode-electrolyte component due to the presence of charged particles (ions) [1]. Characterizing an electrochemical cell may be complex because one has to take into account several variables, such as ionic motion in an electrolyte solution, ion concentration, ionic mobility, the acidity (pH), and electrolyte conductivity. These variables, however, affect the efficiency of electrochemical cells at a macroscale, but it is important to understand the electrochemical events at the solid electrode-electrolyte interface, which is known as the interfacial region, which, in turn, is constituted by an electrical double layer in solution and a space-charge region in the solid. The latter arises due to mobile charges and immobile charges in the interfacial region [2, 3].

N. Perez, *Electrochemistry and Corrosion Science*,
DOI 10.1007/978-3-319-24847-9_2

2.1 Electrostatics

This section deals with a brief review of the physics of electric charges at rest on a particular α-phase being electrically neutral. Consider that an electrical conductor, such as aqueous solutions containing known ions, is **electrically neutral** and that there exists an infinitesimal current flow. Thus, an electric field strength vector $\overrightarrow{E}$ at a point P in space and an inner electric potential ϕ of the system are defined as [1, 3, 4]

$$\overrightarrow{E} = \overrightarrow{F}/Q \tag{2.1}$$

$$\phi = \frac{W}{Q} \tag{2.2}$$

where $\overrightarrow{F}$ = electric force vector (N)
Q = electric charge (C = Coulomb)
W = electric work (J)

In fact, Q depends upon the type of ions and it is related to **Faraday's constant** F. Both F and Q are defined by

$$F = q_e N_A \simeq 96{,}500 \text{ C/mol} \tag{2.3}$$

$$Q = zFX \tag{2.4}$$

where q_e = electron charge = $1.6022x10^{-19}$ C/ions
N_A = Avogadro's number = $6.02213x10^{23}$ ions/mol
X = mole fraction (mol)
z = valence or oxidation number
C/mol = A · s/mol = J/ (mol · V)

The Cartesian electric field strength components in a three-dimensional scheme are defined as follows:

$$\begin{bmatrix} E_x \\ E_y \\ E_z \end{bmatrix} = -\nabla\phi = -\begin{bmatrix} \partial\phi/\partial x \\ \partial\phi/\partial y \\ \partial\phi/\partial z \end{bmatrix} \tag{2.5}$$

$$\begin{bmatrix} E_x \\ E_y \\ E_z \end{bmatrix} = \lambda Q \begin{bmatrix} x^{-2} \\ y^{-2} \\ z^{-2} \end{bmatrix} \tag{2.6}$$

where $\lambda = 1/(4\pi\epsilon_o\epsilon_r) = \left(9x10^9 \text{ N m}^2/\text{C}^2\right)/\epsilon_r$ = constant
$\epsilon_o = 8.854x10^{-12}\text{ C}/\left(\text{N m}^2\right)$ = permittivity of vacuum
ϵ_r = relative permittivity (dielectric constant)

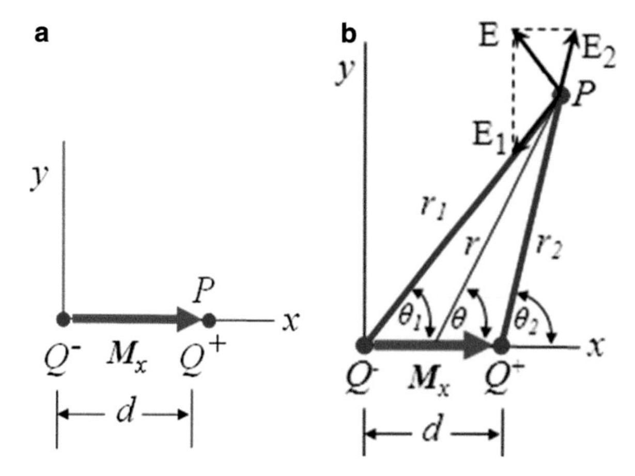

Fig. 2.1 Schematic electric poles in a phase being electrically neutral. Denote the direction of the electric field strength components: E_1 and E_2 at point P. (**a**) Monopole. (**b**) Dipole

Monopole Let ϕ be defined at a point P in space around a charge Q^+ (ion) in an electric monopole Q^+-P path shown in Fig. 2.1a. Integrating Eq. (2.5) yields the inner electric potential for an isolated ion in the x-direction [1, 4]:

$$\phi_{x,iso} = -\int_0^r E_x dx = -\int_0^r \frac{\lambda Q}{x^2} dx = \frac{\lambda Q}{r} + C_x \tag{2.7}$$

$$\phi_{x,iso} = \frac{\lambda Q}{r} = \frac{\lambda Q}{d} \tag{2.8}$$

and that for an ion in solution being influenced by other ions is [4]

$$\phi_{x,sol} = \frac{\lambda Q}{r} \exp(-\kappa r) \tag{2.9}$$

Here, the integration constant is $C_x = 0$ when $\phi_x = 0$ at $r = \infty$. Then Eq. (2.8) gives the definition of the monopole potential due to a point charge Q. On the other hand, when $\phi_x \longrightarrow \infty$ as $r \longrightarrow 0$, an electric potential "singularity" can be established for ϕ_x in the order of r^{-1} as predicted by Eq. (2.8).

Consider a metal as a single electric conductor phase in equilibrium. In this case, Eq. (2.5) gives $E_x = E_y = E_z = 0$, which means that the inner electric potential ϕ is constant in the bulk phase so that $\phi \neq 0$ and the current is $I = 0$. Thus, Eq. (2.5) becomes $\partial\phi/\partial x = 0$, but $\int \partial\phi = 0$ gives $\phi + C_x = 0$ where C_x is the integration constant. Therefore, $\phi \neq 0$, which implies that it is a net electric charge being uniformly distributed over the surface of the phase; otherwise, the repulsive force of like charges will not allow diffusion of charges to the phase surface [1].

Dipole Consider the charge couple shown in Fig. 2.1b, in which the negative and positive charges are separated by a distance $d << r_i$, making an electric dipole for which the electric dipole moment has a magnitude of $M_x = Qd$ and units of Coulomb meter (C m). Denote the direction of the electric field strength Components, E_1 and E_2, in Fig. 2.1b. Nonetheless, the electric dipole moment (M_x) is a measure of the separation of positive and negative electrical charges in a particular α-phase containing many charged particles. It is assumed that α-phase is electrically neutral, and it is not connected to another, say, β-phase. This specific physical condition assures that there are no differences in electric potential between phases: $\Delta\phi = \left|\phi^{\alpha} - \phi^{\beta}\right| = 0$ [1].

The configuration of two closely spaced particles having equal and opposite point charges (Q) making an **oblique triangle** $Q^{-}PQ^{+}$ is shown in Fig. 2.1b. Here, $Q = Q^{+}$ is the positive charge and $Q = Q^{-}$ is negative. These two charged particles form what is known as the electric dipole. This is an extension of the single charge case described by Eq. (2.8). The inner electric potential ϕ of α-phase is the electric potential in the bulk of a dipole at point P. It can be derived using Eq. (2.5) along with Pythagorean theorem ($r_1^2 = d^2 + r_2^2$) and the following relationships: $d = r_1 \cos\theta$, $E = \lambda Q/r^2$, $\cos\theta = d/r$, and $M_x = Qd$. Thus,

$$\phi = -\int_{r_2}^{r_1} E dr = -\int_{r_2}^{r_1} \frac{\lambda Q}{r^2} dr = \lambda Q\left(\frac{1}{r_2} - \frac{1}{r_1}\right) \tag{2.10}$$

$$\phi = \lambda Q\left[\frac{r_1^2 - r_2^2}{r_1 r_2 (r_1 + r_2)}\right] \tag{2.11}$$

Using the law of cosines for the triangle $Q^{-}PQ^{+}$ and assuming that the distance $r_i >> x = d$ yield

$$r_2^2 = r_1^2 + d^2 - 2r_1 d\cos\theta \tag{2.12}$$

which can be approximated as

$$r_2^2 \approx r_1^2 - 2r_1 d\cos\theta \quad \text{since} \quad r_i >> d \tag{2.13}$$

$$r_1^2 - r_2^2 \approx 2r_1 d\cos\theta \tag{2.14}$$

Substituting Eq. (2.12) and $M_x = Qd$ into (2.11) gives

$$\phi \simeq \lambda Q\left[\frac{2d\cos\theta}{r_2 (r_1 + r_2)}\right] \simeq \lambda M_x\left[\frac{2\cos\theta}{r_2 (r_1 + r_2)}\right] \tag{2.15}$$

Now, letting $r \simeq r_1 \simeq r_2$ in Eq. (2.15), the electric potential becomes [1]

$$\phi \simeq \frac{\lambda Q d\cos\theta}{r^2} \simeq \frac{\lambda M_x \cos\theta}{r^2} \quad \text{for } r >> d \tag{2.16}$$

The electric potential ϕ "singularity" is in the order of r^{-2} since $\phi \longrightarrow \infty$ as $r \longrightarrow 0$. Conversely, $\phi \longrightarrow 0$ as $r \longrightarrow \infty$. According to Eqs. (2.8) and (2.16), the electric field strengths are, respectively,

$$E_x = -\frac{\partial \phi}{\partial r} = \frac{\lambda Q}{r^2} \qquad \text{(monopole)} \tag{2.17}$$

$$E_\theta = -\frac{\partial \phi}{\partial r} = \frac{\lambda Q d \cos\theta}{r^3} \qquad \text{(dipole)} \tag{2.18}$$

Notice that the E "singularity" is in the order of r^{-2} for the monopole and r^{-3} for the dipole case.

This brief introduction to the source of the electric field strength of a phase is for the reader to notice that a phase has been treated as an electrically neutral thermodynamic system, which is not the case for a heterogeneous electrochemical cell. For all practical purposes, this section serves the purpose of comparison with the proceeding treatment.

Example 2.1. *Let a monopole contain a net charge (Q) due to 10^{-11} moles of a particle in an electrically neutral system. Calculate the electric potential (ϕ_x), the electric potential strength (E_x), and the electric force (F_x) acting on the particle within a distance of 7 cm from a charged particle in the x-direction. Plot $\phi = f(r)$ to determine how fast ϕ decays from the monopole at a distance r.*

Solution. *From Eq. (2.4),*

$$Q = zFX = (2)(96{,}500 \text{ C/mol}) \left(10^{-11} \text{ mol}\right)$$

$$Q = 1.93x10^{-6} \text{ C} = 1.93 \text{ μC}$$

But,

$$\lambda = \frac{1}{4\pi\epsilon_o} = 9x10^9 \frac{\text{N m}^2}{\text{C}^2}$$

$$\lambda = 9x10^9 \frac{\text{J m}}{\text{C}^2}$$

$$r = 0.07 \text{ m}$$

Inserting these values into Eq. (2.8) gives the magnitude of the electric potential on a monopole system. Thus,

$$\phi_x = \frac{\lambda Q}{r} = \frac{\left(9x10^9 \text{ J m/ C}^2\right)\left(1.93x10^{-6} \text{ C}\right)}{0.07 \text{ m}}$$

$$\phi_x = 2.48x10^5 \text{ J/C} = 2.48x10^5 \text{ N m/C}$$

$$\phi_x = 2.48x10^5 \text{ J/(J/V)} = 2.48x10^5 \text{ V} = 0.25 \text{ MV}$$

which is a high value for the isolated monopole case. The magnitude of the electric potential strength is expected to have a very high magnitude given by Eq. (2.17). Thus,

$$E_x = \frac{\lambda Q}{r^2} = \frac{\phi}{r} = \frac{2.48x10^5\ \mathrm{N\,m/C}}{0.07\ \mathrm{m}}$$

$$E_x = 3.54x10^6\ \mathrm{N/C} = 3.54x10^6\ \mathrm{V/m}$$

Thus, the electric force acting on the particle becomes

$$F_x = QE_x = \left(1.93x10^{-6}\ \mathrm{C}\right)\left(3.54x10^6\ \mathrm{N/C}\right)$$

$$F_x = 6.83\ \mathrm{N}$$

The required plot is given below.

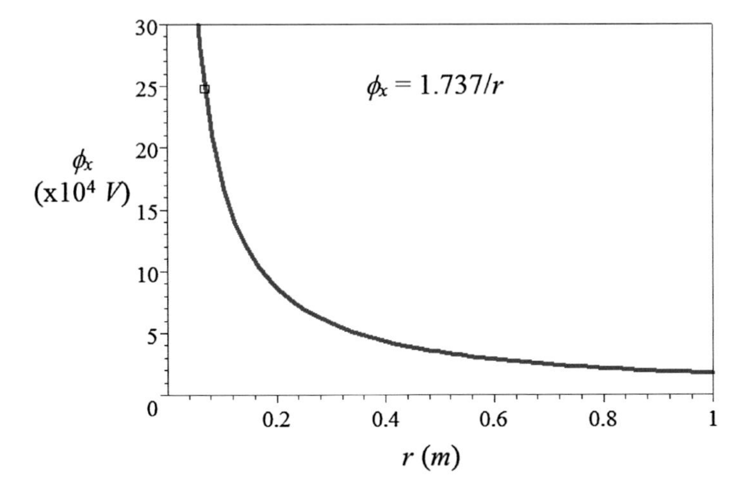

Obviously, the electric potential, $\phi_x = f(r) = 1.737/r$, decays very fast as a function of monopole distance r. According to the above plot, ϕ_x has a negative slope, $d\phi_x/dr = -1.737/r^2$, at any point on the curve, and it is a potential gradient.

A 3D plot reveals a different trend. Thus, $E_x = \phi_x/r$ gives

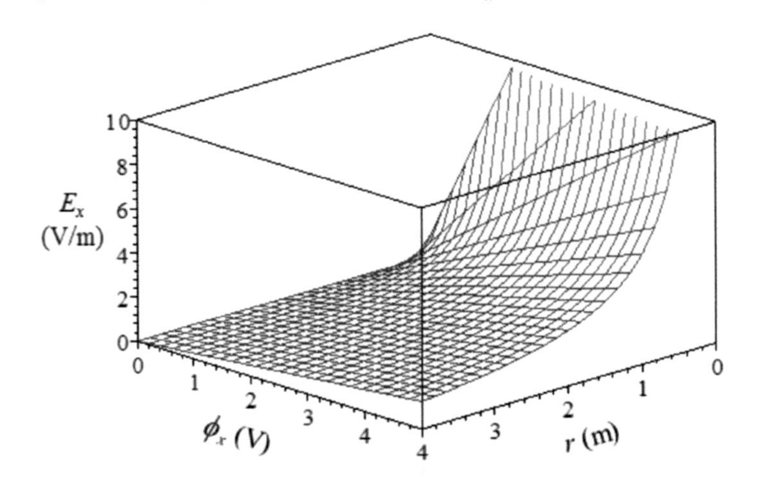

2.2 Electrochemical Cells

Electrochemical cells (systems) are significantly useful as technological tools for producing or consuming electrical energy in ionic mass transport processes. The size and number of polyfunctional electrodes of an electrochemical cell play an important role in technological applications or electrochemical engineering. In fact, electrochemical engineering is an overlap between chemical and electrical engineering fields.

Generally, the transport of charge across the interface between a polyfunctional electrode (electric conductor) and an electrolyte (ionic conductor) depends mainly on temperature, pressure, ion concentration, homogeneity of the electrolyte, electrode surface roughness, the magnitude of the applied or generated electric potential (E), the electric current (I), and so on.

Despite that the electric current flow depends upon several variables, it is principally governed by the mass transport and the electron-transfer processes [5,6]. Specifically, the electric current is a flow of electric charge carried by moving electrons in a wire or by ions in an electrolyte. In particular, the current density (i) is the preferred term because it can be related to oxidation or corrosion rate on an electrode contact surface (A_s) in an electrolyte. Thus,

$$i = I/A_s \tag{2.19}$$

Electrochemical cells can be constructed using metallic electrodes or membrane electrodes. The former produces an electric potential in response to a metallic surface reaction, while the latter cell generates an electric potential in response to ion migration [7,8]. Each type of electrode has its own purpose and limitations. The analysis of membrane electrodes is not included throughout this book.

Unlike the electrostatic system, an electrochemical cell is a heterogeneous thermodynamic system that has at least two solid electrode phases (α-phase and β-phase) in electrical contact. Accordingly, electron flow (e^-) between $\alpha-$ and β-phases and transfer of charged particles (ions) within each phase occur, generating an electric potential change, say,

$$\Delta\phi = \phi^\alpha - \phi^\beta \tag{2.20}$$

Henceforth, ϕ^α and ϕ^β are bulk electric potentials referred to as inner electric potentials. The electrodes are individually immersed into a solution called electrolyte containing charged particles. In general, either the α-phase or β-phase undergoes metal degradation, which is an electrochemical process known as oxidation or corrosion, while the other phase undergoes metal reduction.

Hence, electrochemical cells deal with electron transfer due to these oxidation and reduction reactions on electrode surfaces. Their application is significant in battery and fuel cell production, in corrosion studies as galvanic cells, and in electrodeposition as electrolytic cells. The latter cells can be used for preventing corrosion in metallic structures (Chap. 8) and for recovering metals from solution (Chap. 9).

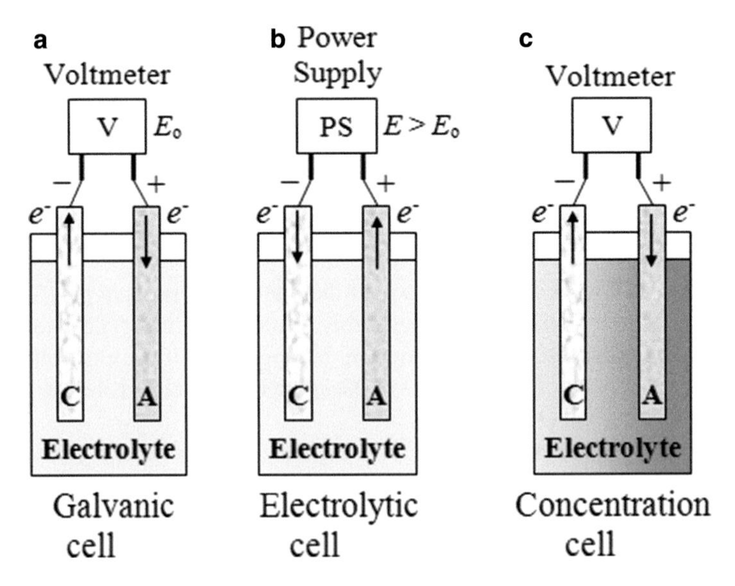

Fig. 2.2 Electrochemical cells. (**a**) Galvanic cell. (**b**) Electrolytic cell. (**c**) Concentration cell

2.2.1 Galvanic Cells

In general, Fig. 2.2 schematically depicts three electrochemical cells called (a) galvanic cell, (b) electrolytic cell, and (c) concentration cell [9]. These types of cells can be classified as a one-electrolyte and two-electrode system. With respect to the concentration cell, it is a galvanic cell with a high initial concentration of an ion at the anode. See Appendix B for other cell arrangements.

Electrochemical cells having two half-cells separated by either a porous diaphragm or a salt bridge are common. Each half-cell contains an electrode phase immersed in an electrolyte solution. If the electrodes are electrically connected using a wire, then the electrochemical cell is complete and ready to work as an energy-producing device (galvanic cell) or as an energy consumption device (electrolytic cell). The application of these devices as technological tools leads to electrochemical engineering, which deals with electroplating, electrowinning, electrorefining, batteries, fuel cells, sensors, and the like.

The types of electrochemical cells are then classified as (1) galvanic cells which convert chemical energy into electrical energy while electron and current flows exist. The electrochemical reactions are spontaneous. (2) Electrolytic cells for converting electrical energy into chemical energy. The electrochemical reactions are not spontaneous. (3) Concentration cells when the two half-cells are equal, but the concentration of a species j is higher in the anode half-cell than in the cathodic counterpart.

Figure 2.3 schematically shows two different galvanic cells and their macro-components. Figure 2.3a contains one electrolyte and two iron (Fe) electrodes

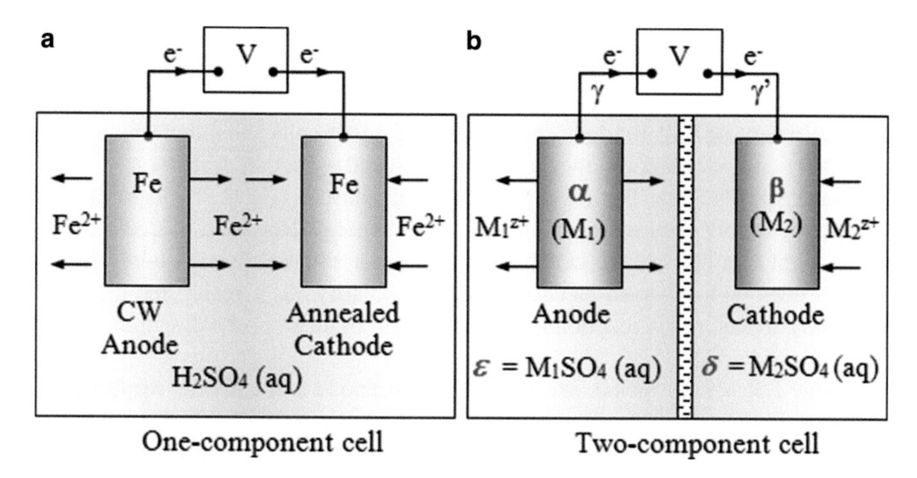

Fig. 2.3 Schematic galvanic cells. (**a**) One-component cell. (**b**) Two-component cell

having different microstructures, despite that the common metallurgical phase is ferrite. The annealed microstructure is constituted by soft equi-axed grains, having insignificant or no microstructural defects and no residual stresses. On the other hand, the cold-worked (*CW*) microstructure has hard elongated grains as a result of the cold-working deformation process, which is a mechanical deformation method that causes a high level of residual stresses due to atomic plane distortion and high dislocation density. Dislocations are line defects in crystalline solid phases, such as cold-worked ferrite. Normally, cold-worked structures exhibit deformed grain structures with significant slip (BCC and FCC metals) or mechanical twining (HCP metals). Consequently, a galvanic cell forms due to differences in microstructures and residual stress levels. In multiphase alloys, phase differences with different electric potentials cause galvanic corrosion.

Furthermore, Fig. 2.3b can be used as the point of departure for developing the thermodynamics of electrochemical cells containing at least two half-cells. In a conventional galvanic cell setup, the cathode is the most positive electrode kept on the right-hand side, while the anode is the most negative electrode on the left-hand side of the cell [10]. Thus, electron flow occurs from left to right through the terminals (Fig. 2.3).

The schematic cell in Fig. 2.3b has two half-cells separated by a porous membrane. The macrocomponents of this galvanic cell are:

- **Anode Half-Cell:** It is an electric conductor containing metallic α-phase called the anodic electrode (plate or cylinder) immersed into an electrolyte ε-phase (liquid, paste, gel, or moist soil). The α-phase is a metal M_1, such as *Zn*.
- **Cathode Half-Cell:** It is also an electric conductor containing metallic β-phase called the cathode electrode (plate or cylinder) immersed into an electrolyte δ-phase (liquid, paste, gel, or moist soil). The β-phase is a metal M_2, such as *Cu*.

- **Wiring System:** Both half-cells are connected using a metallic wire, usually *Cu*, in order to complete an electrical circuit. Denote the direction (arrow) of the electron flow, which produces a direct current (DC).
- **Diaphragm or Salt Bridge**: This separates the half-cells and allows the flow of ions for charge balance between the oxidation and reduction processes.
- **Power Supply (PS):** The above schematic galvanic cell (Fig. 2.3b) does not need a power supply because it produces electrical energy. However, this cell can be made into an electrolytic cell by connecting an external power supply to the electrodes with reversed electrode terminal polarity. As a result, the direction of the galvanic redox reaction is forced to occur in the opposite direction.

Figure 2.3b shows metals M_1 and M_2 as the anode and cathode, respectively. Both are immersed in their own sulfate-base electrolytes, M_1-$M_1SO_4(aq)$ and M_2-$M_2SO_4(aq)$, which are in their ionic state in solution (aq = aqueous). Both metals are externally connected to an electrical circuit in order to measure the potential difference between them. In other words, the electrical circuit is used to measure the galvanic cell potential using a voltmeter (V). This measurable cell potential (E) is current/resistance dependent.

In addition, a galvanic cell is a **bimetallic** cell since it has different metallic electrodes, but electrons flow clockwise. This electrochemical event produces electrical energy from the stored chemical energy in solution. Thus, the electrochemical reactions occur spontaneously without the aid of an external power supply.

The following generalized single reaction represents the electrochemical cell shown in Fig. 2.3b. Assume that the reactions proceed as written in an irreversible manner:

$$M_1 \rightarrow M_1^{z+} + ze^- \quad \text{(anode)} \tag{2.21}$$

$$M_2^{z+} + ze^- \rightarrow M_2 \quad \text{(cathode)} \tag{2.22}$$

$$M_1 + M_2^{z+} \rightarrow M_1^{z+} + M_2 \quad \text{(redox)} \tag{2.23}$$

The metallic cation M_1^{z+} produced at the anodic α-phase (negative terminal) goes into ϵ-phase solution, and the number of electrons ze^- moves along the wires to arrive at the cathode electrode-electrolyte interface (positive terminal). As a result, the cation M_2^{z+} gains ze^- electrons and reduces on the cathode β-phase. Thus, M_2^{z+} cations become atoms that deposit or electroplate on the surface of the cathodic β-phase, specifically at surface defects.

The overall reaction, known as the redox reaction, is the sum of Eqs. (2.21) and (2.22) in which the electrons ze^- cancel out.

One can assume that the oxidation (on the anode) and reduction (on the cathode) processes are spontaneous as written and are the representation of ionic mass transport being dependent upon the strength of the electric field among other factors. In fact, the numbers of the above reactions are in the millions, but a single reaction per electrode serves the purpose of explaining how an electrochemical cell really works.

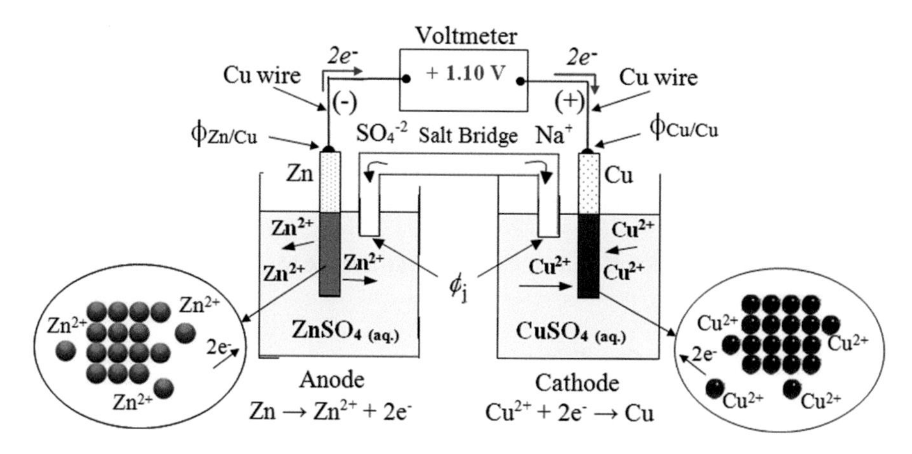

Fig. 2.4 The Zn/Cu galvanic (voltaic) cell, also known as Daniell cell. The *inset* shows the atomic level at the electrode surfaces

Figure 2.4 schematically illustrates a galvanic cell, known as voltaic cell and Daniell cell (invented in 1836), with more details [1, 4, 11–13]. Notice that $\phi_{Zn/Cu}$ and $\phi_{Cu/Cu}$ are contact potentials and ϕ_j is the liquid junction potential. The standard cell potential is defined as the sum of interfacial potentials. Thus,

$$E^o_{cell} = \left(E^o_{Zn} + E^o_{Cu}\right) + \left[\phi_{Zn/Cu} + \phi_{Cu/Cu} + \phi_j\right] \tag{2.24}$$

In real cases, nonstandard conditions exist, and the cell potential can be predicted by

$$E_{cell} = E^o_{cell} + [E_{Nernst} - E_{Ohmic} - \eta] \tag{2.25}$$

where E_{Nernst}, E_{Ohmic}, and $\eta = \eta_{overpotential}$ are potential terms for correcting the standard potential expression, Eq. (2.25). These terms will be dealt with in a later chapter.

This electrochemical coupling (Fig. 2.4) undergoes several events which can be described as follows:

- When zinc (Zn) is immersed into aqueous zinc sulfate, $ZnSO_4(aq)$, the subsystem $Zn\text{-}ZnSO_4(aq)$ becomes the anode half-cell. Similarly, copper (Cu) immersed in aqueous copper sulfate, $Cu\text{-}CuSO_{4(aq)}$, turns into the cathode half-cell.
- Both $ZnSO_{4(aq)}$ and $CuSO_{4(aq)}$ are electrolyte solutions that act as ionic conductors.
- Normally, *Cu* wire is used to connect the *Zn* and *Cu* electrodes so that no potential difference develops between the wires, which are called terminals.
- If a salt bridge is not used, both half-cells become isolated, and electrical neutrality is lost since the *Zn* electrode becomes more negative as Zn^{2+} cations

are produced. Conversely, the Cu electrode becomes more positive as Cu^{2+} are removed from solution and deposited as Cu atoms on the electrode surface. The salt bridge is filled with an agar gel (polysaccharide agar or polymer gel) and concentrated aqueous salt ($NaCl$ or KCl) or sodium sulfate ($NaSO_4$) solution in order to maintain electrical neutrality between the half-cells. The gel allows diffusion of ions, but eliminates convective currents due to turbulent transport of electrical charges [1].
- A voltmeter is used to measure the potential difference (E_{cell}) between the Zn and Cu electrodes.

2.2.1.1 Open-Circuit Condition

This is the case when the voltmeter is removed leaving an open circuit. A summary of events is outlined below for Fig. 2.4.

Due to instability of the galvanic cell, Cu^{2+} cations slowly diffuse into the $ZnSO_4(aq)$ solution through the salt bridge and come in contact with the Zn electrode. Consequently, the zinc anode oxidizes and the copper cations reduce spontaneously. The Zn electrode will go into solution or simply dissolves. The redox reaction results by adding both the anodic and cathodic reactions

$$Zn \longrightarrow Zn^{2+} + 2e^- \quad \text{(anode)} \tag{2.26}$$

$$Cu^{2+} + 2e^- \longrightarrow Cu \quad \text{(cathode)} \tag{2.27}$$

$$Zn + Cu^{2+} \longrightarrow Zn^{2+} + Cu \quad \text{(redox)} \tag{2.28}$$

In order to avoid destruction of the cell, just connect a resistor like a voltmeter to the terminals. This will keep Cu^{2+} cations away from the $ZnSO_4(aq)$ solution due to the electric field force in the $CuSO_4(aq)$ solution.

2.2.1.2 Closed-Circuit Condition

This is the case when a voltmeter is connected to the galvanic cell. Thus, the electrical circuit is complete, and the potential difference between the Zn and Cu electrodes is measured as $E^o_{cell} = +1.10\,\text{V}$ at standard conditions, $P = 1\,\text{atm}$, $T = 25\,°\text{C}$, and 1 mol/l:

- The Cu-Zn junction potential $\phi_{Zn/Cu}$ is lower than the Cu-Cu junction $\phi_{Cu/Cu}$. Electrons will flow from the lowest electric potential reservoir (anode half-cell) to the highest reservoir (cathode half-cell) through the external wires or terminals [1].
- The electrical work is $W = Q\Delta\phi = QE_{cell} > 0$ for electron flow to occur as negative charges from the anode to the cathode. According to Ohm's law, the change in inner potential is $\Delta\phi = IR$.

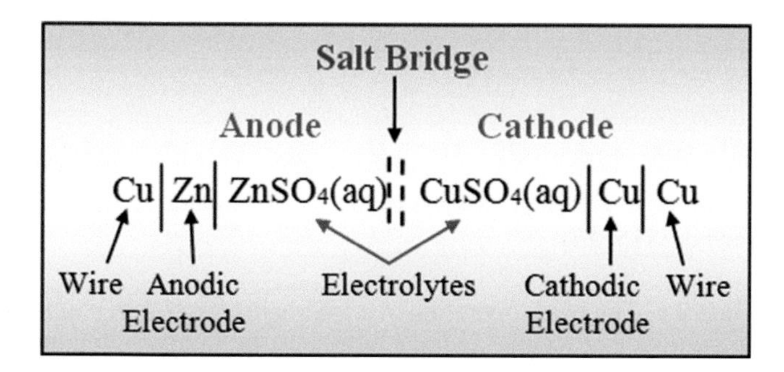

Fig. 2.5 Shorthand notation for the galvanic cell diagram in Fig. 2.4

- Consequently, the redox reaction, $Zn + Cu^{2+} \longrightarrow Zn^{2+} + Cu$, occurs spontaneously. This electrochemical process is called electroplating or electrodeposition.
- In due time, $CuSO_4(aq)$ electrolyte becomes depleted of Cu^{2+} cations, and $ZnSO_4(aq)$ electrolyte becomes enriched with Zn^{2+} cations.
- The maximum potential difference between *Zn-Cu* electrodes in Fig. 2.4 can be defined as

$$E_{cell} = \left(E^o_{Zn} + E^o_{Cu}\right) + \left[\phi_{Zn/Cu} + \phi_{Cu/Cu} + \phi_j\right] \tag{2.29}$$

$$E_{cell} \simeq E^o_{Zn} + E^o_{Cu} \simeq 1.10\ V_{SHE} \tag{a}$$

since $\left[\phi_{Zn/Cu} + \phi_{Cu/Cu} + \phi_j\right] \approx 0$.

For convenience, Fig. 2.4 can be represented in shorthand notation as a cell diagram. This is shown in Fig. 2.5 where vertical lines represent phase boundaries. A salt bridge is symbolized by two vertical lines, and it has two liquid junctions of insignificant potentials.

2.2.2 Application of Galvanic Cells

Lithium Ambient-Temperature Batteries (LAMBS) These are high-energy density devices, in which the *Li* anode is passivated. The solid cathode can be made of $LiCuO$, $LiMnO_2$, LiV_2O_5, or $LiBi_2Pb_2O_5$. Some LAMBS use liquid cathodes, such as $LiSO_2$ [14]. For negative cathodes made of carbon (graphite or coke) in lithium-ion batteries, the ideal electrode reactions are reversed during charging $\left(\underset{\rightarrow}{c}\right)$ and discharging: $\left(\underset{\leftarrow}{d}\right)$

$$C_6 + xLi^+ + xe^- \underset{d}{\overset{c}{\rightleftarrows}} = Li_xC_6 \tag{2.30}$$

The reduction reaction is for charging, and the oxidation is for discharging processes of the lithium-ion cells. Detailed analysis of these cells, including side reactions, can be found elsewhere [15–17].

Lead-Acid Battery The basic operation of a lead-acid (Pb-H_2SO_4) battery is based on groups of positive and negative plates immersed in an electrolyte that consists of diluted sulfuric (H_2SO_4) acid and water. Hence, the mechanism of this type of battery is based on the electron-balanced anodic (−) and cathodic (+) reactions. Hence, the ideal reactions are

$$(+)\text{:}\quad PbSO_4 + 2H_2O \underset{d}{\overset{c}{\rightleftarrows}} PbO_2 + 3H^+ + HSO_4^- + 2e^- \tag{2.31}$$

$$(-)\text{:}\quad PbSO_4 + H^+ + 2e \underset{d}{\overset{c}{\rightleftarrows}} Pb + HSO_4^- \tag{2.32}$$

The redox reaction in lead-acid batteries is the sum of the above half-cell reaction:

$$\text{Redox:}2PbSO_4 + 2H_2O \underset{d}{\overset{c}{\rightleftarrows}} PbO_2 + 3H^+ + 2HSO_4^- + Pb \tag{2.33}$$

According to the above half-cells, the anode is pure lead (Pb), and the cathode is lead dioxide (PbO_2). In addition, both electrodes dissolve in the electrolyte during discharge, forming lead sulfate ($PbSO_4$). However, when the battery is charged, reverse reactions occur. This reversible electrochemical cycle can last for a prolonged time, but in practice batteries has a finite lifetime due to the lead sulfate buildup acting as an insulation barrier.

Most automotive batteries have lead-calcium grids for maintenance-free and have a lifetime from 1 to 5 years; however, longer battery life is possible. Generally, a lead-acid battery is used as a 12-V electrochemical device having six 2-V cells connected in series. The average activity and density of the sulfuric acid solution are in the order of $a = 4.5$ mol/l and $\rho = 1.25\,\text{g/cm}^3$ at 20 °C, respectively.

Dry-Cell Battery This is a common galvanic cell, known as voltaic cell or Daniell cell, which contains a moist ammonium chloride electrolyte. A schematic battery is shown in Fig. 2.6 [18]. The zinc casing and the solid carbon in contact with the electrolyte (electric conductor) develop a potential difference, which in turn produces an electron flow when the zinc and carbon are electrically connected. Consequently, the zinc eventually corrodes galvanically since it provides the electrons to the electrolyte for generating reduction reactions. The electrolyte (moist paste) carries the current from the zinc anode to the carbon cathode.

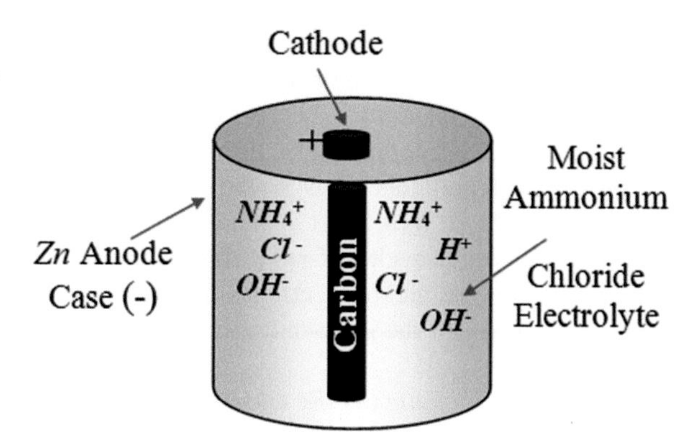

Fig. 2.6 Schematic dry-cell battery known as Daniell cell

This particular electrochemical cell is a well-sealed battery (dry-cell battery) useful for flashlights, portable radios, and the like. Nonetheless, zinc casing oxidizes, $Zn \to Zn^{+2} + 2e^{-}$, at the anode surface, while MnO_2 is reduced at the carbon cathode surface, and ammonia is then produced, decreasing the cell current by forming a gaseous film around the carbon electrode. Logically, the dry-cell battery does not last long when used continuously. However, if this battery is partially used, its design life is slightly extended.

Sintered Nickel Electrode in Alkaline Batteries These batteries are galvanic devices containing a porous Ni matrix that holds the active (anodic) materials. The following reaction is reversed upon discharging [19, 20]:

$$Ni(OH)_2 + OH^- \underset{d}{\overset{c}{\rightleftarrows}} NiOOH + H_2O + e^- \tag{2.34}$$

If the battery is overcharged, then electrolysis (dissociation) of water occurs leading to hydrogen evolution at the cathodes and oxygen evolution at the anodes.

In addition, corrosion of the Ni electrode may occur under unfavorable conditions, leading to loss of electrical continuity due to the following reaction:

$$2Ni + O_2 + 2H_2O \underset{d}{\overset{c}{\rightleftarrows}} 2Ni(OH)_2 \tag{2.35}$$

Other side reactions must be taken into account for characterizing nickel-hydrogen cells. Details on this matter can be found elsewhere [16–20]. For the purpose of clarity, the driving force for current flow through a moist electrolyte and electrochemical corrosion is the potential (voltage) difference between the anode and the cathode electrodes. In fact, a battery is simply an electrochemical device used as an energy storage. The capacity of a battery depends on the amount of stored chemical energy into the electrolyte. Accordingly, a large battery converts more chemical energy into electrical energy than a small one.

Nonetheless, the electrochemical properties can easily be evaluated by controlling the charge–discharge characteristics and determining the coulombic efficiency. The use of cyclic voltammetry seems very promising in this regard.

In general, any battery has a life expectancy based on the application and the diffusion of particular ions in the electrolyte and degree of electrochemical efficiency, which, in turn, must depend on the diffusing ionic phases that form on the charge–discharge cycles. X-ray diffraction is a technique useful for determining ionic phases on the nickel electrodes, but it is out of the scope of this book to determine or reveal if the ions are crystalline or amorphous phases.

Example 2.2. *Calculate* **(a)** *the mass and number of moles of a zinc battery* (Zn) *casing and* **(b)** *the mass and number of moles of the manganese dioxide* (MnO_2) *in the electrolyte if the battery has a stored energy of* $36\,kJ/V$ *and a power of* $3\,W$. **(c)** *Find the time it takes to consume the stored energy if the battery operates at a current of* $2\,A$ *and the potential (voltage). The thickness of the cell is* $x = 1\,mm$. *Other dimensions, such as length* (L) *and radius* (r), *are indicated below. The discharging reaction is* $MnO_2 + H^+ + e \rightarrow MnOOH$.

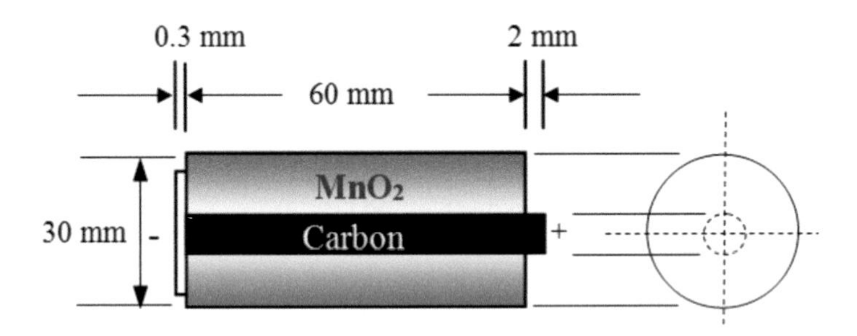

Solution. *The stored energy needs to be converted into units of Coulombs. Hence,*
$Q = 36\,\text{kJ/V} = 36{,}000\,\text{A s} = 36{,}000\,\text{C}$.
The area and volume are
$A_s = 2\pi rL = (2\pi)\,(1.5\,\text{cm})\,(6\,\text{cm}) = 56.55\,\text{cm}^2$
$V = xA_s = (0.1\,\text{cm})\left(56.55\,\text{cm}^2\right) = 5.65\,\text{cm}^3$

(a) *The mass and the moles of Zn are*

$$m_{Zn} = V\rho = \left(5.65\,\text{cm}^3\right)\left(7.14\,\text{g/cm}^3\right) = 40.34\,\text{g}$$

$$X_{Zn} = \frac{m_{Zn}}{A_{w,Zn}} = \frac{40.34\,\text{g}}{65.37\,\text{g/mol}} = 0.62\,\text{mol}$$

(b) *The moles and the mass of MnO_2 are*

$$X_{MnO_2} = \frac{Q}{zF} = \frac{36{,}000\ \text{A s}}{(1)\,(96{,}500\ \text{A s/mol})} = 0.37\ \text{mol}$$

$$m_{MnO_2} = X_{MnO_2} A_{w,MnO_2} = (0.37\ \text{mol})\,(86.94\ \text{g/mol})$$

$$m_{MnO_2} = 32.17\ \text{g}$$

(c) *The time for discharging the stored energy and the potential are*

$$t = \frac{Q}{I} = \frac{36{,}000\ \text{A s}}{2\ \text{A}} = 5\ \text{h}$$

$$E = P/I = (3\ \text{V A})\,/\,(2\ \text{A}) = 1.5\ \text{V}$$

2.2.3 Electrolytic Cells

An electrolytic cell (electrolyzer) is a reversed galvanic cell in which the current flows counterclockwise and the source of power is external. The cell has **bimetallic** or **monometallic** electrodes. This cell consumes power, $P = EI$ where E is the applied external potential and I is DC current, and the cell reactions are driven in the reverse direction as opposed to galvanic reactions; that is, the galvanic reactions are driven backward. This is possible if the applied potential is $E > E_{cell}$. In fact, Faraday's laws of electrolysis are obeyed for generating electrochemical reactions under the principles of electrochemical stoichiometry, and therefore, the electrochemical reactions occur with the aid of an external power supply. The theme of electrolysis will be dealt within Chap. 9 for large-scale production of pure metals.

This type of cell is very useful (1) in the electrometallurgical field for recovering metals from treated oxide ores by electroplating the metal cations on cathodes, (2) in the decomposition of water into hydrogen and oxygen, and (3) in converting bauxite into aluminum and other compounds. Hence, if current flows, then the principles of electrochemical stoichiometry are used for producing electrochemical reactions through the process of electrolysis. This process stands for an electrochemical event in which the electrochemical reactions are forced (non-spontaneous) to occur by passing a direct current (DC) through the electrolyte. Therefore, $E_{electrolytic} > E_{galvanic}$, but the measurable cell potential difference in an electrolytic cell (or galvanic cell) can be influenced by the Ohm's potential ($E = IR$) drop.

Table 2.1 summarizes generalized comparisons between galvanic and electrolytic cells ,including the free energy change (ΔG) not being defined yet, but it determines the direction of a reaction.

Furthermore, the suitable mechanism of electrolysis dictates that the electron transfer from the reduced form of the reactant to the metal electrode may occur at any available empty state in the electronic structure on the metal. Thus, the energy of activation for the electron transfer may be the difference in energy between the final state of the electron and the Fermi level (see section on STM and SECM techniques

Table 2.1 Comparison of electrochemical cells

Galvanic cell	Electrolytic cell
Produces chemical energy	Provides electrical energy
Chemical ⟶electrical energy	Electrical⟶chemical energy
$\Delta G^o < 0$ and $E^o > 0$	$\Delta G^o > 0$ and $E^o < 0$
Spontaneous reactions	Forced reactions
Positive cathode	Negative cathode
Negative anode	Positive anode
Dry batteries	Electroplating, cathodic protection

in Chap. 4). However, the electron can only be transferred if an empty state of that energy is available on the electrode surface [3]. For example, the reduction of copper cations (Cu^{2+}) requires that its ionic electronic configuration must have two empty states, $3d^9$ and $4s$ subshells, which are filled by $2e^-$ electrons coming from the anode. Hence,

$$\begin{aligned} &\mathrm{Cu}^{2+} \ : \ 1s^2 2s^2 2p^6 3s^2 3p^6 3d^9 = [Ar]3d^9 \\ Cu^{2+} + 2e^- &\rightarrow Cu \\ &\mathrm{Cu}: \quad 1s^2 2s^2 2p^6 3s^2 3p^6 3d^{10} 4s^1 = [Ar]3d^{10} 4s^1 \end{aligned} \tag{2.36}$$

and Cu atoms deposit on the cathode electrode surface. The available states for deposition are mainly surface defects, such as vacancies and step edges, where atomic clusters form and eventually an atomic surface layer is produced [21].

Additionally, the fundamental foundation of electrolytic cells is due to **Faraday's laws of electrolysis**:

- **Faraday's first law** states that the mass of a substance produced during electrolysis is proportional to the quantity of electricity passed ($Q = zF$). Also, $Q = It$ where I is the direct current (DC) and t is time, and the product is in Coulomb units (C). For the reduction of copper, $Cu^{2+} + 2e^- \rightarrow Cu$, the amount of charge is $Q = zF = (2)(96{,}500\,\mathrm{C}) = 1.93x10^5\,\mathrm{C}$ of electric charge.
- **Faraday's second law** states that the mass of a substance dissolved or deposited during the passage of an amount of electricity (Q) is proportional to the electrochemical equivalent mass (EEM) of the substance. This is the mass of the cation deposited by passing an amount of charge $Q = zF$. Fundamentally, one Faraday (1 F) of electricity is equal to the charge carried by one mole of electrons. Thus, $1\,\mathrm{F} = 96{,}500\,\mathrm{C}$.

2.2.4 Concentration Cells

In this type of cells, the electric potential arises as a result of direct charge transfer of the electrolyte from the more concentrated solution to the less concentrated solution.

As a matter of fact, a concentration cell acts to dilute the more concentrated electrolyte solution, creating a potential difference between the electrodes as the cell reaches an equilibrium state. This equilibrium occurs when the concentration of the reactant, say, Ag^+ cations, in both cells is equal. This electrochemical process is achieved by transferring the electrons from the half-cell containing the lower cation concentration to the half-cell having the higher concentration. Furthermore, concentration cell corrosion is also a phrase being used to indicate that corrosion occurs when different areas of a metal surface are in contact with different concentrations of the same solution. This particular case can be found in buried steel pipes exposed to varying *pH* of the soil. There are three general types of concentration cells: (1) metal ion concentration cells due to the presence of water, (2) oxygen concentration cells due to dissolved oxygen in solution, and (3) active-passive cells due to a passive film formation for corrosion protection.

2.3 Standard Electric Potential

In general, the reference electrode selected to measure the standard potential (E^o) of a metal has to be reversible since classical thermodynamics applies to all reversible processes. The standard hydrogen electrode (*SHE*) is used for this purpose. The standard potential is also known as the electromotive force (*emf*) under equilibrium conditions: unit activity, 25 °C, and 1 atm (101 kPa) pressure. The E^o measurements are for reducing metallic cations (M^{z+}). Figure 2.7 shows the *SHE* cell diagram, and Fig. 2.8 schematically illustrates the *SHE* electrochemical cell. Table 2.2 lists *emf* values represented as E_M^o for some pure elements.

As can be seen in Fig. 2.8, the electric potential of *M* at the surface is measured against the *SHE*. It is the resultant of several interfacial potentials described by Eq. (2.24), and it is also known as the **interfacial cell potential**, which depends on the chemical potential of the ions into solution at equilibrium. On the other hand, the electrons flow toward the cathode electrode *M*, where the metal cations M^{z+} gain these electrons and enter the electrode lattice.

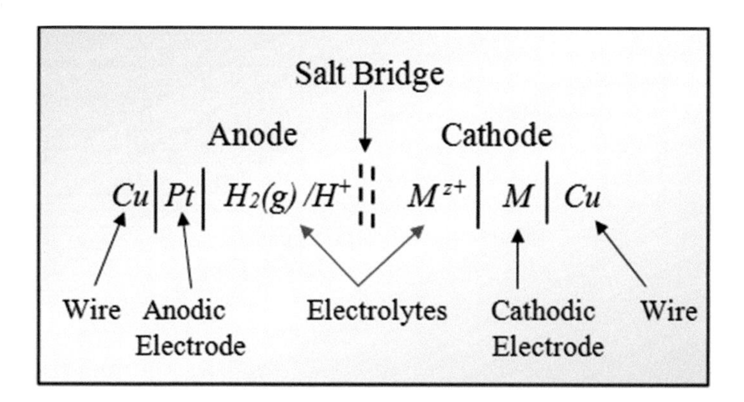

Fig. 2.7 Metal/SHE standard cell diagram

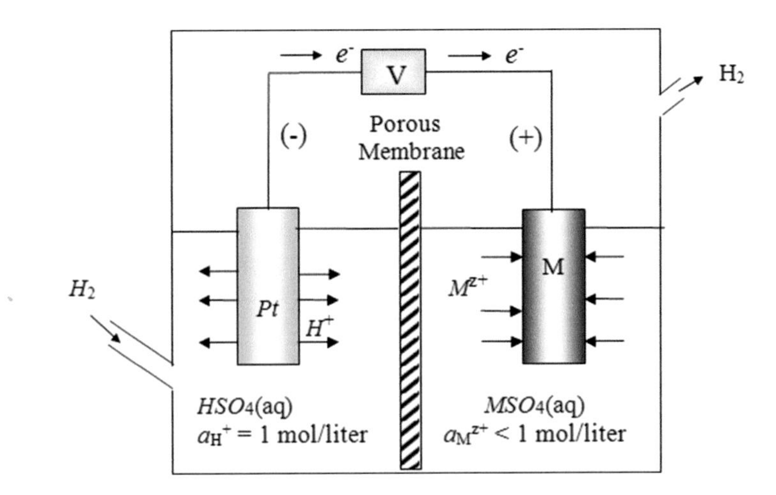

Fig. 2.8 Schematic metal/SHE cell

Table 2.2 Standard potential for metal reduction [22, 23]

Type	Reduction reaction	E^o (V_{SHE})
Noble	$Au^{3+} + 3e^- = Au$	+1.498
↑	$O_2 + 4H^+ + 4e^- = 2H_2O$	+1.229
	$Pt^{2+} + 2e^- = Pt$	+1.200
	$Pd^{2+} + 2e^- = Pd$	+0.987
	$Ag^+ + e^- = Ag$	+0.799
	$Fe^{3+} + e^- = Fe^{2+}$	+0.770
	$Cu^{2+} + 2e^- = Cu$	+0.337
↕	$2H^+ + 2e^- = H_2$	0.000
	$Fe^{3+} + 3e^- = Fe$	−0.036
	$Pb^{2+} + 2e^- = Pb$	−0.126
	$Ni^{2+} + 2e^- = Ni$	−0.250
	$Co^{2+} + 2e^- = Co$	−0.277
	$Cd^{2+} + 2e^- = Cd$	−0.403
	$Fe^{2+} + 2e^- = Fe$	−0.440
	$Cr^{3+} + 3e^- = Cr$	−0.744
	$Zn^{2+} + 2e^- = Zn$	−0.763
	$Ti^{2+} + 2e^- = Ti$	−1.630
	$Al^{3+} + 3e^- = Al$	−1.662
	$Mg^{2+} + 2e^- = Mg$	−2.363
	$Na^+ + e^- = Na$	−2.714
↓	$K^+ + e^- = K$	−2.925
Active	$Li^+ + +e^- = Li$	−3.045

The above standard states (Table 2.2) conform to the convention adopted by the International Union of Pure and Applied Chemistry (IUPAC), which requires that all tabulated electromotive force (*emf*) or standard electrode potential values (E^o) be so with respect to the standard hydrogen electrode (*SHE*) for reduction reactions of pure metals: $M^{z+} + ze^- \rightarrow M$.

The spontaneous redox reaction in Fig. 2.8 is

$$\begin{aligned} H_2 &= 2H^+ + 2e^- && \text{(anode)} \\ M^{z+} + 2e^- &= M && \text{(reduction)} \\ M^{z+} + H_2 &= M + 2H^+ && \text{(redox)} \end{aligned} \tag{2.37}$$

The usefulness of the galvanic series in Table 2.2 is illustrated by the assessment of *Fe*-based couplings to form galvanic cells. Iron (*Fe*) is located between copper (*Cu*) and zinc (*Zn*). Suppose that one steel plate is coated with *Zn* and another with *Cu*. Consequently, *Fe* is the anode for *Cu* and the cathode for *Zn* couplings. In the latter case, *Zn* becomes a sacrificial anode, which is the principle of coupling for galvanized steel sheets and pipes. On the other hand, if *Cu* coating breaks down, steel is then exposed to an electrolyte and becomes the anode, and therefore, it oxidizes. These two cases are schematically shown in Fig. 1.5. The cell potential of each coupling case is determined by adding the standard potentials, but the sign of E^o_{Fe} ($E^o_{Fe} = +0.44$ V) and E^o_{Zn} ($E^o_{Zn} = +0.763$ V) is reversed since they act as anodes. Mathematically, the cell potentials for the cited cases are

$$Fe\text{-}Zn{:}\;\; E^o_{cell} = E^o_{Fe} + E^o_{Zn} = -0.44\,\text{V} + 0.763\,\text{V} = 0.323\,\text{V} \tag{a}$$

$$Fe\text{-}Cu{:}\;\; E^o_{cell} = E^o_{Fe} + E^o_{Cu} = +0.44\,\text{V} + 0.337\,\text{V} = 0.777\,\text{V} \tag{b}$$

For comparison purposes, Table 2.3 illustrates standard potentials for some compounds or alloys.

Nonetheless, the *SHE* is a gas electrode that consists of a platinum foil suspended in sulfuric acid solution having H^+ unit activity ($a_{H^+} = 1$ mol/l) at 1 atm and 25 °C. In order to maintain $a_{H^+} = 1$ mol/l, purified hydrogen (H_2) gas is injected into the anode half-cell in order to remove any dissolved oxygen. The platinum foil is an inert material in this solution, and it allows the hydrogen molecules to oxidize, providing electrons to the metal M^{z+} ions to be reduced on the cathode surface. The concentration of M^{z+} ions are also kept at unit activity. By convention, the standard hydrogen electrode potential is zero: $E^o_H = 0$.

For comparison, Table 2.4 lists some pure metals and common alloys for making galvanic series, and Table 2.5 lists some reference electrodes used for measuring corrosion potentials of metals and alloys in specific applications.

The purpose of the reference electrode (RE) is to provide a stable and reproducible potential (voltage). Essentially, most reference electrodes are suitable over a limited range of conditions, such as *pH* or temperature; otherwise, the electrode behavior becomes unpredictable and inaccurate. Apparently, the mer-

Table 2.3 Standard potentials for compounds [23]

Type	Reduction reaction	E^o (V_{SHE})
Anodic	$O_3 + 2H^+ + 2e^- = O_2 + H_2O$	+2.070
↑	$MnO_2 + 4H^+ + 4e^- = Mn^{2+} + 2H_2O$	+1.230
	$O_2 + 4H^+ + 4e^- = 2H_2O$	+1.229
	$TiO_2 + 4H^+ + 4e^- = Ti + 2H_2O$	+0.860
	$NiO_2 + 2H_2O + 2e^- = Ni(OH)_2 + 2(OH^-)$	+0.760
	$O_2 + 2H_2O + 4e^- = 4(OH^-)$	+0.401
	$Ag_2O + H_2O + 2e^- = 2Ag + 2(OH^-)$	+0.342
	$Pt(OH)_2 + 2e^- = Pt + 2(OH^-)$	+0.400
	$Pd(OH)_2 + 2e^- = Pd + 2(OH^-)$	+0.100
↕	$2H^+ + 2e^- = H_2$	0.000
	$Cu(OH)_2 + 2e^- = Cu + 2(OH^-)$	−0.224
	$PbSO_4 + 2e^- = Pb + SO_4^{2-}$	−0.359
	$Ni(OH)_2 + 2e^- = Ni + 2(OH^-)$	−0.660
	$Co(OH)_2 + 2e^- = Co + 2(OH^-)$	−0.730
	$Fe(OH)_3 + 3e^- = Fe + 3(OH^-)$	−0.771
	$2H_2O + 2e^- = H_2 + 2(OH^-)$	−0.828
↓	$Fe(OH)_2 + 2e^- = Fe + 2(OH^-)$	−0.877
Cathodic	$Mn(OH)_2 + 2e^- = Mn + 2(OH^-)$	−1.470

Table 2.4 Galvanic series of metal and alloys in seawater [24]

Noble	Zinc (Zn)
↑	Low-carbon steels
	Alloy steels
	Cast iron
	Stainless steels
	α-Brass
	Red brass
	Copper (Cu)
	Al-bronze
	Cu-Ni alloys
	Nickel (Ni)
↓	Inconel
Active	Titanium

cury/mercurous chloride (Hg_2Cl_2, calomel), the copper/copper sulfate ($Cu/CuSO_4$), and silver/silver chloride ($Ag/AgCl$) reference electrodes are common in industrial applications, such as corrosion and cathodic protection (cathodic polarization) on buried and submerged structures.

The use of Table 2.5: If the measured potential is $E = -0.541\ V_{SCE}$, then convert V_{SCE} to V_{SHE}. Thus, $E = E_{measure} + E_{Table\ 2.5}$, and

$$E = (-0.541 + 0.241)\ V_{SHE} = -0.30\ V_{SHE}$$

Table 2.5 Secondary reference electrode potentials [22]

Name	Half-cell reaction	E (V) vs. SHE
Mercury sulfate	$HgSO_4 + e^- = Hg + SO_4^{2-}$	0.615
Copper sulfate	$CuSO_4 + 2e^- = 2Cu + SO_4^{2-}$	0.318
Saturated calomel	$Hg_2Cl_2 + 2e^- = 2Hg + 2Cl^-$	0.241
Silver chloride	$AgCl + e^- = Ag + Cl^-$	0.222
Standard hydrogen	$2H^+ + 2e^- = H_2$	0.000

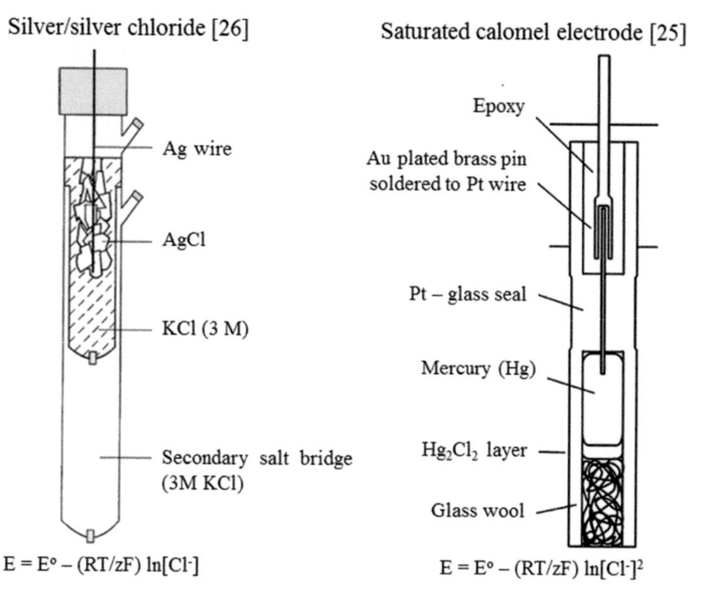

Fig. 2.9 Schematic reference electrodes

A reference electrode is just a stable half-cell electrode used to measure the electric potential difference against a half-cell working electrode. In other words, the potential of a working electrode is measured relative to that of the reference electrode. As a result, meaningful measurements are carried out by knowing that the reference electrode potential is kept constant during current flow. The "Handbook of Reference Electrodes" [25] should be consulted for details on this matter. Only a brief introduction of the type of reference electrodes is included in this section.

The main features of a reference electrode are that (1) it is thermodynamically stable since a reversible reaction takes place between the working electrode and its electrolyte (environment) and that (2) its reversible reaction follows the Nernst equation, which linearly depends on the testing temperature and logarithmically on the concentration ratio.

Figure 2.9 exhibits two schematic stable reference electrodes along with their corresponding half-cell potential equations (Nernst equations).

These reference electrodes can be constructed in a laboratory or purchased from a suitable vender.

Moreover, meaningful electrochemical electric potential measurements are achieved using a stable reference electrode (RE). This implies that RE must be constructed so that its composition remains constant and its response is stable over time. Hence, the measured cell potential undergoing changes must be attributed to the electrolyte depletion of the ion concentration of species j.

With respect to Fig. 2.9, both the Ag/AgCl and SCE reference electrodes are common stable electrochemical half-cells since the half-cell potential does not change over time or with temperature to a reasonable extent. However, contact junctions of the half-cells may degrade over time, and environmental species may be transported into the electrolyte, causing contamination or poisoning. This problem is preventable if the design application stage and subsequent fabrication are carried with caution. The most common problem is human mishandling carelessness.

2.4 Fuel Cells

Fuel cells are like batteries, except that they are continuously receiving an external supply of hydrogen gas (H_2) as the fuel and oxygen (O_2) or air as the oxidant to form water (H_2O). Figure 2.10 schematically shows a simplified fuel cell known as **photon exchange membrane fuel cell (PEMFC)**, where the membrane thickness is $<100\ \mu m$ [4]. Other types of fuel cells and their descriptions can be found elsewhere [4, 26–28].

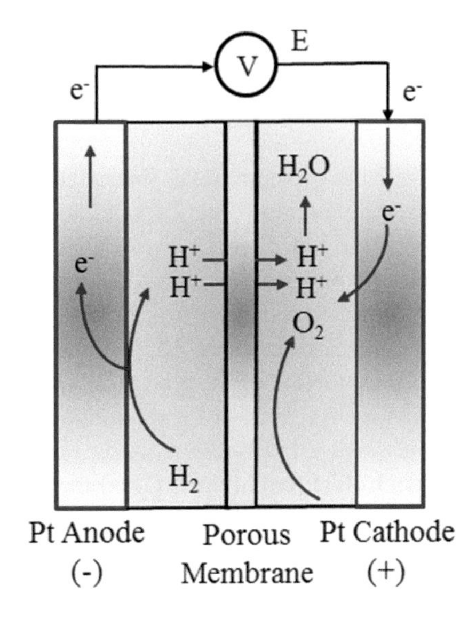

Fig. 2.10 Schematic photon exchange membrane fuel cell (PEMFC)

Nonetheless, the anode and cathode have properly designed channels that transport H_2 and O_2 gases to the Pt catalyst-coated polymeric membrane. These gases are the main reactants and H_2O is the main product.

The electrochemical half-cell reactions that occur in the PEMFC are

$$H_2 \rightarrow 2H^+ + 2e^- \quad E^o = 0 \tag{2.38a}$$

$$2H^+ + \frac{1}{2}O_2 + 2e^- \rightarrow H_2O \qquad E^o = 1.229\ \text{V} \tag{2.38b}$$

Hydrogen cation (H^+) in the anode half-cell crosses the Pt catalyst-coated porous membrane to react with oxygen within the cathode half-cell to form water, which is carried away through design (not shown) channels. This simple mechanism is powerful enough to produce energy for implantable devices, portable electronic devices, and the like [4].

According to Engel and Reid [4], an individual cell can be designed to have a stack of Pt-based catalyst-membrane sequence so that the potential across the stack is a multiple of the individual cell potential. In fact, the goal of any fuel cell is to convert the chemical energy in a fuel into electrical energy as long as fuel is supplied. For comparison, a battery is an energy storage device that ceases to produce electrical energy when the chemical reactants are consumed (discharged).

The fuel cell thermodynamics indicates that the cell potential can be theoretically calculated using Nernst equation. Hence, for the overall reaction, Eq. (2.38b),

$$E = E^o - \frac{RT}{zF}\ln\left(K_{sp}\right) \tag{2.39a}$$

$$E = E^o - \frac{RT}{zF}\ln\left(\frac{[H_2O]}{[H^+]^2[O_2]^{1/2}}\right) \tag{2.39b}$$

where $[H_2O] = [O_2] = 1$ mol/l and $E^o = 1.1229\,\text{V}$ (Table 2.2). From kinetics standpoint, Butler–Volmer equation predicts the fuel cell current density as

$$i = i_o \exp\left[\frac{\alpha zF\left(E - E^o\right)}{RT}\right] \tag{2.40a}$$

$$i = i_o \exp\left[\alpha \ln\left(K_{sp}\right)\right] = i_o K_{sp} \exp\left(-\alpha\right) \tag{2.40b}$$

where $\alpha = 0.5$ and $i_o = 10^{-6}\,\text{A/cm}^2$:

$$n_{th,\max} = \frac{\Delta G}{\Delta H_f} \tag{2.41}$$

Here, $\Delta H_f = -285.84\,\text{kJ/mol}$ is the enthalpy change for the formation of liquid water. For a single PEMFC with $[H^+] = 0.5\,\text{mol/l}$, $z = 2$ at $T = 298\,\text{K}$,

$$E = 1.229 - \frac{8.314 * 298}{2 * 96500} \ln\left(\frac{1}{0.5^2}\right) = 1.21 \text{ V}$$

$$\Delta G = W_e = -zFE = -2 * 96.500 * 1.21 = -233.53 \text{ kJ/mol}$$

$$i = \left(10^{-6}\right) \exp\left(\frac{0.5 * 2 * 93500\,(1.21 - 1.229)}{8.314 * 298}\right) = 4.88 \times 10^{-7} \text{ A/cm}^2$$

$$\eta_{th,\max} = \frac{\Delta G}{\Delta H_f} = \frac{-116.77}{-285.84} * 100 = 81.70\,\%$$

2.5 Problems/Questions

2.1. Define the galvanic electrode potential for the cell shown in Fig. 2.4 using the interfacial potentials involved in the electrochemical process.

2.2. Describe what happens in the galvanic cell shown in Fig. 2.4 when electrons leave the Cu terminal at the Cu-Zn interface.

2.3. Explain why an electrode potential difference occurs in Fig. 2.4.

2.4. Why are Cu^{+2} cations electroplated on the *Cu* electrode surface in Fig. 2.4?

2.5. Calculate the standard potential for the formation of ferric hydroxide $Fe(OH)_3$ (brown rust).

2.6. Explain why care must be exercised in using the galvanic series illustrated in Table 2.2?

2.7. If zinc and copper rods are placed in saltwater, a direct chemical reaction may slightly corrode zinc. Why?

2.8. Is the cell potential a surface potential? If so, why?

2.9. If a monopole contains a net charge (Q) due to 10^{-12} moles of a species j in a electrically neutral system, then calculate the electric potential (ϕ_x), the electric potential strength (E_x), and the electric force (F_x) acting on a particle within a distance of 9 cm from another charged particle in the x-direction. [Solution: $Q = 0.193\,\mu\text{C}$, $F_x = 0.04\,\text{N}$.]

2.10. A battery (Example 2.2) containing 0.4 moles of MnO_2 delivers 1.5 V. For 4-h operation, calculate the electric current and the power (in watts). [Solution: $I = 2.68\,\text{A}$ and $P = 4.02\,\text{W}$.]

2.11. Write down the reversible hydrogen reaction for a standard electrode, and determine its standard electric potential.

2.12. Calculate the E^o values for each of the following reactions as written, and determine which one is the cathode and anode:

$$2Fe^{3+} + Cd \rightarrow Cd^{2+} + 2Fe^{2+}$$

References

1. I.N. Levine, *Physical Chemistry* (McGraw-Hill Book Company, New York, 1978), pp. 350–396
2. C.M.A. Brett, A.M.O. Brett, *Electrochemistry Principles, Methods and Applications* (Oxford University Press, New York, 1994), pp. 56–58, 78–81, 116–117
3. W. Schmickler, *Interfacial Electrochemistry* (Oxford University Press, New York, 1996), pp. 12, 72
4. T. Engel, P. Reid, *Physical Chemistry*, Chap. 11 (Pearson Education, Boston, 2013)
5. A.J. Bard, L.R. Faulkner, *Electrochemica Methods: Fundamentals and Applications*, 2nd edn., Chaps. 1 and 2 (Wiley, New York, 2001)
6. D.W. Rogers, *Concise Physical Chemistry*, Chap. 14 (Wiley, New York, 2011)
7. L.I. Boguslavsky, Electron transfer effects and the mechanism of the membrane pot, in *Modern Aspects of Electrochemistry*, vol. 18, ed. by R.E. White, J. O'M. Bockris, B.E. Conway (Plenum Press, New York, 1986), pp. 113–114
8. D.G. Hall, Ion-selective membrane electrodes: a general limiting treatment of interference effects. J. Phys. Chem **100**, 7230–7236 (1996)
9. O.C. Mudd, *Control of Pipeline Corrosion - A Manual*, Part 1. Corrosion, vol. 1 (1945), pp. 192–201
10. D.L. Reger, S.R. Goode, E.E. Mercer, *Chemistry: Principles & Practice*, 2nd edn. (Saunders College Publishing, New York, 1997), p. 795
11. D.W. Ball, *Physical Chemistry*, Chap. 8 (Brooks/Cole, a division of Thomson Learning, Inc., Pacific Grove, United States, 2003)
12. V.S. Bagotsky, *Fundamentals of Electrochemistry*, 2nd edn., Chaps. 2–3 (Wiley, New Jersey, 2006)
13. A.J. Bard, L.R. Faulkner, *Electrochemical Methods - Fundamentals and Applications*, 2nd edn., Chaps. 1–2 (Wiley, New York, 2001), p. 548
14. J.P. Gabano, Lithium battery systems: an overview, in *Lithium Batteries*, ed. by J.P. Gabano (Academic, New York, 1983), pp. 1–12
15. P. Arora, R.E. White, M. Doyle, J. Electrochem. Soc. **145**(10), 3647–3667 (1998)
16. P. Arora, M. Doyle, R.E. White, J. Electrochem. Soc. **146**(10), 3543–3553 (1999)
17. P. Arora, B.N. Popov, R.E. White, J. Electrochem. Soc. **145**(3), 807–814 (1998)
18. M.G. Fontana, *Corrosion Engineering* (McGraw-Hill Book Company, New York, 1986)
19. P. De Vidts, J. Delgado, B. Wu, D. See, K. Kosanovich, R.E. White, J. Electrochem. Soc. **145**(11), 3874–3883 (1998)
20. B. Wu, R.E. White, Modeling of a nickel-hydrogen cell: phase reactions in the nickel active material. J. Electrochem. Soc. **148**(6), A595–A609 (2001)
21. D.M. Kolb, G.E. Engelmann, J.C. Ziegler, On the unusual electrochemical stability of nanofabricated copper clusters. Angew. Chem. Int. Ed. **39**(6), 1123–1125 (2000)
22. D.A. Jones, *Principles and Prevention of Corrosion* (Macmillan Publishing Company, New York, 1992), pp. 39–70
23. G.F. Carter, *Principles of Physical and Chemical Metallurgy* (American Society For Metals, Metals Park, OH, 1979), pp. 316–317

24. A.L. Rouff, *Materials Science* (Prentice-Hall, New Jersey, 1973), p. 697
25. G. Inzelt, A. Lewenstam, F. Scholz (eds.), *Handbook of Reference Electrodes* (Springer, Berlin, Heidelberg, New York, 2013)
26. H. Kahlert, *Electroanalytical Methods: Guide to Experiments and Applications*, 2nd edn., ed. by F. Scholz (Springer, Berlin, 2010), p. 305
27. K. Sundmacker, Fuel cell engineering: toward the design of efficient electrochemical power plants. Ind. Eng. Chem. Res. **49**, 10159–10182 (2010)
28. Fuel Cell Handbook, 7th edn. EG&G Technical Services, Inc., Under Contract No. DE-AM26-99FT40575, U.S. Department of Energy, Office of Fossil Energy (November 2004)

3 Thermodynamics of an Electrochemical Cell

In principle, most solids can corrode, to an extent, under the influence of an electrode electric potential difference in an electrolyte having a certain pH value at a specific temperature T. Thus, a full characterization of an electrochemical cell must include theoretical concepts of thermodynamics and kinetics of the electrochemical processes.

In fact, electrochemistry is the discipline that takes into account the chemical phenomena associated with homogeneous charge transfer in an electrolyte due to ionic transport and heterogeneous charge transfer on electrode surfaces due to the electric wire system. This leads to electroneutrality due to, at least, two half-cell reactions in opposite directions. However, the electrochemical heterogeneity associated with oxidation and reduction reactions on the corresponding electrode surfaces is located at the interfacial region between the electrolyte and electrode surface, and its characteristics, including the free energy change, are influenced by the ions in this region and the electrode electric potential difference [1].

In this chapter, thermodynamics is the relevant theoretical approach for describing an electrochemical cell, especially galvanic cells containing a pair of electrodes. Multielectrode cells are dealt with in a later chapter. Fundamentally, the thermodynamics of an electrochemical cell is detailed using Pourbaix diagrams, which do not include fundamental aspects of kinetics related to an electrochemical cell.

The subsequent analytical procedure leads to the derivation of the Gibbs free energy change and the Nernst equation, which is suitable for determining the cell electric potential when ion activities are less than unity as nonstandard conditions.

3.1 Electrochemical Cell Phases

This section deals with the phases present in an electrochemical cell as shown in Fig. 3.1. Treat each half-cell as an isolated system. Let metal M_1 rod be the α-phase being immersed in an aqueous solution (electrolyte) ϵ-phase, $MSO_4(aq)$, so that z^+

N. Perez, *Electrochemistry and Corrosion Science*,
DOI 10.1007/978-3-319-24847-9_3

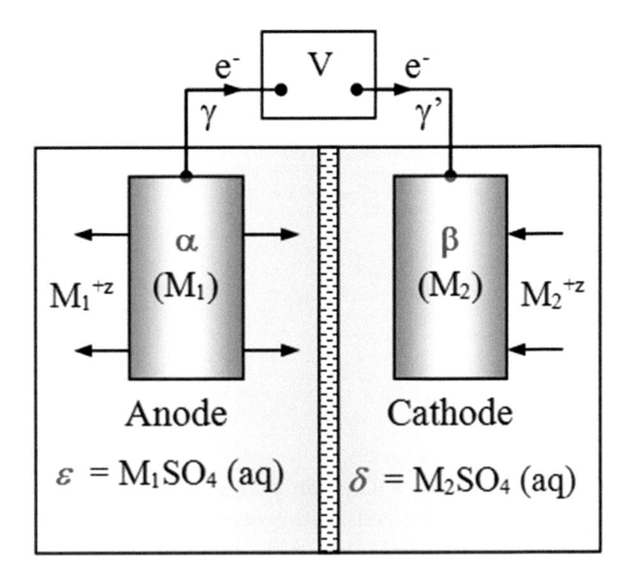

Fig. 3.1 Galvanic cell and its phase components

and z^- are the valence of the cation (M^{z+}) and the anion (SO_4^{2-}), respectively. Then, ϕ_α is the electric potential of the metallic α-phase, and ϕ_ϵ is the electric potential of the ϵ-phase. Once the α-phase is in contact with the ϵ-phase, an exchange of charged particles between α and ϵ phases causes a change in the total energy of the system, which consists of the negatively charged α-phase as electrons are detached from the metallic α-phase. The altered total energy is due to the change in electrostatic potential energy of the phases involved, specifically at the metal-electrolyte interface where a strong electric field exists and an electrical double layer of charged particles disturbs the charge on the interface. This electrical double layer is a normal balance (equal number) of positive and negative ions, which diffuse to the interface causing such as disturbance in energy [2]. Eventually, the system will reach the state of equilibrium within the electrolyte.

The electrochemical reactions for the above cell are

$$(M_1)_\alpha = \left(M_1^{z+}\right)_\epsilon + (ze^-)_\gamma \qquad \text{(Anode)} \tag{3.1}$$

$$\left(M_2^{z+}\right)_\delta + (ze^-)_{\gamma'} = (M_2)_\beta \qquad \text{(Cathode)} \tag{3.2}$$

$$(M_1)_\alpha + \left(M_2^{z+}\right)_\delta = \left(M_1^{z+}\right)_\epsilon + (M_2)_\beta \qquad \text{(Redox)} \tag{3.3}$$

where $(ze^-)_\gamma = (ze^-)_{\gamma'}$ for electron balance and γ and γ' are the wire phases, usually *Pt* or *Cu* metals.

The corresponding electrochemical potential equation for each half-cell is

$$\left(\mu_{M_1}\right)_\alpha = \left(\mu_{M_1^{z+}}\right)_\epsilon + (z\mu_{e^-})_\gamma \tag{3.4}$$

$$\left(\mu_{M_2}\right)_\beta = \left(\mu_{M_2^{z+}}\right)_\delta + (z\mu_{e^-})_{\gamma'} \tag{3.5}$$

For example, if $\alpha = Zn$ and $\beta = Cu$, then Eqs. (3.4) and (3.5) become

$$Zn = Zn^{2+} + 2e^{-} \quad \text{(Anode)} \tag{3.6}$$

$$Cu^{2+} + 2e^{-} = Cu \quad \text{(Cathode)} \tag{3.7}$$

$$Zn + Cu^{2+} = Zn^{2+} + Cu \quad \text{(Redox)} \tag{3.8}$$

The mathematical procedure for deriving Eqs. (3.4) and (3.5) can be found elsewhere [3]. Nonetheless, the electrical work for the redox reaction is

$$W_e = FE = F\left(\phi_{\gamma} - \phi_{\gamma'}\right) \tag{3.9}$$

$$FE = \left(\mu_{e^-}\right)_{\gamma} - \left(\mu_{e^-}\right)_{\gamma'} \tag{3.10}$$

where $E = \left(\phi_{\gamma} - \phi_{\gamma'}\right)$ is the electric potential difference between α and β phases being connected through voltmeter V and F is the Faraday's constant.

Multiplying Eq. (3.10) by z and combining the resultant expression with Eqs. (3.4) and (3.5) yields

$$zFE = \left(\mu_{M_1}\right)_{\alpha} - \left(\mu_{M_1^{z+}}\right)_{\epsilon} - \left(\mu_{M_2}\right)_{\beta} + \left(\mu_{M_2^{z+}}\right)_{\delta} \tag{3.11}$$

Thus, the Gibbs free energy change for the redox reaction, Eq. (3.3), takes the form

$$\Delta G = -\left[\left(\mu_{M_1}\right)_{\alpha} - \left(\mu_{M_1^{z+}}\right)_{\epsilon} - \left(\mu_{M_2}\right)_{\beta} + \left(\mu_{M_2^{z+}}\right)_{\delta}\right] \tag{3.12}$$

Combining Eqs. (3.11) and (3.12) at standard and nonstandard conditions gives

$$\Delta G^{o} = -zFE^{o} \quad \text{(standard)} \tag{3.13}$$

$$\Delta G = -zFE \quad \text{(nonstandard)} \tag{3.14}$$

This expression, Eq. (3.14), is used for determining if $\Delta G > 0$ for a non-spontaneous process, $\Delta G < 0$ for a spontaneous process, and $\Delta G = 0$ for equilibrium. Similarly, ΔG^{o} in Eq. (3.13) undergoes the same conditions.

Actually, ΔG^{o} and E^{o} are extensive (dependent on the amount of a species) and intensive (independent of the amount of a species) properties, respectively [4].

The standard Gibbs free energy change can be defined in a general form using the chemical potential (energy related to diffusion of ions) of the product and reactants. Thus,

$$\Delta G = \sum \nu_j \mu_j = \left(\sum \nu_j \mu_j\right)_{product} - \left(\sum \nu_j \mu_j\right)_{reactants} \tag{3.15}$$

where ν_j = number of ions or molecules in a reaction and
$j = 1, 2, 3, \ldots$ = species involved in the reaction.

Combining Eqs. (3.14) and (3.15) yields the standard electric potential (E^o) in terms of standard chemical potential (μ^o)

$$E^o = -\frac{\sum \nu_j \mu_j}{zF} \tag{3.16}$$

Applying Eq. (3.16) to the redox reaction given by Eq. (3.3) yields

$$E^o = -\frac{1}{zF}\left\{\left[(\mu_{M^{z+}})^{\epsilon} + \left(\mu_{M_2}\right)^{\beta}\right] - \left[\left(\mu_{M_1}\right)^{\alpha} + (\mu_{M^{z+}})^{\delta}\right]\right\} \tag{3.17}$$

The standard cell electric potential using Table 3.1 is

$$E^o_{M^{z+}/M} = \frac{1}{zF}\left\{\begin{array}{c}\left[(1)\mu^o_M + (2)\mu^o_{H+}\right] \\ -\left[(1)\mu^o_{M^{2+}} + (1)\mu^o_{H_2}\right]\end{array}\right\} \tag{3.18}$$

$$E^o_{M^{z+}/M} = \frac{\mu^o_{M^{z+}}}{zF} \quad \text{since } \mu^o_M = \mu^o_{H+} = \mu^o_{H_2} = 0 \tag{3.19}$$

Table 3.1 illustrates standard chemical potential (μ^o) for some species [5, 6].

According to Pourbaix classical lecture, the transformation of water gives [8]

$$2H_20(g) = 2H_2(g) + O_2 \tag{3.19a}$$

$$2\mu_{H_20(g)} = 2\mu_{H_2(g)} + \mu_{O_2}(g) \tag{3.19b}$$

From Table 3.1 under standard conditions, $\mu_{H_20(g)} = \mu^o_{H_20(g)} = -237.19\,\text{kJ/mol}$ since $\mu_{H_2(g)} = \mu_{O_2}(g) = 0$.

Table 3.1 Standard chemical potentials [5, 7]

Name	μ^o (kJ/mol)	Name	μ^o (kJ/mol)
Al	0	Cu	0
Al_2O_3	−1576.41	CuO	−127.19
$Al(OH)_3$	−1137.63	$Cu(OH)_2$	−356.90
Al^{3+}	−481.16	$CuSO_4$	−661.91
AlO_2^-	−839.77	Cu^{2+}	+64.98
Cr	0	Fe	0
$Cr(OH)_2$	−587.85	$Fe(OH)_2$	−483.54
$Cr(OH)_3$	−900.82	$Fe(OH)_3$	−694.54
Cr^{3+}	−215.48	$FeSO_4$	−829.69
CrO_2^-	−835.93	Fe^{2+}	−84.94
$H_2(g)$	0	Fe^{3+}	−10.59
H^+	0	Zn	0
OH^-	−157.30	$Zn(OH)_2$	−559.09
H_2O	−237.19	Zn^{2+}	−147.21

Example 3.1. *Assume that the cell in Fig. 3.1 is in equilibrium at room temperature and that the cell phases are* $M_1^\alpha = Zn^\alpha$ *with* $z_1 = +2$, $\epsilon = ZnSO_4(aq)$, $M_2^\beta = Cu^\beta$ *with* $z_2 = +2$, *and* $\delta = CuSO_4(aq)$. *Will the redox reaction occur? Why or why not?*

Solution. *From Table 3.1 along with* $M_1^{z+} = Zn^{2+}$, $M_2^{z+} = Cu^{2+}$, *and Eq. (3.19),*

$$E^o_{Zn/Zn^{2+}} = \frac{\mu^o_{Zn^{z+}}}{zF} = \frac{+147.21\ \text{kJ/mol}}{(2)(96.50\ \text{kJ/mol V})} = 0.763\ \text{V}$$

$$E^o_{Cu^{2+}/Cu} = \frac{\mu^o_{Cu^{z+}}}{zF} = \frac{+64.98\ \text{kJ/mol}}{(2)(96.50\ \text{kJ/mol V})} = 0.337\ \text{V}$$

which are exactly the results given in Table 2.2. Then, the standard cell potential is

$$E^o = E^o_{Zn} + E^o_{Cu} = +0.763\ \text{V} + 0.337\ \text{V} = 1.10\ \text{V}$$

Using Eq. (3.13) along with and $z_1 = z_2 = z = 2$ *yields the standard Gibbs free energy change*

$$Zn^\alpha + \left(Cu^{2+}\right)^\delta = \left(Zn^{2+}\right)^\epsilon + Cu^\beta$$

$$\Delta G^o = -zFE^o = (2)(96.50\ \text{kJ/mol V})(1.10\ \text{V})$$

$$\Delta G^o = -212.30\ \text{kJ/mol}$$

For each half-cell reaction,

$$\Delta G^o_{Zn} = -zFE^o_{Zn} = (2)(96.50\ \text{kJ/mol V})(0.763\ \text{V}) = -147.26\ \text{kJ/mol}$$

$$\Delta G^o_{Cu} = -zFE^o_{Cu} = (2)(96.50\ \text{kJ/mol V})(0.337\ \text{V}) = -65.041\ \text{kJ/mol}$$

Then,

$$\Delta G^o = \Delta G^o_{Zn} + \Delta G^o_{Cu} = (-147.26 - 65.041)\ \text{kJ/mol} = -212.30\ \text{kJ/mol}$$

Therefore, the redox reaction occurs spontaneously because $\Delta G^o < 0$. In this particular galvanic cell, the spontaneous redox reaction drives the cell to produce an electric potential since the change in Gibbs free energy is negative. This means that the oxidation-reduction reaction generates an electric current and hence the name of the cell, galvanic or voltaic cell. Also, calculated standard potential (E^o) of this cell must do work by driving an electric current through a wire system (not shown). Nonetheless, E^o fundamentally measures the strength of the driving force behind the electrochemical redox reaction.

Example 3.2. *Let the anodic phases in Fig. 3.1 be* $(M_1)_\alpha = Cr$ *with* $z = +3$ *in* $\epsilon = CrSO_4(aq)$*, whereas the cathodic phases are* $(M_2)_\beta = Cu$ *with* $z = +2$ *in* $\delta = CuSO_4(aq)$*. **(a)** Determine the redox reaction and **(b)** calculate the Gibbs free energy change and the cell electric potential. Will the redox reaction occur?*

Solution. (a) *The half-cell reactions from Table 2.2 must be balanced, but the standard cell potentials do not change. Thus,*

$$\begin{aligned} 2Cr &\rightarrow 2Cr^{3+} + 6e^- && E^o_{Cr} = 0.744\ \text{V} \\ 3Cu^{2+} + 6e^- &\rightarrow 3Cu && E^o_{Cu} = 0.337\ \text{V} \\ 2Cr + 3Cu^{2+} &\rightarrow 2Cr^{3+} + 3Cu && E^o_{cell} = E^o_{Cr} + E^o_{Cu} = 1.081\ \text{V} \end{aligned}$$

(b) *The Gibbs free energy change for the* Cr/Cr^{3+} *half-cell is*

$$\Delta G^o_{Cr/Cr^{3+}} = -zFE^o_{Cr/Cr^{3+}} = -(6)\,[96.50\ \text{kJ}/(\text{mol V})]\,(0.744\ \text{V})$$
$$\Delta G^o_{Cr/Cr^{3+}} = -430.78\ \text{kJ/mol}$$

that for Cu^{2+}/Cu *is*

$$\Delta G^o_{Cu^{2+}/Cu} = -zE^o_{Cu^{2+}/Cu} = (6)\,[96.50\ \text{kJ}/(\text{mol V})]\,(0.337\ \text{V})$$
$$\Delta G^o_{Cu^{2+}/Cu} = -195.12\ \text{kJ/mol}$$

For the entire cell,

$$\Delta G^o_{Cr/Cu} = \Delta G^o_{Cr/Cr^{3+}} + \Delta G^o_{Cu^{2+}/Cu}$$
$$\Delta G^o_{Cr/Cu} = -430.78\ \text{kJ/mol} - 195.12\ \text{kJ/mol}$$
$$\Delta G^o_{Cr/Cu} = -625.90\ \text{kJ/mol}$$

In addition, $\Delta G^o_{Cr/Cu}$ *can also be determined in the following way*

$$\Delta G^o_{Cr/Cu} = \Delta G^o_{Cr/Cr^{3+}} + \Delta G^o_{Cu^{2+}/Cu} = -zFE^o_{Cr/Cr^{3+}} - zE^o_{Cu^{2+}/Cu}$$
$$\Delta G^o_{Cr/Cu} = -zF\left(E^o_{Cr/Cr^{3+}} + E^o_{Cu^{2+}/Cu}\right) = -zFE_{cell}$$
$$\Delta G^o_{Cr/Cu} = -(6)\,[96.50\ \text{kJ}/(\text{mol V})]\,(1.081\ \text{V})$$
$$\Delta G^o_{Cr/Cu} = -625.90\ \text{kJ/mol}$$

Therefore, the redox reaction occurs spontaneously because $\Delta G^o < 0$*, and in principle,* $\Delta G^o = -625.90\,kJ/mol$ *is the amount of the driving force for generating the electric potential* $E^o_{cell} = 1.081\,V$ *between the metal electrodes and for allowing spontaneous electrochemical reactions to occur. Again,* $E^o_{cell} = 1.081\,V$ *fundamentally measures the strength of the driving force behind the electrochemical redox reaction.*

3.2 The Nernst Equation

The activity $a_j = [j]$ of a dissolved species j in a electrolyte solution is fundamentally the concentration in such a solution. Its mathematical definition is based on chemical potential (μ_j) of the species

$$a_j = \exp\left[\frac{\mu_j - \mu^o}{RT}\right] \tag{3.20}$$

Table 3.2 illustrates some important definitions, which are needed to solve electrochemical problems at standard conditions, $T = 25\,°\text{C}$, $P_o = 1$ atm and $a_j = 1$ mol/ l.

From the second law of thermodynamics, the change in free energy is [3]

$$dG = -SdT + VdP \tag{3.21}$$

For an isothermal (T = constant) and isometric (V = constant) system, Eq. (3.21) yields

$$\int_{\Delta G_j^o}^{\Delta G_j} dG = V \int_{P_o}^{P} dP \tag{a}$$

$$\Delta G_j - \Delta G^o = V(P - P_o) \tag{3.22}$$

Dividing this equation by RT and using the natural exponential function gives

$$\exp\left[\frac{\Delta G - \Delta G^o}{RT}\right] = \exp\left[\frac{V(P - P_o)}{RT}\right] \tag{3.23}$$

where P = pressure
P_o = standard pressure = 1 atm (101 kPa)
$R = 8.314510\,\text{J mol}^{-1}\,\text{K}^{-1}$ (universal gas constant)
T = absolute temperature (K)

Table 3.2 Useful definitions in the science of interactions

a_j = activity (mol/l)	$a_j = [j] = \gamma_j X_j$(Henry's law)
C_j = concentration (g/l)	$C_j = a_j A_{w,j}$
V_j = molar volume (l)	$a_j = P_j/P_o = \gamma\left[C_j/C^o\right]$
P_j = pressure (MPa)	$H_2O = H^+ + OH^-$
P_o = 1 atm = 101 kPa	$K_w = \left[H^+\right]\left[OH^-\right] = 10^{-14}$
γ_j = activity coefficient	$pH = -\log\left[H^+\right]$
A_w = atomic weight (g/mol)	$pH = 14 + \log\left[OH^-\right]$
X_j = mole fraction	μ^o = standard chemical potential

It is understood henceforth that dissolved species (substances) j are at unit activity of one molarity (1 M = mol/l).

Now, the equilibrium constant K_e (reaction quotient) for a redox reaction such as Eq. (3.3) can be defined in terms of activities as follows:

$$K_e = \frac{\sum a_{\text{Product}}}{\sum a_{\text{Reactants}}} = \frac{\sum [\text{Product}]}{\sum [\text{Reactants}]} \tag{3.24}$$

$$K_{e,\alpha-\epsilon} = \frac{\left[\left(M_1^{z+}\right)^{\epsilon}\right]}{\left[M_1^{\alpha}\right]} \quad \text{(anode half-cell)} \tag{3.25}$$

$$K_{e,\beta-\delta} = \frac{\left[M_2^{\beta}\right]}{\left[\left(M_2^{z+}\right)^{\delta}\right]} \quad \text{(cathode half-cell)} \tag{3.26}$$

and for the redox reaction

$$K_e = K_{e,\alpha-\epsilon} K_{e,\beta-\delta} = \frac{\left[\left(M_1^{z+}\right)^{\epsilon}\right] \cdot \left[M_2^{\beta}\right]}{\left[M_1^{\alpha}\right] \cdot \left[\left(M_2^{z+}\right)^{\delta}\right]} \tag{3.27}$$

For the dissolution of a compound ($AgCl$ salt), the reaction constant is called solubility product (K_{sp}) instead of equilibrium constant. From Eq. (3.23), K_{sp} for a simple redox reaction is

$$K_{sp} = \exp\left(\frac{\Delta G - \Delta G^o}{RT}\right) \tag{3.28}$$

Solving Eq. (3.28) for the non-equilibrium Gibbs energy change ΔG gives

$$\Delta G = \Delta G^o + RT \ln\left(K_{sp}\right) \tag{3.29}$$

At standard conditions, $\Delta G = 0$ and Eq. (3.29) becomes [9]

$$\Delta G^o = -RT \ln\left(K_{sp}\right) \tag{3.30}$$

Thus, the equilibrium constant K_{sp} can be determined from measurements of the electric potential (E^o) or chemical potential (μ^o). Substituting Eqs. (3.13) and (3.14) into (3.29) yields the **Nernst Equation** for calculating the non-equilibrium electric potential (E) of an electrochemical cell. Hence,

$$E = E^o - \frac{RT}{zF} \ln\left(K_{sp}\right) \tag{3.31}$$

The interpretation of the Nernst equation suggests that the current resulting from the change of oxidation state of the metals M_1 and M_2 is known as **Faradaic current**, which is a measure of the rate of redox reaction. The term zF in Eq. (3.31) stands for the number of Coulombs needed for reacting 1 mol of an electroactive species. Therefore, Eq. (3.31) embodies the Faraday's law of electrolysis since the redox reaction involves electron transfer between phases having different electric potentials. For instance, electrochemical equilibrium is achieved if current flow ceases in a reversible galvanic cell (Fig. 3.1). Despite that the Nernst equation is derived assuming thermodynamic equilibrium of a reversible cell, K_{sp} in Eq. (3.31) is not exactly the same as the chemical equilibrium constant due to a current flow. The Nernst equation gives the open-circuit potential difference for a reversible galvanic cell and excludes liquid junction potentials, which are always present but small in magnitude.

From a macroscale, the Nernst equation, Eq. (3.31), is graphically depicted in Fig. 3.2, which elucidates a simple calculus plot. The intercept is just the standard potential (E^o) at temperature T and the slope, $b = -RT/zF$, is the energy/charge ratio.

Example 3.3. *Apply the preceding procedure to the galvanic cell shown in Fig. 3.1 at a temperature of 25 °C for determining the solubility product K_{sp} and the Nernst cell potential E. Let $\alpha = Zn$ with $\left[Zn^{2+}\right] = 0.0002\,mol/l$ and $\beta = Cu$ with $\left[Cu^{2+}\right] = 0.04\,mol/l$ at 298 K.*

Solution.

$$\begin{array}{ll} Zn \longrightarrow Zn^{2+} + 2e^- & E^o_{Zn} = 0.763\,\text{V} \\ Cu^{2+} + 2e^- \longrightarrow Cu & E^o_{Cu} = 0.337\,\text{V} \\ \hline Zn + Cu^{2+} \longrightarrow Zn^{2+} + Cu & E^o = E^o_{Zn} + E^o_{Cu} = 1.10\,\text{V} \end{array}$$

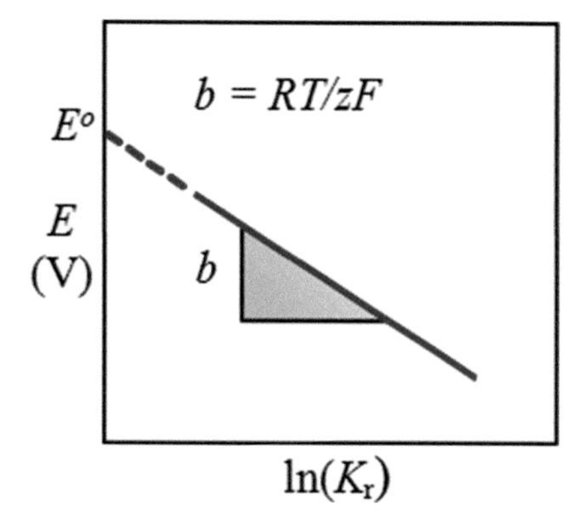

Fig. 3.2 Nernst plot

From Eq. (3.24),

$$K_{sp} = \frac{[Cu]\left[Zn^{2+}\right]}{\left[Cu^{2+}\right][Zn]} = \frac{\left[Zn^{2+}\right]}{\left[Cu^{2+}\right]} = \frac{0.0002}{0.04} = 0.005$$

since $[Cu] = [Zn] = 1\,mol/l$*. Thus, the cell potential defined by Eq. (3.31) along with* $z = 2$ *and* $E^o = 1.10\,V$ *gives*

$$E = E^o - \frac{RT}{zF}\ln K_{sp}$$

$$E = 1.10\text{ V} - (0.0296\text{ V})\ln(0.005) = 1.26\text{ V}$$

3.3 Properties of a Species

An electrochemical cell can be used to measure the entropy change (ΔS) and the enthalpy change (ΔH) of a reversible reaction [4, 9]. This can be accomplished by measurements of the Nernst potential (E) as a function of temperature. At constant pressure, the free energy change (ΔG), ΔS, and ΔH are defined by

$$\Delta G = -zFE \tag{3.14}$$

$$\Delta S = -\left(\frac{\partial \Delta G}{\partial T}\right)_p \tag{3.32}$$

$$\Delta H = \Delta G + T\Delta S \tag{3.33}$$

Differentiating Eq. (3.14) with respect to the temperature T yields

$$\left(\frac{\partial \Delta G}{\partial T}\right)_p = -zF\left(\frac{\partial E}{\partial T}\right)_p \tag{3.34}$$

Then, the entropy change becomes

$$\Delta S = zF\left(\frac{\partial E}{\partial T}\right)_p \tag{3.35}$$

From the second law of thermodynamics, the thermal energy (heat transfer) is defined by

$$Q = -T\Delta S = -zFE - \Delta H \tag{3.36}$$

Now, substituting Eqs. (3.14) and (3.35) into (3.33) yields

$$\Delta H = zF\left[T\left(\frac{\partial E}{\partial T}\right)_p - E\right] \tag{3.37}$$

The heat capacity (C_p) and its change (ΔC_p) at constant pressure (P) are defined by

$$C_p = T\left(\frac{\partial S}{\partial T}\right)_P \tag{3.38}$$

$$\Delta C_p = T\left(\frac{\partial \Delta S}{\partial T}\right)_P \tag{3.39}$$

Substitute Eq. (3.35) into (3.39) to get

$$\Delta C_p = zFT\left(\frac{\partial^2 E}{\partial T^2}\right)_P \tag{3.40}$$

Using an electrochemical cell to measure the cell potential, Eq. (3.31), as a function of temperature at constant pressure leads to curve fitting an empirical function, usually a polynomial [9, 10]. One particular polynomial is of the form [9]

$$E = a + b(T - T_o) + c(T - T_o)^2 + d(T - T_o)^3 \tag{3.41}$$

$$\frac{\partial E}{\partial T} = b + 2c(T - T_o) + 3d(T - T_o)^2 \tag{3.42}$$

$$\frac{\partial^2 E}{\partial T^2} = 2c + 6d(T - T_o) \tag{3.43}$$

where T_o = reference temperature (K) and a, b, c, d = curve fitting coefficients.

Now, the above thermodynamic state functions can be defined in terms of temperature. Thus,

$$\Delta G = -zF\left[a + b(T - T_o) + c(T - T_o)^2 + d(T - T_o)^3\right] \tag{3.44}$$

$$\Delta S = zF\left[b + 2c(T - T_o) + 3d(T - T_o)^2\right] \tag{3.45}$$

$$\Delta C_p = zFT \cdot [2c + 6d(T - T_o)] \tag{3.46}$$

Example 3.4. *Levine [9] applied Eq. (3.41) as a continuous function in the closed temperature interval [0 °C, 90 °C] or [273.15 K, 363.15 K] for plating silver (Ag) from silver chloride (AgCl) according to the following redox (overall) reaction:*

$$2AgCl + H_2 = 2Ag + 2HCl(aq)$$

The curve fitting coefficients at $P = 1$ atm and $T_o = 273.15$ K are

$$a = 0.2366 \text{ V} \qquad b = -4.8564x10^{-4} \text{ V/K}$$
$$c = -3.4205x10^{-6} \text{ V/K}^2 \quad d = 5.8690x10^{-9} \text{ V/K}^3$$

Calculate E, ΔG, ΔS, ΔH, Q, and C_P at $T = 25\,°C$ (298 K) and $P = 1$ atm.

Solution. *If* $T = 25\,°C = 298.15\,K$, $(T - T_o) = 25\,K$, $z = 2$, *and* $F = 96{,}500\,C/mol = 96{,}500\,J/mol\,V$, *then the following results*

From Eq. (3.41),	$E = 0.222\,V$
From Eq. (3.44),	$\Delta G = -21.463\,kJ/mol$
From Eq. (3.45),	$\Delta S = -0.062\,kJ/mol\,K$
From Eq. (3.37),	$\Delta H = -40.04\,kJ/mol$
From Eq. (3.36),	$Q = 61.50\,kJ/mol$
From Eq. (3.46),	$\Delta C_P = -0.17\,kJ/mol\,K$

3.4 Pourbaix Diagrams

A Pourbaix diagram is like a phase diagram having ion boundaries represented by lines. In essence, electric potential-pH plots are known as Pourbaix diagrams since they were first made by Pourbaix in 1938. A compilation of these diagrams is available in the *Atlas of Electrochemical Equilibria in Aqueous Solutions* [11]. In practice, Pourbaix diagrams are suitable for studies of corrosion, electrowinning, electroplating, hydrometallurgy, electrolysis, electrical cells, and water treatment since they are electrochemical maps indicating the domain of stability of ions, oxides, and hydroxides [11]. This map provides the oxidizing power in an electrochemical field measured as potential and the acidity and alkalinity of species measured as pH. Thus, any reaction involving hydroxyl (OH^-) ions should be written in terms of H^+ ion concentration, which in turn is converted into $pH = -\log[H^+]$ (Table 3.2). Besides the possible reactions that may occur in an electrochemical system, a simplified Pourbaix diagram gives important areas for designing and analyzing electrochemical systems. These areas are known as corrosion, passivation, and immunity. However, a Pourbaix diagram does not include corrosion rate, which is essential in kinetic studies.

3.4.1 Diagram For Water and Oxygen

The construction of the *E-pH* Pourbaix diagram for water and oxygen is straightforward. For instance, consider the hydrogen evolution reaction at standard conditions with $T = 25\,°C$, $P_o = 1$ atm

$$2H^+ + 2e^- = H_2 \tag{3.47}$$

for which the reaction constant is

$$K_{sp} = \frac{a_{H_2}}{a_{H^+}} = \frac{[H_2]}{[H^+]^2} = \frac{P_{H_2}/P_o}{[H^+]^2} \tag{3.48}$$

$$\log\left[K_{sp}\right] = \log\left(P_{H_2}/P_o\right) - 2\log\left[H^+\right] \tag{3.49}$$

From Eq. (3.31), the Nernst equation gives the hydrogen potential as

$$E_{H_2} = E^o_{H^+/H_2} - \frac{2.303RT}{zF} \log\left[K_{sp}\right] \quad (3.50)$$

$$E_{H_2} = \frac{b}{2}\left[2pH - \log\left(P_{H_2}/P_o\right)\right] \quad (3.51)$$

where $b = slope = -2.303RT/F = -0.0592\,\text{V}$, $Z = 2$, $E^o_{H^+/H_2} = 0$, and $pH = -\log\left[H^+\right]$.

Figure 3.3 shows the line labeled H_2 given by Eq. (3.51) which is the lower limit of stability of water. Below line H_2 water is reduced to hydrogen.

For reducing dissolved oxygen in an acid solution, the cathodic reaction is

$$O_2 + 4H^+ + 4e^- = 2H_2O \quad (3.52)$$

where $z = 4$ and

$$K_{sp} = \frac{[H_2O]^2}{[O_2]\,[H^+]^4} \quad (3.53)$$

$$K_{sp} = \frac{1}{[P_{O_2}/P_o]\,[H^+]^4} \quad (3.54)$$

Recall that P_o is the standard pressure (1 atm).

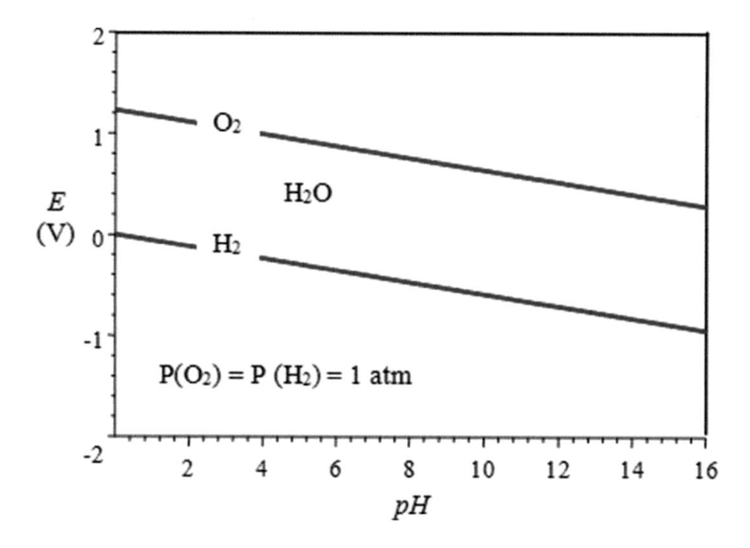

Fig. 3.3 Purbaix diagram domains of hydrogen and oxygen and water at 25 °C

Thus,

$$\log\left[K_{sp}\right] = -\log\left(P_{O_2}/P_o\right) - 4\log\left[H^+\right] \tag{3.55}$$

$$\log\left[K_{sp}\right] = -\log\left(P_{O_2}/P_o\right) + 4pH \tag{3.56}$$

Using Eq. (3.31) along with $Z = 4$ yields

$$E_{O_2} = E^o_{O_2} - \frac{2.303RT}{zF}\log\left[K_{sp}\right] \tag{3.57}$$

$$E_{O_2} = 1.229\,\text{V} + \frac{b}{4}\left[4pH - \log\left(P_{O_2}/P_o\right)\right] \tag{3.58}$$

Accordingly, plotting Eq. (3.58) yields a straight line labeled O_2 as shown in Fig. 3.3, and above line O_2, water is oxidized to oxygen. Notice that the slope $b = -0.0592\,\text{V}$ does not change if P_{H_2} and P_{O_2} change.

Usually, these straight lines are superimposed on a metal M-H_2O diagram for comparison purposes and determining the phase that has an influence on the formation of a metallic ion or an oxide phase.

Mathematically, these lines can be shifted above or below according to the above electric potentials defined by Eqs. (3.51) and (3.58). This can be accomplished by letting $P_{O_2} \neq P_o$ and $P_{H_2} \neq P_o$ in these equations.

Analysis of Fig. 3.3

Line O_2: At this position and above, oxygen gas (O_2) is evolved on the surface of an immersed electrode in water.

Line H_2: At this position and below, hydrogen gas (H_2) is evolved on the surface of an immersed electrode in water.

Between lines O_2 and H_2: Water is stable.

@ $P_{H_2} < 1$ atm, water is unstable and decomposes to hydrogen gas H_2. Thus, $H_2O + 2e^- = H_2 + 2\left(OH^-\right)$.
@ $P_{H_2} = 1$ atm, Eq. (3.51) becomes $E_H = -0.0592\,(pH)$ and establishes the limit of predominance between H_2O and H_2, but H_2 predominates.
@ $P_{H_2} > 1$ atm, water is stable.

@ $P_{O_2} < 1$ atm, water is stable.
@ $P_{O_2} = 1$ atm, Eq. (3.58) becomes $E_{O_2} = 1.229\,\text{V} - 0.0592\,\text{V}\,(pH)$ and establishes the limit of predominance between O_2 and H_2O, but O_2 predominates.
@ $P_{O_2} > 1$ atm, O_2 predominates.

Example 3.5. *Drive Eq. (3.58) for an alkaline solution and compute the E_H and E_{O_2} for a $pH = 10$ under standard conditions.*

Solution.

$$O_2 + 2H_2O + 4e^- = 4(OH^-) \tag{a}$$

from which

$$K_{sp} = \frac{[OH^-]^4}{[O_2]} = \frac{[OH^-]^4}{P_{O_2}/P_o} \tag{b}$$

$$\log\left(K_{sp}\right) = 4\log\left[OH^-\right] - \log\left(P_{O_2}\right) \tag{c}$$

From Table 3.2,

$$pH = 14 + \log\left[OH^-\right] \tag{d}$$

Then, Eq. (c) becomes

$$\log\left(K_{sp}\right) = 4\left(pH - 14\right) - \log\left(P_{O_2}\right) \tag{e}$$

From Eq. (3.31) with $b = -2.303RT/F = -0.0592\ V$ and $Z = 4$, E_{O_2} becomes

$$E_{O_2} = E^o_{O_2} - \frac{2.303RT}{zF}\log\left(K_{sp}\right) \tag{f}$$

$$E_{O_2} = 0.401\ \text{V} + \frac{b}{4}\left[4\left(pH - 14\right) - \log\left(P_{O_2}\right)\right] \tag{g}$$

$$E_{O_2} = 1.229\ \text{V} + \frac{b}{4}\left[4pH - \log\left(P_{O_2}\right)\right] \tag{3.58}$$

3.4.2 Pourbaix Diagram for a Metal

Below is a brief description on how to construct a Pourbaix diagram for a metal $M\text{-}H_2O$ electrochemical system, in which the lines represent predominant ion boundaries enclosing areas of stable phases in an aqueous environment. Figure 3.4 shows a schematic and yet simple $M\text{-}H_2O$ Pourbaix diagram, which includes the water and oxygen lines for comparison purposes. The construction of the diagram consists of determining the potential E for lines 1 and 3 using the Nernst equation and the pH for lines 2 and 4 using known equilibrium constant K_{sp}. The range of metal ion concentration can be $10^{-6}\ \text{mol/l} \le a\left(M^{z+}\right) \le 1\ \text{mol/l}$, but $a\left(M^{z+}\right) = 10^{-6}\ \text{mol/l}$ is chosen hereafter in order to simplify the procedure for constructing the diagram. Otherwise, several vertical and incline lines would be generated. If $a\left(M^{z+}\right) = 1\ \text{mol}/\ \text{l}$, the electrochemical corrosion process is unrealistic [12]. The procedure is as follows:

1. Horizontal line 1 with $[M] = 1$ mol/l:

$$M = M^{z+} + ze^{-} \tag{a}$$

$$K_{sp} = \frac{\left[M^{z+}\right]}{[M]} = \left[M^{z+}\right] \tag{b}$$

This reaction involves electron transfer only. Thus, the Nernst equation, Eq. (3.31), with $\left[M^{z+}\right] = 10^{-6}\ \text{mol}/\ \text{l}$ and $T = 298$ K becomes

$$E_1 = E^{o}_{M/M^{z+}} - \frac{2.303RT}{zF} \log\left[M^{z+}\right] \tag{c}$$

$$E_1 = E^{o}_{M/M^{z+}} - \frac{0.3548}{z} \tag{3.61}$$

The value for the standard potential $E^{o}_{M/M^{z+}}$ is given in Table 2.2, provided that the metal M is known. Thus, Eq. (3.61) plots a horizontal line (line 1) as shown in Fig. 3.4.

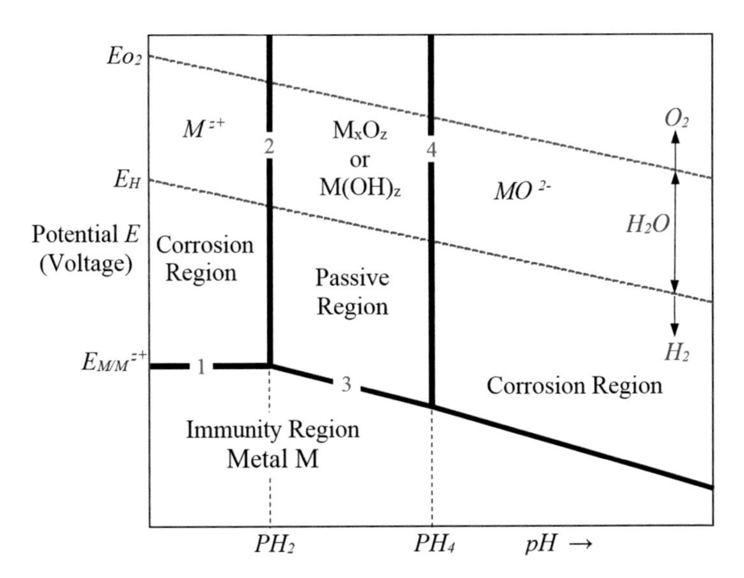

Fig. 3.4 Schematic Purbaix diagram for a metal M showing theoretical domains of corrosion, immunity, and passivation of aluminum at 25 °C. The hydrogen and oxygen linear trends and the stable water region are included as references

2. Vertical line 2:

$$xM^{z+} + zH_2O = M_xO_z + 2zH^+ \tag{3.62}$$

$$K_{sp} = \frac{\left[H^+\right]^{2z}}{\left[M^{z+}\right]^x} \tag{a}$$

This reaction involves a fixed hydrogen concentration only; the equilibrium constant K_{sp} for this reaction must be known so that

$$\log\left(K_{sp}\right) = 2z\log\left[H^+\right] - x\log\left[M^{z+}\right] \tag{b}$$

$$\log\left(K_{sp}\right) = -2z\left(pH_2\right) - 6x \tag{c}$$

$$pH_2 = -\frac{1}{2z}\left[\log\left(K_{sp}\right) + 6x\right] \tag{3.63}$$

Therefore, a straight line is drawn starting at E_1-pH_2 point. This is shown in Fig. 3.4 as line 2. It should be mentioned that the metal oxide M_xO_z can be replaced by a metal hydroxide $M(OH)_2$ or $M(OH)_3$ depending on the chosen metal. Thus, Eq. (3.62) should be balanced accordingly. Hence,

3. Incline line 3:

$$M + zH_2O = M_xO_z + 2zH^+ + 2ze^- \tag{3.64}$$

$$K_{sp} = \left[H^+\right]^{2z} \tag{a}$$

Notice that this reaction involves both hydrogen and electron transfer. Then, the Nernst equation becomes

$$E_3 = E^o_{M/M_xO_z} - 0.0592\left(pH\right) \tag{3.65}$$

The standard potential E^o_{M/M_xO_z} must be known. In Fig. 3.4, line 2 and line 3 intersect at a pH_2 value given by Eq. (3.63). Above this pH, a passivation process takes place on the metal surface according to the following reaction

$$M + zH_2O = M_xO_z + 2zH^+ \tag{3.66}$$

where M_xO_z is a passive film compound. This reaction is reversed below the pH given by Eq. (3.63).

4. Vertical line 4:

$$M_xO_z + H_2O = xMO^-_{z-1} + 2H^+ \tag{3.67}$$

$$K_{sp} = a\left(H^+\right)^2 a\left(MO^-_{z-1}\right)^x \tag{a}$$

Letting $\left[MO_{z-1}^{-}\right] = 10^{-6}\,\mathrm{mol/l}$ in Eq. (a) yields

$$\log\left(K_{sp}\right) = 2\log a\left(H^{+}\right) + x\log a\left(MO_{z-1}^{-}\right) \tag{b}$$

$$\log\left(K_{sp}\right) = -2\left(pH_4\right) - 6x \tag{c}$$

Solving for pH gives

$$pH_4 = -0.5\left[6x + \log\left(K_{sp}\right)\right] \tag{3.68}$$

Subsequently, a vertical line 4 is drawn at this pH_4 value.

As an example, Fig. 3.5 illustrates the E-pH equilibrium diagram or Pourbaix diagram for the Fe-H_2O system containing 10^{-6} mol/l of Fe at 25 °C and 1 atm pressure [11]. This particular diagram shows the sensitivity of the potential (E) to changes in pH and the thermodynamic stability of species as a function of potential and pH. Only three iron oxide species are considered, and not all of the possible thermodynamic species are shown.

In addition, the lines in the Pourbaix diagram (Fig. 3.5) represent the equilibrium conditions for species from two different regions. That is, the activities a_i are equal. For instance, $a_{Fe^{2+}} = a_{Fe^{3+}}$ at slightly below $0.8\,V$ at $-2 \leq pH < 5$. This diagram is fundamentally a two-dimensional representation of the thermodynamic states

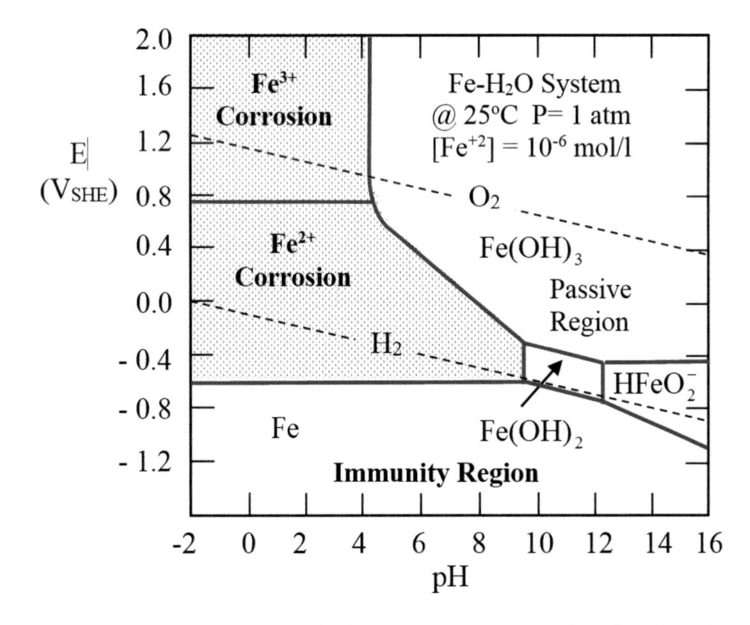

Fig. 3.5 Simplified Pourbaix diagram for iron-water system showing domains of corrosion, immunity, and passivation of iron (Fe) at 25 °C [11]

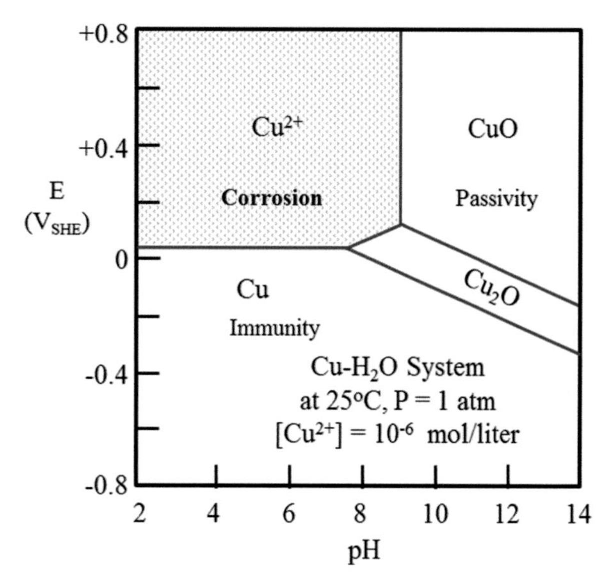

Fig. 3.6 Simplified Pourbaix diagram for copper-water system showing domains of corrosion, immunity, and passivation of copper (CU) at 25 °C [11]

of pure iron in water. The region boundaries are susceptible to be displaced with changes in concentration.

Despite that a Pourbaix diagram is thermodynamically important in predicting the formation of a species, it does not include any information on the corrosion kinetics. Fortunately, corrosion kinetics is provided by a polarization diagram by plotting the cell electric potential vs. current or electric potential vs. current density. Polarization diagrams are dealt with in a latter chapter.

Figure 3.6 depicts the Pourbaix diagram for copper-water system as additional reference. The diagram shows stable species at standard conditions. Notice that only two copper oxides, CuO and Cu_2O, form in alkaline water environments.

Moreover, a Pourbaix diagram is a two-dimensional map in the metallic corrosion space, summarizing the thermodynamic state for immunity, corrosion, and passivity. The term passivity represents a region where metal oxide or hydroxide forms on a metal substrate as a thin coating for protection against metal oxidation.

3.5 Problems/Questions

3.1. Consider the following electrochemical cell $\left|Cu^{2+}, Cu\right| \left|H_2, H^+\right| Pt$ to calculate the maximum activity of copper ions Cu^{2+} in solution due to oxidation of a copper strip immersed in sulfuric acid H_2SO_4 at 25 °C, 101 kPa, and $pH = 2$. [Solution: $a_{Cu^{2+}} = 4.06x10^{-16}$ mol/l.]

3.2. Consider a galvanic cell $|\ Zn\ |0.10 M ZnSO_4|\ |0.10\ M\ NiSO_4|\ Ni\ |$. Calculate the cell potential using the Nernst equation at $T = 25$ °C. [Solution: $E = 0.513$ V.]

3.3. The solubility product (K_{sp}) of silver hydroxide, $AgOH$, is $1.10x10^{-4}$ at 25 °C in an aqueous solution. **(a)** Write down the chemical reaction and **(b)** determine the Gibbs free energy change ΔG^o. [Solution: (b) $\Delta G^o \approx -22.58$ kJ/mol.]

3.4. Calculate **(a)** the concentration of $AgOH$ and Ag^+ in g/l at 25 °C and **(b)** the pH if $[OH^-] = 4.03x10^{-3}$ mol/l *and* $K_{sp} = 1.10x10^{-4}$. [Solutions: (a) $C_{AgOH} =$ 18.73 g/l and (b) $pH = 11.61$.]

3.5. For an electrochemical copper reduction reaction at 25 °C and 101 kPa, calculate **(a)** the Gibbs free energy change ΔG^o as the driving force, **(b)** the equilibrium constant K_e and **(c)**, the copper ion concentration $C_{Cu^{2+}}$ in g/l. [Solution: $C_{Cu^{2+}} = 2.52 \times 10^{-10}$ g/l.]

3.6. Use the electrochemical cell $| \; Cd \left| 0.05 \, M \, Cd^{2+} \right| \left| 0.25 \, M \, Cu^{2+} \right| Cu \; |$ to calculate **(a)** the temperature for an electric potential of $0.762 \, V_{SHE}$ and **(b)** ΔS, ΔG, ΔH, and Q. [Solution: (a) $T = 44.32$ °C, (b) $Q = -4.25$ J/mol.]

3.7. Pure copper and pure cobalt electrodes are separately immersed in solutions of their respective divalent ions, making up a galvanic cell that yields a cell potential of $0.65 \, V_{SHE}$. Calculate **(a)** the cobalt activity if the activity of copper ions is 0.85 mol/l at 25 °C and **(b)** the change in entropy ΔS^o and Gibbs free energy ΔG^o. [Solution: $\Delta G = 1125.45$ kJ/mol.]

3.8. Show that the activities are $\left[M_1^{2+}\right] = \left[M_2^{2+}\right]$ in a galvanic cell, provided that the potentials is $E = E^o$.

3.9. Show that $\Delta S = -R \ln K_{sp}$ in a galvanic cell at constant pressure.

3.10. **(a)** Derive an expression for $\ln \left[Cu^{2+}\right] = f(T)$ for the given cells below. Assume that the current ceases to flow and that the cells are not replenished with fresh solutions. **(b)** Plot the expressions for a temperature range of $0\,°C \le T \le 60\,°C$ and draw some suitable conclusions to explain the electrochemical behavior based on the resultant trend.

References

1. C.M.A. Brett, A.M.O. Brett, *Electrochemistry Principles, Methods and Applications* (Oxford University Press, New York, 1994), pp. 1–7
2. J.W. Evans, L.C. De Jonghe, *The Production of Inorganic Materials* (Macmillan Publishing Company, New York, 1991), pp. 1497–149
3. R.T. Dehoff, *Thermodynamics in Materials Science* (McGraw-Hill, New York, 1993), pp. 458–484
4. T. Engel, P. Reid, *Physical Chemistry*, Chap. 11 (Pearson Education, Boston, 2013)
5. M. Pourbaix, *Enthalpies Libres de Formations Standards*. Cebelcor Report Technique, 87 (1960)

6. L.L. Sheir, R.A. Jarman, G.T. Burstein (eds.), *Corrosion, Vol. 2, Corrosion Control* (Butterworth-Heinemann, Boston, 1994), pp. 21:12–21:24
7. The NBS tables of chemical thermodynamic properties. J. Phys. Chem.Ref. Data, National Bureau of Standards, Washington, DC **11**(2) (1982)
8. M. Pourbaix, *Lectures on Electrochemical Corrosion*, translated by J.A.S. Green, translation edited by R.W. Staehle, foreword by J. Kruger (Plenum Press, New York, London, 1973)
9. I.N. Levine, *Physical Chemistry*, vol. 152 (McGraw-Hill Book Company, New York, 1978), pp. 350–396
10. W. Xing, F. Li, Z.-F. Yan, G.Q. Lu, Synthesis and electrochemical properties of mesoporous nickel oxide. J. Power Sources **134**, 324–330 (2004)
11. M. Pourbaix, *Atlas of Electrochemical Equilibria in Aqueous Solutions* (NACE, Houston, TX, 1974)
12. D.A. Jones, *Principles and Prevention of Corrosion* (Macmillan Publishing Company, New York, 1992), pp. 39–70

4 Nano-electrochemistry

Recent advances at a nanoscale make the atomic world very fascinating since images of nanoparticles or even atom clusters can be revealed on the surface of a substrate. Specifically considered hereafter is the scanning tunneling microscope (STM) for revealing atomic events and the electrochemical scanning tunneling microscope (ESTM) for characterizing electrochemical reactions at the probe tip and substrate. These techniques are briefly described using models and selected atomic images due to the lack of space in this textbook. Among several techniques, STM and ESTM are useful for studying fundamental electrochemical problems through nanoscale science. The studied electrochemical events are commonly conducted in a nanogap between a very sharp metallic tip and a substrate (electrode). The nanogap ranges from 1 nm to less than 10 nm, and it represents a potential energy barrier.

4.1 Electrical Double Layer

In general, when a solid surface is brought into contact with an aqueous medium (polar medium), it acquires charges through ionization, ion adsorption, or ion dissolution. Subsequently, ions with opposite charge (counter ions) are attracted toward the charged surface, and those ions of like charge are repelled away from the surface [1, 2]. In particular, when a metal M is immersed in a suitable electrolyte, initially its atoms oxidize at a relatively high rate, and subsequently, the oxidation (dissolution) process gradually ceases and eventually stops due to a negative charge buildup on the metal surface. Thus, dynamic equilibrium is attained, provided that no other complex electrochemical reactions take place. In this process, metal atoms are removed from their lattice sites to ionize as cations (M^{z+}) into the electrolyte, forming a negatively charged metal surface. If polar water (H_2O) molecules and hydrogen cations (H^+) are in the electrolyte, they are attracted by the negative metal surface to form an ionic structure known as **electrical double layer** (EDL), which is an interfacial boundary (solid–liquid interface) that prevents other ions from the

N. Perez, *Electrochemistry and Corrosion Science*,
DOI 10.1007/978-3-319-24847-9_4

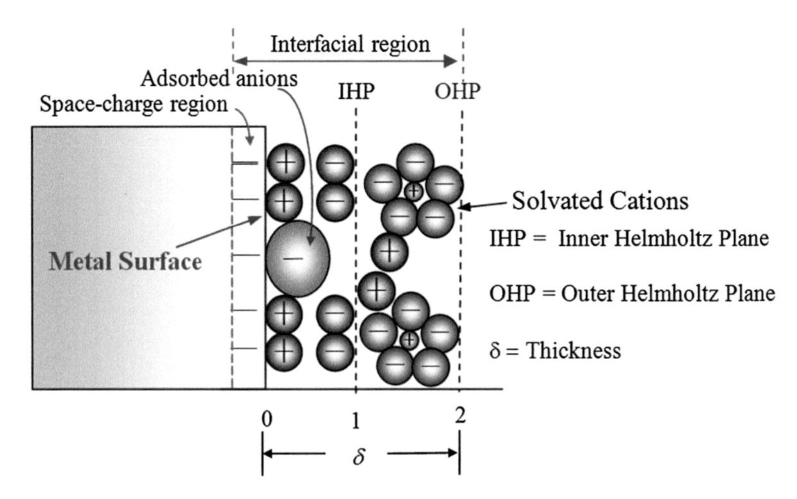

Fig. 4.1 Schematic electrolyte double layer and its interfacial structure having a thickness δ and space-charge region

bulk solution to be part of it. Figure 4.1 shows a schematic electrolyte double-layer model [1–3]. This region is part of the interfacial region known as electrical double layer, which prevents easy charge transfer and limits electrochemical reactions on the electrode surface [3, 4]; it acts as double-layer capacity.

There are different models for describing the electrolyte double layer. The aim of a particular model is to explain experimental observations [3]. The classical electrolyte double-layer models are known as [3] Helmholtz model (1879); Gouy–Chapman model (1910–1913); Stern model (1924); Grahame model (1947); Bockris, Devanathan, and Muller model (1963); Trasatti–Buzzanca model (1971); Conway model (1975–1980); and Marcus model (1992).

This structure is a complex ionic arrangement in an electric field and has a limited thickness. Eventually, metal dissolution ceases if there is no external current flow, but it proceeds until the ionic structure is dense enough to protect and prevent the metal from reacting any further until equilibrium is reached.

One finds in the literature [2, 5–8] different models for explaining the interfacial structure of the electrical double layer. The classical Helmholtz model suggests that the surface charge is neutralized by opposite sign counter ions placed at a distance δ from the charged surface. This model indicates that some negatively charged ions are adsorbed on the metal electrode surface, and polar water covers the rest of this surface, forming a protective layer. This divides the electrical double layer into two planes. The **inner Helmholtz plane** (IHP) is an ionic layer that consists of adsorbed dipole H_2O molecules. The majority of the anions do not penetrate this layer, some do as indicated in Fig. 4.1. The inner potential or the interfacial potential on the boundary of this ionic plane is ϕ_1. The **outer Helmholtz plane** (OHP) consists of a plane of adsorbed ions due to chemical forces in contact with a diffuse ionic layer at a inner potential ϕ_2. Moreover, the electrical double layer seems to limit the

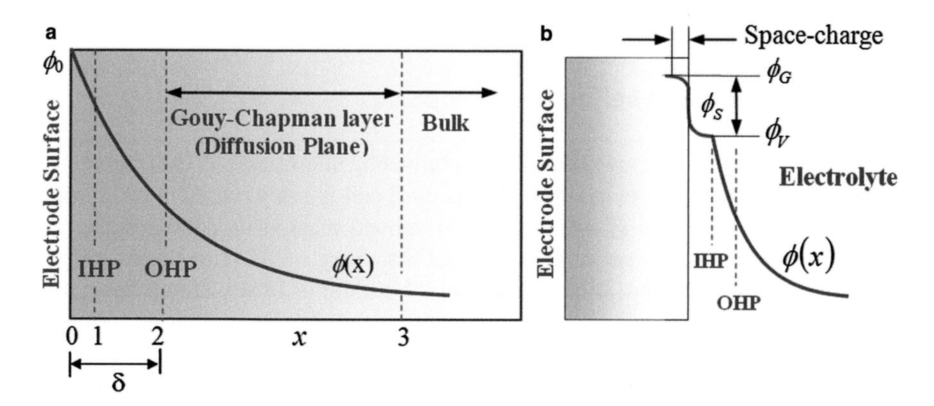

Fig. 4.2 Schematic inner electric potential distribution across the electrical double layer. (**a**) Helmholtz planes and (**b**) space-charge along with surface potential difference (ϕ_s)

continuous electrochemical reactions at the metal surface due to the lack of charge transfer, and consequently, an electric potential decay develops.

Furthermore, Fig. 4.2a shows the combined Helmholtz and Gouy–Chapman models, and Fig. 4.2b depicts a modern model for the electric potential $\phi(x)$ within a half-cell. The latter model takes into account the a diffusion layer of the electrolyte solution called Gouy–Chapman layer. In general, the schematic potential distribution shown in Fig. 4.2a can be approximated as an exponential decay function of the form [9, 10]

$$\phi(x) = \phi_o \exp(-\lambda x) \tag{4.1}$$

where ϕ_o is a constant known as the potential drop across the diffusion layer, λ is another constant related to a characteristic thickness of the diffusion layer [10], and x is the distance from the electrode surface. The diffusion layer is a thick layer located in a region of ions in contact with the *OHP* plane and the bulk of solution at a potential range of $\phi_2 . \le \phi_d \le \phi_3$ (Fig. 4.2a).

Figure 4.2b shows a potential decay, but it also includes a more realistic electric potential distribution within a half-cell. The electric potential distribution is attributable to the Galvanic potential (inner potential) ϕ_G, Volta potential (outer potential) ϕ_V, and the diffusion layer decay potential $\phi(x)$ [3]. Thus, the potential difference, $\phi_S = \phi_G - \phi_V$, is the electrode surface potential related to electron motion from the Fermi energy level (E_F), which corresponds to the free energy of electrons. In fact, E_F is an important theory for evaluating photovoltaic and photoelectrochemical devices because it shows the energy levels occupied by electrons [11].

In addition, analyzing Eq. (4.1) yields

$$\phi(x) = \phi_o \longrightarrow \phi_{\max} \text{ when } x \longrightarrow 0 \tag{4.2}$$

$$\phi(x) = \phi_b \longrightarrow 0 \quad \text{ when } x \longrightarrow \infty \tag{4.3}$$

4.2 Microscopic Techniques

4.2.1 Scanning Tunneling Microscope

This section includes the powerful scanning tunneling microscope (STM) technique for revealing atomic events in an electrochemical cell in real time. This is referred to as in situ STM which allows one to explore and manipulate solid surfaces at an atomic scale. This is possible due to the discovery of STM (a breakthrough technique in 1981) by Gerd Binnig and Heinrich Rohrer of IBM's Zurich Research Center. These researchers were awarded the 1986 Nobel Prize in Physics for their design of the STM, and subsequently, IBM became a pioneer in nanotechnology. Nonetheless, the STM is a device that allows electrical measurements, and it provides means to obtain topographical details of a substrate surface (surface morphology). Figure 4.3 shows a picture of an IBM STM.

Figure 4.4 schematically shows an electrochemical STM cell. The basic principle of STM is based on tunneling of electrons in a gap having a vertical distance in the range $0 < z < 5$ nm between an electrically conductive sharp probe tip (STM tip, typically made out of *Pt–Ir* or tungsten, *W*, wire) and a working electrode (WE) or substrate [10, 12].

When the STM cell is excited in an ultrahigh vacuum (UHV) by an applied bias voltage (ϕ_{bias}), electrons flow across the gap generating a measurable tunneling current (I_t), which is a variable to be controlled during In situ STM work. However, if $\phi_{bias} = 0$, then there is no net tunneling current [13, 14].

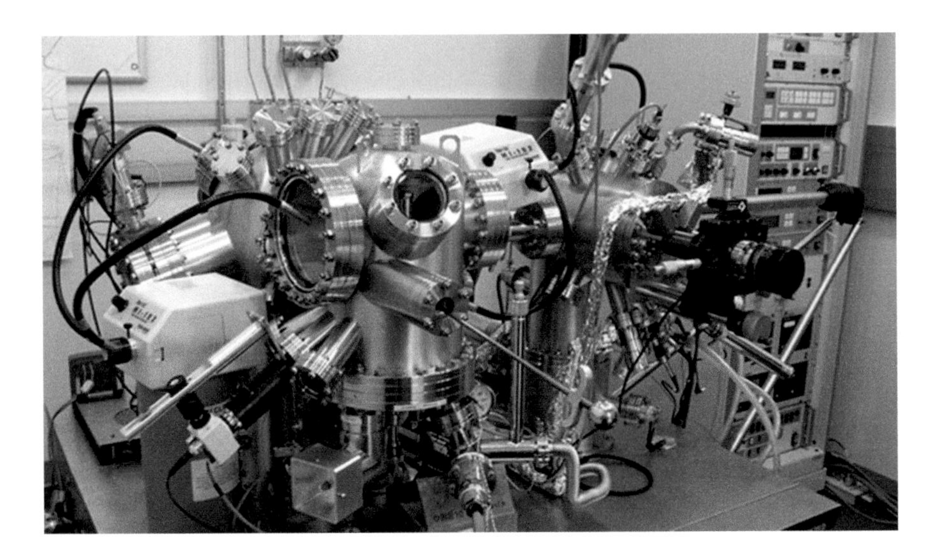

Fig. 4.3 IBM STM picture from www.google.com images

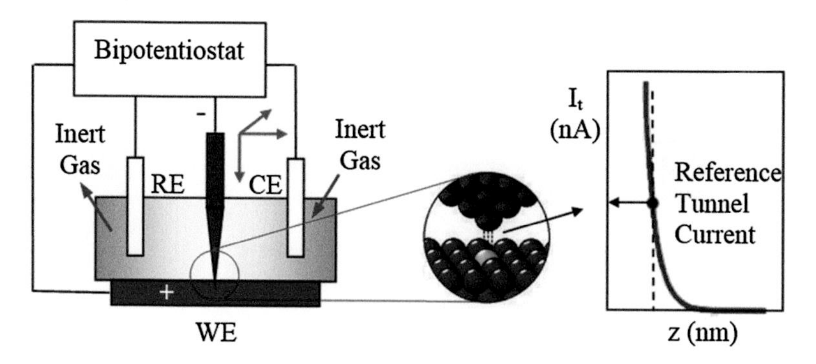

Fig. 4.4 Schematic electrochemical cell for in situ STM as per reference [27]. *Tip insert*: "Figure, Michael Schmid, TU Wien". The schematic plot is based on reference [13] data

At a discrete vertical distance z, electrons can quantum mechanically tunnel through the gap between the STM tip and the substrate surface generating a tunneling current (I_t), which behaves like an exponential decay function of the form [13]

$$I_t = I_o \exp(-\sigma z) = \phi_{bias}/R_t \tag{4.4}$$

Here, R_t is the effective resistance of the tunneling gap, typically $10^9 \leq R_t \leq 10^{11}\ \Omega$ [10]. At this time, I_o and β are constants that must be determined through nonlinear curve fitting using experimental data for $I_t = f(z)$. An schematic plot for this function is at the right-hand side of Fig. 4.4. The behavior of this function with respect to the reference tunneling current, however, implies that a slight change in z causes a significant large change in I_t. Therefore, I_t is very sensitive to the vertical distance (z) or gap between the STM tip and the WE. In fact, only the tunneling current is of principal interest in STM [10].

Furthermore, from a physics point of view, an applied positive potential known as a bias voltage ($\phi_{bias} > 0$) on the STM tip causes electrons to tunnel from the tip to the WE. As a result of this electrochemical event, the potential in the gap decreases, and the energy level of the WE decreases as well. These electrons have the tendency to flow toward empty (unfilled) states of the sample surface. Conversely, an applied negative potential ($\phi_{bias} < 0$) on the tip causes electrons to tunnel into the tip from the sample. Consequently, the potential in the gap and the energy level of the WE increase [15, 16].

Figure 4.5 also illustrates schematic events at the STM tip [17] and depicts a real image of copper deposition on a gold (Au) substrate [18].

Copper cations (Cu^{2+}) in an electrically conductive electrolyte solution are collected around the STM tip, and subsequently, the STM tip is brought down to contact the sample, leaving a cluster of Cu atoms on terraces. Then, Cu is reduced, $Cu^{2+} + 2e^- \rightarrow Cu$, at the STM tip, which moves when attached to a 3D piezoelectric scanner. The scanner is subjected to be electrically charged for adjusting or controlling the position of the STM tip.

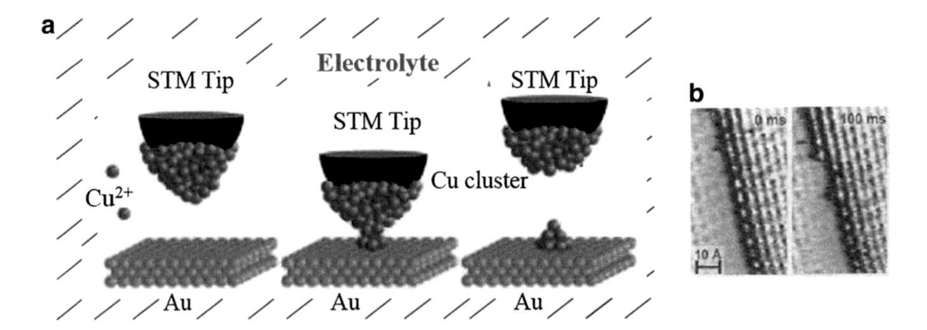

Fig. 4.5 Tip-induced metal deposition of (**a**) schematic atoms [17] and (**b**) real copper (Cu) atoms on gold (Au) surface [18]

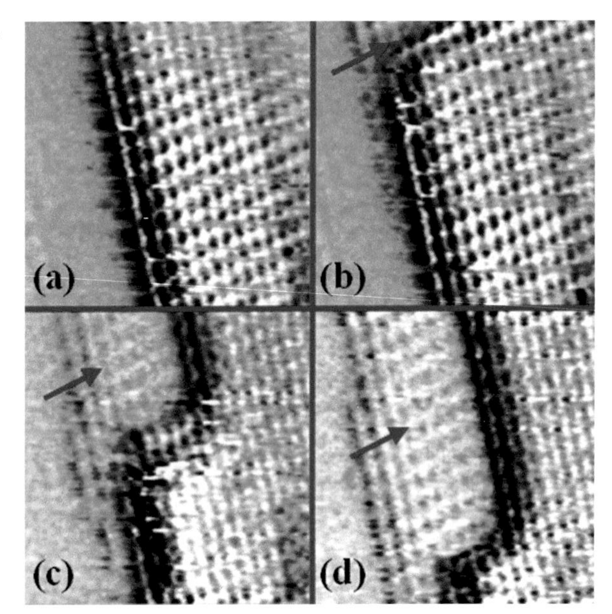

Fig. 4.6 Atomic scale for the dynamic behavior of copper, Cu(100), dissolution in an 0.01 M HCl solution at −0.23 V vs. SCE [18]. Parts (**a**) through (**d**) indicate the Cu dissolution areas marked with arrows. Taken from Prof. Dr. Olaf Magnussen video-STM available online, ttp://www.atomic-movies.uni-kiel.de. These images (size: 5 nm × 5 nm) were captured using Windows media

Additionally, Fig. 4.6 depicts a sequence of real images for copper dissolution on a gold (*Au*) surface [17, 18]. Observe the areas being marked with an arrow. This area increased during *Cu* oxidation at times less than 500 ms. These images were captured from the atomic movie library or video-STM [18] using windows media player. Apparently, *Cu* dissolution (oxidation) occurs in a particular crystallographic [001] direction [2].

Despite the complexity of a STM electrochemical cell, an important application of the STM technique is to observe the shape and size of atom clusters (islands) and their spatial distributions on the atomic scale [19–23]. Nowadays, it is evident that STM images, such as the ones in Fig. 4.6, do exhibit an organized spatial distribution forming a compact atomic structure. It is clear in these images that metal dissolution, as well as deposition, is a step-edge process.

Most of STM work to date concentrates on the crystalline surface structures of face-centered cubic (FCC) and hexagonal close-packed (HCP) materials [14, 20, 24, 25]. If the STM work is for metal deposition, then an atomic reorganization initially forms a monolayer on the substrate. This monolayer can be defined as a pseudomorphic structure since the atom being deposited has to adopt a lattice space of the substrate. For example, in the $Cu/Au(001)$ atomic coupling, Cu settles in the lattice spacing of the Au substrate in the (001) crystallographic plane.

4.2.2 Electrochemical STM

The STM cell being briefly described is for viewing the substrate atomic structure in a small gap. If a steady-state tunneling current, ($I_{t,\infty} >> I_t$ at $z \to \infty$) exists, then STM cell becomes an electrochemical STM (ESTM) technique for characterizing electrochemical reactions at the probe tip and substrate. These reactions are the source of the measurable current due to electron flow [10].

Connecting a bipotentiostat to a ESTM cell allows independent control of the electrochemical potential of the tip and substrate relative to a reference electrode [10, 26]. Figure 4.4 depicts the conventional three-electrode electrochemical cell. The working electrode (WE) is the metal, sample, or the substrate under investigation. The reference electrode (RE) provides the reference potential with insignificant electric current flow. Normally, the saturated calomel electrode (SCE) is used as the reference electrode. The auxiliary electrode (AE) supplies the current required by the working electrode without limiting the measured response of the cell. In the case of ESTM, the probe tip is the fourth electrode [27]. One major advantage of ESTM is that mass transport characteristics can be described for a variety of different ultramicroelectrode (UME) geometries (disk, conical or ring shaped) having at least one dimension smaller than 25 μm [28–31].

The steady-state tunneling current, $I_{t,\infty}$, (referred to as Faradaic current) for a UME-disk with radius r is given by [10, 29, 30]

$$I_{t,disk} = 4zFCDr \tag{4.5}$$

The distance between the tip and the substrate in STM and ESTM affects the current, but in ESTM, the electrochemical reactions can lead to oxidation of the tip, $O + ze^- \to R$, and then back to reduction, $O + ze^- \leftarrow R$. Thus, the ESTM technique can be in the **feedback mode** or **collective mode** [10].

Additionally, Fig. 4.7 shows an ESTM image of copper minigrid scanned in ruthenium hexamine solution.

This image was captured using a 5-μm tip radius at $z = 10\,\mu\text{m}$. Denote the effect of I_t on the background. The higher I_t, the darker the image [3, 32].

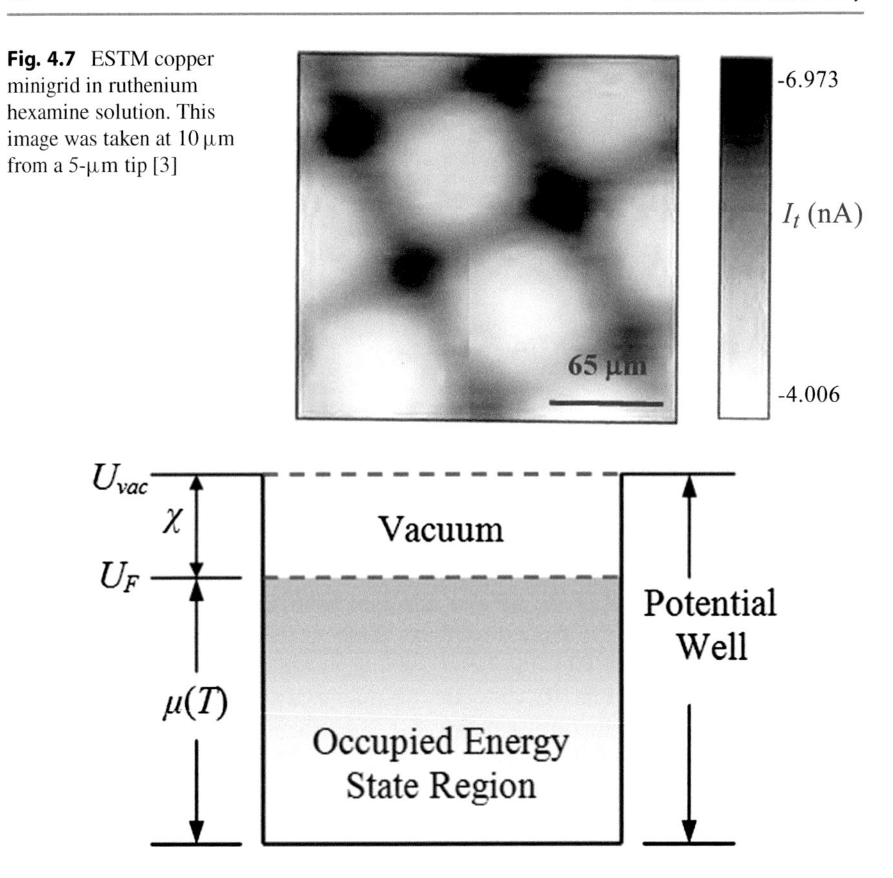

Fig. 4.7 ESTM copper minigrid in ruthenium hexamine solution. This image was taken at 10 μm from a 5-μm tip [3]

Fig. 4.8 Potential well model for free electron motion and energy levels: vacuum (E_{vac}), work function (χ), Fermi (E_F), and chemical (μ) energies

4.3 Fermi Energy Level

The Fermi energy (E_F) is fundamentally the chemical potential $\mu(T)$ of occupied electron energy states at temperature T. Figure 4.8 shows the potential well model indicating the energy level for $\mu(T)$, the Fermi energy, the work function χ, the vacuum energy U_{vac}, and the space for free electron (n) motion.

The functionality of the work function χ (also called ionization energy) is to act as a potential barrier for electron motion, and it is the minimum amount of energy required to remove an electron at the Fermi level from the metal surface into vacuum. Thus, electrons are bound inside metals by the work function. The work function, $\chi = U_{vac} - U_{F,}$ is affected by the structure of the solid surface, but it keeps the electrons into the metal by attractive forces. On the other hand, the Fermi energy U_F level represents the energy of an occupied state in the electron valence band. Essentially, the electron energy levels are mostly filled up to U_F, and U_{vac} is the maximum energy level of vacuum in the potential well model [33].

The Fermi energy and the electron density equations are defined by [33]

$$U_F = \frac{h^2}{2m_e}\left(\frac{3}{8\pi}\rho_e\right)^{2/3} \tag{4.6}$$

$$\rho_e = n_e \rho N_A / A_w \tag{4.7}$$

where $h = 6.6261 * 10^{-34}$ J s $= 4.1357x10^{-15}$ eV s (Plank's constant)
$m_e = 9.11 * 10^{-31}$ kg (mass of an electron)
$\rho_e =$ electron density $(1/\mathrm{m}^3)$
$n_e =$ number of electrons
$\rho =$ bulk density of the metal (kg/m^3)
$N_A = 6.02 * 10^{23}$ atom/mol = Avogadro's number
$A_w =$ atomic weight (g/mol)
1 eV $= 1.60 * 10^{-19}$ J

Figure 4.9 schematically shows the STM tip-substrate setup (Fig. 4.9a) and the Fermi energy level (Fig. 4.9b) for the tip and substrate. Now, assume that the substrate (sample) is biased positively with respect to the tip. This process lowers the Fermi energy (U_F^S) of the substrate by a bias voltage (ϕ_{bias}), which leads to a bias energy $U_{bias} = q_e \phi_{bias}$. Here, q_e is the electron charge, and ϕ_{bias} is the applied external potential in voltage. As a result, electron tunneling occurs from the tip to the sample unoccupied states [33].

Furthermore, the velocity of an electron (v_e), known as the Fermi velocity, participating in the tunneling process can be predicted using its kinetic energy,

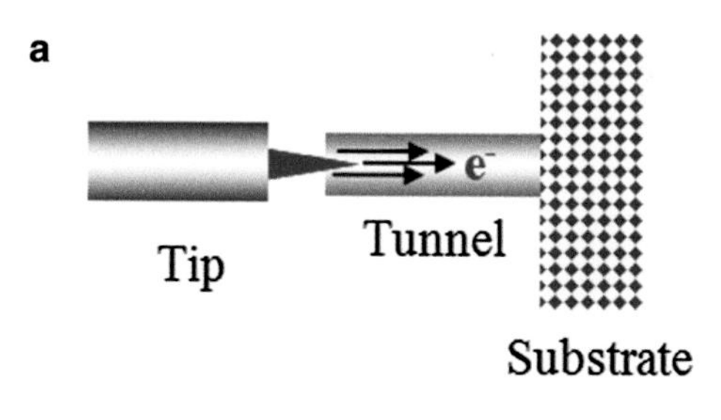

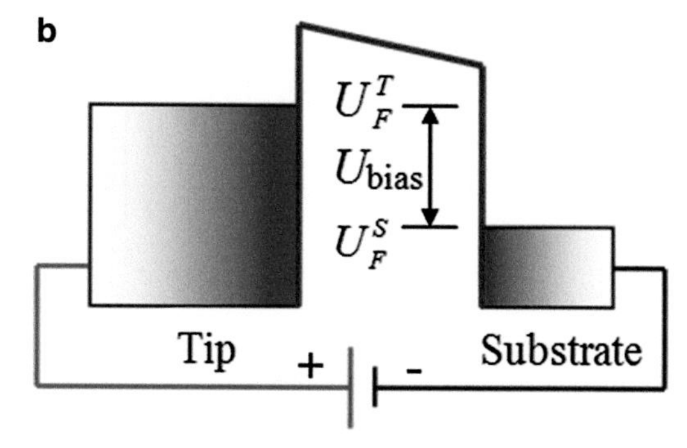

Fig. 4.9 (**a**) Schematic STM tip and substrate for quantum tunneling and (**b**) energy level of electrons in the tip and substrate

$U_F = 0.5 m_e v_e^2$, and Eq. (4.6). Thus,

$$v_e = \sqrt{\frac{2U_F}{m_e}} = \left(\frac{h}{2m_e}\right)\rho_e^{1/3} = \left(3.64x10^{-4}\,\frac{m^2}{s}\right)\rho_e^{1/3} \tag{4.8}$$

Assume that the W-tip and the Au-substrate are used in a STM cell. If the cell is not electrically connected, then the offset in the Fermi energies of the substrate and tip becomes the work function, $\chi = U_{F,Au}^S - U_{F,W}^T$.

Now, if the cell is electrically connected and a bias voltage (ϕ_{bias}) is applied at low temperature, then a tunneling current develops. Figure 4.9b schematically depicts the offset in the Fermi energies.

In addition, the atom density can be defined by

$$n_a = \frac{\rho N_A}{A_w} \tag{4.9}$$

which is a useful expression since the number of free electrons can then be determined as $n_e = \rho_e / n_a$.

Example 4.1. *Determine **(a)** the Fermi energy and **(b)** the velocity of the electron for a gold (Au) substrate and a tungsten (W) tip in an ultrahigh vacuum STM cell at temperature T.*

Solution. **(a)** *The Fermi energy: For gold, $Au \rightarrow Au^{1+} + e^-$, with one free electron per reaction. The electron density, Eq. (4.7), is*

$$\rho_{e,Au} = [n_e \rho N_A / A_w]_{Au}$$

$$\rho_{e,Au} = (1)\left(\frac{19.32 \times 10^3\ \text{kg/m}^3}{196.97\ \text{kg/kmol}}\right)\left(\frac{6.02 \times 10^{23}\ \text{atom}}{10^{-3}\ \text{kmol}}\right)$$

$$\rho_{e,Au} = 5.9048 \times 10^{28}\ 1/\text{m}^3$$

Thus, the Fermi energy as per Eq. (4.6) is

$$U_{F,Au} = \frac{h^2}{2m_e}\left(\frac{3}{8\pi}\rho_e\right)^{2/3}$$

$$U_{F,Au} = \frac{\left(6.63 \times 10^{-34}\ \text{J s}\right)^2}{2 * 9.11 \times 10^{-31}\ \text{kg}}\left[\left(\frac{3}{8\pi}\right)\left(5.90x \times 10^{28}\ 1/\text{m}^3\right)\right]^{2/3}$$

$$U_{F,Au} = 8.8689 \times 10^{-19}\ \frac{\text{J}^2\,\text{s}^2}{\text{kg m}^2} = 8.8689 \times 10^{-19}\ \text{J}$$

$$U_{F,Au} = \frac{8.8689 \times 10^{-19}\ \text{J}}{1.60 \times 10^{-19}\ \text{J/eV}} = 5.54\ \text{eV}$$

The units: $J^2\,s^2/\left(kg\,m^2\right) = J^2/(N\,m) = J$. *Similarly, for tungsten,* $W \rightarrow W^{4+} + 4e^-$, *with four free electrons, Eqs. (4.6) and (4.7) give*

$$\rho_{e,W} = [n_e \rho N_A / A_w]_W$$

$$\rho_{e,W} = (4)\left(\frac{19.3\times 10^3}{183.85}\right)\left(\frac{6.02\times 10^{23}}{10^{-3}}\right) = 2.5278\times 10^{29}\,1/m^{-3}$$

$$U_{F,W} = \frac{\left(6.63\times 10^{-34}\,J\,s\right)^2}{2 * 9.11\times 10^{-31}\,kg}\left(\frac{3\times 1.694\times 10^{29}}{8\pi}\frac{1}{m^3}\right)^{2/3} = 2.3383\times 10^{-18}\,J$$

$$U_{F,W} = \frac{2.3383\times 10^{-18}\,J}{1.60\times 10^{-19}\,J/eV} = 14.61\,eV$$

Obviously, $U_{F,W} > U_{F,Au}$ *due to the differences in the amount of free electrons and the electron densities between gold and tungsten metals.*

(b) *The velocity of electrons: Using Eq. (4.8) yields*

$$v_e = \sqrt{\frac{2U_F}{m_e}}$$

$$v_{e,Au} = \sqrt{\frac{2\,(8.8689\times 10^{-19}\,J)}{9.11\times 10^{-31}\,kg}} = 1.40\times 10^6\sqrt{\frac{N\,m}{kg}}$$

$$v_{e,Au} = 1.40\times 10^6\sqrt{\frac{kg\,m}{s^2}\frac{m}{kg}} = 1.40\times 10^6\,m/s$$

and for tungsten

$$v_{e,W} = \sqrt{\frac{(2)\,(2.3383\times 10^{-18}\,J)}{9.11\times 10^{-31}\,kg}}$$

$$v_{e,W} = 2.27\times 10^6\,m/s = 2270\,km/s$$

Therefore, an electron travels very fast in the vacuum gap. Let us determine the atom density using Eq. (4.9)

$$n_{a,Au} = \frac{\rho N_A}{A_w} = \frac{\left(19.32\times 10^3\,kg/m^3\right)}{196.97\,kg/kmol}\left(\frac{6.02\times 10^{23}\,atoms}{10^{-3}\,kmol}\right)$$

$$n_{a,Au} = 5.9048x10^{28}\,atoms/m^3$$

$$n_{a,W} = \frac{\rho N_A}{A_w} = \frac{\left(19.30\times 10^3\,kg/m^3\right)}{183.85\,kg/kmol}\left(\frac{6.02\times 10^{23}\,atoms}{10^{-3}\,kmol}\right)$$

$$n_{a,W} = 6.3196\times 10^{28}\,atoms/m^3$$

Verify the number of free electrons being used. Thus,

$$n_{e,Au} = \frac{5.9048 \times 10^{28}}{5.9048 \times 10^{28}} = 1$$

$$n_{e,W} = \frac{\rho_e}{n_a} = \frac{2.5278 \times 10^{29}}{6.3196 \times 10^{28}} = 4$$

According to the above calculations, the amount of energy (offset) provided to the electrons during the tunneling process is

$$\Delta U = U_{F,W} - U_{F,Au} = 14.61\ \text{eV} - 5.54\ \text{eV}$$
$$\Delta U = 9.07\ \text{eV}$$

This is the energy that causes the electrons to acquire a specific velocity in the tunneling gap (electrical conduction region).

4.4 Electrochemical Interfaces

When an electrode (metal) is immersed in an electrolyte, an interface forms, and the solid surface properties are susceptible to undergo electrochemical changes. This is a fundamental concept related to surface science since the electrode is in contact with a liquid at a temperature T and pressure P (atmospheric or ultrahigh vacuum) forming a localized and heterogeneous charge distribution at the interface [34]. As a result, a surface potential between the outer potential (Volta potential) and the inner surface potential (Galvanic potential) is accomplished, $\phi_S = \phi_G - \phi_V$ (Fig. 4.2b).

Let us use the generalized half-cell reaction (known as redox reaction in the semiconductor field) in the following form for iron (Fe):

$$O + ze^- \rightleftarrows R \tag{4.10}$$

$$Fe^{2+}\,(aq) + 2e^- \rightleftarrows Fe\,(s) \quad E^o = -0.44\ V_{SHE} \tag{4.11}$$

where O and R are the oxidized (Fe^{2+}) and reduced (Fe) ions, respectively. The Nernst potential equation for the above reaction is

$$E = E^o - \frac{RT}{zF}\ln K_{sp} = E^o - \frac{RT}{zF}\ln\frac{[R]}{[O]} \tag{4.12}$$

$$E = -0.44\ V_{SHE} - \frac{RT}{zF}\ln\frac{[Fe]}{[Fe^{2+}]} \tag{4.13}$$

This simple electrochemical half-cell reaction indicates that ferrous iron (Fe^{2+}) is reduced to atomic iron (Fe) on the cathode electrode surface at available unfilled atomic sites. The measurable cell potential, $E = E_{cell}^{half} = E_{redox}^{half}$, at a macroscale

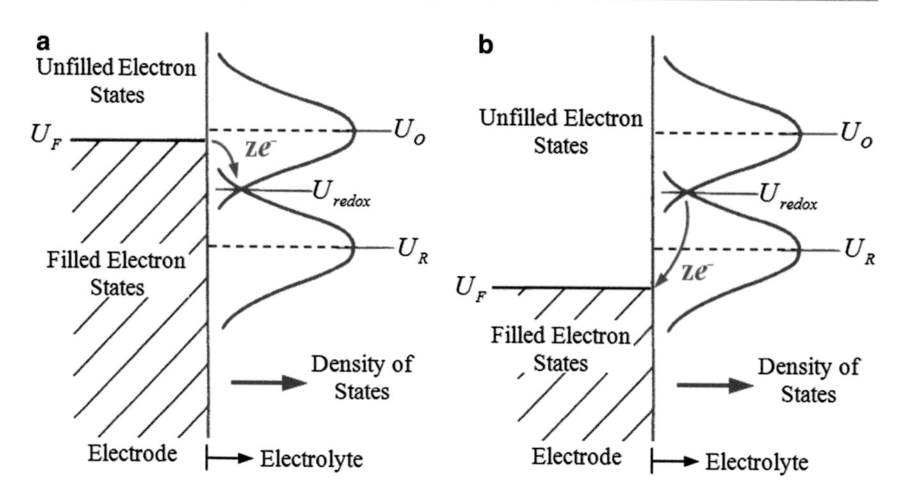

Fig. 4.10 Energy distribution of a redox couple of species O and R on a metallic surface electrode [3, 34]. (**a**) Reduction. (**b**) Oxidation

must be associated with the redox half-cell electronic energy (U). The distribution of this electronic energy is schematically shown in Fig. 4.10 for oxidation and reduction processes [3, 34].

Figure 4.10a indicates that $U_O > U_F > U_R$ and O (Fe^{3+}) can be reduced, $O + ze^- \rightarrow R$, when the electrons at the Fermi energy level are removed at U_{redox} energy level [3]. On the other hand, Fig. 4.10b elucidates the reduction process due to $U_{redox} > U_F$ so that electrons are released by R (Fe) at U_{redox} energy level. The electrons move to available unfilled atomic sites so that $O + ze^- \leftarrow R$ occurs.

As a crude approximation, the redox half-cell energy may be define as [3]

$$U_{redox} = U_F + q_e\phi_S \tag{4.14}$$

$$U_{redox} = U_F + U_S \tag{4.15}$$

Here, $U_S = q_e\phi_S$ may be defined as the surface energy of the electrode. Recall that the surface potential (U_S) is simply the potential difference between the interface and electrode immediate surface, but it cannot be measured.

The measurable potential is the potential difference between electrodes as predicted by Eq. (4.12). If this equation is multiplied by the electron charge, $q_e = 1.6022x10^{-19}$ C, and inserting $F = q_eN_A$ and $\kappa = R/N_A = 8.62x10^{-5}$ eV/K, gives the half-cell redox energy relationship

$$U = U^o - \frac{\kappa T}{z} \ln \frac{[R]}{[O]} \tag{4.16}$$

For the reduction of ferrous iron at 298 K, Eq. (4.13) becomes

$$U = -0.44\ \text{eV} - (0.013\ \text{eV}) \ln \frac{[Fe]}{[Fe^{2+}]} \tag{4.17}$$

since 1 V = 1 J/C (Volts = Joules/Coulomb)

Example 4.2. *If* $[Fe] = 1\,mol/l$ *and* $\left[Fe^{2+}\right] = 0.05\,mol/l$, *then calculate the theoretical half-cell redox energy for the reduction reaction, Eq. (4.11).*

Solution. *From Eq. (4.17),*

$$U = -0.44\ \text{eV} - (0.013\ \text{eV}) \ln \frac{1}{0.05}$$

$$U = -0.48\ \text{eV}$$

This analogy is simply a crude approximation scheme for determining the energy needed to reduce ferrous iron (Fe^{2+}).

4.5 Tunneling Probability

Figure 4.10 depicts an ideal quantum-mechanical system that describes the tunneling event of a particle from region 1 to region 3. Quantum tunneling is used in scanning tunneling microscopes, resonant tunneling diodes, quantum well lasers, etc. where a beam of identical particles having the same wave function $\psi(x)$ will ideally tunnel through the barrier (region 2). Realistically, one needs to determine the tunneling probability to predict the amount of particles that will tunnel and the amount that will reflect at $x = 0$.

According to quantum-mechanical tunneling theory, the motion of a particle can be described by a wave function $\psi(x)$, which is oscillatory with fixed wavelength (λ) and wave amplitudes, A in region 1 (left) and F in region 3 (right). The wave function $\psi_1(x)$ enters the barrier at $x_1 = 0$ and leaves the barrier at $x_1 = w$. Thus, $\psi_1(x)$ is ideally transmitted or tunneled to region 3. However, when $E < U_o$ (Fig. 4.10), all incident particles impinge on a potential barrier because U_o is higher than the particles energy E. Consequently, the wave function $\psi_1(x)$ is partially transmitted (tunneled) having a behavior described by $\psi_2(x)$, while $U(x)$ is more likely to follow the path shown in Fig. 4.9b.

It can be assumed that (1) transmission through the barrier does not alter a particle's energy E and that (2) there must exist a tunneling probability P_T for a particle to strike the barrier from region 1 (left) and eventually tunnel through region 2 and escape to region 3 (right) as indicated in Fig. 4.11. This process is a complex phenomenon known as quantum-mechanical tunneling, which is also referred to as the tunnel effect [35–39].

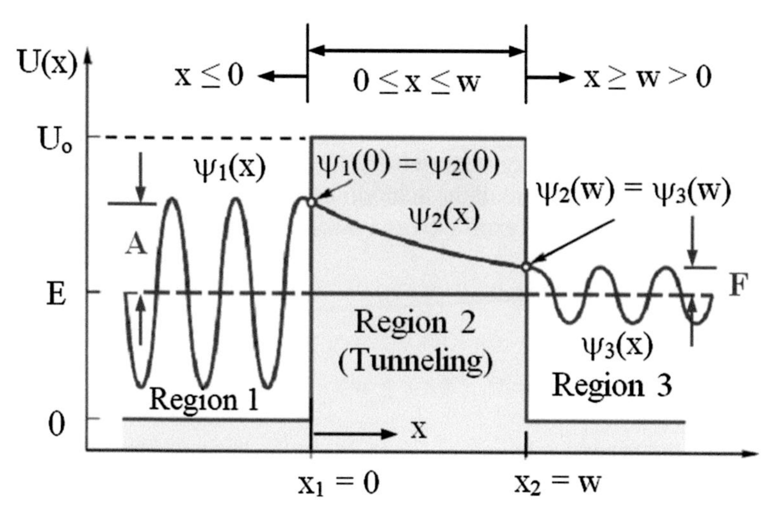

Fig. 4.11 Schematic quantum tunneling process. Region 2 is the classically forbidden barrier. Sketch taken from reference [35]

Accordingly, there should be a non-zero probability that a particle will tunnel through region 2, which is the idealized energy barrier or tunneling gap. Conversely, the classical physics approach predicts a zero probability for the quantum tunneling through region 2, which is known as the classically forbidden region [2, 35].

The energy balance for the quantum tunneling process is of the form

$$\text{Total energy} = \text{potential energy - kinetic energy} \tag{a}$$

$$E = U - K \tag{b}$$

For one-dimensional analysis, the time-independent Schrodinger equation is based on Eq. (b) and the wave function $\psi(x)$ describing the motion of the particle. Thus,

$$E\psi(x) = U(x)\,\psi(x) - \frac{\hbar^2}{2m}\frac{d^2\psi(x)}{dx^2} \tag{c}$$

which is recast as

$$\frac{d^2\psi(x)}{dx^2} + \frac{2m}{\hbar^2}\left[U(x) - E\right]\psi(x) = 0 \tag{4.18}$$

where m is the mass of the particle (electron, proton, or alpha decay), $E > 0$ is the particle energy, $U(x) > 0$ is the barrier potential energy for $0 \le x \le w$, $\hbar = h/2\pi$, and $\psi(x)$ is the general wave function.

4.5.1 Step Potential (SP) Method

The general solutions of Eq. (4.18) for a traveling particle from left to right, as per model in Fig. 4.11, are divided into regions (incident, reflected, and transmitted) with **constant potential energy barrier** U_o. Accordingly, the step potential $U(x)$ needed for finding the solutions of the Schrodinger equation, as sets of superposition of the left and right waves, is expressed as a piecewise function [2, 35–39]:

$$\begin{aligned} U(x) &= 0 && \text{for } x \leq 0 && \text{Region 1} \\ U(x) &= U_o && \text{for } 0 \leq x \leq w && \text{Region 2} \\ U(x) &= 0 && \text{for } x \geq w && \text{Region 3} \end{aligned} \tag{4.19}$$

For $E > U_o$, the general solutions of Eq. (4.18) in the three regions (Fig. 4.11) are all wave functions:

$$\begin{aligned} \psi_1(x) &= A\exp(ik_1x) + B\exp(-ik_1x) && \text{for } x \leq 0 \\ \psi_2(x) &= C\exp(ik_2x) + D\exp(-ik_2x) && \text{for } 0 \leq x \leq w \\ \psi_3(x) &= F\exp(ik_1x) + G\exp(-ik_1x) && \text{for } x \geq w \end{aligned} \tag{4.20}$$

where the wave vectors k_1 and k_2 are defined as

$$k_1 = \frac{2\pi}{h}\sqrt{2mE} \tag{4.21}$$

$$k_2 = \frac{2\pi}{h}\sqrt{2m[E - U(x)]} \tag{4.22}$$

For $E < U_o$, only $\psi(x)_2$ changes to an exponential function:

$$\begin{aligned} \psi_1(x) &= A\exp(ik_1x) + B\exp(-ik_1x) && \text{for } x \leq 0 \\ \psi_2(x) &= C\exp(k_2x) + D\exp(-k_2x) && \text{for } 0 \leq x \leq w \\ \psi_3(x) &= F\exp(ik_1x) + G\exp(-ik_1x) && \text{for } x \geq w \end{aligned} \tag{4.23}$$

For $E = U_o$, only $\psi(x)_2$ changes to a linear function:

$$\begin{aligned} \psi_1(x) &= A\exp(ik_1x) + B\exp(-ik_1x) && \text{for } x \leq 0 \\ \psi_2(x) &= C + Dx && \text{for } 0 \leq x \leq w \\ \psi_3(x) &= F\exp(ik_1x) + G\exp(-ik_1x) && \text{for } x \geq w \end{aligned} \tag{4.24}$$

The arbitrary probability amplitudes are constants in the above wave functions and are real or complex. Thus, A = incident wave amplitude, B = reflected wave amplitude, C = incident wave amplitude from $x = 0$, D = reflected wave amplitude from $x = w$, F = transmitted wave amplitude at $x_2 \geq w$, and G = reflected wave amplitude from $x = \infty$. Eventually, $G = 0$ for $x = \infty$ since no particle comes from the right.

These wave amplitudes can be related to each other, provided that $\psi(x)$ and $d\psi(x)/dx$ are continuous $[\psi(x) = d\psi(x)/dx]$ at $x = 0$ and $x = w$; otherwise, $d^2\psi(x)/dx^2$ would not exist in Eq. (4.18) [2, 39].

For region 2, k_2 is the decay length (tunneling length), and it indicates how quickly the exponential function decays. From the k_2 definition, Eq. (4.22), one can define the tunneling or penetration length α, the wave length λ, and the momentum $P(x)$ as

$$\alpha = 1/k_2 \tag{4.25}$$

$$\lambda = 2\pi/k_2 \tag{4.26}$$

$$P(x) = \sqrt{2m[U(x) - E]} \quad \text{(Real)} \tag{4.27}$$

$$P(x) = i\sqrt{2m[E - U(x)]} \quad \text{(Imaginary)} \tag{4.28}$$

In quantum mechanics, the probability density (P_T) for a particle to emerge to region 3 (Fig. 4.10) can be defined as the absolute square of the complex probability amplitude $P_T = |\psi(x)|^2$, but $P_T = |F/A|^2$ is the preferred definition since F and A are the transmission and incident wave amplitudes, respectively [2]. In general, P_T is the tunneling probability for finding a particle at a position $x > 0$ within a potential energy barrier $U(x)$. This can be accomplished by applying boundary conditions ($BC's$) on the wave functions at the edges of the barrier. For continuous $\psi(x)$ and $d\psi(x)/dx$ functions at $x = 0$ and $x = w$, the $BC's$ are

$$\begin{aligned} \psi_1(0) &= \psi_2(0) & \psi_2(0) &= \psi_3(0) \\ \frac{d\psi_1(0)}{dx} &= \frac{d\psi_2(0)}{dx} & \frac{d\psi_2(0)}{dx} &= \frac{d\psi_3(0)}{dx} \end{aligned} \tag{4.29}$$

and

$$\begin{aligned} \psi_1(w) &= \psi_2(w) & \psi_2(w) &= \psi_3(w) \\ \frac{d\psi_1(w)}{dx} &= \frac{d\psi_2(w)}{dx} & \frac{d\psi_2(w)}{dx} &= \frac{d\psi_3(w)}{dx} \end{aligned} \tag{4.30}$$

For $E < U_o$, Eq. (4.23) along with the $BC's$ yields

$$A + B = C + D \tag{4.31}$$

$$ik_1(A + B) = k_2(C + D) \tag{4.32}$$

$$C\exp(k_2 w) + D\exp(-k_2 w) = F\exp(ik_1 w) \tag{4.33}$$

$$k_2[C\exp(k_2 w) - D\exp(-k_2 w)] = ik_1 F\exp(ik_1 w) \tag{4.34}$$

There are four simultaneous equations which can be solve for the B/A, C/A, D/A, and F/A. But, the algebraic manipulation takes a lot of space and it is tedious. Therefore, the most relevant quotients are derived as [39]

$$\frac{F}{A} = \frac{4ik_1k_2 \exp(-ik_1w)}{(k_2 + ik_1)^2 \exp(-k_2w) - (k_2 - ik_1)^2 \exp(k_2w)} \tag{4.35}$$

$$\frac{B}{A} = \frac{F}{A}\frac{k_1^2 + k_2^2}{2ik_1k_2} \cosh(k_2w) \tag{4.36}$$

Hence, the probability densities, P_T for transmission and P_R for reflection, are defined as

$$P_T = \left|\frac{F}{A}\right|^2 = \frac{F^*F}{A^*A} \tag{4.37}$$

$$P_R = \left|\frac{B}{A}\right|^2 = \frac{B^*B}{A^*A} \tag{4.38}$$

where F^* is the conjugate of F, B^* is the conjugate of B, and A^* is the conjugate of A and $P_T + P_R = 1$. For example, the conjugate of $\exp(-ik_1w)$ is $\exp(ik_1w)$, and $(k_2 - ik_1)^2$ is the conjugate of $(k_2 + ik_1)^2$.

For transmission, Eqs. (4.35) and (4.37) yield the exact tunneling probability for $E < U_o$:

$$P_T = \frac{16k_1^2k_2^2}{(k_1 + k_2)^2 e^{-2k_2w} + (k_1 + k_2)^2 e^{2k_2w} + 12k_1^2k_2^2 - 2\left(k_2^4 + k_1^4\right)} \tag{4.39}$$

But, $(k_1 + k_2)^2 e^{-2k_2w} + (k_1 + k_2)^2 e^{2k_2w} >> 12k_1^2k_2^2 - 2\left(k_2^4 + k_1^4\right)$ and Eq. (4.39) reduces to

$$P_T = \frac{16k_1^2k_2^2}{(k_1 + k_2)^2 e^{-2k_2w} + (k_1 + k_2)^2 e^{2k_2w}} \tag{4.40}$$

If $e^{2k_2w} >> e^{-2k_2w}$, then Eq. (4.40) becomes a suitable approximation equation of the form

$$P_T = \frac{16k_1^2k_2^2}{(k_1 + k_2)^2} \exp(-2k_2w) \quad \text{for } E < U_o \tag{4.41}$$

$$P_T = \left(\frac{4k_1k_2}{k_1 + k_2}\right)^2 \exp(-2k_2w) \quad \text{for } E < U_o \tag{4.42}$$

For $E > U_o$, a similar analytical procedure gives [2]

$$P_T = \frac{4k_1k_2}{(k_1 + k_2)^2} \tag{4.43}$$

and for $E = U_o$,

$$P_T = \frac{1}{1 + 4mE\,(\pi w)^2\,/h^2} \tag{4.44}$$

In fact, P_T is also known as the transmission coefficient or the transmission probability. For particle tunneling, the barrier having a fixed potential energy height $U_2\,(x) = U_0$ and a finite width w must be comparable to the particle wave length (λ) through k_2. The counterpart of P_T is known as the reflection probability (P_R) since a particle may emerge into region 2 and return to region 1 due to some perturbations including external electric fields, magnetic fields, or electromagnetic radiation at the edge of the barrier at position x_1 [35]. Despite that $P_T + P_R = 1$, $P_R > P_T$ in most cases.

4.5.2 WKB Method

The WKB is a method for approximating the solution of a differential equation. This is known as the Wentzel, Kramers, and Brillouin (1926) method, **WKB method**, or **WKB approximation** [35]. For a slowly **changing barrier potential energy** $U = U(x)$, the solution of the Schrodinger equation within the barrier (region 2) can be written in a general form

$$\psi\,(x) = A \exp\left[i\theta\,(x)\right] \tag{4.45}$$

where $\theta\,(x)$ is the phase of $\psi\,(x)$ that will determine specific discrete values (quantization condition). Omitting the analytical procedure for proving that Eq. (4.45) is a solution of Eq. (4.18), only the results for $\theta\,(x)$ and P_T are given as

$$\theta\,(x) = \frac{1}{\hbar} \int_o^w P\,(x)\,dx \tag{4.46}$$

$$P_T = \exp\left(-2k_2 \int_o^w dx\right) \tag{4.47}$$

For $E < U_o$,

$$P_T = \exp\left(-2k_2 w\right) \quad \text{for small } k_2 \tag{4.48}$$

For $E \to U_o$ [38],

$$P_T = \left[1 + \exp\left(2k_2 w\right)\right]^{-1} \quad \text{for large } k_2 \tag{4.49}$$

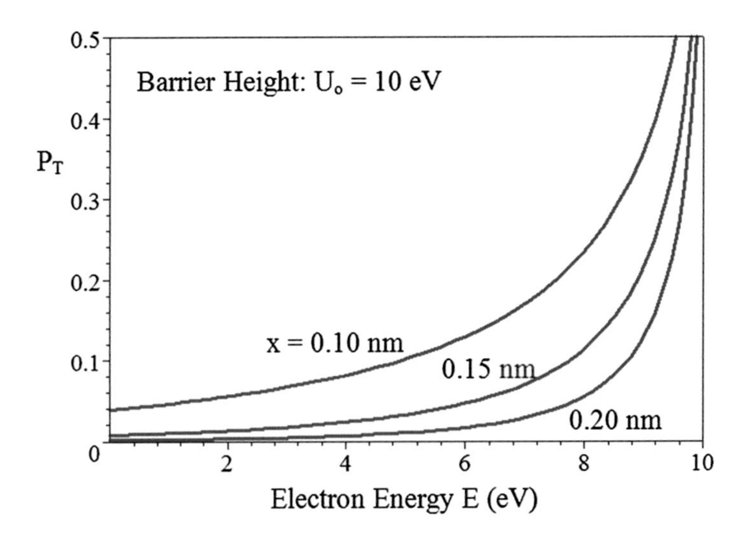

Fig. 4.12 Tunneling probability in a barrier height of 10 eV and varying tunneling distance x

Moreover, if the tunneling particle is an electron, then Eqs. (4.21) and (4.22) become

$$k_1 = \left(5.1268\,\text{nm}^{-1}\,\text{eV}^{-1/2}\right)(U_o)^{1/2} \tag{4.50}$$

$$k_2 = \left(5.1268\,\text{nm}^{-1}\,\text{eV}^{-1/2}\right)(U_o - E)^{1/2} \tag{4.51}$$

$$k_2 = \left(5.1268\,\text{nm}^{-1}\,\text{eV}^{-1/2}\right)\chi^{1/2} \tag{4.52}$$

where $\chi = U_o - E$ is in electron volt (eV) units. Then, as an example, Eq. (4.48) becomes

$$P_T = \exp\left[-\left(10.254\,\text{nm}^{-1}\,\text{eV}^{-1/2}\right)(U_o - E)^{1/2}\,w\right] \tag{4.53}$$

Figure 4.12 depicts the tunneling effect on tunneling probability for $U_o = 10\,\text{eV}$ and tunneling width x as per Eq. (4.53) [2].

Example 4.3. *This example reflects the significance of the tunneling currents through a thin metal oxide. The conductor-oxide-conductor system shown below for transmission tunneling can be modeled as per Fig. 4.11. For a 1.2-nm thick barrier, barrier height $U_o = 10\,eV$, electron energy $E = 8\,eV$, and current $I = 1\,mA$ (incident current), calculate **(a)** the number of electrons per second impinging on the oxide barrier, **(b)** the fraction of transmitted electrons per second, and **(c)** the transmitted current. Use the SP and the WKB methods and compare results for the $E < U_o$ case.*

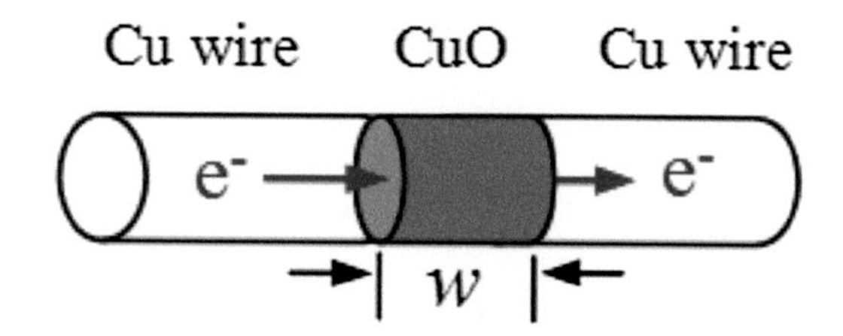

Solutions. **(a)** *The number of electrons per second impinging on the oxide barrier*

$$N_e = I/q_e = \left(10^{-3}\ \mathrm{A}\right) / \left(1.6022 \times 10^{-19}\ \mathrm{A\,s}\right) \simeq 6.241 \times 10^{15}\ \text{electrons/s}$$

$$N_e = 6241 \times 10^{12} = 6241\ \text{tera electrons/s}\quad \textit{(Impinged)}$$

(b) *The fraction of transmitted electrons per second*

SP method: *From Eqs. (4.50) and (4.51),*

$$k_1 = \left(5.1268\ \mathrm{nm}^{-1}\,\mathrm{eV}^{-1/2}\right)(E)^{1/2} = (5.1268)\,(8)^{1/2} = 14.501\ \mathrm{nm}^{-1}$$

$$k_2 = \left(5.1268\ \mathrm{nm\,eV}^{-1/2}\right)(U_o - E)^{1/2}$$

$$k_2 = \left(5.1268\ \mathrm{nm}^{-1}\,\mathrm{eV}^{-1/2}\right)(10\ \mathrm{eV} - 8\ \mathrm{eV})^{1/2} = 7.2504\ \mathrm{nm}^{-1}$$

From Eq. (4.39),

$$P_T = \frac{16k_1^2k_2^2}{(k_1+k_2)^2 e^{-2k_2w} + (k_1+k_2)^2 e^{2k_2w} + 12k_1^2k_2^2 - 2\left(k_2^4 + k_1^4\right)}$$

$$P_T = 1.0364 \times 10^{-5}$$

and from Eq. (4.41),

$$P_T = \frac{16k_1^2k_2^2}{(k_1+k_2)^2} e^{-2k_2w} = 1.0364 \times 10^{-5}$$

Then,

$$N_x^{SP} = N_e P_T^{SP} = \left(6.241 \times 10^{15}\ \text{electrons/s}\right)\left(1.0364 \times 10^{-5}\right) = 6.468 \times 10^{10}$$

$$N_x^{SP} \simeq 65 \times 10^9\ \text{electrons/s} = 65\ \text{giga electrons/s}\ \textit{(Transmitted)}$$

Only 65 *giga electrons/s out of* 6241 *tera electrons/s are transmitted a distance*

$$\alpha^{SP} = 1/k_2 = 1/7.2504 \simeq 0.14\ \mathrm{nm}$$

WKB method: *From Eq. (4.48),*

$$P_T^{WKB} = \exp(-2k_2 w) = \exp\left[-(2)\left(7.2504\ \text{nm}^{-1}\right)(1.2\ \text{nm})\right]$$
$$P_T^{WKB} = 2.7724x10^{-8}$$

and

$$N_x^{WKB} = N_e P_T^{WKB} = \left(6.241 \times 10^{15}\ \text{electrons/s}\right)\left(2.7724 \times 10^{-8}\right)$$
$$N_x^{WKB} = 1.7303 \times 10^{8}\ \text{electrons/s}$$
$$N_x^{WKB} \simeq 173 \times 10^{6}\ \text{electrons/s} \quad \textit{(Transmitted)}$$

The penetration distance is

$$\alpha^{SP} = \alpha^{WKB} = 1/k_2 = 1/7.2504 \simeq 0.14\ \text{nm}$$

Therefore, only 173 *millions out of* 6241 *tera electrons were transmitted a distance of* 0.14 *nm within the barrier.*

(c) *From SP method, the transmitted current is*

$$I_T^{SP} = N_x^{SP} q_e = I P_T^{SP} = \left(10^{-3}\ \text{A}\right)\left(1.0364 \times 10^{-5}\right)$$
$$I_T^{SP} = 1.0364 \times 10^{-8}\ \text{A}$$
$$I_T^{SP} = 10.36\ \text{nA}$$

From the WKB method, the transmitted current is

$$I_T^{WKB} = N_x^{WKB} q_e = I P_T^{WKB} = \left(10^{-3}\ \text{A}\right)\left(2.7724 \times 10^{-8}\right)$$
$$I_T^{WKB} = 2.772 \times 10^{-11}\ \text{A}$$
$$I_T^{WKB} = 27.72\ \text{pA}$$

Therefore, only a small current was transmitted through the barrier. The SP and WKB methods differ in current values by 99 %. The SP method will always give higher P_T results since the factor $16k_1^2 k_2^2 / \left((k_1 + k_2)^2\right) = 373.82$ in this example. Consequently, one has to be cautious in choosing a particular equation to carry out calculations that may lead to a significant error.

4.6 Problems/Questions

4.1. In an electrochemical cell, the bulk electrolyte has uniform and constant ion densities, but the electrical double layer is an inhomogeneous fluid. Why?

4.2. The interface region between two bulk phases may contain a complex distribution of electric charge. Name at least two causes for this phenomenon.

4.3. Explain how it is possible for a scanning tunneling microscope (STM) to image atoms on the surface of a sample. Recall that one of the metals is the sample and the other is the probe.

4.4. For a finite depth energy well, the wave function $\psi(x)$ within the barrier of width w, the tunneling current I_t through the barrier, and the decay length are related as shown below:

$$\psi(x) = \psi_o \exp(-k_2 x)$$

$$I_t \propto |\psi_o|^2 \exp(-k_2 x)$$

$$k_2 = \frac{2\pi}{h}\sqrt{2m(U_o - E)} = \frac{2\pi}{h}\sqrt{2m\chi}$$

If the work function and the barrier width are $\chi = 4.32\,\text{eV}$ and $w = 0.151\,\text{nm}$ in a STM experiment, then determine I_t for a barrier of width $2w$. Explain. [Solution: $I_{t,2w} = 0.2 I_{t,w}$].

4.5. Determine **(a)** the Fermi energy and **(b)** the velocity of the electron for a copper (Cu) substrate and a tungsten (W) tip in a ultrahigh vacuum STM cell at temperature T. [Solution: (a) $U_{F,Cu} = 6.59\,\text{eV}$, (b) $\chi = -8.02\,\text{eV}$].

4.6. For the reaction $O + e^- \rightleftarrows R$ in an electrochemical cell, the cyclic voltammetry method provided the formal potential (standard potential) $\phi^o = 133\,\text{mV}$ against a reference electrode. If the activity ratio is $[O]/[R] = 0.2$, then calculate **(a)** the applied potential ϕ at 25 °C and **(b)** the required energy for the reaction to proceed to the right as written. [Solution: $U = 0.092\,\text{eV}$].

4.7. Show that $P_T = \exp(-2k_2 x)$, Eq. (4.48), using the following equations:

$$\psi = A \exp(i\theta)$$

$$\frac{d^2\psi}{dx^2} = -\frac{\sqrt{2m[U-E]}}{\hbar^2}\psi = -\frac{P^2}{\hbar^2}\psi$$

where P is the moment, $x = w$, and $\hbar = h/2\pi$. Hint: extract the resultant real and imaginary parts for solving this problem.

4.8. Below is a sketch for the principle of the STM technique. If the work function (χ) and the tunneling gap column (z) are 4 eV and 0.3 nm, respectively, calculate the probability for an electron to tunnel from the probe tip to the metal sample. [Solution: $P_T = 0.21\,\%$].

4.9. Calculate the electron energy (E) if the tunneling probability (P_T) is 0.106 and the barrier height (U_o) 4.5 eV. Assume a barrier width of 0.4 nm. [Solution: $E = 4.2\,\text{eV}$].

4.10. Assuming that there is no wave reflection, Eq. (4.20), at the boundary $x = 0$, derive $P_R = |B/A|^2$ and P_T for $E > U_o$. Use the continuity of the wave function and its derivative at the origin.

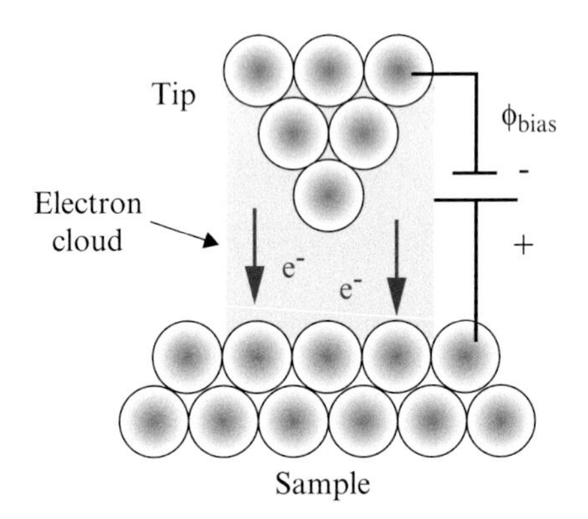

4.11. Assuming that the wave amplitude, Eq. (4.23), $C = 0$ when $x \to \infty$, derive $P_R = |B/A|^2$ and P_T for $E < U_o$. Use the continuity of the wave function and its derivative at the origin. Explain.

4.12. For$E < U_o$, use Eq. (4.23) to show that

$$\frac{F}{A} = \frac{4ik_1k_2 \exp(-ik_1w)}{(k_2 + ik_1)^2 \exp(-k_2w) - (k_2 - ik_1)^2 \exp(k_2w)}$$

$$P_T = \frac{16k_1^2k_2^2}{\left[(k_1 + k_2)^2 e^{-2k_2w} + (k_1 + k_2)^2 e^{2k_2w}\right] + \left[12k_1^2k_2^2 - 2\left(k_2^4 + k_1^4\right)\right]}$$

If $k_1 = 14$, $k_2 = 7$ and $w = 1.2\,\text{nm}$, then prove that

$$P_T = \frac{16k_1^2k_2^2}{(k_1 + k_2)^2} \exp(-2k_2w) = 1.7619 \times 10^{-5}$$

References

1. T.F. Yen, *Environmental Chemistry: Essentials of Chemistry for Engineering Practice*. Prentice Hall Environmental, vol. 4A (Prentice Hall PTR, New York, 1999)
2. T. Engel, P. Reid, *Physical Chemistry* (Pearson Education, Boston, 2013), pp. 279–286, 367–380
3. C.M.A. Brett, A.M.O. Brett, *Electrochemistry Principles, Methods and Applications* (Oxford University Press, New York, 1994), pp. 39–44
4. D.A. Jones, *Principles and Prevention of Corrosion* (Macmillan Publishing Company, New York, 1992), p. 40
5. J. O'M. Bockris, M.A.V. Devanathan, K. Muller, On the structure of charge interfaces. Proc. R. Soc. Lond. Ser. A **274**(1356), 55–79 (1963)
6. C.A. Barlow Jr., The electrical double layer, in *Physical Chemistry: An Advanced Treatise. Vol. IXA/Electrochemistry*, ed. by H. Eyring (Academic, New York, 1970)
7. T. Erdey-Gruz, *Kinetics of Electrode Processes* (Wiley-Interscience/Wiley, New York, 1972), p. 442
8. H. Wang, L. Pilon, Accurate simulations of electric double layer capacitance of ultramicroelectrodes. J. Phys. Chem. C **115**, 16711–16719 (2011)
9. J.W. Evans, L.C. De Jonghe, *The Production of Inorganic Materials* (Macmillan Publishing Company, New York, 1991)
10. A.J. Bard, L.R. Faulkner, *Electrochemical Methods - Fundamentals and Applications*, 2nd edn., Chaps. 1–2, 16 (Wiley, New York, 2001), pp. 548, 670
11. M. Jakob, H. Levanon, P.V. Kamat, Charge distribution between UV-irradiated TiO(2) and gold nanoparticles: determination of shift in the Fermi level. Nano Lett. **3**(3), 353–358 (2003)
12. J. McBreen, Physical methods for investigation of electrode surfaces, in *Fundamentals of Electrochemistry*, 2nd edn. (Wiley, New Jersey, 2006), p. 485
13. M.-B. Song, J.-M. Jang, C.-W. Lee, Electron tunneling and electrochemical currents through interfacial water inside an STM junction. Bull. Korean Chem. Soc. **23**(1), 71–74 (2002)
14. S. Duffe, T. Irawan, M. Bieletzki, T. Richter, B. Sieben, C. Yin, B. von Issendorff, M. Moseler, H. Hövel, Softlanding and STM Imaging of Ag561 Clusters on a C_{60} Monolayer. Eur. Phys. J. D 1–8 (2007)
15. K. von Bergmann, Iron nanostructures studied by spin-polarised scanning tunneling microscopy, Ph.D. thesis, Universität Hamburg, 2004
16. C.J. Chen, *Introduction to Scanning Tunneling Microscopy*, 2nd edn. (Oxford University Press, New York, 2008), p. 24
17. N.D. Kolb, M.A. Schneeweiss, Scanning tunneling microscopy for metal deposition studies. Electrochem. Soc. Interface 26–30 (1999)
18. O.M. Magnussen, L. Zitzler, B. Gleich, M.R. Vogt, R.J. Behm, In-situ atomic-scale studies of the mechanisms and dynamics of metal dissolution by high-speed STM. Electrochem. Acta **46**(24), 3725–3733 (2001)
19. D.M. Kolb, G.E. Engelmann, J.C. Ziegler, On the unusual electrochemical stability of nanofabricated copper clusters. Angew. Chem. Int. Ed. **39**(6), 1123–1125 (2000)
20. Y.-F. Liu, K. Krug, P.-C. Lin, Y.-D. Chiu, W.-P. Dow, S.-L. Yau, Y.-L. Leea, In situ STM study of Cu electrodeposition on TBPS-modified Au(111) electrodes. J. Electrochem. Soc. **159**(2), D84–D90 (2012)
21. D.M. Kolb, An atomistic view of electrochemistry. Surf. Sci. **500**, 722–740 (2002)
22. R.Q. Hwang, M.C. Bartelt, Scanning tunneling microscopy studies of metal on metal epitaxy. Chem. Rev. **97**, 1063–1082 (1997)
23. A. Lehnert, P. Buluschek, N. Weiss, J. Giesecke, M. Treier, S. Rusponi, H. Brune, High resolution in situ magneto-optic Kerr effect and scanning tunneling microscopy setup with all optical components in UHV. Rev. Sci. Instrum. **80**, 023902, 1–7 (2009)
24. W. Auwärter, M. Muntwiler, J. Osterwalder, T. Greber, Defect lines and two-domain structure of hexagonal boron nitride films on Ni(111). Surf. Sci. **545**, L735–L740 (2003)

25. S. Speller, T. Rauch, A. Postnikov, W. Heiland, Scanning tunneling microscopy and spectroscopy of S on Pd(111). Phys. Rev. B **61**(11), 7297–7300 (2000)
26. A.J. Bard, G. Denuault, H. Oongmolke, D. Mandler, D.O. Wipf, Scanning electrochemical microscopy: a new technique for the characterization and modification of surfaces. Acc. Chem. Res. **23**, 357–363 (1990)
27. J. Zhang, J. Ulstrup, Oxygen-free in-situ scanning tunneling microscopy. J. Electroanal. Chem. **599**, 213–220 (2007)
28. A.J. Bard, F.R.F. Fan, V.M. Michael, Scanning electrochemical microscopy, in *Electroanalytical Chemistry: A Series of Advances*, vol. 18, ed. by A.J. Bard (Marcel Dekker, New York, 1994), pp. 243–373
29. S. Bruckenstein, J. Janiszewska, Diffusion currents to (ultra) microelectrodes of various geometries: ellipsoids, spheroids and elliptical 'disks. J. Electroanal. Chem. **538–539**, 3–12 (2002)
30. M. Fleischmann, S. Pons, D. Rolison, P.P. Schmidt, *Ultramicroelectrodes* (Datatech Systems, Morgantown, NC, 1987)
31. A. Szabo, Theory of the current at microelectrodes: application to ring electrodes. J. Phys. Chem. **91**, 3108–3111 (1987)
32. G. Nagy, L. Nagy, Scanning electrochemical microscopy: a new way of making electrochemical experiments. Fresenius J. Anal. Chem. **366**, 735–744 (2000)
33. K.W. Kolasinski, *Surface Science: Foundations of Catalysis and Nanoscience*, 2nd edn. (Wiley, West Sussex, 2008), pp. 36–38
34. W. Schmickler, *Interfacial Electrochemistry*, vol. 12 (Oxford University Press, New York, 1996), pp. 72–76
35. E.F. Schubert, *Physical Foundations of Solid-State Devices*, 2006 edition, Chap. 9-5, 10-1 (Rensselaer Polytechnic Institute, Troy, NY)
36. A. Beiser, *Concepts of Modern Physics*, 6th edn. (McGraw-Hill Companies, Boston, 2003), pp. 193–196
37. A.D. Polyanin, A.I. Chernoutsan (eds.), *A Concise Handbook of Mathematics, Physics, and Engineering Sciences* (CRC Press, Taylor and Francis Group, LLC, Boca Raton, 2011), p. 577, 627
38. E.C. Kemble, *The Fundamental Principles of Quantum Mechanics with Elementary Applications* (Dover, New York, 1958)
39. A.I.M. Rae, *Quantum Mechanics*, 4th edn. (IOP Publishing Ltd., London, 2002)

Kinetics of Activation Polarization 5

Polarization is an electrochemical process induced by deviation of the electrochemical equilibrium potential (E_{corr}) due to an electric current passing through the electrochemical cell.

Activation polarization is an electrochemical process induced by an applied overpotential to deviate an electrochemical cell from its equilibrium state during current flow through the electrochemical cell. A positive overpotential causes oxidation reactions to take place. Conversely, a negative overpotential induces reduction reactions.

For comparison purposes, concentration polarization is an electrochemical process induced by mass transport due to a concentration gradient in the electrolyte solution. This type of polarization is induced by diffusion mechanism which is significantly detailed in Chap. 6.

Electrochemical reaction kinetics is essential in determining the rate of corrosion of a metal M exposed to a corrosive medium (electrolyte). On the other hand, thermodynamics predicts the possibility of corrosion, but it does not provide information on how slow or fast corrosion occurs. Thus, a reaction rate strongly depends on the rate of electron flow to or from an electrode interfacial region. If the electrochemical system is at equilibrium, then the net rate of reaction is zero. In comparison, reaction rates are governed by chemical kinetics, while corrosion rates are primarily governed by electrochemical kinetics [1].

Electrochemical kinetics of a corroding metal can be characterized by determining the corrosion current density (i_{corr}) and the corrosion potential (E_{corr}). Then the corrosion behavior can be disclosed by a polarization curve (E vs. $\log i$). Evaluation of these parameters leads to the determination of the polarization resistance (R_p) and the corrosion rate as i_{corr}, which is often converted into Faradaic corrosion rate C_R having units of mm/y.

N. Perez, *Electrochemistry and Corrosion Science*,
DOI 10.1007/978-3-319-24847-9_5

5.1 Energy Distribution

For a polarized electrode under steady-state current flow, the generalized reaction given by Eq. (1.4) can be used to derive the Butler–Volmer equation, which involves energy barriers known as activation energies. Only the activation energy change is used for the forward (ΔG_f) (reduction) and reverse (ΔG_r) (oxidation) reactions. For example, the hydrogen reaction, $2H^+ + 2e^- = H_2$ at equilibrium, requires that the rate of discharge of H^+ ions in the forward direction (reduction) must be exactly equal to the rate of ionization of H_2 molecule in the reverse direction (oxidation). Thus, the rate of this simple reaction is known as the exchange current density (i_o). At equilibrium, $i_o = i_a = -i_c$ and $i_{net} = i_a - i_c = 0$ represent a convenient way for defining electrochemical equilibrium. However, if deviation from the equilibrium state occurs, an overpotential develops, and consequently, the electrochemical cell polarizes, and the activation energies become dependent on i_o. These energies are depicted in Fig. 5.1 in which the **activation state** is at the maximum point (saddle point).

This figure represents the Boltzmann or Maxwell–Boltzmann distribution law for the energy distribution of the reacting species (ions) [2–4]. This schematic energy distribution is for reversible electrodes. If these are polarized by an overpotential under steady-state conditions, the rate of reactions are not equal, $R_f \neq R_r$.

In general, the electrochemical and chemical rates of reactions due to either anodic or cathodic overpotentials can be predicted using both Faraday's and

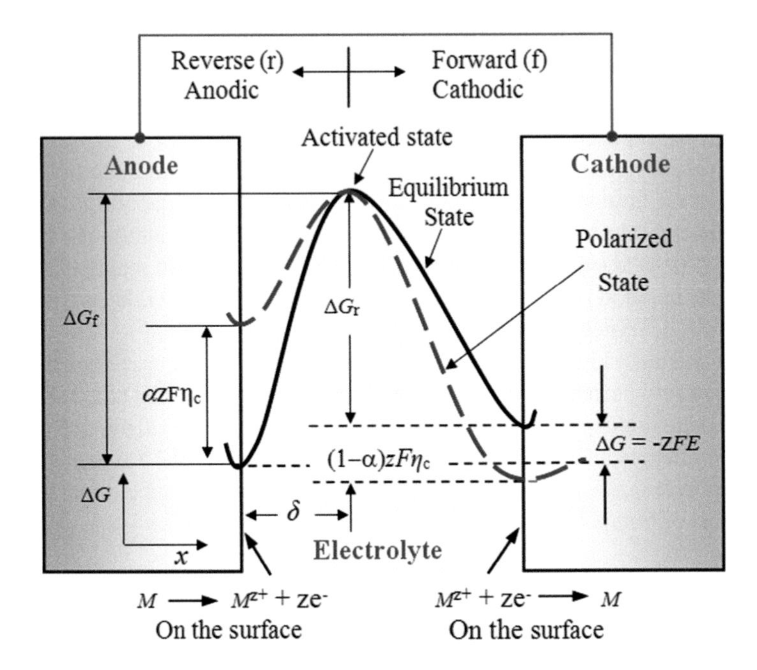

Fig. 5.1 Schematic activation free energy distribution. After Jones [2]

Arrhenius equations, respectively:

$$R_F = \frac{iA_{w,j}}{zF} \tag{5.1}$$

$$R_A = \gamma_a \exp\left(-\frac{\Delta G^*}{RT}\right) \tag{5.2}$$

where i = applied current density $\left(\mathrm{A/cm^2}\right)$
$A_{w,j}$ = atomic weight of species j (g/mol)
$A_{w,alloy} = \sum f_j A_{w,j}/z_j$ in units of g/mol
z = oxidation state or valence number
z_j = valence number of element j
f_j = weight fraction of element j
γ_a = chemical reaction constant
ΔG^* = activation energy or free energy change (J/ mol)

At equilibrium, Faraday's and Arrhenius rate equations become equal ($R_F = R_A$), and consequently, the current density becomes

$$i = \gamma_o \exp\left(-\frac{\Delta G^*}{RT}\right) \tag{5.3}$$

Here, the term $\gamma_o = zF\gamma_a/A_w$ may be defined as the electrochemical rate constant having a unit of current density $\left(\mathrm{A/cm^2}\right)$. For a reversible electrode at equilibrium, the current density in Eq. (5.3) becomes the exchange current density; that is, $i = i_o$. In addition, Table 5.1 gives typical experimental values for i_o in some solutions.

Denote that values of *Pb* and *Pt* in 1 *N HCl* solution. Thus, $i_{o,Pt}/i_{o,Pb} = 5x10^9$ means that hydrogen evolution, $2H^+ + 2e^- = H_2$, is 5 trillion times faster on the *Pt* surface electrode than on the *Pb* surface. If *Pt-Pb* are galvanically coupled in

Table 5.1 Exchange current density data at 25 °C [3]

Electrode	Electrolyte	Reaction	i_o (A/cm^2)
Al	2 N H_2SO_4	$2H^+ + 2e^- = H_2$	$1x10^{-10}$
Au	1 N *HCl*	$2H^+ + 2e^- = H_2$	$1x10^{-6}$
Cu	0.1 N *HCl*	$2H^+ + 2e^- = H_2$	$2x10^{-7}$
Fe	2 N H_2SO_4	$2H^+ + 2e^- = H_2$	$1x10^{-6}$
Ni	1 N *HCl*	$2H^+ + 2e^- = H_2$	$4x10^{-6}$
Ni	05 N $NiSO_4$	$Ni = Ni^{2+} + 2e^-$	$1x10^{-6}$
Pb	1 N *HCl*	$2H^+ + 2e^- = H_2$	$2x10^{-13}$
Pt	1 N *HCl*	$2H^+ + 2e^- = H_2$	$1x10^{-3}$
Pt	0.1 N *NaOH*	$O_2 + 4H^+ + 4e^- = 2H_2O$	$4x10^{-13}$
Pd	06 N *HCl*	$2H^+ + 2e^- = H_2$	$2x10^{-4}$
Sn	1 N *HCl*	$2H^+ + 2e^- = H_2$	$1x10^{-8}$

1 N HCl solution, Pt becomes the cathode where hydrogen evolution (H_2) occurs and Pb the anode for providing the electrons. In addition, i_o is an experimental kinetic parameter since there is no accurate theoretical background for its derivation at this time.

On the other hand, if an electrode is polarized by an overpotential under steady-state conditions, then the rates of reactions are not equal ($R_F \neq R_A$), and consequently, the forward (cathodic) and reverse (anodic) current density components must be defined in terms of the free energy change ΔG^* deduced from Fig. 5.1. Letting $\gamma_o = k_f'$ for forward and $\gamma_0 = k_r'$ for reverse reactions gives

$$i_f = k_f' \exp\left[-\frac{\Delta G_f^*}{RT}\right] \quad \text{(cathodic)} \tag{5.4}$$

$$i_r = k_r' \exp\left[-\frac{\Delta G_r^*}{RT}\right] \quad \text{(anodic)} \tag{5.5}$$

where $\Delta G_f^* = \Delta G_f - \alpha z F \eta_c$
$\Delta G_r^* = \Delta G_r + (1-\alpha) z F \eta_a$
$\alpha =$ symmetry coefficient (transfer coefficient).

For a cathodic case, the net current, $i = i_f - i_r$ or $i = i_c - i_a$, and the overpotential, $\eta_c < 0$, are, respectively,

$$i = k_f' \exp\left(-\frac{\Delta G_f}{RT}\right) \exp\left(\frac{\alpha z F \eta}{RT}\right) - k_r' \exp\left(-\frac{\Delta G_r}{RT}\right) \exp\left[-\frac{(1-\alpha) z F \eta}{RT}\right] \tag{5.6}$$

from which the exchange current density is deduced as

$$i_o = k_f' \exp\left(-\frac{\Delta G_f}{RT}\right) = k_r' \exp\left(-\frac{\Delta G_r}{RT}\right) \tag{5.7}$$

Substituting Eq. (5.7) into (5.6) for one-step reaction yields the well-known **Butler–Volmer equation**, without mass transfer effects, for polarizing an electrode from the open-circuit potential E_o under steady-state conditions:

$$i = i_o \left\{ \exp\left[\frac{\alpha z F \eta}{RT}\right]_f - \exp\left[-\frac{(1-\alpha) z F \eta}{RT}\right]_r \right\} \tag{5.8}$$

where $\eta = E - E_o =$ overpotential (Volts)
$E =$ applied potential (Volts)
$i_o =$ exchange current density (A/cm^2)

It is clear that overpotential depends on the applied current density; therefore, $\eta = f(i)$. In addition, the exchange current density (i_o) can be defined as the rate of oxidation and reduction reactions at equilibrium. Specifically, i_o is the current density at which the rate of oxidation and rate of reduction are equal in a state of equilibrium. Thus, i_o is just the reversible reaction rate at equilibrium, and it is a kinetic parameter, whereas ΔG_f and ΔG_r are thermodynamic parameters. In fact, i_o is very sensitive to electrode surface condition, and it is temperature-dependent as indicated by Eq. (5.8).

Inherently, the Butler–Volmer equation, Eq. (5.8), defines two electrochemical cases: high activation polarization for high current density and low activation polarization for low current density studies. In other words, Eq. (5.8) measures the current density response due to an applied overpotential potential on an electrode surface.

In addition, Eq. (5.8) can be rearranged as a hyperbolic sine function along with $\alpha = 0.5$. Thus,

$$i = 2i_o \sinh\left(\frac{zF\eta}{2RT}\right) \tag{5.9}$$

from which the overpotential becomes

$$\eta = \left(\frac{2RT}{zF}\right)\sinh^{-1}\left(\frac{i}{2i_o}\right) \tag{5.10}$$

Furthermore, one can generalize the Arrhenius type equation, Eq. (5.8), for a series of electrochemical reactions as [2]

$$i_{o,T,r} = i_{o,T_o,r}\exp\left[-\frac{\Delta G^*}{R}\left(\frac{1}{T} - \frac{1}{T_o}\right)\right] \tag{5.11}$$

where $i_{o,T,r}$ = exchange current density at temperature T
$i_{o,T_o,r}$ = reference exchange current density at temperature T_o
$r = 1, 2, 3, 4, \ldots$ = number of reactions

In addition, Eq. (5.8) is the general nonlinear rate equation in terms of current density that most electrochemical cells obey under charge-transfer control. Thus, $\eta = \eta_a = (E - E_o) > 0$ and $\eta = \eta_c = (E - E_o) < 0$ are the anodic and cathodic overpotentials, respectively, which represent a deviation from the half-cell potential E^o, and E is the applied potential. In summary, the applied anodic current density is $i = i_a - i_o$ at $\eta_a > 0$, and the cathodic current density is $i = i_c - i_o$ at $\eta_c < 0$. At equilibrium conditions, the potential is $E = E_o$ and $\eta = 0$, and Eq. (5.8) yields $i_a = i_c = i_o$. This implies that the rates of metal dissolution (oxidation) and deposition (reduction) are equal. However, a deviation from **equilibrium of the half-cells** is the normal case in real situations.

Furthermore, if one is interested in determining the total ionic molar flux J due to the mass transfer of a species j, then combine $i = zFJ$ and Eq. (5.8) to get

$$J = \frac{i}{zF}\left\{\exp\left[\frac{\alpha zF\eta}{RT}\right]_f - \exp\left[-\frac{(1-\alpha)\,zF\eta}{RT}\right]_r\right\} \tag{5.12}$$

This ionic molar flux will be defined in details in Chap. 6. The reaction rate in terms of current density i given by Eq. (5.8) is valid for one-step reaction. Hence, the controlling reaction may be part of a series of reaction steps, but the slowest reaction step is the rate controlling. For any reaction, Eq. (5.8) can be approximated by letting $x = \alpha z F \eta / RT$ and $y = (1-\alpha) z F \eta / RT$. Expanding the exponential functions according to Taylor's series yields

$$\exp(x) = \sum_{k=0}^{\infty} \frac{x^k}{k!} \simeq 1 + x \tag{5.13}$$

$$\exp(-y) = \sum_{k=0}^{\infty} \frac{(-y)^k}{k!} \simeq 1 - y \tag{5.14}$$

Thus, Eq. (5.8) can be expressed as a linear approximation when the overpotential is small in magnitude [5]. Inserting Eqs. (5.13) and (5.14) into (5.8) yields

$$i = i_o\,(x + y) = [\alpha z F \eta / RT + (1-\alpha)\, z F \eta / RT] \tag{5.15}$$

$$i \simeq i_o \left(\frac{z F \eta}{RT} \right) \tag{5.16}$$

These expressions, Eqs. (5.15) and (5.16), can be used for small values of overpotential like $\eta < 0.01$ V [4]. However, one can take a different approach to approximate the current density by considering the following inequality:

$$\exp\left[\frac{\alpha z F \eta}{RT}\right]_f >> \exp\left[-\frac{(1-\alpha)\, z F \eta}{RT}\right]_r \tag{5.17}$$

If this is the case, then Eq. (5.8) yields

$$i = i_o \exp\left(\frac{\alpha z F \eta}{RT}\right) \tag{5.18}$$

Solving Eq. (5.18) for the overpotential yields the **Tafel equation**

$$\eta = a + b \log(i) \tag{5.19}$$

where

$$a = -\frac{2.303 RT}{\alpha z F} \log(i_o) \tag{5.20}$$

$$b = \frac{2.303 RT}{\alpha z F} \tag{5.21}$$

Let R_f and R_r be the forward (reduction) and reverse (oxidation) reaction rates, respectively, and let $K = K_f/K_r$ be the equilibrium constant so that

$$\ln(K) = \ln\left(K_f\right) - \ln(K_r) \tag{5.22}$$

Using from Eq. (5.2) along with $\gamma_a = K_f$ or K_r yields

$$\ln\left(K_f\right) = \ln\left(R_f\right) + \frac{\Delta G_f}{RT} \tag{5.23}$$

$$\ln(K_r) = \ln(R_r) + \frac{\Delta G_r}{RT} \tag{5.24}$$

Substituting Eqs. (5.23) and (5.24) into (5.22) yields

$$\ln(K) = \ln\left(R_f\right) + \frac{G_f}{RT} - \ln(R_r) - \frac{G_r}{RT} \tag{5.25}$$

Differentiating Eq. (5.25) with respect to the temperature T gives

$$\frac{d\ln(K)}{dT} = \frac{\Delta G_r}{RT^2} - \frac{\Delta G_f}{RT^2} = \frac{\Delta G^*}{RT^2} \tag{5.26}$$

Rearranging Eq. (5.26) yields the activation energy as [6]

$$\Delta G^* = \left(RT^2\right)\frac{d\ln(K)}{dT} \tag{5.27}$$

According to Faraday's law of electrolysis, the quantity of charge transferred (Q) at a time t and known current (I) or current density (i) is

$$Q = It = iA_s t \tag{5.28}$$

The quantity of moles for **zF charge transfer** at t and i is

$$N = \frac{Q}{zF} \tag{5.29}$$

Now, the Faraday's mass loss or gain during the electrochemical process called electrolysis can be deduced by multiplying the number of moles, Eq. (5.29), by the atomic weight of metal M:

$$m = NA_w = \frac{ItA_w}{zF} = \frac{ItA_w N}{Q} \tag{5.30}$$

$$A_{w,alloy} = \sum \frac{f_j A_{w,j}}{z_j} \tag{5.31}$$

Here, f_j is the weight fraction of an element j in the alloy. When the mass of a substance being altered at an electrode surface, Eq. (5.30), the electrochemical reaction rate is quantified by the rate of electron flow being measured as current [3].

Furthermore, the current distribution within an electrochemical cell must ideally be uniform under steady conditions; otherwise, short-circuit problems may arise, inducing low cell efficiency and short-life design. In electrodeposition of metals such as electroplating, the metal deposited (metallic coating) on the cathodes must be well adhered as homogeneous as possible having uniform thickness. In electrowinning and electroplating, a homogeneous current distribution induces uniform and smooth plated metals.

In order to optimize an electrochemical cell, thermodynamic and kinetic numerical or experimental assessments are important in understanding the evolved electrochemical behavior of a cell being designed to convert chemical to electrical energy (battery) or the opposite as in electrodeposition (electroplating, electrowinning, and electrorefining).

Example 5.1. *Two identical nickel rods are exposed to* $1\,N$ *HCl-base electrolyte. One rod is immersed at* $25\,°C$ *for* $2\,h$*, while the other rod is at* $50\,°C$ *for* $1\,h$*.* ***(a)*** *Calculate the time of exposure of a third identical rod immersed into the electrolyte at* $30\,°C$*. Assume that nickel oxidizes in the temperature range* $25\,°C \leq T \leq 50\,°C$*. Use this temperature range to determine* ***(b)*** *the solubility constant* (K_{sp}) *for the Ni oxidation reaction and* **(c)** *the current density when one identical rod is polarized from the exchange current density* $(4 \times 10^{-6}\,A/cm^2)$ *at an anodic overpotential of* $0.05\,V_{SHE}$*. This implies that* $n_a = (E - E^o) > 0$ *and* $K_{sp} < 1$. Use a symmetry coefficient of $\alpha = 0.5$.

Solution. (a) *The nickel oxidation reaction is*

$$Ni \longrightarrow Ni^{2+} + 2e^-$$

which requires a chemical activation energy to proceed in the written direction. The nickel rods are not polarized through an external circuit. Thus, the Arrhenius equation, Eq. (5.2), yields

$$R_1 = 1/t_1 = \gamma_a \exp\left(-\frac{\Delta G^*}{RT_1}\right) \tag{a}$$

$$R_2 = 1/t_2 = \gamma_a \exp\left(-\frac{\Delta G^*}{RT_2}\right) \tag{b}$$

Dividing these equations gives

$$\frac{t_2}{t_1} = \exp\left[-\frac{\Delta G^*}{R}\left(\frac{1}{T_1} - \frac{1}{T_2}\right)\right]$$

Then, at $T_1 = 25\,°C = 298\,K$ *and* $t_1 = 2\,h$ *with* $R = 8.314510\,J\,mol^{-1}\,K^{-1}$ *and at* $T_2 = 50\,°C = 323\,K$ *and* $t_2 = 1\,h$*, the reaction rates are*

$$R_1 = 1/t_1 = 0.50\,\text{h}^{-1}$$
$$R_2 = 1/t_2 = 1.00\,\text{h}^{-1}$$

Combining Eqs. (a) and (b) gives the chemical activation energy

$$\Delta G^* = -R\left(\frac{1}{T_1} - \frac{1}{T_2}\right)^{-1} \ln\left(\frac{t_2}{t_1}\right)$$

$$\Delta G^* = -\left(8.314510\,\text{J}\,\text{mol}^{-1}K^{-1}\right)\left(\frac{1}{298\text{ K}} - \frac{1}{323\,\text{K}}\right)^{-1} \ln\left(\frac{1\text{ h}}{2\,\text{h}}\right)$$
$$\Delta G^* = 22,189.16\,\text{J/mol}$$

Substituting this energy into Eq. (a) and solving for γ_a *yield*

$$\gamma_a = \frac{1}{2\,h}\exp\left[\frac{22,189.16\,\text{J/mol}}{\left(8.314510\,\text{J}\,\text{mol}^{-1}\,\text{K}^{-1}\right)(298\,\text{K})}\right]$$
$$\gamma_a = 3875.05\,\text{h}^{-1}$$

At $T_3 = 30\,°\text{C} = 303\,K$*, the rate of reaction equation is*

$$R_3 = 1/t_3 = \gamma_a \exp\left(-\frac{\Delta G^*}{RT_3}\right)$$

Hence, the time and the rate of reaction are

$$t_3 = \frac{1}{\gamma_a}\exp\left(+\frac{\Delta G^*}{RT_3}\right)$$
$$t_3 = \frac{1}{3875.05\,h^{-1}}\exp\left[\frac{22,189.16\,\text{J/mol}}{\left(8.314510\,\text{J}\,\text{mol}^{-1}\,\text{K}^{-1}\right)(303\,\text{K})}\right]$$
$$t_3 = 1.73\,\text{h}$$
$$R_3 = 1/t_3 = 0.58\,\text{h}^{-1}$$

Therefore, the reaction rates are $R_1 < R_3 < R_2$.

(b) *Determine the solubility constant at 30 °C. Using the Nernst equation gives the required overpotential expression as*

$$E = E^o - \frac{RT}{zF} \ln\left(K_{sp}\right)$$

$$\eta_a = E - E^o = -\frac{RT}{zF} \ln\left(K_{sp}\right)$$

$$K_{sp} = \exp\left(-\frac{zF\eta_a}{RT}\right)$$

Then,

$$K_{sp} = \exp\left(-\frac{2 * 96,500 * 0.05}{8.31451 * 298}\right) = 2.03 \times 10^{-2} \;\; at\; 25\,°\mathrm{C}$$

$$K_{sp} = \exp\left(-\frac{2 * 96,500 * 0.05}{8.31451 * 323}\right) = 2.75 \times 10^{-2} \;\; at\; 50\,°\mathrm{C}$$

(c) *From Eq. (5.8), the anodic current density for pure anodic polarization becomes*

$$i_a = i_o \exp\left(\frac{\alpha zF\eta_a}{RT}\right)$$

Hence,

$$i_a = 2.804 \times 10^{-5}\,\mathrm{A/cm^2} = 28.04\,\mu\mathrm{A/cm^2} \;\; at\; 25\,°\mathrm{C}$$

$$i_a = 2.412 \times 10^{-5}\mathrm{A/cm^2} = 24.12\,\mu\mathrm{A/cm^2} \;\; at\; 50\,°\mathrm{C}$$

Example 5.2. *Given the following data for a hypothetical metal undergoing chemical oxidation in a aqueous solution,* ***(a)*** *plot the data (Arrhenius plot) and determine the activation energy from the plot and* **(b)** *calculate the chemical rate (Arrhenius rate of reaction) R_A at 308 °C.*

T (K)	R_A $(x10^{-2}\,h^{-1})$
298	50.01
300	50.26
305	50.88
310	51.48
315	52.08

Solution. (a) *Using Eq. (5.2) to curve fit the given data yields the resultant curve fitting equation*

$$\ln\left(R_A\right) = 4.3046 \times 10^{-2}\,\mathrm{h}^{-1} - (219.20\,\mathrm{K})/T$$

where the intercept and the slope are $\ln(\gamma_a) = 4.3046 \times 10^{-2}\, h^{-1}$ *and* $S = 219.20\, K$*, respectively. The required Arrhenius plot is shown below and typical Arrhenius equation becomes*

$$R_A = \left(3.1455\,\mathrm{h}^{-1}\right) \exp\left(-\frac{219.20\,\mathrm{K}}{T}\right)$$

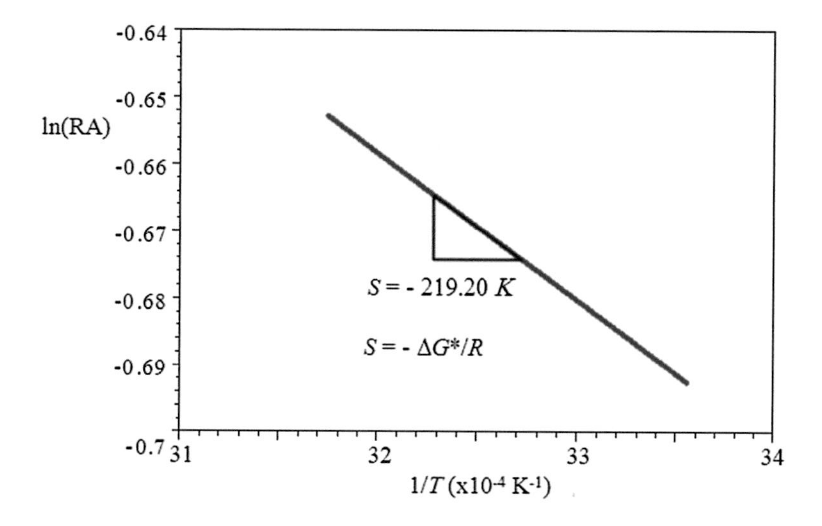

(b) *If* $T = 308\, K$*, then* $\ln(R_A) = -0.66863$ *and* $R_A \simeq 0.51\, h^{-1}$*. The activation energy is*

$$\Delta G^* = -RS = -\,(8.314\mathrm{J/mol\,K})\,(-219.20\,\mathrm{K}) = 1822.40\,\mathrm{J/mol}$$

5.2 Polarization Diagram

This section partially treats the copious literature for supporting the mathematical models used to characterize the kinetics of charge-transfer mechanism involved in an electrochemical system. Thus, electrode reactions are assumed to induce deviations from equilibrium due to the passage of an electric current through an electrochemical cell causing a change in the working electrode (WE) potential. This electrochemical phenomenon is referred to as polarization. In this process, the deviation from equilibrium causes an electric potential difference between the polarized and the equilibrium (unpolarized) electrode potential known as overpotential (η) [7].

Figure 5.2 shows a partial polarization diagram and related kinetic parameters. For instance, both Evans and Stern diagrams [8] are superimposed in order for the reader to understand the significance of the electrochemical behavior of a polarized

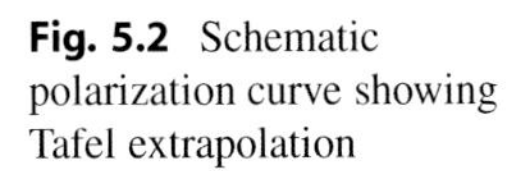
Fig. 5.2 Schematic polarization curve showing Tafel extrapolation

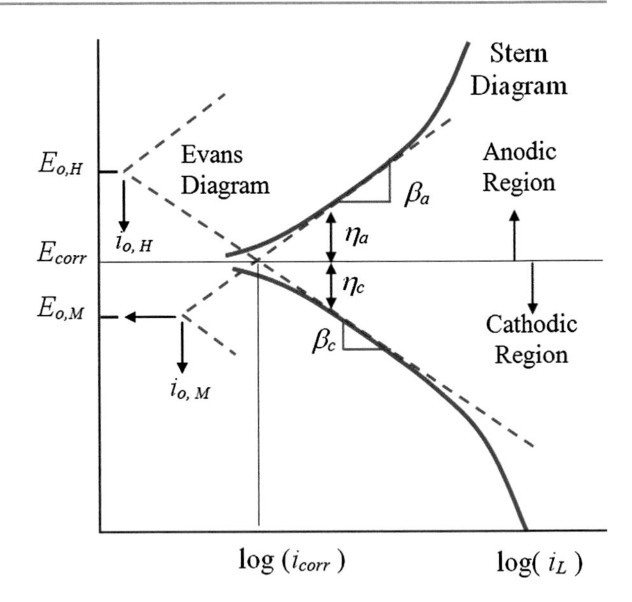

metal (M) electrode in a hydrogen-containing electrolyte. From Fig. 5.2, $E_{o,H}$ and $E_{o,M}$ are the open-circuit potentials for hydrogen and metal M, respectively, $i_{o,H}$ $i_{o,M}$ are the exchange current densities, and i_L is the limiting current density.

For a reversible electrode, Evans diagram allows the determination of the corrosion point where both the hydrogen cathodic and the metal anodic line intercept.

On the other hand, the irreversible electrochemical behavior denoted by the cathodic and anodic Stern diagram is also used for determining the corrosion point by simply extrapolating the linear portions of both curves until the intercept as indicated in Fig. 5.2. The latter diagram is very common in electrochemical studies of pure metals and their alloys. Therefore, both Evans and Stern polarization diagrams provide the corrosion potential (E_{corr}) and the corrosion current density (i_{corr}). The other parameters illustrated in Fig. 5.2 will become relevant as the electrochemical analysis advances in this chapter. With respect to the Stern diagram, this represents a polarization behavior that can be determined experimentally using potentiostatic and potentiodynamic methods, and it will be referred to as a polarization curve. Thus, a polarization curve is the result of polarizing from the corrosion potential anodically or cathodically, and it is a very common experimental output in electrochemical studies on corroding electrodes.

For instance, anode surfaces can be polarized by formation of a thin, Impervious, and stable oxide film through a chemisorption process. This film is a barrier separating the metal surface from the environment. When the oxidation and reduction reactions are completely polarized, the metal is in the passive state because there is no potential difference between anodic and cathodic electrodes. At this point, the anodic and cathodic currents are equal, the net current is zero, and corrosion ceases. In fact, corrosion is an oxidation process, and it is the representation of the electrochemical deterioration of a material due to interactions with the surrounding environment.

Further analysis of Fig. 5.2 yields the following summary:

- The solid curve can be obtained statically or dynamically.
- This nonlinear curve is divided into two parts. If $E > E_{corr}$, the upper curve represents an anodic polarization behavior for oxidation of the metal M. On the contrary, if $E < E_{corr}$ the lower curve is a cathodic polarization for hydrogen reduction as molecular gas (hydrogen evolution). Both polarization cases deviate from the electrochemical equilibrium potential (E_{corr}) due to the generation of anodic and cathodic overpotentials, which are arbitrarily shown in Fig. 5.2 as η_a and η_c, respectively.
- Both anodic and cathodic polarization curves exhibit small linear parts known as Tafel lines, which are used for determining the Tafel slopes β_a and β_c. These slopes can be determined using either the Evans or Stern diagram.
- Extrapolating the Tafel or Evans straight lines until they intersect the (E_{corr}, i_{corr}) point.
- The disadvantage in using the Evans diagram is that the exchange current density (i_o), the open-circuit potential E_o (no external circuit is applied), and the Tafel slopes for the metal and hydrogen have to be known quantities prior to determining the (E_{corr}, i_{corr}) point.
- The advantage of the Stern diagram over the Evans diagram is that it can easily be obtained using the potentiodynamic polarization technique at a constant potential sweep (scan rate) and no prior knowledge of the above kinetics parameter is necessary for determining the (E_{corr}, i_{corr}) point. The resultant curve is known as a potentiodynamic polarization curve.
- In conclusion, E_{corr} and i_{corr} can be determined from an Evans diagram for an **unpolarized metal** since $i_{corr} = i_a = -i_c$ at $E = E_{corr}$. On the other hand, if the metal is **polarized**, then the Stern diagram can be used for determining E_{corr}, i_{corr}, β_a, and β_c. In addition, E_{corr} is a reversible potential also known as a mixed potential.

Evaluation of corrosion behavior is normally done through a function that depends on kinetic parameters depicted in Fig. 5.2. The current density function for polarizing an electrode irreversibly from the corrosion potential is similar to Eq. (5.8). Hence, for one-step reaction under steady-state condition and no mass transport effects, the current density function as per Butler–Volmer equation is

$$i = i_{corr}\left\{\exp\left[\frac{\alpha zF\eta}{RT}\right]_f - \exp\left[-\frac{(1-\alpha)\,zF\eta}{RT}\right]_r\right\} \tag{5.32}$$

where $\eta = E - E_{corr}$

E = applied potential (V)

β_a = Tafel anodic slope (V) (see Fig. 5.2)

β_c = Tafel cathodic slope (V) (see Fig. 5.2)

The anodic polarization caused by an overpotential $\eta_a = (E - E_{corr}) > 0$ is referred to as an electrochemical process in which a metal surface oxidizes (corrodes) by losing electrons. Consequently, the metal surface is positively charged due to the loss of electrons. This electrochemical polarization is quantified by η_a. On the other hand, cathodic polarization requires that electrons be supplied to the metal surface at an negative overpotential, $\eta_c = (E - E_{corr}) < 0$, which implies that $E < E_{corr}$.

In general, the activation polarization is basically an electrochemical phenomenon related to a charge-transfer mechanism, in which a particular reaction step controls the rate of electron flow from a metal surface undergoing oxidation. This is the case in which the rate of electron flow is controlled by the slowest step in the half-cell reactions [3, 7, 9, 10].

Despite that Eq. (5.22) is a generalized expression, it represents a measure of anodic polarization for corrosion studies and indicates that $\eta_a = 0$ for an unpolarized and $\eta_a \neq 0$ for a polarized electrode surface. For the latter case, the reaction rate for activation polarization depends on the charge-transfer overpotential as in metal oxidation due to electrons loss, the diffusion overpotential as in mass transport of ions, the reaction overpotential due to rate determining chemical reaction mechanism, the crystallization overpotential as in metal deposition in which atoms are incorporated into the electrode crystal surface lattice, and the ohmic overpotential due to a resistance at the electrode-terminal junctions [9].

Furthermore, the anodic metal undergoes a succession of reaction steps prior to dissolve into the electrolyte by liberating z electrons. For hydrogen evolution, Fontana's idealized model [1] for activation polarization is depicted in Fig. 5.3, which shows the succession of the reaction steps that may take place after the hydrogen cations are adsorbed (attached) on the electrode surface.

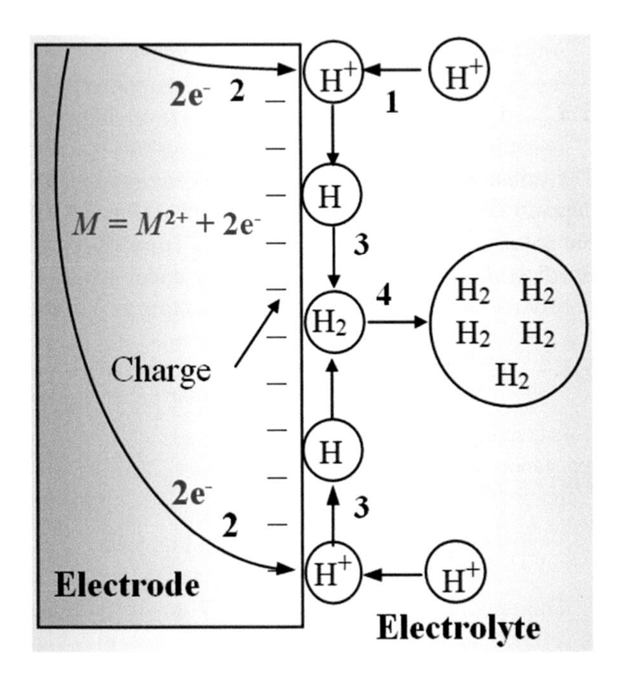

Fig. 5.3 Activation polarization model after Fontana [1]

This succession is hypothetically shown by the following idealized metal oxidation reactions during activation polarization:

$$M_{lattice} \longrightarrow M^{z+}_{surface} \longrightarrow M^{z+}_{solution} \tag{5.33}$$

This succession indicates that the metal M looses ze^- electrons on its surface and eventually the metal cation M^{z+} goes into solution. If the metal is silver undergoing oxidation, then the cation is just Ag^+. For iron, the succession may be $Fe \rightarrow F^+ \rightarrow Fe^{2+} \rightarrow Fe^{3+}$.

Hence, the possible reaction steps are

$$H^+ + e^- \longrightarrow H \quad \text{(atomic hydrogen)} \tag{5.34a}$$

$$H + H \longrightarrow H_2 \quad \text{(hydrogen molecule)} \tag{5.34b}$$

$$H_2 + H_2 \longrightarrow 2H_2 \quad \text{(gas bubble)} \tag{5.34c}$$

Thus, only one step in Eq. (5.33) and one in Eq. (5.34) controls the charge transfer for activation polarization.

The above activation polarization model depicted in Fig. 5.3 suggests that this electrochemical process is based on an electron-transfer mechanism, which depends on the activation energy (driving force) and the working temperature. As shown in Fig. 5.3, the electrochemical reactions for hydrogen evolution are inherent to kinetics of electron transfer at the electrode–electrolyte interface. This interfacial region has specific characteristics described by different electric double-layer models being introduced in Sect. 4.1.

5.3 Polarization Methods

The polarization resistance (R_p) of a metal-electrolyte system and the pitting or breakdown potential (E_b) can be determined using at least two-electrode system. Subsequently, the rate of metal dissolution or corrosion rate is calculated using a function of the form $i_{corr} = f\left(\beta, R_p\right) > i_o$.

5.3.1 Linear Polarization

This technique is useful for deriving the Tafel slopes β_a and β_c and the polarization resistance R_p. Linear polarization (LP), as schematically shown in Fig. 5.4, covers both anodic and cathodic portions of the potential E versus current density i curve.

Now, β_a, β_c, and R_p can be defined by manipulating Eq. (5.8) to yield the anodic (i_a) and the cathodic (i_c) current densities as

$$i_a = i_o \exp\left[\frac{\alpha z F \eta_a}{RT}\right] \qquad \text{(for } i_a >> i_c,\ \eta_a >> \eta_c\text{)} \tag{5.35}$$

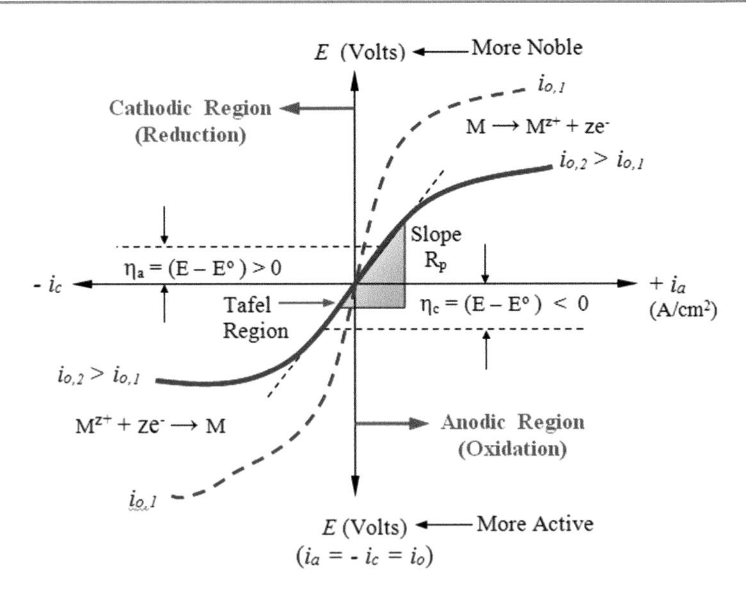

Fig. 5.4 Schematic linear polarization curves for two different exchange current densities. The anodic and cathodic overpotentials and the slope are the polarization parameters

$$i_c = -i_o \exp\left[-\frac{(1-\alpha)\,zF\eta_c}{RT}\right] \quad (\text{for } i_c >> i_a,\, \eta_c >> \eta_a) \tag{5.36}$$

Recall that α is the symmetry factor or charge-transfer coefficient.

Solving Eqs. (5.35) and (5.36) for the overpotentials gives

$$\eta_a = \beta_a \log\left(\frac{i_a}{i_o}\right) \tag{5.37}$$

$$\eta_c = -|\beta_c| \log\left(\frac{i_c}{i_o}\right) \tag{5.38}$$

where α is the symmetry coefficient and Tafel slopes become

$$\beta_a = \frac{2.303RT}{\alpha zF} = \frac{(1-\alpha)\,\beta_c}{\alpha} \tag{5.39}$$

$$\beta_c = \frac{2.303RT}{(1-\alpha)\,zF} = \frac{\alpha\beta_a}{(1-\alpha)} \tag{5.40}$$

For activation polarization, the linear polarization is confined to a small magnitude of the overpotentials η_a and η_c, respectively, using linear coordinates. This technique allows the determination of i_{corr} using a potential range of ± 10 mV from the corrosion potential E_{corr} [10]. Prior to determining i_{corr}, the polarization resistance R_p is estimated from the linear slope of the curve shown in Fig. 5.4 [8]:

$$R_p = \frac{dE}{di} = \frac{\Delta E}{\Delta i} = \frac{\eta}{\Delta i} \tag{5.41}$$

This equation excludes potential gradients due to ionic concentration gradients near the metal surfaces and any potential drop (ohmic effect) due to current flowing through the electrolyte. In fact, R_p measurements include both the actual polarization resistance (R_a) and the solution resistance (R_s). Thus,

$$R_p = R_a + R_s \tag{5.42}$$

The corresponding corrosion current density depends on kinetic parameters since $i_{corr} = f\left(\beta, R_p\right)$. Thus, the simple linear relationship that defines the corrosion current density is of the form [7, 8, 11]

$$i_{corr} = \frac{\beta}{R_p} \tag{5.43}$$

$$\beta = \frac{\beta_a \left|\beta_c\right|}{2.303\left(\beta_a + \left|\beta_c\right|\right)} \tag{5.44}$$

Here, the Tafel slopes, $\beta_a < 1\,\text{V}$ and $\beta_c < 1\,\text{V}$, are taken as positive kinetic parameters for determining i_{corr}. Denote that Eq. (5.43) predicts that the corrosion current density is very sensitive to changes in the polarization resistance (R_p). Conversely, the magnitude of the polarization resistance is mainly controlled by the corrosion current density. The constant $\beta < 1\,\text{V}$ will be derived in the next section.

5.3.2 Tafel Extrapolation

The Tafel extrapolation technique (TE) takes into account the linear parts of the anodic and cathodic curves for determining R_p. This method involves the determination of the Tafel slopes β_a and β_c as well as E_{corr} and i_{corr} from a single polarization curve as shown in Fig. 5.2. This curve is known as the Stern diagram (nonlinear polarization) based on Eq. (5.32). The Evans diagram (linear polarization) is also included in order to show that both diagrams have a common ($E_{corr} \cdot i_{corr}$) point. This figure illustrates a hypothetical electrochemical behavior of a metal M immersed in an electrolyte containing one type of oxidizer, such as H^+ ions.

Further interpretation of the polarization curves can be extended using Pourbaix graphical work (Pourbaix diagram or E_H-pH diagram) depicted in Fig. 5.5 for pure iron (Fe). The resultant plots represent the functions $E = f\,(i)$ and E vs. $f\,[\log(i)]$ for an electrolyte containing $C_{Fe^{2+}} = 0.01\,\text{g/l} = 1.79x10^{-4}\,\text{mol/l} = 1.49x10^{-7}\,\text{mol/cm}^3$ at $pH = 0$. Additionally, the reactions depicted in Fig. 5.5 and some related kinetic parameters are listed in Table 5.2 for convenience.

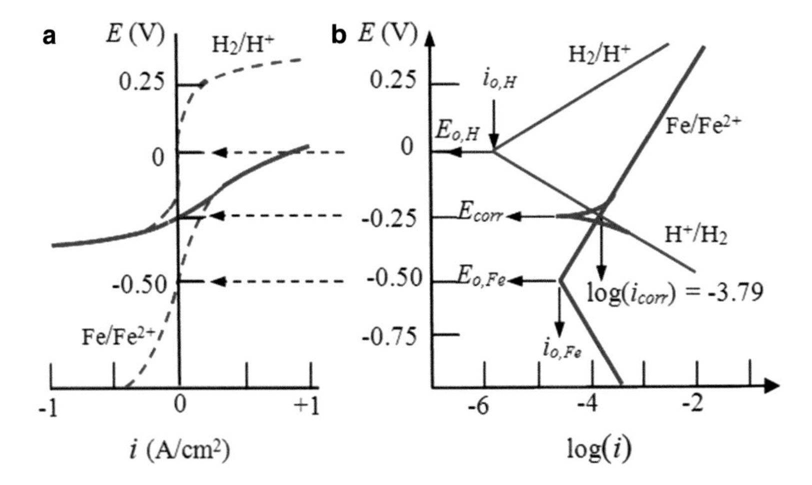

Fig. 5.5 Corrosion of iron (0.01 g/l) with hydrogen evolution at $pH = 0$ (**a**) Linear plot and (**b**) linear-log plot [46]

Table 5.2 Kinetic parameters for iron (0.01 g/l) and hydrogen reactions at $pH = 0$ [8]

Parameters	$Fe = Fe^{2+} + 2e^-$	$2H^+ + 2e^- = H_2$
E^o (V_{SHE})	-0.50	0
i_o (A/cm^2)	$3.16x10^{-5}$	$1.41x10^{-6}$
β_a (V)	$+0.328$	$+0.123$
β_c (V)	-0.328	-0.123

For electrochemical systems containing several oxidizers, determining the corrosion point is more complex using the Evans diagram, but the Stern diagram would provide a similar polarization curve as shown in Fig. 5.2, from which both E_{corr} and i_{corr} are easily determined by extrapolating the Tafel anodic and cathodic linear parts until they intersect as straight lines. Also included in Fig. 5.2 are the exchange current densities, $i_{o,H}$ and $i_{o,M}$, and their counterpart potentials, $E_{o,H}$ and $E_{o,M}$, for hydrogen evolution and metal oxidation, respectively. These potentials, $E_{o,H}$ and $E_{o,M}$, are known as open-circuit potentials. Furthermore, the limiting current density i_L for cathodic polarization is included as an additional information one can extract from a cathodic polarization curve. The latter term will be dealt with in Chap. 6.

The use of the Ohm's law gives the cell and the inner potentials, respectively:

$$E = E_{corr} + \eta_a + |\eta_c| + \phi_s = IR_x \tag{5.45}$$

$$\phi_s = IR_s \tag{5.46}$$

where I = current (A)
R_x = external resistance (Ohm = V/ A)
R_s = solution resistance (Ohm = V/ A)
ϕ_s = internal potential (V)

Solving Eq. (5.45) for I yields

$$I = -\frac{E_{corr} + \eta_a + |\eta_c| + \phi_s}{R_x} \tag{5.47}$$

The current in Eq. (5.47) strongly depends on the magnitude of the external resistance. A slight decrease in R_x increases the current I. Hence, if $R_x \longrightarrow 0$, then $I \longrightarrow \infty$ and $I \longrightarrow 0$ when $R_x \longrightarrow \infty$. In addition, $IR_x \longrightarrow 0$ as $E \longrightarrow 0$.

With regard to Eq. (5.46), ϕ_s can be neglected due to its small contribution to the cell potential. However, the electrolyte conductivity is of significance in determining the governing current expression. For instance, when $IR_x >> IR_s$, the electrolyte has a high conductivity, and if $IR_x << IR_s$, the electrolyte has a low conductivity. Hence, from Eq. (5.47), the governing current expressions are

$$I = \frac{E_{corr} - \eta_a - \eta_c}{R_x} \quad \text{for } IR_x >> IR_s \tag{5.48}$$

$$I = \frac{E_{corr} - \eta_a - \eta_c}{R_s} \quad \text{for } IR_s >> IR_x \tag{5.49}$$

Nowadays, sophisticated instrumentation, such as a potentiostat/galvanostat, is commercially available for conducting electrochemical experiments for characterizing the electrochemical behavior of a metal or an alloy in a few minutes. Nevertheless, a polarization diagram or curve is a **potential-control** technique. This curve can experimentally be obtained statically or dynamically. The latter approach requires a linear potential scan rate to be applied over a desired potential range in order to measure the current response.

On the other hand, a galvanostat can be used as the **current control** source for determining the potential response on a electrode surface. However, the potential-control approach is common for characterizing electrochemical behavior of metallic materials. The potential can be applied uniformly or in a stepwise manner using a waveform. The former case generates a steady-state current response, while the latter provides a transient current response.

It should be mentioned that the principal resourcefulness of electrochemistry for characterizing oxidation of metals at room temperature is the brilliant work of Marcel Pourbaix [8], who ingeniously and effectively used thermodynamic principles in general corrosion of metals immersed in aqueous solutions, specifically in water.

One important observation is that both anodic and cathodic Tafel slopes, β_a and β_c, respectively, are equal numerically, and consequently, Fig. 5.5b has an inflection point at (i_{corr}, E_{corr}). This electrochemical situation is mathematically predicted and discussed in the next section using a current density function for a mixed-potential system.

In fact, Fig. 5.5 compares both linear and extrapolation results for iron. The interpretation of this figure is based on complete activation control in the absence of diffusion and external current [8]. Thus, it can be deduced from this Fig. 5.5 that:

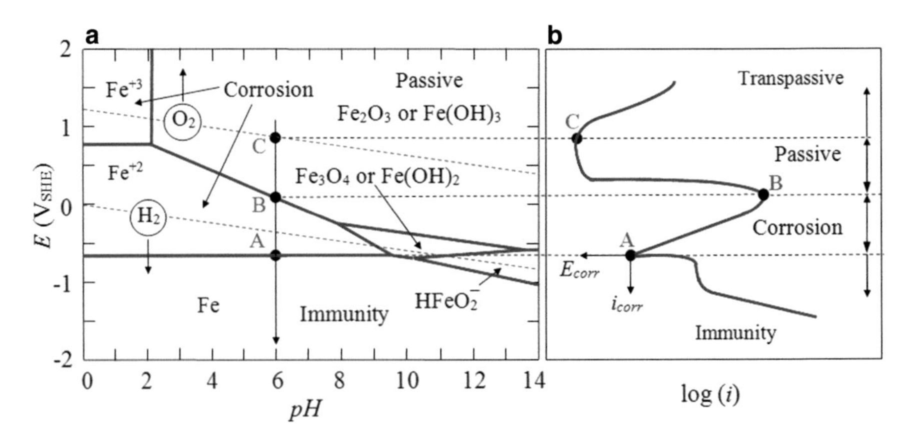

Fig. 5.6 Comparison of Pourbaix ($E - pH$) diagram and polarization curve for iron or steel in water containing $[Fe^{+2}] = 10^{-6}$ mol/liter at $T = 25\,°C$, $pH = 6$, and $P = 101$ kPa. (**a**) Partial Pourbaix diagram. (**b**) Schematic Polarization curve

- The net current density ($i_{net} = i_a - i_c$) is zero at equilibrium, but the corrosion current density is $i_{corr} = 0.1622\,mA/cm^2$ at the corrosion potential, $E_{corr} = -0.25\ V_{SHE}$.
- The polarization curves are reversible in nature; however, if irreversibility occurs, the heavy line in Fig. 5.5a would follow the vertical axis at a potential range $E_{o,c} \leq E \leq E_{o,a}$ when $i_{net} = 0$.
- The open-circuit potentials, $E_{o,H}$ and $E_{o,Fe}$, can be estimated using the Nernst equation.

One important fact is that the solution must be continuously stirred to keep a uniform concentration of the species in solution; otherwise, the concentration at the electrodes becomes uneven, and the open-circuit potential $E_{c,Fe}$ increases during oxidation, and $E_{c,H}$ decreases during reduction. At equilibrium, the overpotential becomes $\eta = E - E_o = 0$ since $E = E_o$, and $i = i_o$, respectively. However, if $i > i_{o,Fe}$, the iron reaction is irreversible because it corrodes by liberating electrons.

Combining the Pourbaix diagram and the polarization curve for iron in water at $pH = 6$ yields interesting information on the electrochemical states shown in Fig. 5.6, which includes the hydrogen and oxygen evolution lines for comparison [8, 12].

Observe the correspondence of potential between both diagrams. For instance, point A corresponds to the corrosion potential (E_{corr}) and corrosion current density (i_{corr}). Corrosion occurs along line AB, and passivation takes place along the line BP, and iron is protected anodically. Point B is a transitional potential at the critical current density, and point C corresponds to oxygen evolution and to the iron pitting potential (E_p). Thus, iron stays passive along the line BC.

For example, both E-pH diagram and the polarization curve show the corrosion region for iron or carbon steel at a $pH = 6$, but only the latter exhibits the corrosion point (A) from which the corrosion rate (C_R) is determined using Faraday's law of electrolysis. This topic is dealt with in the next section, and the passivation phenomenon is discussed in great details in Chap. 7 using potentiodynamic polarization curves. In addition, Fig. 5.6b shows the regions for cathodic protection (CP) and anodic protection (AP). These types of protection are dealt with in later chapters.

Moreover, black magnetite (Fe_3O_4) and dark brown maghemite (Fe_2O_3) are magnetic and thermodynamically metastable phases with cubic (isometric) structures. These phases constitute the passive oxide film on iron or plane carbon steel. It has been reported [13] (cited in [14]) that Fe_2O_3 (also known as γ-Fe_2O_3) may form on Fe_3O_4, resulting in a bilayer passive oxide film containing multiple crystallographic defects. This bilayer film may form at $pH > 8$; however, it is clear from this Fig. 5.6a that Fe_2O_3 may form by the oxidation of Fe_3O_4.

At the chosen $pH = 6$ in Fig. 5.6a, the corrosion potential, $E_A \leq E < E_B$, can lie anywhere between points A and B, and hydrogen evolution is likely to occur at $E < E_{H_2}$.

In general, passivation may occur in low- and high-oxygen contents according to the following reactions at 25 °C, respectively, [15, 16] (cited in [17]):

$$2Fe + 3H_2O \rightarrow Fe_2O_3 + 6H^+ + 6e^- \tag{5.50}$$

$$2Fe^{2+} + 3O^{-2} \rightarrow 2Fe_2O_3 \tag{5.51}$$

Unexceptionably, corroded steel reinforcement embedded in chloride contaminated concrete has been studied by many researchers. As a result, many papers have been published on relevant data obtained under different environmental conditions. For instance, a mixture of complex iron oxide phases have been reported [16] to consists of an inner bilayer, Fe_3O_4 and Fe_2O_3, and an outer porous layer of corrosion product, $Fe(OH)_3$ and $Ca(OH)_2$. The latter is attributed to a low-oxygen content in concrete [16]. Moreover, other iron oxide phases, γ-$FeOOH$ and α-$FeOOH$, may form on steels exposed to oxygen-rich atmospheres [18].

Putting the kinetics of activation polarization into another prospective, it is appropriate to conclude this section by restating activation polarization and comparing it with concentration polarization given in the next chapter.

Activation polarization is the electrochemical process associated with a positive deviation from thermodynamic equilibrium (E_{corr}, i_{corr}) point so that anodic overpotential, $\eta_a = |E - E_{corr}| > 0$, is the potential difference between electrodes in the cell. Thus, η_a is called activation overpotential for an activation polarization of the redox reaction event. The generalized half-cell redox reaction for oxidation-reduction is

$$O + ze^- \rightleftarrows R \tag{5.52}$$

where O is the oxidized species, R is the reduced species, and ze^- is the number of electron transfer.

The activation energy for this redox reaction under an applied anodic overpotential η_a, which is referred to as bias voltage in scanning tunneling microscope (STM), is of Fermi energy level, crudely approximated as

$$U_a = q_e \eta_a \;\; \& \;\; U_c = q_e \eta_c \tag{5.53}$$

Assume that the oxidation of iron, $Fe \rightarrow Fe^{2+} + 2e^-$, occurs at $\eta_a = 0.2\,\mathrm{V}$. Then, Eq. (5.53) gives

$$U_a = q_e \eta_a = \left(1.6x10^{-19}\,\mathrm{C}\right)(0.2\,\mathrm{V}) = 3.2 \times 10^{-20}\,\mathrm{J} = 0.2\,\mathrm{eV} \tag{a}$$

Therefore, 0.2 eV is needed for iron oxidation. This calculation serves the purpose of illustrating the possible magnitude of a redox reaction. Conversely, **concentration polarization** is the electrochemical process that requires a negative deviation from equilibrium so that $\eta_c = (E - E_{corr}) < 0$. If $\eta_c = -0.2\,\mathrm{V}$ for the reduction of iron, $Fe^{2+} + 2e^- \rightarrow Fe$, then $U_c = -0.2\,\mathrm{eV}$.

5.3.3 Impedance Spectroscopy

Polarization impedance of an electrode involves the determination of basic parameters such as polarization resistance (R_p) and capacitance (C_{dl}). Subsequently, the impedance of the working electrode is determined as a useful kinetic reference.

Fundamentally, when an electrode is immersed in an electrolyte, metal and electrolyte ions combine to form a ionic layer inducing a charge distribution at the interfacial region with unique properties related to the half-cell inner potential and the electric double-layer impedance. This implies that the electrode-electrolyte interface is subjected to a current flow perturbation.

Theoretically, this interface is disturbed by the magnitude of the current flowing through it and the type of ionic reaction that may take place within the interface thickness. The ionic behavior within the interfacial dimension may be modeled as per the conceptual electric double-layer introduced in Chap. 4. However, a polarization impedance experiment is also based on conceptual electrical circuit models that take into account the mechanism of charge transfer or ionic diffusion for determining the polarization impedance, which, in turn, is used to estimate the corrosion rate of a certain electrode in a particular electrolyte.

Electrochemical impedance spectroscopy (*EIS*) requires an alternating current (*AC*), and the output is a Nyquist plot for charge-transfer or diffusion-control process, which can be used to determine R_p, which, in turn, is inversely proportional to the corrosion current density i_{corr}.

This method is very useful in characterizing an electrode corrosion behavior. The electrode characterization includes the determination of the polarization resistance (R_p), corrosion rate (C_R), and electrochemical mechanism [3, 7, 19–29]. The usefulness of this method permits the analysis of the alternating current (*AC*) impedance data, which is based on modeling a corrosion process by an electrical circuit. Several review papers address the electrochemical impedance technique based on the AC circuit theory [23–25, 30, 31].

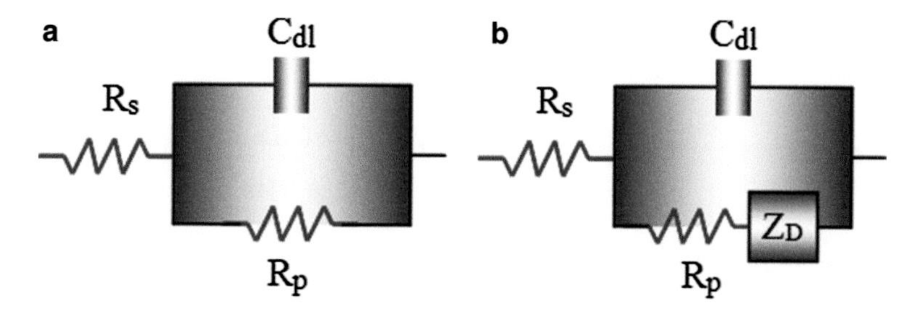

Fig. 5.7 Schematic electrochemical impedance circuits. (**a**) Charge control. (**b**) Diffusion control

The *EIS* technique is based on a transient response of an equivalent circuit for an electrode-solution interface. The response can be analyzed by transfer functions due to an applied small-amplitude potential ($\pm 10\,\text{mV}$) excitation at varying signals or sweep rates. In turn, the potential excitation yields a current response and vice verse. In impedance methods, a sine-wave perturbation of small amplitude is employed on a corroding system being modeled as an equivalent circuit (Fig. 5.7) for determining the corrosion mechanism and the polarization resistance. Thus, a complex transfer function takes the form

$$T = \frac{Output}{Input} \tag{5.54}$$

The transfer function depends on the angular frequency, and it is expressed as impedance $Z(\omega)$ or admittance $Y(\omega)$. It should be emphasized that $Z(\omega)$ is the frequency-dependent proportionality factor of the transfer function between the potential excitation and the current response.

Thus, for a sinusoidal current perturbation, the transfer function is the system impedance $Z(\omega)$, and for a sinusoidal potential perturbation, the transfer function is the system admittance $Y(\omega)$. Hence,

$$Z(\omega) = \frac{E(t)}{I(t)} = Z'(\omega) + jZ''(\omega) \tag{5.55}$$

$$Y(\omega) = \frac{I(t)}{E(t)} = Y'(\omega) + jY''(\omega) \tag{5.56}$$

where $E(t)$ = time-dependent potential (V)
$I(t)$ = time-dependent current (A)
$\omega = 2\pi f$ = angular frequency (Hz)
f = signal frequency (Hz)
$Z'(\omega)$, $Y'(\omega)$ = real parts
$Z''(\omega)$, $Y''(\omega)$ = imaginary parts
t = time (s)
$j = \sqrt{-1}$ = imaginary operator
$j^2 = -1$

In addition, Ohm's law can be viewed in two different current imposition cases as per ASTM G-106 standard testing method. Hence,

$$E = IR \qquad \text{For } DC, f = 0\,\text{Hz} \tag{5.57}$$

$$E(t) = I(t)Z(\omega) \qquad \text{For } AC, f \neq 0\,\text{Hz} \tag{5.58}$$

Ohm's law described by Eq. (5.57) defines the ideal resistance R as the potential/current ratio, $R = E/I$, for a single circuit element, and it is independent of angular frequency since the alternating current (AC) and the potential (voltage) signals are inphase (phase angle $\theta = 0$). However, real circuit elements of an electrochemical cell are complex and depend on the angular frequency and time. In such a case, the impedance $Z(\omega)$ defines the general circuit resistance and becomes analogous to Ohm's resistance R. Thus, the analogous expression to Ohm's law is defined by Eq. (5.55).

Moreover, $Z(\omega)$ is the magnitude of the impedance containing elements of an equivalent circuit, such as capacitors and inductors. Capacitors oppose or impede the current flow. In modeling an electrochemical system as an electrochemical circuit, a potential waveform is applied across the circuit, and a current response to the frequency signal generates impedance data. Thus, the impedance data is related to a phase shift angle and a variation in potential and current amplitudes. This technique is a straightforward approach for analyzing the corrosion behavior of a metal [25]. Figure 5.7 shows two schematic electrochemical circuit models. For a charge-transfer control (Fig. 5.7a), only the solution resistance (R_s), polarization resistance $\left(R_p\right)$, and the capacitor (C_{dl}) are needed in a simple circuit. On the other hand, if the electrochemical system is diffusion control (Fig. 5.7b), a diffusion impedance (Z_D) is incorporated in the circuit.

The potential excitation and its current response are schematically shown in Fig. 5.8 as sinusoidal excitations. The electrochemical impedance spectroscopy method is conducted according to the ASTM G-106 standard practice, in which a range of small-amplitude sinusoidal potential perturbation is applied to the Electrode-solution interface at discrete frequencies. These frequencies cause an out of phase current response with respect to the applied sinusoidal potential waveform.

If a sinusoidal potential excitation is applied to the electrode-solution interface, the potential, current, and impedance can be predicted as per Barn and Faulkner mathematical models [5, 19]. Thus,

$$E(t) = I(t)\,Z(\omega) = E_o \sin(\omega t) \tag{5.59}$$

$$I(t) = I_o \sin(\omega t + \theta) \tag{5.60}$$

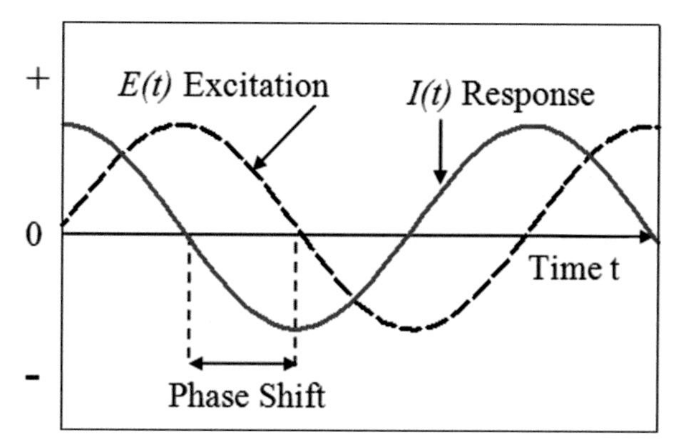

Fig. 5.8 Schematic sinusoidal potential excitation for impedance measurements

where E_o, I_o = constants
θ = phase shift angle between $E(t)$ and $I(t)$
$\omega = 2\pi f$ and $10\,Hz \leq f \leq 100\,\text{Hz}$ (intermediate)

The magnitude of $Z(\omega)$ and $Y(\omega)$ are, respectively,

$$Z(\omega) = \sqrt{\left[Z(\omega)'\right]^2 + \left[Z(\omega)''\right]^2} \tag{5.61}$$

$$Y(\omega) = \sqrt{\left[Y(\omega)'\right]^2 + \left[Y(\omega)''\right]^2} \tag{5.62}$$

and the phase shift angle is defined as

$$\theta = \tan^{-1}\left[Z(\omega)''/Z'(\omega)\right] \tag{5.63}$$

The fundamental characteristic of an AC signal in a simple electrochemical circuit is described by the impedance of the form [23, 25]

$$Z(\omega) = \left[R_s + \frac{R_p}{1 - \omega^2 C^2 R_p^2}\right] - j\left[\frac{\omega C R_p^2}{1 - \omega^2 C^2 R_p^2}\right] \tag{5.64}$$

Here, C is the interfacial capacitance of an electrical double layer at the electrode surface (Farad/cm^2).

For low- and high-frequency amplitudes, Eq. (5.64) yields

$$Z(\omega)_o = R_s + R_p \quad \text{for } \omega = 0 \tag{5.65}$$

$$Z(\omega)_\infty = R_s \quad \text{for } \omega = \infty \tag{5.66}$$

Combining Eqs. (5.65) and (5.66) yields the polarization resistance as

$$R_p = Z(\omega)_o - R_s = Z(\omega)_o - Z(\omega)_\infty \tag{5.67}$$

which is the sought output in electrochemical impedance measurements. Equating Eqs. (5.43) and (5.67) yields the corrosion current density in terms of impedance

$$i_{corr} = \frac{\beta}{Z(\omega)_o - Z(\omega)_\infty} \tag{5.68}$$

The equation of a circle for charge-control mechanism becomes [23]

$$\left[Z'(\omega) - \left(R_s + \frac{1}{2}R_p\right)\right]^2 + \left[Z''(\omega)\right]^2 = \left[\frac{1}{2}R_p\right]^2 \tag{5.69}$$

The locus of Eq. (5.69) is experimentally shown in Fig. 5.9 for AISI 1030 steel immersed in phosphoric acid (H_3PO_4) containing butanol and some concentrations in molality (M) of thiosemicarbonate (TSC) inhibitor at room temperature [26]. This figure is known as the **Nyquist plot**, from which the maximum phase shift angle and polarization resistance become

$$\tan(\theta) \simeq \frac{Z(\omega)}{R_p/2} \tag{5.70}$$

$$R_p \simeq \frac{2Z(\omega)}{\tan(\theta)} \tag{5.71}$$

These data is for a charge-control mechanism with and without the *TSC* inhibitor [27]. Notice from this Fig. 5.9 that the Nyquist impedance semicircles increase with increasing content of the *TSC* inhibitor. As a result, the polarization resistance (R_p) increases, and both double-layer capacitance (C_{dl}) and corrosion rate (C_R) decrease with additions of this inhibitor.

Table 5.3 gives relevant experimental data extracted from the complex Nyquist plot elucidated in Fig. 5.9 and calculated C_R and C_{dl} values. The double-layer capacitance (C_{dl}) listed in Table 5.3 can be calculated using the following expression [32]:

$$C_{dl} = 1/[\omega R_p] \tag{5.72}$$

Table 5.3 Impedance data for AISI 1030 steel [27]

TSC (x10^{-4} M)	R_p (Ohm cm^2)	C_{dl} (μF/cm^2)	C_R (mm/year)
0	4.80	130.80	63.25
0.10	14.50	121.00	20.85
0.50	31.30	77.49	9.65
1.00	33.50	74.08	9.02
5.00	77.90	45.45	3.89
10.00	119.50	39.11	2.52

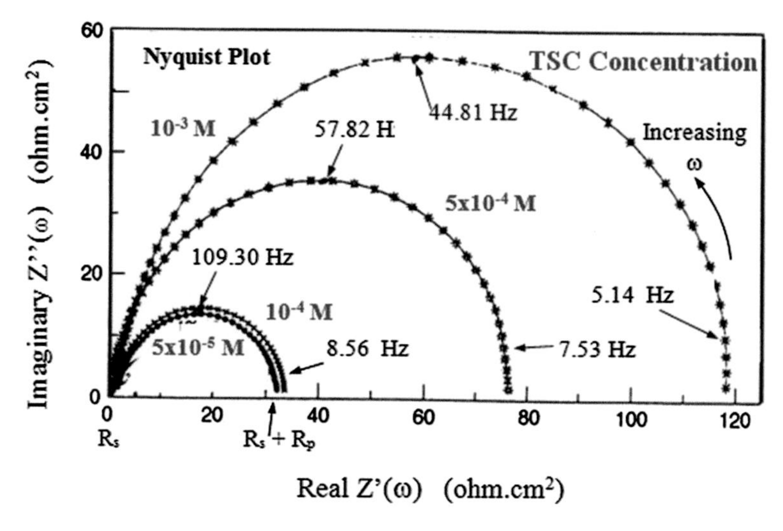

Fig. 5.9 Experimental Nyquist plots for AISI 1030 carbon steel in 35 % H_3PO_4 +6 %butanol + TSC inhibitor at room temperature [26]

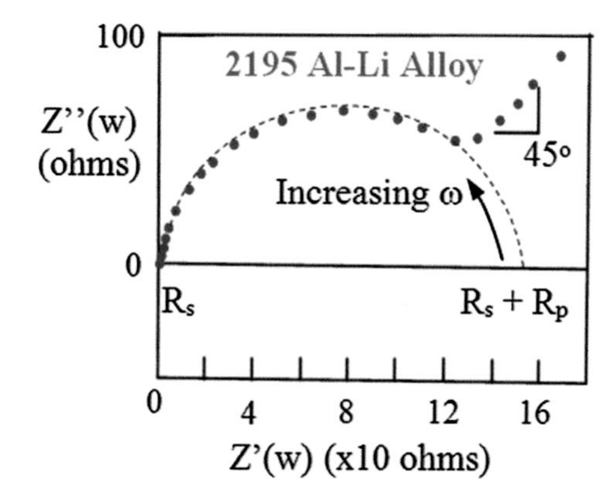

Fig. 5.10 Nyquist–Warburg plot for 2195 Al-Li alloy. (1) Aged at 190 °C for 0.5 h and (2) tested in 3.5 % NaCl aerated solution. Tests done by Laura Baca (graduate student, 2003)

When an electrochemical process is controlled by diffusion or film adsorption, the electrochemical system can be modeled using the ideal circuit shown in Fig. 5.7b. In this case, a diffusion impedance (Z_D) is included in the circuit series, and it is known as **Warburg impedance**.

Notice that Z_D and R_p are connected in series. An experimental Nyquist–Warburg plot is shown in Fig. 5.10. The interpretation of this figure indicates that the 45° portion of the line corresponds to a low angular frequency range. In this case, the kinetics of the electrochemical system is limited by a diffusion-control process (concentration polarization). Also, extrapolating the semicircle (dashed curve) to intercept the real impedance axis $Z(\omega)'$ graphically defines the polarization resistance R_p.

Notice that the semicircle in Fig. 5.10 is depressed due to a diffusion process, which is confirmed by the 45° line. The polarization resistance for this alloy is $R_p = 140\,\Omega$ in an aerated solution containing 3.5 % $NaCl$ at 21.5 °C.

A particular case for using Z_D is when diffusion through a surface film is the rate controlling of carbon steel immersed in concentrated sulfuric acid (H_2SO_4). Thus, iron (Fe) oxidizes and forms a $FeSO_4$ film [26]. The diffusion impedance, $Z_D(\omega)$, and the Warburg impedance coefficient (σ_w) expressions due to the formation of a thin oxide film are define by [32, 33]

$$Z_D(\omega) = \sigma_w \omega^{-1/2} - j\sigma_w \omega^{-1/2} \coth\left[\delta_x \left(\frac{j\omega}{D}\right)^{1/2}\right] \tag{5.73}$$

$$\sigma_w = RT / \left(zFAC_x\sqrt{2D}\right) \tag{5.74}$$

where δ_x = thickness of the diffusion film (cm)

C_x = concentration of the species in the diffusion film (mol/cm^3)

A = exposed area (cm^2)

D = diffusivity or diffusion coefficient (see Chap. 6), (cm^2/s)

In addition, the Cole–Cole impedance formula [34–37] is used to find $Z(\omega)$ due to surface roughness, dielectric heterogeneities, and diffusion:

$$Z(\omega) = \frac{R_p}{1 + \left(jR_pC_x\right)^n} \quad \text{for } 0 < n < 1 \tag{5.75}$$

5.4 Corrosion Rate

During corrosion (oxidation) process, both anodic and cathodic reaction rates are coupled together on the electrode surface at a specific current density known as i_{corr}. This is an electrochemical phenomenon which dictates that both reactions must occur on different sites on the metal-electrolyte interface. For a uniform process under steady-state conditions, the current densities at equilibrium are related as $i_a = -i_c = i_{corr}$ @ E_{corr}. Assume that corrosion is uniform and there is no oxide film deposited on the metal electrode surface; otherwise, complications would arise making matters very complex. The objective at this point is to determine both E_{corr} and i_{corr} either using the Tafel extrapolation or linear polarization techniques. It is important to point out that i_{corr} cannot be measured at E_{corr} since $i_a = -i_c$ and current will not flow through an external current-measuring device [10].

When polarizing from the corrosion potential with respect to anodic or cathodic current density, the overpotential expressions given by Eqs. (5.37) and (5.38) become

$$\eta_a = \beta_a \log\left(\frac{i_a}{i_{corr}}\right) \tag{5.76}$$

$$\eta_c = -|\beta_c| \log\left(\frac{i_c}{i_{corr}}\right) \tag{5.77}$$

Recall that $\eta_a = \Delta E = (E - E_{corr}) > 0$ and $\eta_c = \Delta E = (E - E_{corr}) < 0$ now represent the potential changes from the steady-state corrosion potential E_{corr}. Solving Eqs. (5.76) and (5.77) for the anodic and cathodic current densities yields, respectively,

$$i_a = i_{corr} \exp\left[\frac{2.303\,(E - E_{corr})}{\beta_a}\right] \tag{5.78}$$

$$i_c = i_{corr} \exp\left[-\frac{2.303\,(E - E_{corr})}{|\beta_c|}\right] \tag{5.79}$$

Assume that the applied current density is $i = i_a - i_c$, and substituting Eqs. (5.78) and (5.79) into this expression yields the **Butler–Volmer equation** that quantifies the kinetics of the electrochemical corrosion:

$$i = i_{corr}\left\{\exp\left[\frac{2.303\,(E - E_{corr})}{\beta_a}\right] - \exp\left[-\frac{2.303\,(E - E_{corr})}{|\beta_c|}\right]\right\} \tag{5.80}$$

This expression resembles Eq. (5.32), but it is a convenient expression at this point since the inverse polarization resistance is easily obtainable by deriving Eq. (5.80) with respect to the applied potential E. Thus,

$$\left(\frac{di}{dE}\right) = 2.303 i_{corr}\left\{\begin{array}{c} \beta_a^{-1} \exp\left[2.303\,(E - E_{corr})\,/\beta_a\right] \\ -\left|\beta_c^{-1}\right| \exp\left[-2.303\,(E - E_{corr})\,/\beta_c\right] \end{array}\right\} \tag{5.81}$$

The second derivative of Eq. (5.80) provides useful information on the cell potential. Thus,

$$\left(\frac{d^2 i}{dE^2}\right) = 5.3038 i_{corr}\left\{\begin{array}{c} \beta_a^{-2} \exp\left[2.303\,(E - E_{corr})\,/\beta_a\right] \\ -\left|\beta_c^{-2}\right| \exp\left[-2.303\,(E - E_{corr})\,/\beta_c\right] \end{array}\right\} \tag{5.82}$$

Let us set some conditions for Eq. (5.82) such that

$$\left(\frac{d^2 i}{dE^2}\right) \Rightarrow \left\{\begin{array}{ll} < 0 & \text{for } E = E\max > E_{corr} \\ = 0 & \text{for an inflation point: } E = E_{corr} \\ > 0 & \text{for } E = E_{\min} < E_{corr} \end{array}\right. \tag{5.83}$$

Evaluating Eq. (5.82) at the inflation point yields

$$\left(\frac{d^2 i}{dE^2}\right)_{E=E_{corr}} = 5.3038 i_{corr}\left(\beta_a^{-2} - \left|\beta_c^{-2}\right|\right) \tag{5.84}$$

which clearly indicates that the inflection point is achieved if and only if $\beta_a = \beta_c$. This condition was pointed out by Oldham and Mansfield [1] as the mathematical proof for the $E_{corr} \cdot i_{corr}$ point depicted in Fig. 5.2.

Additionally, evaluating Eq. (5.81) at $E = E_{corr}$ yields the polarization resistance:

$$\left(\frac{di}{dE}\right)_{E=E_{corr}} = 2.303 i_{corr} \left(\frac{\beta_a + |\beta_c|}{\beta_a |\beta_c|}\right) \tag{5.85}$$

$$R_p = \left(\frac{di}{dE}\right)^{-1}_{E=E_{corr}} \tag{5.86}$$

Thus,

$$R_p = \frac{\beta_a |\beta_c|}{2.303 i_{corr} (\beta_a + |\beta_c|)} = \frac{\beta}{i_{corr}} \tag{5.87}$$

Here, β is the proportionality constant, and R_p is inversely proportional to i_{corr} as indicated by Eq. (5.87). For convenience, Eq. (5.87) can be linearized as

$$\log\left(R_p\right) = \log(\beta) - \log(i_{corr}) \tag{5.88}$$

Figure 5.11 depicts the R_p profiles as per Eqs. (5.87) and (5.88). Now, dividing Faraday's rate of reaction, Eq. (5.1), by the metal density ρ defines the corrosion rate (rate of metal dissolution) as

$$C_R = \frac{R_F}{\rho} \tag{5.89}$$

$$C_R = \frac{i A_w}{zF\rho} = \frac{i_{corr} A_w}{zF\rho} \tag{5.90}$$

In addition, corrosion rate can be represented by the rate of weight loss [7] or rate of penetration [3], but Eq. (5.90) is a mathematical model convenient for determining the metal dissolution in terms penetration per year in units of mm/year. These are common units to the engineers or designers.

Combining Eqs. (5.67), (5.87), and (5.90) gives the corrosion rate as

$$C_R = \frac{\beta A_w}{\left[Z(\omega)_o - Z(\omega)_\infty\right](zF\rho)} \tag{5.91a}$$

$$C_R = \frac{\beta A_w}{R_p (zF\rho)} \tag{5.91b}$$

Moreover, oxidation reactions on metal surfaces can be treated as corrosion reactions, and in general, corrosion is classified as wet corrosion that is due to

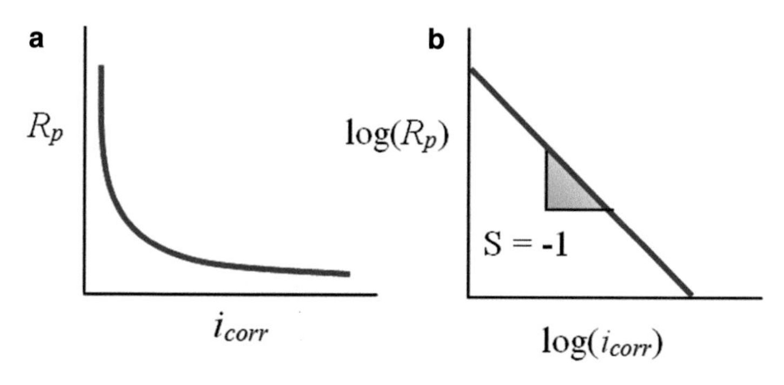

Fig. 5.11 Polarization resistance **(a)** Nonlinear and **(b)** semi-log plots

metal-liquid interface interactions, where the liquid is an electrolyte at low or high temperature. Conversely, dry corrosion is due to metal-gas interface interactions, where the gas, such as steam, is at high temperatures.

Corrosion rate (C_R) is just the rate of metal oxidation, and traditionally, it quantifies the rate of mass loss (dm/dt) induced by electrochemical reactions that involve electron transfer and hence the phrase corrosion reactions. The electron transfer produces a instantaneous current flow named corrosion current, and according to Faraday's law, the rate of corrosion is characterized in terms of rate of mass loss:

$$\frac{dm}{dt} = \frac{IA_w}{zF} \tag{a}$$

where dm/dt is the rate of mass loss (g/s) known as the weight loss.

Henceforth, mass loss is referred to as weight loss in the literature. Dividing Eq. (a) by the product $A_s\rho$, where A_s is the exposed area and ρ is the metal density, gives the engineering rate of penetration as the corrosion rate:

$$C_R = \frac{1}{A_s\rho}\frac{dm}{dt} = \frac{IA_w}{zFA_s\rho} \tag{5.92}$$

$$C_R = \frac{iA_w}{zF\rho} \tag{5.90}$$

Here, $i = I/A_s$.

Example 5.3. *Calculate* ***(a)*** *the corrosion rate C_R in units of mm/year,* ***(b)*** *the electrochemical rate constant in $\mu g/cm^2\, s$ and mol/cm² s for a low-carbon steel plate (1 cmx1 cmx4 cm) immersed in seawater, and* ***(c)*** *the exchange current density i_o. Given data: $n_a = 0.05\,V$, $I_a = 110\,\mu A$ (current), $\rho = 7.87\,g/cm^3$, and $A_w = 55.85$ g/mol.*

Solution. *Since the carbon steel is basically iron (Fe), the oxidation reaction involved in this problem is* $Fe \longrightarrow Fe^{2+} + 2e^-$ *with* $z = 2$ *and* $F = 96,500\,A\,s/mol$, *and the exposed area is* $A = 1\,cmx1\,cm = 1\,cm^2$. *Then, the current density is* $i_a = I_a/A = 110x10^{-6}\,A/cm^2$. *Using Eq. (5.90), the corrosion rate is*

$$C_R = \frac{i_a A_w}{zF\rho} = \frac{\left(110x10^{-6}\,\text{A/cm}^2\right)(55.85\,\text{g/mol})}{(2)(96,500\,\text{A s/mol})\left(7.87\,\text{g/cm}^3\right)}$$

$$C_R = 4.04x10^{-9}\,\text{cm/s} = 1.28\,\text{mm/year}$$

From Eq. (5.1), the electrochemical rate constant is

$$R_F = \frac{i_a A_w}{zF} = \frac{\left(110\,\mu\text{A/cm}^2\right)(55.85\,\text{g/mol})}{(2)(96,500\,\text{A s/mol})}$$

$$R_F = 0.0318\,\mu\text{g}/\left(\text{cm}^2\,\text{s}\right)$$

Eliminating the atomic weight in Eq. (5.1) yields

$$R_F = \frac{i_a}{zF} = \frac{\left(110\,\mu\text{A/cm}^2\right)}{(2)(96,500\,\text{A s/mol})}$$

$$R_F = 5.70x10^{-10}\,\text{mol}/\left(\text{cm}^2\,\text{s}\right)$$

From Eq. (5.10),

$$i_o = \frac{i_a}{2\sinh\left[zFn/(2RT)\right]}$$

$$\sinh\left[zFn/(2RT)\right] = \sinh\frac{(2)(96,500\,J/mol.V)(0.05\,\text{V})}{2\,(8.3145\,J/mol.K)(298\,K)} = 3.4342$$

$$i_o = \frac{110\times10^{-6}\,\text{A/cm}^2}{(2)(3.4342)} = 1.60\times10^{-5}\,\text{A/cm}^2$$

From Eq. (5.16),

$$i_o = \frac{i_a RT}{zFn} = \frac{\left(1.10\times10^{-4}\,\text{A/cm}^{-6}\right)(8.3145\,\text{J/mol K})(298\,\text{K})}{(2)(96,500\,\text{J/mol/V})(0.05\,\text{V})}$$

$$i_o = 2.82\times10^{-5}\,\text{A/cm}^2$$

Therefore, there is approximately 24 % error in the i_o *calculated results.*

Example 5.4. *Assume that the rates of oxidation of Zn and reduction of H^+ are controlled by activation polarization. Use the data given below to* ***(a)*** *plot the appropriate polarization curves and determine E_{corr} and i_{corr} from the plot. Calculate* ***(b)*** *both E_{corr} and i_{corr},* ***(c)*** *the corrosion rate C_R in mm/year, and the polarization resistance R_p in Ω cm^2. Data:*

$Zn = Zn + 2e^- \Longrightarrow$	$E_{Zn} = -0.80\,V$	$i_{o,Zn} = 10^{-7}\,A/cm^2$
	$\beta_a = 0.10\,V$	$\rho = 7.14\,g/cm^3$
	$A_w = 65.37\,g/mol$	
$2H^+ + 2e^- = H_2 \Longrightarrow$	$E_H = 0.10\,V$	$i_{o,H} = 10^{-10}\,A/cm^2$
	$\beta_c = -0.10\,V$	

Determine ***(d)*** *the anodic current density i_a if the cell is polarized at an overpotential of* $0.01\,V$ *and* ***(e) the anodic resistance*** dE/di.

Solutions. (a) *E vs.* log(*i*) *plot:*

(1) Plot $(i_{o,H}, E_H) = \left(\log 10^{-10}\,A/cm^2, 0.10\,V\right)$.

$(i_{o,Zn}; E_{o,Zn}) = \left(\log 10^{-7}\,A/cm^2, -0.80\,V\right)$

(2) *Draw the hydrogen reduction and zinc oxidation lines with β_c and β_a slopes, respectively.*

(3) *The corrosion point $(i_{corr}, E_{corr}) = \left(10^{-4}\,A/cm^2, -0.50\,V\right)$ is for the unpolarized Zn metal. Also, $i_{corr} = i_a = i_c$ at $E = E_{corr}$ and $E_{o,Zn}$ and $E_{o,H}$ are the open-circuit potentials, which can be determined using the Nernst equation, Eq. (3.31).*

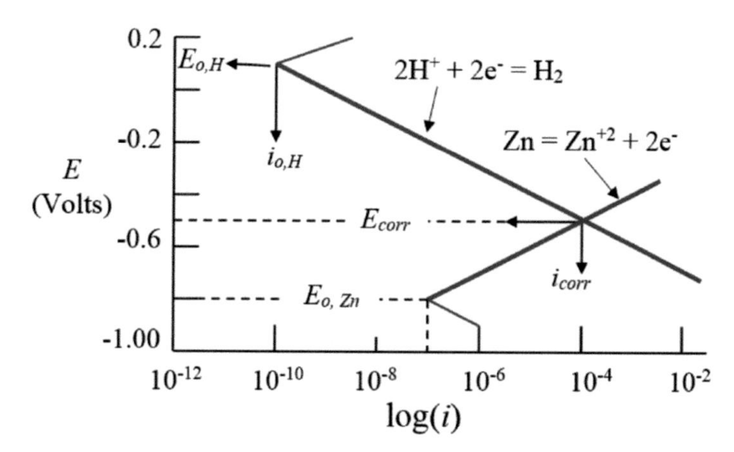

(b) *Using Eqs. (5.37) and (5.38) with* $i_a = i_c = i_{corr}$, $\eta_a = E - E_{o,Zn}$ *and* $\eta_c = E - E_{o,H}$ *yields*

$$E = E_{o,Zn} + \beta_a \log\left(i_{corr}/i_{o,Zn}\right) \tag{a}$$

$$E = E_{o,H} - |\beta_c| \log\left(i_{corr}/i_{o,H}\right) \tag{b}$$

Equating Eqs. (a) and (b) and solving for i_{corr} *yield the corrosion current density:*

$$\log i_{corr} = \frac{1}{\beta_a + |\beta_c|}\left[E_{o,H} - E_{o},_{Zn} + \beta_a \log i_{o,Zn} + |\beta_c| \log i_{o,H}\right] \tag{c}$$

$$\log i_{corr} = \frac{0.1 + 0.8 + 0.1 \log 10^{-7} + 0.1 \log 10^{-10}}{0.2}$$

$$\log i_{corr} = -4$$

$$i_{corr} = 10^{-4}\,\text{A/cm}^2$$

Letting $E = E_{corr}$ *and using Eq. (a) or (b) gives* $E_{corr} = -0.50\,\text{V}$

(c) *From Eq. (5.90)*

$$C_R = \frac{i_{corr} A_w}{zF\rho} = \frac{\left(10x10^{-4}\,\text{A/cm}^2\right)(65.37\,\text{g/mol})}{(2)(96,500\,\text{A s/mol})\left(7.14\,\text{g/cm}^3\right)}$$

$$C_R = 1.50\,\text{mm/year}$$

From Eq. (5.44),

$$\beta = \frac{\beta_a |\beta_c|}{2.303\left(\beta_a + |\beta_c|\right)}$$

$$\beta = \frac{(0.10\,\text{V})(0.10\,\text{V})}{2.303\,(0.10\,\text{V} + 0.10\,\text{V})} = 0.0217\,\text{V}$$

and from Eq. (5.87),

$$R_p = \frac{\beta}{i_{corr}} = \frac{0.0217\,\text{V}}{10^{-4}\,\text{A/cm}^2}$$

$$R_p = 217\,\Omega\,\text{cm}^2$$

(d) *From Eq. (5.78), the anodic current density i_a is*

$$i_a = i_{corr} \exp\left(\frac{2.303 n_a}{\beta_a}\right)$$

$$i_a = \left(10 \times 10^{-4}\,\mathrm{A/cm^2}\right) \exp\left(\frac{(2.303)\,(0.01\,\mathrm{V})}{0.1\,\mathrm{V}}\right)$$

$$i_a = 1.26 \times 10^{-3}\,\mathrm{A/cm^2} > i_{corr}$$

(e) *From Eq. (5.81), the anodic resistance di/dE is*

$$di/dE = \frac{2.303 i_{corr}}{\beta_a} \exp\left(\frac{2.303 n_a}{\beta_a}\right)$$

$$di/dE = \frac{2.303 i_a}{\beta_a} = \frac{(2.303)\left(1.26 \times 10^{-3}\,\mathrm{A/cm^2}\right)}{0.1\,\mathrm{V}}$$

$$di/dE = 2.9018 \times 10^{-2}\,\mathrm{A/cm^2\,V}$$

$$dE/di = 34.46\,\mathrm{cm^2\,V/A} = 34.46\,\Omega\,\mathrm{cm^2}$$

Example 5.5. *A copper surface area, $A = 100\,cm^2$, is exposed to an acid solution. After 24 h, the loss of copper due to corrosion (oxidation) is $15x10^{-3}$ g. Calculate* ***(a)*** *the current density i in $\mu A/cm^2$,* ***(b)*** *the corrosion rate C_R in mm/year,* ***(c)*** *the number of reactions per unit time, and* ***(d)*** *the number of electrons per unit time. Use the data for pure copper: $A_w = 63.55\,g/mol$, $\rho = 8.94\,g/cm^3$, and $q_e = 1.6022x1o^{-19}\,C/reaction$.*

Solution. *First of all, the corrosion process of copper can be represented by the following anodic reaction:*

$$Cu \longrightarrow Cu^{2+} + 2e^-$$

where the valence is $Z = +2$. Thus,

(a) *According to Faraday's law the current density is*

$$i = \frac{I}{A_s} = \frac{zFm}{A_s t A_w}$$

$$i = \frac{(2)\,(96,500\,\mathrm{A\,s/mol})\,\left(15x10^{-3}\,\mathrm{g}\right)}{(100\,\mathrm{cm^2})\,(24x3600\,\mathrm{s})\,(63.54\,\mathrm{g/mol})}$$

$$i = 5.27\,\mu\mathrm{A/cm^2}$$

(b) *From Eq. (5.90),*

$$C_R = \frac{iA_w}{zF\rho}$$

$$C_R = \frac{\left(5.27x10^{-6}\ \text{A/cm}^2\right)(63.54\,\text{g/mol})}{(2)\,(96,500\,\text{A s/mol})\left(8.94\,\text{g/cm}^3\right)}$$

$$C_R = 1.941 \times 10^{-10}\,\text{cm/s}$$

$$C_R = 0.06\,\text{mm/year}$$

(c) *Recall that* $1\ C = A\,s$ *so that* $q_e = 1.6022x10^{-19} A\,s/reactions$ *or* $q_e = 1.6022x10^{-19} A\,s/electrons$*. Thus, the number of reactions per time in seconds is*

$$r = \frac{I}{zq_e} = \frac{iA}{zq_e}$$

$$r = \frac{\left(5.27x10^{-6}\,\text{A/cm}^2\right)\left(100\,\text{cm}^2\right)}{(2)\,(1.6022x10^{-19}\,\text{A s/reactions})} = 1.64x10^{15}$$

$$r \simeq 2 \times 10^{15}\,\text{reactions/s}$$

(d) *Electrons per time*

$$r = \frac{IA}{q_e} = \frac{\left(5.27x10^{-6}\,\text{A/cm}^2\right)\left(100\,\text{cm}^2\right)}{(1.6022x10^{-19}\,\text{A s/electrons})}$$

$$r = 3 \times 10^{15}\,\text{electrons/s}$$

5.5 Ionic Current Density

The ionic current density distribution in the near field or far field of a corroding electrode is difficult to determine in precise details due to the interactions between positive and negative charged ions within the interface region adjacent to the active electrode surface. However, in principle, the current density distribution may not be uniformly distributed within the electrolyte and on the electrode surface due to a temperate gradient within the electrolyte and microstructural as well as mechanical defects of the electrode surface. This, then, is a current density inhomogeneity treated as an asymmetric current density distribution.

Henceforward, a simple mathematical method is adopted in this section in order to generalize the ionic current density distribution in an electrolyte, assuming a constant temperature a smooth electrode surface.

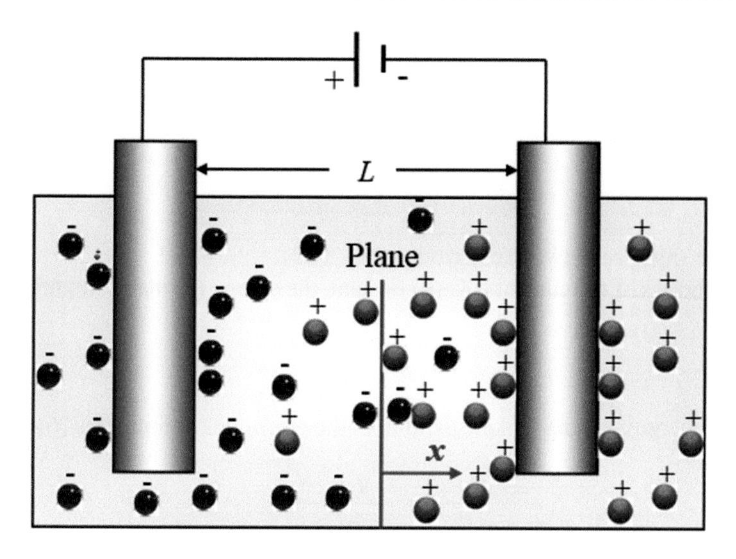

Fig. 5.12 Schematic cell showing the ionic travel distance

Assume that an electrolyte solution contains positively and negatively charged ions (particles) with oxidation number (valence) z^+ and z^-. If the solution has different type of ions, the number of cations (z^+) and anions (z^-) are

$$N^+ = \sum N_j^+ = \sum y_j^+ N_A \tag{5.93}$$

$$N^- = \sum N_j^- = \sum y_j^- N_A \tag{5.94}$$

where j = represents species or ions of different valance
y_j^+, y_j^- = number of moles

Figure 5.12 illustrates the ideal distribution of ions in an electrolyte. Let an applied electric field disturb the thermodynamical equilibrium so that the charge carriers (ions) migrate a distance x from their initial position toward the electrode surface.

Hence, an electrical mobility B_j of ions generates in the solution, and the distance traveled by the ions to the opposite charged electrode (Fig. 5.12) is [6]

$$x_j^+ = v_j^+ dt \tag{5.95}$$

$$x_j^- = v_j^- dt \tag{5.96}$$

where v_j^+ = velocity of cations (cm/s)
v_j^- = velocity of anions (cm/s)
dt = infinitesimal change of time (s)

The number of ions that travel a distance x_j can be defined as

$$N^{+} = \sum \frac{N_j^{+} x_j^{+}}{L} \tag{5.97}$$

$$N^{-} = \sum \frac{N_j^{-} x_j^{-}}{L} \tag{5.98}$$

where $L =$ distance between electrodes (Fig. 5.12).

Nevertheless, if the current is kept constant, the charge (q) that flows through the circuit is

$$dq = Idt \tag{5.99}$$

The total charge crossing a plane parallel to the electrodes in time dt is [6]

$$\frac{dq^{+}}{dt} = \frac{z_j^{+} q_e v_j^{+} N_j^{+}}{L} \tag{5.100}$$

$$\frac{dq^{-}}{dt} = \frac{\left|z_j^{-}\right| q_e v_j^{-} N_j^{-}}{L} \tag{5.101}$$

where $q_e =$ electron charge $= 1.602x10^{-19}$ C.

In addition, the current I and the current density i are

$$I = \frac{dq}{dt} \tag{5.102}$$

$$i = \frac{1}{A}\frac{dq}{dt} \tag{5.103}$$

Recall that electrical conduction is a mass transfer phenomenon, in which electrons and ions carry the electric charge due to their particular mobility within the electrically charged system. Hence, the charge (q) that passes through the cross section of the electrolyte solution (conductor) at time dt is related to the electric current (I). It should be mentioned that the electrons carry the current through the wires and electrodes, while the ions carry the current through the solution.

This phenomenon implies that electrochemical reactions occur at the electrode-solution interface by transferring electrons from or to the electrode. For instance, if $Cu^{2+} + 2e^{-} \Longrightarrow Cu$ occurs, then one mole of Cu is deposited on the electrode surface, while 2 mol of electrons flow through the circuit.

Combining Eq. (5.100), (5.101), and (5.103) for all ions into solution gives

$$i^{+} = \sum \frac{z_j^{+} q_e v_j^{+} N_j^{+}}{V} \tag{5.104}$$

$$i^{-} = \sum \frac{\left|z_j^{-}\right| q_e v_j^{-} N_j^{-}}{V} \tag{5.105}$$

where $V = AL$ is the volume $\left(\text{cm}^3\right)$ and A is the cross-sectional area of the solution. The concentration (activity) of ions is

$$C_j^+ = \frac{1}{V}\sum y_j^+ \quad (5.106)$$

$$C_j^- = \frac{1}{V}\sum y_j^- \quad (5.107)$$

Combining Eqs. (5.93)–(5.94), (5.104)–(5.105), and (5.106)–(5.107) along with Eq. (2.3), $F = q_e N_A,$ yields

$$i^+ = \sum z_j^+ F v_j^+ C_j^+ \quad (5.108)$$

$$i^- = \sum z_j^- F v_j^- C_j^- \quad (5.109)$$

The total current density becomes $i = i^+ + i^-$ as in electrolysis. Hence,

$$i = \sum z_j^+ F v_j^+ C_j^+ + \sum \left| z_j^- \right| F v_j^- C_j^- \quad (5.110)$$

The ionic velocity v_j for migration toward the electrode is small when compared with the random velocity (drift velocity) v_r based on the translational kinetic energy in the absence of an electric field [6]. This is mathematically shown below

$$\frac{3}{2}kT = \frac{1}{2}mv_{r,j} \quad (5.111)$$

From Eqs. (5.108), (5.109), and (5.111), in general, one can deduced that

$$v_{r,j}^+ = \sqrt{\frac{3kT}{m_j^+}} >> v_j^+ = \frac{i^+}{z_j^+ F C_j^+} \quad (5.112)$$

$$v_{r,j}^- = \sqrt{\frac{3kT}{m_j^+}} >> v_j^- = \frac{i^-}{z_j^- F C_j^-} \quad (5.113)$$

where m_j^+ = mass of cations (species).

5.6 Ionic Conductivity

In this section, the electrical conductivity of a conductor can be defined as (1) the ionic conductivity (K_c) because it is a measure of an electrolyte's ability to conduct an electric current due to the motion of positively and negatively charged ions through defects in the a crystal lattice or aqueous electrolyte or (2) as the electronic conductivity due to the motion of electrons. The electrical conductivity is emphasized on charge (q), current (I), and potential (E) as the macroscale variables and an electrolyte containing ions and electrons as the nanoscale charge carriers, which are capable of migrating under the influence of an electrical field.

Of particular interest in this section is the definition of current. It is defined as the rate of charge that flows through a cross-sectional area (A_c) in a specific direction (Fig. 5.12). If current is due to electrons, then one may take into account the random motion of electrons in the presence of an electric field and the use of the Drude model to describe electron behavior or use quantum mechanics to treat the electrons as particles moving or tunneling through a gap to reach a conduction band.

Mathematically, electroneutrality is achieved when the rate of charge for cations and anions are equal; $dq^+/dt = dq^-/dt$. If the electrolyte is treated as an aqueous electrical conductor, then its conductivity (K_c) and resistivity (ρ_c) are related as

$$K_c = 1/\rho_c = i/E_x \tag{5.114}$$

where ρ_c has units of Ω cm, i (electric current density) in A/cm^2, and $E_x = d\phi/dx$ (magnitude of the electric potential gradient in the x-direction) in V/cm.

Moreover, the effectiveness of ion migration can be characterized by the ionic mobility (B_j) in the solution when an electrical force field acts on the ions. This implies that the ionic mobility is related to the electrolyte conductivity. This can be assessed by substituting Eq. (5.110) into (5.114)

$$K_c = \frac{1}{F_x}\left(\sum z_j^+ F v_j^+ C_j^+ + \sum |z_j^-| F v_j^- C_j^-\right) \tag{5.115}$$

$$K_c = \sum z_j^+ F B_j^+ C_j^+ + \sum |z_j^-| F B_j^- C_j^- \tag{5.116}$$

since the ionic or electrical mobility is defined as

$$B_j = v_j/F_x \tag{5.117}$$

which has units of cm^2/ (V s) $=$ mol cm^2/ (J s). Also, F_x is the electric force gradient.

In general, the electrolyte conductivity can be defined by

$$K_c = z_j F B_j C_j \tag{5.118}$$

Then, B_j and i become

$$B_j = \frac{K_c}{zFC_j} \tag{5.119}$$

$$i = zFv_jC_j \tag{5.120}$$

Thus, electric current ($I = iA_e$) flow causes energy dissipation as heat in the electrolyte since the ions must overcome frictional forces during their motion through the medium [38].

From this crude approximation, one can determine the diffusion coefficient or diffusivity of the metal cation M^{z+} as follows. Multiply both sides of Eq. (5.119) by the product κT, where $\kappa = 8.62x10^{-5}$ eV/K is the Boltzmann constant and T is the absolute temperature. Hence,

$$\kappa TB_j = \kappa T \frac{K_c}{zFC_j} \tag{5.121}$$

Notice that κTB_j is the Nernst–Einstein diffusion coefficient D, which is the rate of swept area of cations [39]. Hence, the **Nernst–Einstein equation** is

$$D = \kappa TB_j \tag{5.122}$$

$$D = \kappa T \left(\frac{K_c}{zFC_j} \right) \tag{5.123}$$

The subject of diffusion is well documented [38,40–47], and it is a manifestation of continuous atom or ion motion at random. The K_c expression given by Eq. (5.114) is redefined as [45]

$$K_c = \frac{L}{A_e} \frac{1}{R_s} = \frac{\lambda}{R_s} \tag{5.124}$$

where $\lambda = L/A_e =$ cell constant $\left(\text{cm}^{-1}\right)$
$L =$ distance between electrodes (cm)
$A_e =$ cross-sectional area of the electrolyte conducting path $\left(\text{cm}^2\right)$
$R_s =$ electrolyte electric resistance (Ω)

Moreover, K_c can also be defined as

$$K_c = \frac{1}{\rho_c} = \frac{I}{EL} \tag{5.125}$$

Clearly, K_c is strongly dependent on L. Here, E is the applied electric potential (volts) between two electrodes, and I is the electric current response. The ionic conductivity of a solution is usually measured using an alternating current (AC) to limit any electrodeposition on the cathodic electrode. The conductivity cell contains two electrodes having well-defined areas, and it is connected to a Wheatstone bridge for adjusting an external resistance R_x until the internal resistances R_1 and R_2 are equal for zero current flow [48]. As a result, the resistance of the solution becomes

$$R_s = \frac{R_2 R_x}{R_1} \tag{5.126}$$

Then, Eq. (5.124) is given by

$$K_c = \frac{L}{A_e}\frac{1}{R_s} = \frac{\lambda R_1}{R_2 R_x} \tag{5.127}$$

In addition, dividing K_c by the molar concentration (C) of a species j yields the molar conductivity of a strong or weak electrolyte

$$K_m = \frac{K_c}{C} \tag{5.128}$$

If the molar concentration of the electrolyte has units of mol/cm^3, then K_m is in $\Omega^{-1}\,\text{cm}^2\,\text{mol}^{-1}$ or $S\,\text{cm}^2\,\text{mol}^{-1}$ where $S = \Omega^{-1}$ is called siemens.

Example 5.6. *Show that the electric current flows in an electrolyte (conductor) only if there is an electric potential gradient, $d\phi/dx$, within the electrolyte.*

Solution. *Substituting $E_x = -d\phi/dx$ Eq. (2.5) and $i = I/A_e$, Eq. (2.19), into Eq. (5.114) yields*

$$K_c = i/E_x = -\frac{I}{A_e d\phi/dx}$$

from which

$$I = -K_c A_e \frac{d\phi}{dx}$$

$$\frac{dq}{dt} = -K_c A_e \frac{d\phi}{dx}$$

where $d\phi/dx = \Delta\phi/\Delta x$ so that $|\Delta\phi| = IR_s$ (Ohm's law) and $\Delta x = L$.

5.7 Problems/Questions

5.1. What are the three conditions for galvanic corrosion to occur?

5.2. If the state of equilibrium of an electrochemical cell is disturbed by an applied current density i_x, then $i_a = -i_c = i_{corr}$ no longer holds. Let i_x be a cathodic current density and the slope of the corresponding polarization curve be $dE/d\log(i_x)$, which increases approaching the Tafel constant β_c. Determine **(a)** the value of i_x when $i_c = 10^{-3}\,\text{A/cm}^2$ and $i_a = 10^{-9}\,\text{A/cm}^2$ and **(b)** the value of β_c when $i_{corr} = 10^{-5}\,\text{A/cm}^2$ and $\eta = -0.20\,\text{V}$. [Solutions: (a) $i_x = 10^{-3}\,\text{A/cm}^2$ and (b) $\beta_c = 0.10\,\text{V}$.]

5.3. According to the Stockholm convention cell $Fe\left|Fe^{2+}\right|\left|H^{+},H_2\right|Pt$, the corrosion potential of iron (Fe) is $-0.70\ V_{SCE}$ at $pH = 4.4$ and 25 °C in a deaerated (no oxygen is involved) acid solution. Calculate **(a)** the corrosion rate in mm/year when

$$i_{o,H} = 10^{-6}\ \text{A/cm}^2\ \beta_c = -0.10\ V_{SHE}$$
$$i_{o,Fe} = 10^{-8}\ \text{A/cm}^2\ E_{Fe} = -0.50\ V_{SHE}$$

(b) and the Tafel anodic slope β_a and (c) draw the kinetic diagram E vs. $\log i$. [Solutions: (a) $C_R = 1.16\ \text{mm/year}$, (b) $\beta_a = 0.010\ V_{SHE}$.]

5.4. Let the following anodic and cathodic reactions be, respectively, $M = M^{+2} + 2e^-$ $E_M = -0.941\ V_{SCE}$ $i_{o,M} = 10^{-2}\ \mu\text{A/cm}^2$ $2H^+ + 2e^- = H_2$ $E_H = -0.200\ V_{SHE}$ $i_{o,H} = 0.10\ \mu\text{A/cm}^2$ where $i_{corr} = 10^2\ \mu\text{A/cm}^2$ and $E_{corr} = -0.741\ V_{SCE}$. **(a)** Construct the corresponding kinetic diagram and **(b)** determine both β_a and β_c from the diagram and from the definition of overpotential equations. [Solutions: (b) $\beta_a = 0.05\ V_{SHE}$ and $\beta_c = 0.10\ V_{SHE}$.]

5.5. A steel tank is hot dipped in a deaerated acid solution of $5x10^{-4}\ \text{mol/cm}^3$ molality zinc chloride ($ZnCl_2$) so that a 0.15-mm zinc coating is deposited on the steel surface. This process produces a galvanized steel tank. Calculate the time it takes for the zinc coating to corrode completely at a $pH = 4$. Data: $E_{Zn} = -0.8\ V$, $i_{o,Zn} = 10\ \mu\text{A/cm}^2$, $i_{o,H} = 10^{-3}\ \mu\text{A/cm}^2$, $\beta_a = 0.08\ \text{V}$, and $\beta_c = 0.12\ \text{V}$, $T = 25\ °\text{C}$. [Solution: $t \simeq 0.40$ years.]

5.6. Calculate the activity and the corrosion rate of iron (Fe) immersed in an aerated aqueous solution of $pH = 9$. The dissociation constant for ferrous hydroxide, $Fe\,(OH)_2$, is $1.64x10^{-14}$. Given data:

$$i_{o,Fe} = 10^{-2}\mu\text{A/cm}^2 \qquad E_{corr} = -0.30\ \text{V} \qquad \beta_a = 0.10\ \text{V}$$
$$A_{w,Fe} = 55.85\ \text{g/mol} \qquad \rho_{Fe} = 7.86\ \text{g/cm}^3$$

[Solution: $\left[Fe^{2+}\right] = 1.64x10^{-4}$ mol/l and $C_R = 0.0473\ \mu\text{m/year}$.]

5.7. Plot the anodic data given below and determine the polarization resistance (R_p) and the anodic Tafel slope (β_a) for a metal M. Use $\beta_c = 0.07\ V$ and $i_{corr} = 0.019\ \text{A/cm}^2$. The reactions are

$$M = M^{2+} + 2e^-$$
$$2H^+ + 2e^- = H_2$$

$i\ (\mu A/cm^2)$	$E\ (V_{SHE})$
0.08	−0.32
0.10	−0.30
0.15	−0.25
0.18	−0.22
0.20	−0.20

[Solution: $R_p = 1\ \Omega\ cm^2$ and $\beta_a = 0.12\ V$.]

5.8. Why does a pearlitic steel corrode rapidly in an acidic solution?

5.9. Why will the tip and the head of an iron nail behave as anodes relative to the shank? See Fig. 1.6.

5.10. It is known that the standard electrode potential (E^o) for pure crystalline zinc is $-0.763\ V$. Will this value change by cold working and impurities? Explain.

5.11. An electrolyte contains a very low activity ($8x10^{-9}$ mol/l) of silver cations (Ag^+) and an unknown concentration of copper cations. If the cell potential difference between the copper anode and the silver cathode is −0.04 V, determine **(a)** which cation will be reduced (electroplated) on the cathode and **(b)** the concentration of $[Cu^{2+}]$ in g/l at 40 °C. Neglect the effects of other ions that might react with silver. [Solution: (a) Ag will be reduced and (b) $[Cu^{2+}] = 60$ g/l.]

5.12. Suppose that a cold-worked copper electrode has 8000 J/mol stored energy and it dissolves in an aqueous electrolyte. Calculate the overpotential.

5.13. What is the significant differences between the overpotential (η) and the ohmic potential (E)?

5.14. If a sheet of zinc (Zn) is placed in hydrochloric acid (HCl), Zn goes into solution. Why? [Solution: $\Delta G = -147.26\ kJ/mol$.]

5.15. Calculate the equilibrium constant at 25 °C for the electrochemical cell shown in Fig. 2.4. [Solution: $K_e = 1.64x10^{37}$].

5.16. Show that $\Delta G = zF\beta_a \log(i_a/i_{corr})$ where the local potential E can be defined by the Nernst equation.

5.17. If an electrochemical cell operates at 10 A, 25 °C, and 101 kPa for 30 min, calculate **(a)** the number of moles and **(b)** the weight of copper that would be produced. [Solution: (a) $N = 0.093$ mol and (b) $W = 5.93$ g.]

5.18. **(a)** Derive the Arrhenius equation from the general definition of the activation energy of any rate reaction process and **(b)** determine the Arrhenius constants B and A_e using the derived equation and the following data for the dissociation constants: $K_{sp@300} = 10^{11}$, $K_{sp@350} = 3K_{sp@300}$ between 300 and 350 K. It is assumed that the activity coefficients approach 1 in diluted electrolyte solution and that K_{sp} follows an Arrhenius type trend. Recall that K_{sp} can be expressed in terms of concentrations. **(c)** Plot the given data below and draw the trend line in the close interval [200,400] for the temperature axis and [0,4]x10^{11} for the reaction constant K_{sp} axis. [Solutions: (b) $K_{sp} = 2.1869 \times 10^{14}$, $\Delta G^* = 19.181$ kJ/mol.]

T (K)	K_{sp}
200	$0.02x10^{11}$
250	$0.22x10^{11}$
280	$0.50x10^{11}$
300	$1.00x10^{11}$
330	$1.95x10^{11}$
350	$3.00x10^{11}$
360	$3.65x10^{11}$

5.19. Plot the given conductivity data vs. the aqueous $Cu^{2+}SO_4^-$ concentration at 25 °C and 1 atm. (The data was taken from [5]). Conduct a nonlinear regression analysis on these data and explain the resultant curve.

C $\left(\text{mol/cm}^3\right)$ $x10^{-6}$	K_c $(\Omega^{-1} \cdot \text{cm}^{-1})$ $x10^{-3}$
0	0
25.00	3.00
50.00	5.10
75.00	7.40
100.00	8.50
102.50	10.00
105.00	12.00
107.50	14.00
200.00	15.00
202.50	16.00

5.20. An electrochemical cell operates at a small overpotential, and the corroding metal is exposed to a H^+ ion-containing electrolyte. Use the given data

$\eta = 0.005\,\text{V}$	$T = 25\,°\text{C}$
$i = 3.895 \times 10^{-8}\,\text{A/cm}^2$	$\rho = 7.14\,\text{g/cm}^3$
$i_o = 10^{-7}\,\text{A/cm}^2$	$D = 10^{-5}\,\text{cm}^2/\text{s}$
$C_R = 5.8277x10^{-4}\,\text{mm/year}$	

in order to determine **(a)** the corroding metal by calculating the atomic weight A_w and determining its anodic reaction. Assume that there is a linear relationship between the current density and the overpotential. **(b)** The activity of the corroding metal if $a_{H^+} = 1\,mol/l$, **(c)** the free energy change ΔG. Will the reaction you have chosen occur? **(d)** The ionic velocity of the corroding metal. As a crude approximation, neglect the ionic velocity of other ions in solution. **(e)** The ionic mobility B and **(f)** electrolyte conductivity neglecting other ions present in the electrolyte. [Solutions: (b) $6.77x10^{-4}$ mol/cm^3, (d) $v_j \simeq 2.98x10^{-10}$ cm/s.]

5.21. If an electrochemical copper reduction process is carried out at 5 A. for 20 min, determine **(a)** the electric charge (amount of Coulombs of electricity), **(b)** the number of electrons if there are $1/\left(1.6022x10^{-19}\right) = 6.24x10^{18}\,electron/C$, **(c)** the number of moles, **(d)** Faraday's weight reduced on a cathode, and the reduction rate. Data: $A_w = 63.55$ g/mol and $T = 35\,°\text{C}$. [Solutions: (a) $Q = 6000\,\text{C}$, (b) $N_e = 4x10^{22}$ electrons, (c) $N = 0.0311$ mol, (d) $W = 1.98$ g, and (e) $P_R = 5.94$ g/h.]

5.22. Calculate the overpotential that causes hydrogen evolution on a flat platinum (Pt) electrode surface immersed in an acid solution, when the applied cathodic and exchange current densities are $6000\,\mu\text{A/cm}^2$ and $100\,\mu\text{A/cm}^2$, respectively. Assume a symmetry factor of 0.50 at room temperature (25 °C). [Solution: $\eta_c = -0.21$ V.]

5.23. Consider a discharge (chemical desorption) mechanism as the rate determining for $Ni^{2+} + 2e^- = Ni$ at 25 °C in a nickel battery. Calculate the cathodic Tafel slope per decade if the symmetry factor is 0.50, the exchange current density is constant, and the cathodic overpotential is $\eta_c < 0$. [Solution: $\beta_c = -0.06$ V/decade.]

5.24. Plot the normalized current profile as a function of both symmetry factor (α) and overpotential; that is, $i/i_o = f(\alpha, \eta)$ when the oxidation state is defined by $z = 2$ and $\alpha = 0.35$, 0.50 and 0.75 at room temperature (25 °C). Explain the polarization behavior very succinctly.

5.25. Calculate the number of moles and mass of **(a)** the battery zinc (Zn) casing and **(b)** the manganese dioxide (MnO_2) in the electrolyte if the battery has a stored energy of 36 kJ/V and a power of 3 W. Calculate **(c)** the time it takes to consume the stored energy if the battery operates at a current of 2 A and **(d)** the potential (voltage). The thickness of the cell casing is $x = 1$ mm, and other dimensions are indicated below. The discharging reaction is

$$MnO_2 + H^+ + e \rightarrow MnOOH$$

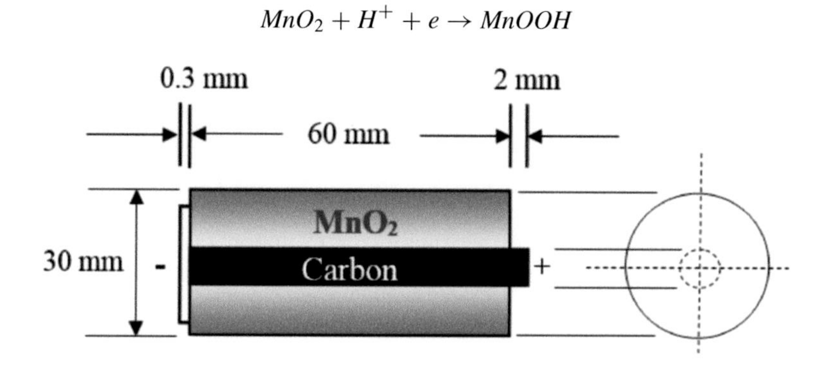

[Solutions: (a) $M_{Zn} = 40.34$ g and $X_{Zn} = 0.62$ mol, (b) 0.37 mol and 32.17 g, and (c) 5 h and 1.5 V.]

5.26. A plate of pure nickel (*Ni*) oxidizes in an electrochemical cell containing an acid solution at 25 °C. The total surface area of the nickel plate is 100 cm^2. If $2x10^{16}$ electrons per second are relieved on the plate surface, then calculate **(a)** the corrosion rate in mm/year and **(b)** the mass of nickel being lost in a year. [Solution: (a) $C_R = 0.35\,\text{mm/year}$ and (b) $m = 31.15$ g.]

5.27. Use the data listed in Table 5.3 to perform a least squares analysis and, subsequently, determine the polarization proportionality constant β. Let the atomic weight and the density of the steel be $A_{w,steel} \simeq A_{w,Fe} = 55.85$ g/mol and $\rho = 7.85$ g/cm^3, respectively. [Solution: $\beta = 0.03$ V.]

5.28. Equal amounts of $CuSO_4$ and $NiSO_4$ are dissolved with water to make up an electrolyte. Hypothetically, the ion velocities and concentrations are

$$\begin{aligned} v_{Cu^{2+}} &= 0.22\,\text{cm/s} \qquad & C_{Ni^{2+}} &= 10^{-5}\,\text{mol/cm}^3 \\ v_{SO_4^{-2}} &= 0.1\,\text{cm/s} \qquad & C_{SO^{-2}} &= 10^{-5}\,\text{mol/cm}^3 \\ C_{Cu^{2+}} &= 10^{-5}\,\text{mol/cm}^3 \end{aligned}$$

If the current density is 1 A/cm^2, calculate the velocity of the nickel ions $\left(Ni^{2+}\right)$. [Solution: 0.20 cm/s.]

5.29. An electrochemical cell operates at $10\,A$, $R_x = 0.25\,\Omega$, and $\omega = 50$ Hz. Determine **(a)** the electrolyte resistance (R_s), **(b)** the potential E_x when the external resistance and the capacitance are $R_x = 0.25\,\Omega$ and $C_x = 20$ A s/V at 30°C, and

(c) the electrolyte conductivity K_c. The distance between electrodes is $L = 15\,\text{cm}$, and the effective electrode surface is $A_s = 8000\,\text{cm}^2$. [Solution: (a) $Z(w)_x = 0.25\,\Omega$, (b) $E_x = 2.5\,\text{V}$, and (c) $K_c = 7.50x10^{-3}\,\Omega^{-1}\,\text{cm}^{-1}$.]

5.30. Determine and analyze the impedance profile by varying the angular frequency for fixed $R_x = 0.25\,\Omega$ and $C_x = 20\,\text{A s/V}$.

5.31. Assume that an electrolytic cell is used for recovering magnesium from a solution containing $10^{-4}\,mol/cm^3$ of Mg^{+2} at 35 °C. The nickel ionic mobility and the electric field strength are $B = 55x10^{-5}\,\text{cm}^2\text{V}^{-1}\,\text{s}^{-1}$ and $F_x = 10\,\text{V/cm}$ [Taken from [5]], respectively; calculate **(a)** the ionic velocity (v), **(b)** the solution electric conductivity (K_c), and the electric resistivity (ρ_c). [Solution: (a) $v = 0.0055\,\text{cm/s}$, (b) $K_c = 0.0106\,\Omega^{-1}\,\text{cm}^{-1}$, and $\rho_c = 94.21\,\Omega\text{cm}$.]

5.32. It is known that current flows when there exists a gradient of electric potential ($d\phi/dx$) within an electric conductor, such as an electrolyte. Consider a current-carrying homogeneous conductor with constant cross-sectional area (A_c) so that the electric field strength (E_x) is constant at every point in the conductor. Derive an expression for the current as a function of the gradient of electric potential. In this particular problem, "x" stands for direction, and Δx is the length of the electric conductor. Start with the following current density definition $i = K_c E_x$, where K_c is the electric conductivity ($\Omega^{-1}\,\text{cm}^{-1}$).

References

1. M.G. Fontana, *Corrosion Engineering*, 3rd edn. (McGraw-Hill, New York, 1986)
2. D.A. Jones, *Principles and Prevention of Corrosion* (Macmillan, New York, 1992)
3. T.N. Andersen, H. Eyring, Principles of electrode kinetics, in *Physical Chemistry: An Advanced Treatise*, vol. IXA, ed. by H. Eyring. Electrochemistry (Academic Press, New York, 1970)
4. D.W. Shoesmith, Kinetics of aqueous corrosion, in *Corrosion*, 9th edn. ASM Handbook, vol. 13 (ASM International, Materials Park, 1987)
5. I.N. Levine, *Physical Chemistry* (McGraw-Hill, New York, 1978), pp. 454–459, 498–500
6. A.J. Bard, L.R. Faulkner, *Electrochemical Methods* (Wiley, New York, 1980), p. 316
7. L.L. Shreir, Outline of electrochemistry, in *Corrosion*, vol. 2, ed. by L.L. Shreir, R.A. Jarman, G.T. Burstein (eds.) Corrosion Control (Butterworth-Heinemann, Boston, 1994)
8. M. Stern, A.L. Geary, Electrochemical polarization I: a theoretical analysis of the shape of polarisation curves. J. Electrochem. Soc. **104**(1), 56–63 (1957)
9. U.R. Evans, The distribution and velocity of the corrosion of metals. J. Frankl. Inst. **208**, 45–58 (1929)
10. U.R. Evans, *The Corrosion and Oxidation of Metals* (Arnold, London, 1961)
11. M. Stern, Electrochemical polarization ii: ferrous-ferric electrode kinetics on stainless steel. J. Electrochem. Soc. **104**(9), 559–563 (1957)
12. L.L. Shreir, Corrosion in aqueous solutions, *in Corrosion*, vol. 1, ed. by L.L. Shreir, R.A. Jarman, G.T. Burstein. Metal/Environment Reactions (Butterworth-Heinemann, Boston, 1994)
13. J.R. Scully, Electrochemical methods of corrosion testing, in *Corrosion*, vol. 13, 9th edn. (ASM International, Materials Park, 1987), p. 213

14. H.P. Hack, Evaluation of galvanic corrosion, in *Corrosion*, vol. 13, 9th edn. (ASM International, Materials Park, 1987), p. 236
15. J.R. Maloy, Factors affecting the shape of current-potential curves. J. Chem. Educ. **60**, 285–289 (1983)
16. J. Wang, *Analytical Electrochemistry*, 2nd edn. (Wiley-VCH, Winheim, 2000), pp. 4–7
17. J.W. Evans, L.C. De Jonghe, *The Production of Inorganic Materials* (Macmillan, New York, 1991), p. 149
18. T. Erdey-Cruz, *Kinetics of Electrode Processes* (Wiley-Interscience, New York, 1972), p. 442
19. D.D. McDonald, An impedance interpretation of small amplitude cyclic voltammetry: I. Theoretical analysis for a resistive-capacitive system. J. Electrochem. Soc. **125**(9), 1443–1449 (1978)
20. A.C. Makrides, Some electrochemical methods in corrosion research. Corrosion **18**(9), 338t–349t (1962)
21. A.C. Makrides, Dissolution of iron in sulfuric acid and ferric sulfate solutions. J. Electrochem. Soc. **107**(11), 869–877 (1960)
22. F. Mansfeld, Recording and analysis of AC impedance data for corrosion studies. I. Background and methods of analysis, corrosion. Corrosion **36**(5), 301–307 (1981)
23. F. Mansfeld, M.W. Kendig, S. Tsai, Recording and analysis of AC impedance data for corrosion studies. II. Experimental approach and results. Corrosion **38**(11), 570–579 (1982)
24. J.R. Scully, Polarization resistance method for determination of instantaneous corrosion rates. Corrosion **56**(2), 199–218 (1999/2000)
25. J.H. Wang, F.I. Wei, H.C. Shih, Electrochemical studies of the corrosion behavior of carbon and weathering steels in alternating wet/dry environments with sulfur dioxide gas. Corrosion **52**(8), 600–608 (1996)
26. E. Khamis, M.A. Ameer, N.M. Al-Andis, G. Al-Senani, Effect of thiosemicarbazones on corrosion of steel in phosphoric acid produced by wet process. Corrosion **56**(2), 127–138 (2000)
27. P. Pierson, K.P. Bethune, W.H. Hartt, P. Anathakrishnan, A new equation for potential attenuation and anode current output projection for cathodically polarized marine pipelines and risers. Corrosion **56**(4), 350–360 (2000)
28. D.W. Law, S.G. Millard, J.H. Bungey, Galvanostatic pulse measurements of passive and active reinforcing steel in concrete. Corrosion **56**(1), 48–56 (2000)
29. D.D. McDonald, M.H. McKubre, in *Modern Aspects of Electrochemistry*, vol. 14, ed. by J.O'M. Bockris, B.E. Conway, R.E. White (Plenum Press, New York, 1982), p. 61
30. K. Hladky, L.M. Callow, J.L. Dawson, Corrosion rates from impedance measurements: an introduction. Br. Corros. J. **15**(1), 20–27 (1980)
31. O.C. Ho, D. Raistrick, R.A. Huggins, Application of A-C techniques to the study of lithium diffusion in tungsten trioxide thin films. Electrochem. Acta **127**(2), 343–350 (1980)
32. D.R. Franceschetti, J.R. McDonald, Small-signal A-C response theory for electrochromic thin films. Electrochem. Acta **129**(8), 1754–1756 (1982)
33. M.W. Kendig, E.T. Allen, F. Mansfeld, Optimized collection of AC impedance data. J. Electrochem. Soc. **131**(4), 935–936 (1984)
34. R. de Levie, L. Pospisil, On the coupling of interfacial and diffusional impedances, and on the equivalent circuit of an electrochemical cell. J. Electroanal. Chem. Interfacial Electrochem. **22**(3), 277–290 (1969)
35. L. Nyikos, T. Pajkossy, Electrochemistry at fractal interfaces: the coupling of AC and DC behaviour at irregular electrodes. Electrochem. Acta **35**(10), 1567–1572 (1990)
36. K.S. Cole, R.H. Cole, Dispersion and absorption in dielectrics I. Alternating current characteristics. J. Chem. Phys. **9**, 341–352 (1941)
37. P.G. Shewmon, *Diffusion in Solids* (McGraw-Hill, New York, 1963)
38. B. Wu, R.E. White, Modeling of a nickel-hydrogen cell: phase reactions in the nickel active material. J. Electrochem. Soc. **148**(6), A595-A609 (2001)
39. R.J. Weiss, *Physics of Materials* (Hemisphere Publishing, New York, 1990), pp. 360–362

40. A.G. Guy, J.J. Hren, *Physical Metallurgy*, 3rd edn. (Addison-Wesley, Reading, MA, 1974), pp. 367–399
41. C.A. Wert, R.M. Thompson, *Physics of Solids*, 2nd edn. (McGraw-Hill, New York, 1970), pp. 54–73
42. J. Crank, *The Mathematics of Diffusion*, 2nd edn. (Oxford University Press, New York, 1990)
43. J. Braunstein, G.D. Robbins, Electrolytic conductance measurements and capacitive balance. J. Chem. Educ. **48**(1), 52–59 (1971)
44. F. Kohlrausch, Ueber Platinirte Elecktroden und Widerstandsbestimmung (German) [About platinized electrodes and resistance determination (English)]. Weid. Ann. **60**, 315–332 (1897)
45. G.H. Geiger, D.R. Poirier, *Transport Phenomena in Metallurgy* (Addison-Wesley, Reading, MA, 1973), pp. 434, 443, 449–451
46. M. Pourbaix, *Lectures on Electrochemical Corrosion* (Plenum Press, New York, 1973), pp. 234–253
47. J.I. Munro, W.W. Shim, Anodic protection - its operation and applications. Mater. Perform. **40**(5), 22–25 (2001)
48. T. Engel, P. Reid, *Physical Chemistry* (Pearson Education, Boston, 2013), pp. 900–904

Mass Transport by Diffusion and Migration

6

Whenever the concentration of a species j like $\left[H^{+}\right]$ becomes the rate controlling, an electrochemical cell is cathodically polarized. This is attributed to the cathodic reactions due to ionic mass transfer [1]. In this case, the concentration of the species j at the cathode electrode surface is lower than the bulk concentration ($C_s < C_b$), and therefore, the electrochemical process is controlled by the rate of mass transfer (flux) rather than charge transfer. In general, the mass transfer may be due to diffusion, migration of charged species under the influence of an electric field, and convection due to fluid flow by stirring, rotation, or vibration. In the latter case, the convective flux strongly depends on the velocity of the mass transfer and concentration of the species. Indiscriminating the source of the flux, steady-state or transient conditions must be taken into account in the analysis of an electrochemical process at the electrode-electrolyte interface. Henceforth, an electrolyte is an electrolytic solution containing positively and negatively charged ions (species).

The theory of concentration polarization deals with the accumulation of ions forming a thin layer on an electrode surface, as in the case of ion deposition. Initially, the ion concentration at localized sites depends on the diffusivity of the species, and consequently, the fundamental properties of the solution may depend on ionic interactions.

Any deviation from electrochemical equilibrium due to an overpotential (η) causes the electrochemical system to be polarized cathodically, and consequently, ions deposit or plate out on the cathode surface. If an undesirable boundary layer forms at the electrode surface, it may interfere the deposition process.

6.1 Mass Transfer Modes

The fundamental understanding of electrochemical corrosion requires a thorough analysis of the electrochemical processes involving metal oxidation of a particular species j and the simultaneous reduction of other species. This leads to the

N. Perez, *Electrochemistry and Corrosion Science*,
DOI 10.1007/978-3-319-24847-9_6

characterization of ionic mass transport within an electrochemical system. For metal oxidation, the ionic mass transport is induced by natural diffusion of at least one species j from the electrode surface into the electrolyte solution. However, if the natural state of an electrochemical system is disturbed by an external electrical energy through the application of a bias voltage (E), the ionic mass transport is controlled by the migration mechanism due to a potential gradient (dC/dx) within an electrical field. The third mechanism for ionic mass transport is by convection when the electrolyte flow is induced by some electromechanical device (rotating electrode or pump). In fact, mass transport can be driven by an external perturbation of an electrochemical system due to a deviation from the equilibrium state. The mass transport of the ions in an electrolyte is very important in characterizing and understanding the working principles of electrochemical cells and other systems. For instance, ionic conductors such as organic and water-based electrolytes have their unique properties and mass transport limitations. The former electrolytes are considered as weak ionic conductors due to their low electrical conductivity capability, while the latter electrolytes have high electrical conductivity. Nonetheless, the mass transport in an electrolyte is the main process that may limits the power that can be drawn from electrochemical cells. Thus, the main role of electrolytes in electrochemical cells is to serve as the medium for the mass transfer of ions and the wiring system for electron flow in connected half-cells under electrochemical stability conditions.

Mass transfer of electrochemical species j, known as ions, can be a complex phenomenon because the solution (fluid) containing the ions may be strongly influenced by turbulent flow and to a lesser extent laminar flow, diffusion, and an electric field. For a stationary electrochemical system, such as a tank, pipe, or battery, as observed by a stationary observer, under internal laminar flow, the mass transfer is quantified by the molar (J) or mass flux (J^*). In general, the term **flux** can be defined as a quantity transferred through a given area in a given time, and as a result, dC/dx must exist in a particular electrolyte. Hence, flux acts in opposition to this concentration gradient [2].

For a one-dimensional treatment in the x-direction, the mass transfer can be characterized as an idealized cylindrical or rectangular (parallelepiped) element of a solution. This is schematically shown in Fig. 6.1 as a simplified model of a three-dimensional analysis [3]. Thus, the molar flux J is a vector quantity perpendicular to a plane of species j [4].

The model indicates that a plane of differential area (dA) containing the species j moves in the x-direction from position 1 to position 2 and then to position 3. This motion is influenced by the modes of mass transfer, such as **diffusion** due to a molar concentration gradient, **migration** due to an electric field, and natural or forced **convection** due to the kinematic velocity or a combination of these modes. Mass transfer of species j can then be quantified by the absolute value of the molar flux J or the mass flux J^*. Notice that J is perpendicular to the moving plane of

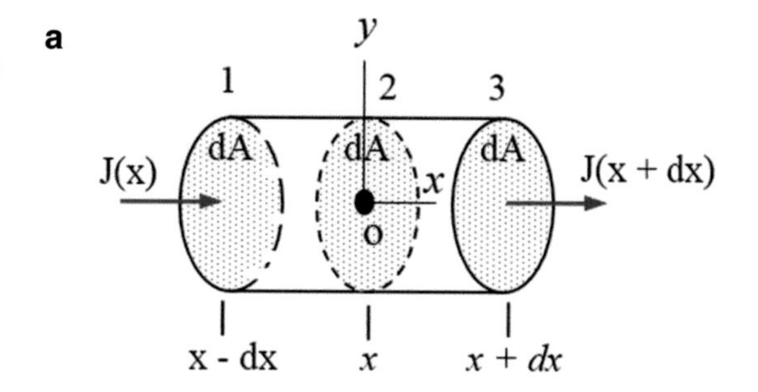

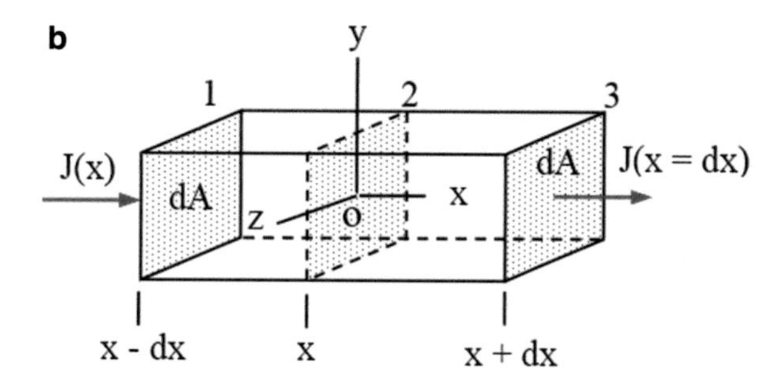

Fig. 6.1 Volume elements for diffusion of a plane along the x-direction. (**a**) Cylindrical element and (**b**) rectangular element

species j and represents the absolute value of the vector molar flux $\vec{J}$. The total flux can be defined as

$$J = \sum J_j \tag{6.1a}$$

$$J^* = \sum J_j A_{w,j} \tag{6.1b}$$

where $A_{w,j}$ = atomic weight of species j

In addition, J represents the number of moles of species j that pass per unit time through a unit area. Thus, the units of J are mol cm^{-2} s^{-1} and that of J^* are g/cm^{-2} s^{-1}. Since the ionic flux is related to mass transfer, it is convenient at this point to introduce the concept of mass transfer of a solute j through an electrolyte. The mass transfer can be driving by on or a combination of the following mechanisms:

- Ionic or molecular diffusion due to a concentration gradient
- Migration of a charged species under the influence of a potential or electric field
- Hydrodynamic velocity due to fluid flow

The general result of diffusion is to balance out differences in concentration of specific ions or species. This means that there must be a concentration gradient $\partial C/\partial x$ which acts as the driving force for diffusion to occur. However, the rate of diffusion depends on the diffusion coefficient (diffusivity) D. Thus, the larger the value of D, the faster the diffusion process will take place.

Mathematically, for a steady-state condition and a stationary x-axis, the concentration rate is $dC_j/dt = 0$, and the total molar flux is defined by the **Nernst-Plank equation** [3, 5–7]:

$$J = -D\frac{\partial C}{\partial x} - \frac{zFDC}{RT}\frac{\partial \phi}{\partial x} + Cv \tag{6.2}$$

$$J_d = -D\frac{\partial C}{\partial x} \tag{6.3}$$

$$J_m = -\frac{zFDC}{RT}\frac{\partial \phi}{\partial x} \tag{6.4}$$

$$J_c = Cv \tag{6.5}$$

where J_d = diffusion molar flux due to dC/dx (mol/cm^2 s)
J_m = migration molar flux due to $d\phi/dx$ (mol/cm^2 s)
J_c = convective molar flux due to fluid flow (mol/cm^2 s)
dC/dx = concentration gradient (mol/cm^4)
$d\phi/dx$ = potential gradient (V/cm)
D = diffusivity (diffusion coefficient) (cm^2/s)
10^{-6} cm^2/s $\leq D < 10^{-4}$ cm^2/s for most cations
C = ionic concentration (mol/cm^3 or mol/l)
v = hydrodynamic velocity (cm/s)

The physical interpretation of Eq. (6.2) is schematically shown in Fig. 6.2 after Maloy's modes of mass transfer [8]. The natural diffusion molar flux J_d arises due to a concentration gradient of ions located at a distance x from the electrode surface. In general, diffusion is the process by which matter moves due to random particle (ion, atom, or molecule) motion. On the other hand, migration molar flux J_m develops due to an electric field, which causes a potential gradient in the solution. The convective molar flux J_c is caused by a moving or flowing solution (electrolyte) due to rotation, stirring, or vibration.

In the absence of an external electrical field, the Faraday's law of electrolysis gives the generalized equation, Eq. (5.30), for the current density in an electrochemical cell:

$$i = zF\left(\frac{m}{A_w t A_s}\right) \tag{6.6}$$

where the rate of reaction process for metal reduction or oxidation is normally represented by the current density (i). Here, m is the mass loss or gain, and A_w is the atomic weight, t is the time, and A_s is the electrode exposed area. The term $m/(A_w tA)$ represents the total ionic molar flux (J) of a particular species j in solution and has units of mol/(cm^2 s). Recall from physics that Faraday's law of electromagnetic induction differs from Faraday's law of electrolysis, despite

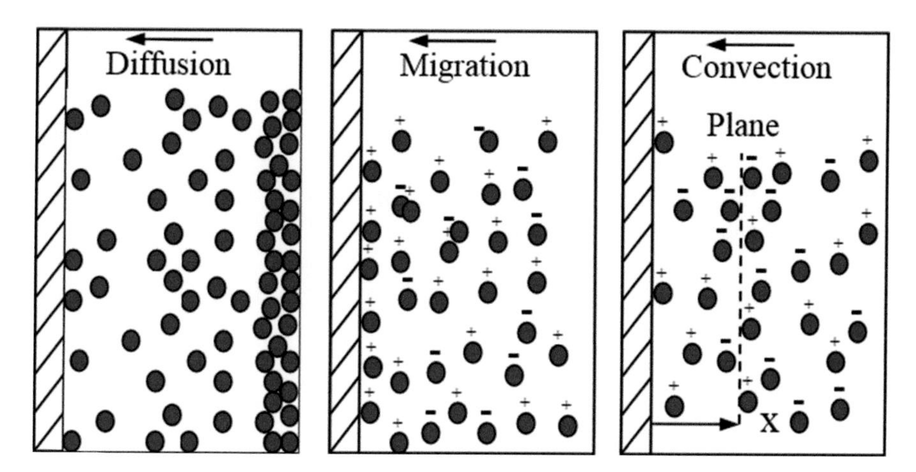

Fig. 6.2 Modes of mass transport

that both are related to an electric field surrounding a conductor of length L. This conductor is the electrolyte in the former law and a solid in the latter law, which in turn is related to a magnetic flux density (magnetic induction) $\overrightarrow{\mathbf{B}}$ produced by currents in a magnetic field.

According to Faraday's law of electrolysis, an electrochemical process requires the passage of one Faraday (1 F) of electricity to transfer one mole of electrons at the cathode and one mole of electrons at the anode. Thus, if one faraday of electricity flows through an external circuit, then $6.022x10^{23}$ electrons (Avogadro's number) pass through the circuit wire system, and subsequently, the well-known Faraday's constant can be calculated using Eq. (2.3). For convenience,

$$F = q_e N_A = \left(1.6022x10^{-19}\text{C/electrons}\right)\left(6.022x10^{23}\text{ electrons/mol}\right) \quad \text{(a)}$$

$$F = 96,484\,\text{C/mol} \simeq 96,500\,\text{C/mol of electricity}$$

For one-dimensional analysis, the diffusion molar flux is

$$J_x = \frac{m}{A_w t A_S} \tag{6.7}$$

For one-dimensional analysis, combining Eqs. (6.6) and (6.7) yields the current density (i) due to species j in the electrolyte:

$$i = +zFJ_x = -zFD\left(\frac{\partial C}{\partial x}\right)_{x=0} \quad \text{(cathodic)} \tag{6.8}$$

$$i = -zFJ_x = zFD\left(\frac{\partial C}{\partial x}\right)_{x=0} \quad \text{(anodic)} \tag{6.9}$$

This expression can be generalized including all three modes of transport of matter in an electrolyte; that is,

$$i = zF\left[J_d + J_m + J_c\right] \tag{6.10}$$

Substituting Eqs. (6.3) through (6.5) into (6.10) yields the total anodic current density for the three mass transfer modes:

$$i = zF\left[-D\frac{dC}{dx} - \frac{zFDC}{RT}\frac{\partial \phi}{\partial x} + Cv\right] \tag{6.11}$$

The potential gradient due to natural diffusion and due to an external electrical source can be estimated as [9, 10]

$$\left(\frac{\partial \phi}{\partial x}\right)_{diffusion} = \frac{RT}{xF} \tag{6.12}$$

$$\left(\frac{\partial \phi}{\partial x}\right)_{electric} = \frac{E}{L} \tag{6.13}$$

where $L =$ distance from the electrode surface.

Usually, J_d and J_m are coupled in the operation of electrochemical cells. This is a coupled mass transfer process referred to as **ambipolar diffusion** [11] or binary system where the migration molar flux J_m evolves due to a chemical potential gradient $d\mu/dx$, which in turn emerges as a result of a potential gradient $d\phi/dx$. However, other factors such as pressure gradient dP/dx and temperature gradient dT/dx are neglected as contributing factors to the total value of J_x expression as defined by Eq. (6.2) and to the total anodic current density, Eq. (6.11).

However, ambipolar diffusion due to the coupling of the electric field across the electrode-electrolyte interfaces occurs since the mass transport of ions in the electrolyte is affected by the transport of electrons in the solid electrodes and vice versa [12, 13]. In the light of this, ambipolar diffusion is usually modeled to take into account a fixed rate of diffusion for both ions and electrons [14].

During bipolar diffusion, the amount of electric charge carried by electrons during mass transport can theoretically be determined by the Faraday's constant, which is an important physical constant in chemistry, electrochemistry physics, and electronics.

Example 6.1. *Calculate Faraday's constant (F) if $I = 250\,mA$, $t = 12\ min$, and $N = 1.8654 \times 10^{-3}\,mols$.*

Solution. *To experimentally determine Faraday's constant F, one needs to know the total amount of charge (Q) in Coulombs and the number of moles of charge (N). Then measure the electric current (I) at time (t). Thus,*

$$F = \frac{Q}{N} = \frac{It}{N}$$

$$F = \left(250 * 10^{-3}\,\text{A}\right)(12 * 60\,\text{s}) \,/\, \left(1.8654 \times 10^{-3}\,\text{mol}\right) = 96,494\,\text{A s/mol}$$

$$F = 96,494\,\text{C/mol} \simeq 96,500\,\text{C/mol}$$

Therefore, this numerical result is in accord with the American Standard Definitions of Electrical Terms which states that "96,500 C/mol is required for an electrochemical reaction involving one chemical equivalent."

6.2 Fick's Laws of Diffusion

Diffusion can be defined as the process in which the mass transfer of matter is due to random motion. In particular, Fick's postulate states that the flux (J) of matter across a given plane is proportional to the concentration gradient across the plane. However, the concentration gradient ($\partial C/\partial x$) of matter (ion or impurity) in a finite volume of a conductive medium, such as an electrolyte or a silicon substrate, decreases when the mass transfer of matter is distributed uniformly within the finite volume. As a result, concentration homogeneity of matter in the finite volume is achievable and obtainable if matter is inactive.

The mathematics of diffusion can be found in Crank's book [15], in which different diffusion processes are analyzed. Let us assume an isotropic and homogeneous medium in which the diffusivity D is constant and the rate of transfer of matter (ions, atom, or molecules) is described by **Fick's first law of diffusion**. Despite that diffusion may be treated as a three-dimensional process, it may be assumed that it occurs in isotropic media.

In general, the conditions for steady state ($\partial C/\partial t = 0$) and for transient state ($\partial C/\partial t \neq 0$) are governed by Fick's laws of diffusion:

$$J_n = -\sum_{n=1}^{3} D_{nn} \frac{\partial C}{\partial x_n} \qquad \text{(first law)} \tag{6.14}$$

$$\frac{\partial C}{\partial t} = \sum_{n=1}^{3} D_{nm} \frac{\partial^2 C}{\partial x_{nm}^2} = -\frac{\partial J_n}{\partial x_n} \qquad \text{(second law)} \tag{6.15}$$

where $\partial C/\partial t$ = concentration rate $\left(\text{mol/cm}^3\,\text{s}\right)$

$n, m = 1, 2, 3$ which stand for $x_1 = x$, $x_2 = y$, and $x_3 = z$

$\partial^2 C/dx_{nm}^2 = \partial^2 C/dx_{mn}^2$ and $D_{nm} = D_{mn.}$ due to symmetry

Fick's laws are significant in measuring diffusivity in isotropic and anisotropic media. Confining a diffusion problem to a one-dimensional treatment suffices most approximations in diffusion. Hence, Eqs. (6.14) and (6.15) can be simplified for diffusion flow along the x-direction in isotropic media:

$$J_x = -D\frac{\partial C}{\partial x} \qquad \text{(first law)} \tag{6.16}$$

$$\frac{\partial C}{\partial t} = D\frac{\partial^2 C}{\partial x^2} = -\frac{\partial J_x}{\partial x} \qquad \text{(second law)} \tag{6.17}$$

These expressions imply that both the molar flux and concentration rate strongly depend on the concentration gradient along the x-direction only since $\partial C/\partial y = \partial C/\partial z = 0$, and that the diffusion flow is perpendicular to the moving plane of a solute (Fig. 6.1). General solutions of Fick's second law equation can be found elsewhere [15, 16] for a few relevant diffusion problems related to a two-component system, such as an electrode sheet immersed or suspended in a liquid solution.

For natural diffusion of ionic mass transport, the current density (i) is one of the kinetic parameters of major importance because it represents the rate of oxidation (corrosion) or reduction (electrodeposition). In this case, substituting Eq. (6.3) into (6.8) and (6.9) along with $J_d = J$ yields the current density as a function of concentration gradient:

$$i_a = +zFD\frac{\partial C}{\partial x} \qquad \text{for oxidation at the anode} \tag{6.18}$$

$$i_c = -zFD\frac{\partial C}{\partial x} \qquad \text{for reduction at the cathode} \tag{6.19}$$

Another kinetic parameter called exchange current density (i_o) is very significant in characterizing an electrochemical process. It is the common current density for oxidation and reduction reactions in a specific electrochemical cell. Fundamentally, i_o is that current density in the absence of an electrical field so that the overpotential is zero ($\eta = 0$). However, it can be approximated by substituting either Eq. (6.18) or (6.19) into (5.16) along with a small overpotential. Hence,

$$i_{o,a} = +\frac{RTD}{\eta}\frac{\partial C}{\partial x} \qquad \text{for oxidation} \tag{6.20}$$

$$i_{o,c} = -\frac{RTD}{\eta}\frac{\partial C}{\partial x} \qquad \text{for reduction} \tag{6.21}$$

which implies that $i_{o,a} = -i_{o,c}$ at equilibrium.

This confirms that the anodic current density is balanced by the cathodic current density when $\eta \to 0$ [17]. Additionally, the diffusivity or diffusion coefficient (D) in the above equations is assumed to be independent of concentration (C), and it may be defined as an Arrhenius (Svante Arrhenius, 1889) type equation:

$$D = D_o \exp\left(-\frac{U^*}{RT}\right) \tag{6.22}$$

where D_0 = diffusion constant (cm^2/s)
U^* = activation energy (J/mol)

The Arrhenius equation was especially for rate constant (k_c) of chemical reactions at absolute temperature T. This equation is generalized as

$$k_c = A\exp\left(-\frac{E_A}{RT}\right) = A\exp\left(-\frac{E_A}{k_B T}\right) \tag{6.22a}$$

where A is the reaction constant known as the pre-factor which depends on the order of the chemical reaction, E_A is the activation energy that represents the driving force for a chemical reaction and is treated as a temperature-independent variable in a suitable temperature range, R is the universal gas constant, and k_B is Boltzmann constant. The Arrhenius equation is applicable in many fields as defined by Eq. (6.22).

6.2.1 Diffusion in a Rectangular Element

Fick's second law of diffusion is for a nonsteady state or transient conditions in which $dC_j/dt \neq 0$. Using Crank's model [15] for the rectangular element shown in Fig. 6.3 yields the fundamental differential equations for the rate of concentration. Consider the central plane as the reference point in the rectangular volume element and assume that the diffusing plane at position 2 moves along the x-direction at a distance x-dx from position 1 and $x + dx$ to position 3.

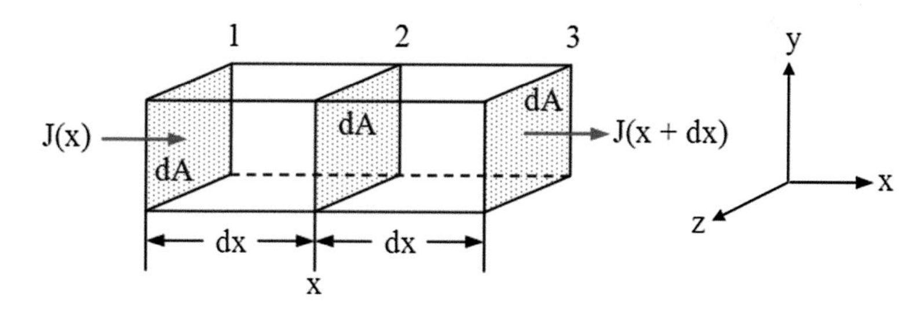

Fig. 6.3 Rectangular element for one-dimensional diffusion in the opposite direction of the concentration gradient (dC/dx)

Thus, the rate of diffusion that enters the volume element at position 1 and leaves at position 3 is

$$R_x = dydz\left[J_x - \frac{\partial J_x}{\partial x}dx\right] - dydz\left[J_x + \frac{\partial J_x}{\partial x}dx\right] \tag{a}$$

$$R_x = -2dxdydz\frac{\partial J_x}{\partial x} = -2dV\frac{\partial J_x}{\partial x} \tag{6.23}$$

Similarly,

$$R_y = -2dxdydz\frac{\partial J_y}{\partial y} = -2dV\frac{\partial J_y}{\partial y} \tag{6.24}$$

$$R_z = -2dxdydz\frac{\partial J_z}{\partial z} = -2dV\frac{\partial J_z}{\partial z} \tag{6.25}$$

The concentration rate is defined as

$$\frac{\partial C}{dt} = \frac{1}{2dV}\sum R_i \tag{6.26}$$

$$\frac{\partial C}{dt} = -\frac{\partial J_x}{\partial x} - \frac{\partial J_y}{\partial y} - \frac{\partial J_z}{\partial z} \tag{6.27}$$

This expression is the equation of continuity for conservation of mass [17]. Combining Eqs. (6.15) and (6.27), with D = constant, yields Fick's second law equation in three dimensions:

$$\frac{\partial C}{dt} = D\left[\frac{\partial^2 C}{\partial x^2} + \frac{\partial^2 C}{\partial y^2} + \frac{\partial^2 C}{\partial z^2}\right] \tag{6.28}$$

Note that "z" has been used as a coordinate in Eqs. (6.23) through (6.28). Hence, for a one-dimensional mass transport, the concentration rate in Eq. (6.28) becomes the most familiar Fick's second law equation:

$$\frac{\partial C}{dt} = D\frac{\partial^2 C}{\partial x^2} \tag{6.29}$$

If D is not constant, then Eq. (6.29) yields the concentration rate in the x-direction as

$$\frac{\partial C}{dt} = \frac{\partial}{dx}\left[D\frac{\partial C}{\partial x}\right] \tag{6.30}$$

$$\frac{\partial C}{dt} = \frac{\partial D}{\partial x}\frac{\partial C}{\partial x} + D\frac{\partial^2 C}{\partial x^2} \tag{6.31}$$

In fact, Eq. (6.29) is the most used form of Fick's second law, and it is the general diffusion expression for one-dimensional analysis under nonsteady state condition.

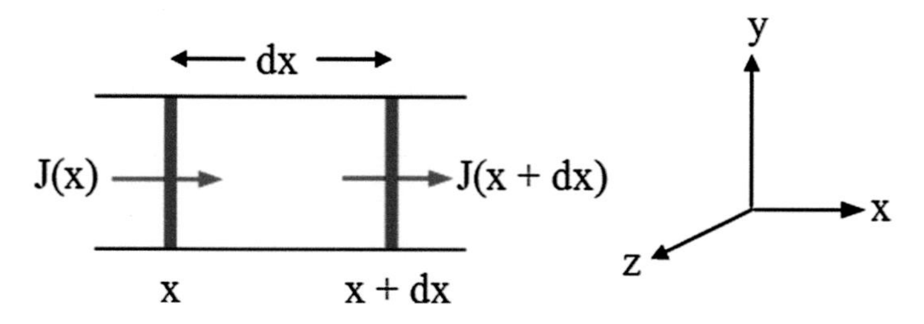

Fig. 6.4 Rectangular planes for one-dimensional diffusion in the opposite direction of the concentration gradient (dC/dx)

In addition, Fick's second law equation can also be derived using Fig. 6.4 and Taylor series on $J(x + dx)$ [17].

Define the rate concentration as [17]

$$\frac{\partial C}{dt} = \frac{J(x) - J(x + dx)}{dx} \tag{6.32}$$

Expanding $J(x + dx)$ so that

$$J(x + dx) = J(x) + \frac{\partial J(x)}{\partial x} dx + \frac{\partial^2 J(x)}{\partial x^2} \frac{dx^2}{2} + \ldots . \tag{6.33}$$

Substituting the first two terms into Eq. (6.32) yields

$$\frac{\partial C}{dt} = \frac{J(x) - J(x) - \frac{\partial J(x)}{\partial x} dx}{dx} \tag{6.34}$$

$$\frac{\partial C}{dt} = -\frac{\frac{\partial J(x)}{\partial x} dx}{dx} = -\frac{\partial J(x)}{\partial x} = -\frac{\partial J_x}{\partial x} \tag{6.35}$$

which is exactly the same as Eq. (6.17) for transient ionic or atomic diffusion in the x-direction.

6.2.2 Solution of Fick's Second Law

For the rectangular element shown in Fig. 6.3, the solution of Eq. (6.29) is subjected to proper initial and boundary conditions given in Appendix A for a particular case, which includes detailed analytical procedures for deriving two suitable solutions. From Appendix A, the normalized concentration expression for the concentration polarization case with $C_\infty = C_b > C_o$ is

$$\frac{C_x - C_b}{C_o - C_b} = 1 - \text{erf}\left(\frac{x}{\sqrt{4Dt}}\right) \qquad \text{for } C_b > C_o \tag{6.36}$$

$$C_x = C_o + (C_b - C_o)\,\mathrm{erf}\left(\frac{x}{\sqrt{4Dt}}\right) \quad \text{for } C_b > C_o \tag{6.37}$$

and for activation polarization with $C_o > C_b = C_\infty$, the resultant expression takes the form

$$\frac{C_x - C_o}{C_b - C_o} = \mathrm{erf}\left(\frac{x}{\sqrt{4Dt}}\right) \quad \text{for } C_o > C_b \tag{6.38}$$

$$C_x = C_o - (C_o - C_b)\,\mathrm{erf}\left(\frac{x}{\sqrt{4Dt}}\right) \quad \text{for } C_o > C_b \tag{6.39}$$

where C_b is the bulk concentration of a species j and C_o is the concentration at the electrode surface.

The error function, erf (y) with $y = x/\sqrt{4Dt}$, is an integral function that is solved numerically by series expansion [17]. It is defined by

$$\mathrm{erf}\,(y) = \frac{2}{\sqrt{\pi}} \int_o^y \exp\left(-y^2\right) dy \tag{6.40}$$

The exponential function may be defined as a power series. Thus,

$$\exp\left(-y^2\right) = \sum_{n=1}^{\infty} \frac{(-1)^{n+1}\, y^{2(n-1)}}{(2n-1)\,(n-1)!} \tag{6.41}$$

Assume that the series is uniformly convergent so that

$$\mathrm{erf}\,(y) = \frac{2}{\sqrt{\pi}} \int_o^y \exp\left(-y^2\right) dy = \frac{2}{\sqrt{\pi}} \int_o^y \sum_{n=1}^{\infty} \frac{(-1)^{n+1}\, y^{2(n-1)}}{(2n-1)\,(n-1)!} dy \tag{6.41a}$$

Integrating gives the error function as a power series

$$\mathrm{erf}\,(y) = \frac{2}{\sqrt{\pi}} \sum_{n=1}^{\infty} \frac{(-1)^{n+1}\, y^{2n-1}}{(2n-1)\,(n-1)!} \tag{6.42}$$

Expanding the series yields

$$\mathrm{erf}\,(y) = \frac{2}{\sqrt{\pi}} \left(y - \frac{y^3}{3} + \frac{y^5}{10} - \frac{y^7}{42} + \cdots\right) \tag{6.42a}$$

The complementary error function is defined by

$$\mathrm{erf}\,(y) + \mathrm{erf}\,c\,(y) = 1 \tag{6.43}$$

Table 6.1 Values of the error function erf (y)

y	erf (y)	y	erf (y)	y	erf (y)
0	0	0.55	0.5633	1.30	0.9340
0.025	0.0282	0.60	0.6039	1.40	0.9523
0.05	0.0564	0.65	0.6420	1.50	0.9661
0.10	0.1125	0.70	0.6778	1.60	0.9763
0.15	0.1680	0.75	0.7112	1.70	0.9838
0.20	0.2227	0.80	0.7421	1.80	0.9891
0.25	0.2763	0.85	0.7707	1.90	0.9928
0.30	0.3286	0.90	0.7970	2.00	0.9953
0.35	0.3794	0.95	0.8209	2.10	0.9970
0.40	0.4284	1.00	0.8427	2.20	0.9981
0.45	0.4755	1.10	0.8802	2.30	0.9989
0.50	0.5205	1.20	0.9103	2.40	0.9993

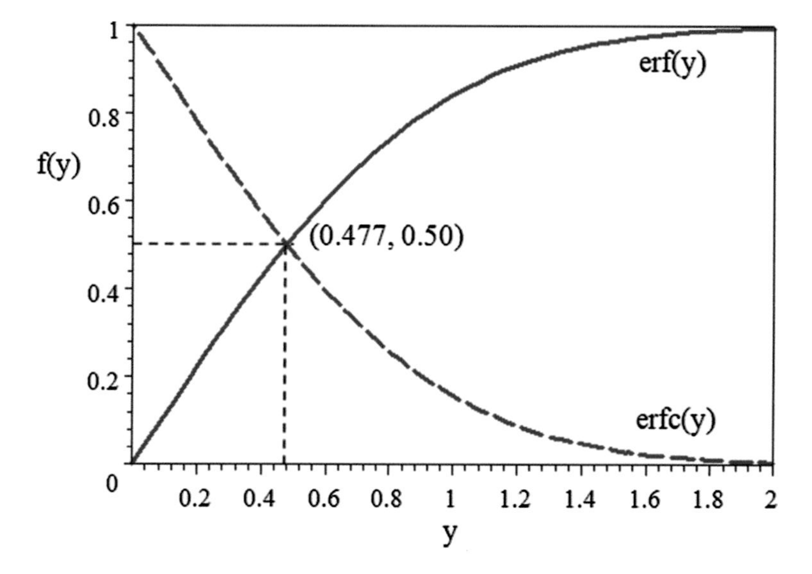

Fig. 6.5 Error and complement function profiles

$$\operatorname{erf} c\,(y) = \frac{2}{\sqrt{\pi}} \int_{y}^{\infty} \exp\left(-y^2\right) dy \tag{6.44}$$

The function erf (y) values are illustrated in Table 6.1, and Fig. 6.5 depicts both erf (y) and erf c (y) profiles.

The graphical application of Eq. (6.36) is schematically shown in Fig. 6.6. Note that the curves are displaced upward with increasing time and downward with increasing distance x.

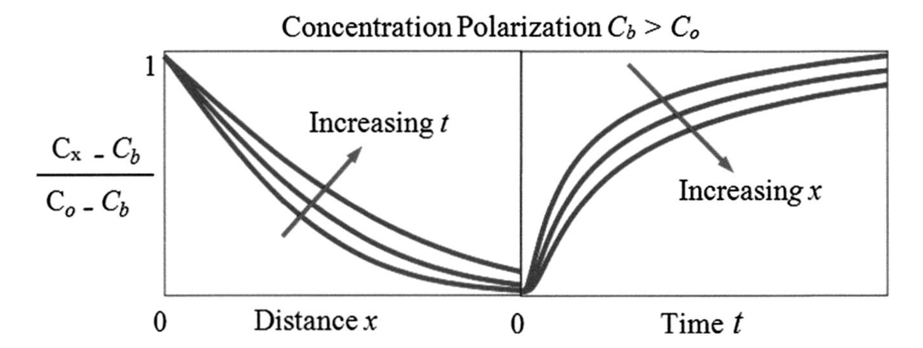

Fig. 6.6 Normalized concentration for concentration polarization as per Eq. (6.36)

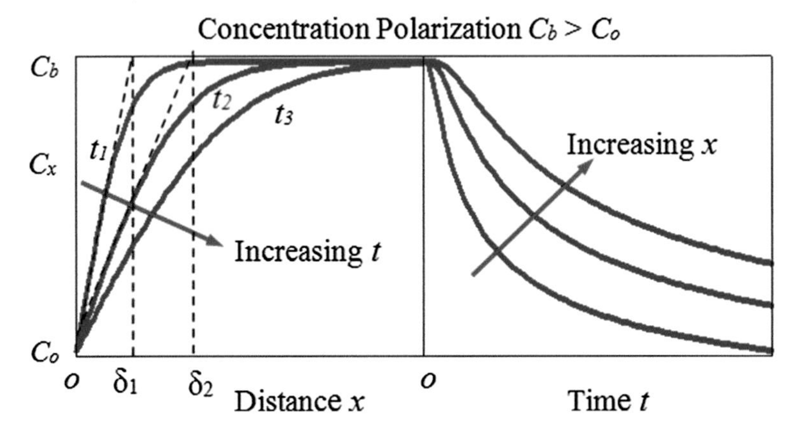

Fig. 6.7 Schematic concentration profiles for concentration polarization as per Eq. (6.37)

Furthermore, Eq. (6.37) is plotted in Fig. 6.7 for graphically defining the diffusion layer δ, which increases as the curve is displaced to the right with increasing time t [18–20].

In addition, the concentration gradient dC/dx for both concentration polarization and activation polarization can be determined from Eqs. (6.37) and (6.39).

For concentration polarization with $C = C_x$, the diffusion of the concentration gradient toward the electrode surface becomes

$$\frac{\partial C}{\partial x} = -(C_o - C_b)\frac{\partial}{\partial x}\left[\frac{2}{\sqrt{\pi}}\int_y^{\infty}\exp\left(-y^2\right)dy\right] \quad (6.45)$$

$$\frac{\partial C}{\partial x} = -\frac{(C_o - C_b)}{\sqrt{\pi D t}}\exp\left(-\frac{x^2}{4Dt}\right) \quad \text{for } C_b > C_o \quad (6.46)$$

$$\left.\frac{\partial C}{\partial x}\right|_{x=0} = -\frac{(C_o - C_b)}{\sqrt{\pi D t}} \approx +\frac{C_b}{\sqrt{\pi D t}} \quad \text{for } C_b >> C_o \quad (6.47)$$

Conversely, the diffusion of the concentration gradient with $C = C_x$ away from the electrode surface is

$$\frac{\partial C}{\partial x} = -(C_o - C_b)\frac{\partial}{\partial x}\left[\frac{2}{\sqrt{\pi}}\int_y^{\infty}\exp\left(-y^2\right)dy\right] \quad (6.48)$$

$$\frac{\partial C}{\partial x} = -\frac{(C_o - C_b)}{\sqrt{\pi Dt}}\exp\left(-\frac{x^2}{4Dt}\right) \quad \text{for } C_o > C_b \quad (6.49)$$

$$\left.\frac{\partial C}{\partial x}\right|_{x=0} = -\frac{(C_o - C_b)}{\sqrt{\pi Dt}} \approx -\frac{C_o}{\sqrt{\pi Dt}} \quad \text{for } C_o >> C_b \quad (6.50)$$

Substituting Eq. (6.47) into (6.18) yields the **Cottrell equation** for the current density due to diffusion mass transfer [6]:

$$i = \frac{zFDC_b}{\sqrt{\pi Dt}} = \frac{zFDC_b}{\delta} \quad (6.51)$$

$$\delta = \sqrt{\pi Dt} \quad (6.52)$$

where δ = thickness of the diffusion double layer. Experimentally, this parameter is $0.3\,\text{mm} \leq \delta \leq 0.5\,\text{mm}$ [21].

Furthermore, consider the species O being transported by diffusion to an electrode surface, where it reduces according to the reaction $O + ze^- = R$ at a temperature T [3]. This implies that the number of electrons being transferred to the electrode surface is proportional to the number of species O. As a result, the current density due to O mass transport is related to the diffusion flux under steady-state conditions. In this particular case, combining Eqs. (6.19) and (6.46) gives the current density as

$$i_c = -zFJ_x = -zFD\frac{C_o - C_b}{\sqrt{\pi Dt}}\exp\left(-\frac{x^2}{4Dt}\right) \quad (6.52a)$$

Evaluating Eq. (6.52a) on the electrode surface with $x = 0$ gives

$$i_c = -zFD\frac{C_o - C_b}{\sqrt{\pi Dt}} \quad (6.52b)$$

which resembles Cottrell expression, Eq. (6.51).

If more than one species is being transported to the electrode surface, then the current density becomes

$$i_c = -zF\sum_{j=1}^{N} J_{x,j} = -zF\sum_{j=1}^{N} D_j\frac{C_{j,o} - C_{j,b}}{\sqrt{\pi D_j t}}\exp\left(-\frac{x^2}{4D_j t}\right) \quad (6.52c)$$

$$i_c = -zF \sum_{j=1}^{N} J_{x,j} = -zF \sum_{j=1}^{N} D_j \frac{C_{j,o} - C_{j,b}}{\sqrt{\pi D_j t}} \text{ at } x = 0 \tag{6.52d}$$

Therefore, all species j contribute to the total current density.

Example 6.2. *Consider the application of Fick's second law to solve an activation polarization problem. An electrochemical cell contains a bulk concentration and an electrode surface concentration of copper ions Cu^{2+} equal to $4x10^{-4}\,mol/cm^3$ and $10^{-2}\,mol/cm^3$, respectively. If the diffusivity of copper ions is $1.5x10^{-6}\,cm^2/s$, calculate **(a)** the concentration of Cu^{2+} ions at $0.12\,mm$ from the electrode surface after 10 minutes. Plot $C_x = f(t)$ in interval $[0, 200]\,sec$ at $x = 0.12\,mm$. How does this function behave? Plot $C_x = f(x)$ in interval $[0, 0.2]\,mm$ at $t = 180\,sec$. How does this function behave now? Determine **(b)** the time for a concentration of $10^{-3}\,mol/cm^3$ of Cu^{2+} ions at $0.10\,mm$ from the electrode surface. The anodic reaction for copper is $Cu \rightarrow Cu^{2+} + 2e^-$.*

Data: $D = 1.5x10^{-6}\,cm^2/s$ $\quad C_b = 4x10^{-4}\,mol/cm^3$ @ $x = \infty$

$z = 2$ $\quad C_o = 10^{-2}\,mol/cm^3$ @ $x = 0$

(a) $C_x = ?$ @ $x = 0.12\,mm$, $t = 10$ min $= 600\,s$

$$y = x/\sqrt{4Dt} = (0.012\,\text{cm})/\sqrt{(4)\,(1.5x10^{-6}\,\text{cm}^2/\text{s})\,(600\,\text{s})} = 0.20$$

From Table 6.1,

$$\text{erf}\,(y) = \text{erf}\,(0.20) = 0.22$$

From Eq. (6.39) with $C_o > C_b$,

$$C_x = C_o - (C_o - C_b)\,\text{erf}\left(x/\sqrt{4Dt}\right)$$

$$C_x = 10^{-2}\text{mol/cm}^3 - \left(10^{-2}\,\text{mol/cm}^3 - 4x10^{-4}\,\text{mol/cm}^3\right)(0.22)$$

$$C_x = 7.89x10^{-3}\text{mol/cm}^3$$

For plotting $C_x = f(t)$ in the interval $[0, 200]\,s$ at $x = 0.12\,mm$,

$$C_x = 10^{-2} - \left(10^{-2} - 4 * 10^{-2}\right)\text{erf}\left(\frac{0.12}{\sqrt{4 * 1.5 * 10^{-5} t}}\right) \; in \text{ mol/cm}^3$$

$$C_x = 10 + 30.0\,\text{erfc}\left(\frac{0.12}{\sqrt{0.000\,06t}}\right) \; with \; (x10^{-3}\text{mol/cm}^3)$$

The plot is is given below. For plotting $C_x = f(x)$ in the interval $[0, 0.2]\,mm$ at $t = 180\,s$,

$$C_x = 10^{-2} - \left(10^{-2} - 4 * 10^{-2}\right) \operatorname{erf}\left(\frac{x}{\sqrt{4 * 1.5 * 10^{-5}\,(180)}}\right) \quad in\ \mathrm{mol/cm^3}$$

$$C_x = 10 + 30.0\,\operatorname{erfc}\left(\frac{x}{\sqrt{0.000\,06\,(100)}}\right) \quad with\ (x10^{-3}\mathrm{mol/cm^3})$$

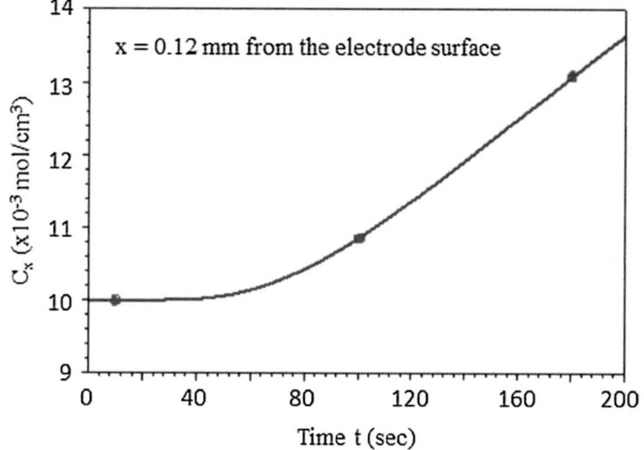

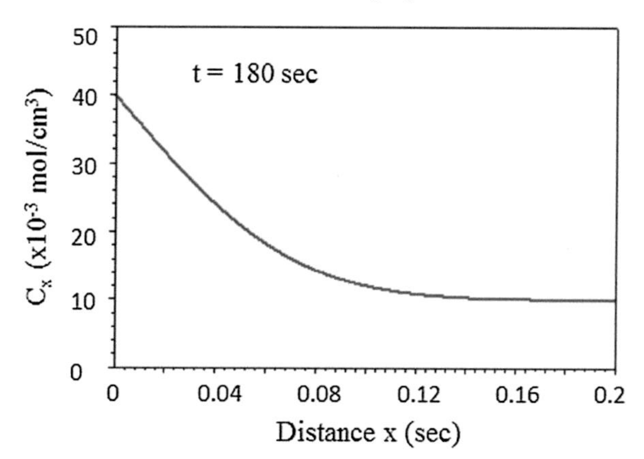

(b) $t = ?$ @ $C_x = 10^{-3}\,mol/cm^3$, $x = 0.10\,mm$ *and from Eq. (6.38)*

$$(C_x - C_o)\,/\,(C_b - C_o) = \operatorname{erf}\left(x/\sqrt{4Dt}\right)$$

$$(C_x - C_o)\,/\,(C_b - C_o) = \left(10^{-3} - 10^{-2}\right)/\left(4x10^{-4} - 10^{-2}\right)$$

$$(C_x - C_o)\,/\,(C_b - C_o) = 0.94 = \operatorname{erf}(y)$$

From Table 6.1,

$$y = 1.33 = x/\sqrt{4Dt}$$

Then

$$t = \frac{1}{4D}\left(\frac{x}{y}\right)^2 = \left\{1/\left[(4)\left(1.5x10^{-6}\,\text{cm}^2/\text{s}\right)\right]\right\}(0.010\,\text{cm}/1.33)^2$$
$$t = 9.42\,\text{s}$$

Therefore, 9.12 s is not sufficient time for a significant diffusion.

Example 6.3. *Assume that two columns of ionic solutions are connected by a barrier, such as a diaphragm. If the barrier is suddenly removed at time $t = 0$, free diffusion takes place. Derive an expression for* **(a)** *the concentration C_x at $x = 0$ and* **(b)** *the molar flux J_x in the x-direction at $x = 0$. Given data: $D = 10^{-5}\,cm^2/s$, $C_o = 10^{-4}\,mol/cm^3$.*

Solution. **(a)** *From Appendix A along with $y = x/\sqrt{4Dt}$*

$$\int_o^{C_o} dC = B\int_{-\infty}^{\infty}\exp\left(-y^2\right)dy$$

But, $B = C_o/\sqrt{\pi}$ (given)

$$\int_o^{C_x} dC = B\int_y^{\infty}\exp\left(-y^2\right)dy$$
$$C_x = \left(C_o/\sqrt{\pi}\right)\left(\sqrt{\pi}/2\right)\text{erf}\,c(y)$$
$$C_x = (C_o/2)\,\text{erf}\,c(y)$$
$$C_x = (C_o/2)\,[1 - \text{erf}(y)]$$

Then,

$$C_x = (C_o/2)\left[1 - \text{erf}(x/\sqrt{4Dt})\right]$$
$$C_x = C_o/2 \quad @\; x = 0$$

(b) *From part (a),*

$$C_x = (C_o/2)\,\text{erf}\,c(y)$$
$$C_x = -\frac{C_o}{2}\frac{2}{\sqrt{\pi}}\int_y^{\infty}\exp\left(-y^2\right)dy$$

But, $y = x/\sqrt{4Dt}$ & $dy = dx/\sqrt{4Dt}$

$$dC_x/dx = -\left(\frac{C_o}{\sqrt{4\pi Dt}}\right)\exp\left(-x^2/4Dt\right)$$
$$dC_x/dx = -C_o/\sqrt{4\pi Dt} \quad @\; x = 0$$

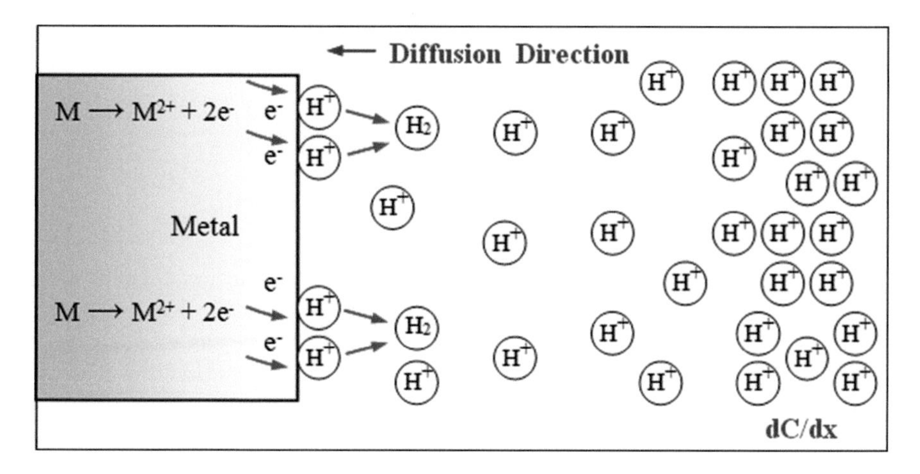

Fig. 6.8 Schematic model for concentration polarization [1]

From Eq. (6.3),

$$J_x = -D\,(dC_x/dx)$$

$$J_x = C_o\sqrt{D/\,(4\pi t)}\quad @\ x = 0$$

Note that the diffusion molar flux J_x is strongly dependent on diffusion time, specifically at very small periods of time.

6.2.3 Concentration Gradient Model

Recall that polarization in the electrochemical field is understood as a deviation of the electrochemical equilibrium potential (E_{corr}). On the other hand, concentration polarization in electrochemistry occurs from changes in the electrolyte concentration during current flow through the electrolyte. As a result, the rate of reaction at the electrode surface is faster than the rate of mass transport by diffusion and by migration.

The goal in this section is to elucidate the mechanism of diffusion due to an existing concentration gradient. Let us use Fontana's classical model shown in Fig. 6.8 for diffusion of hydrogen anions due to a concentration gradient [1].

This model suggests that the ionic concentration in the bulk of solution is higher than that on the electrode surface. Therefore, $C_b > C_o$ meaning that the diffusion direction of hydrogen cations is toward the electrode surface where anodic and cathodic reactions take place. Hence, the diffusion molar flux is determined by Fick's first law equation under steady-state conditions. It can be assumed that no other mass transfer phenomenon is present.

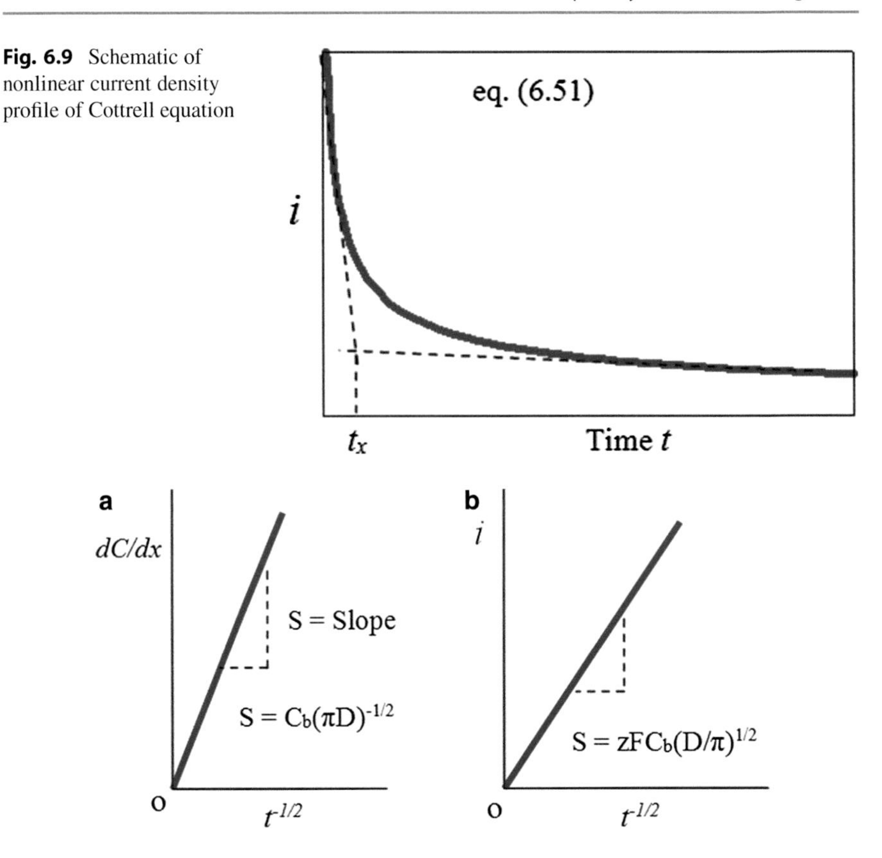

Fig. 6.9 Schematic of nonlinear current density profile of Cottrell equation

Fig. 6.10 Schematic concentration gradient, Eq. (6.47), and current density, Eq. (6.51), on the electrode surface at x = 0. (**a**) Concentration gradient and (**b**) current density

Assuming that all variables in Eq. (6.51) are constant, except time t, one can predict the current density profile as a nonlinear behavior for a particular electrolyte. This is shown in Fig. 6.9.

Linearization of Eqs. (6.47) and (6.51) is shown in Fig. 6.10. For a concentration gradient profile at the electrode surface, where the concentration gradient is $dC/dx = f(x = 0, t)$, the diffusivity D and the valence z can be determined from the slope of Fig. 6.10a and b, respectively.

Furthermore, the area under the curve in Fig. 6.9 represents the electrical charge (q). Hence,

$$q = \int iA_s dt = \int I dt \tag{6.53}$$

where A_s = electrode surface area.

Inserting Eq. (6.51) into (6.53) and solving the integral yield

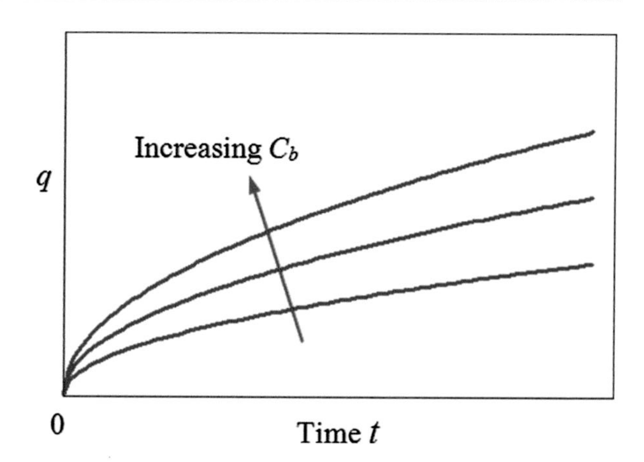

Fig. 6.11 Schematic nonlinear profiles of the electrical charge (q)

$$q = 2zFC_bA_s\sqrt{\frac{Dt}{\pi}} \tag{6.54}$$

If electrons are transferred through a wire due to a potential difference (E) between electrodes, then the electrical work (w_e) is

$$w_e = qE \tag{6.55a}$$

$$w_e = 2zFEC_bA_s\sqrt{\frac{Dt}{\pi}} \tag{6.55b}$$

Figure 6.11 illustrates ideal electrical charge profiles as per Eq. (6.54). The curves are displaced upward with increasing original bulk concentration C_b.

In comparison, species in gases and liquids are always in motion from one place to another, while in solids they oscillate around their lattice positions. Since diffusion in aqueous medium is the main subject in this section, the concentration of a species j can be smoothed out by diffusion since the mass transfer occurs from the high- to low-concentration region as in cathodic processes under concentration polarization.

Maintaining a continuous supply and removal of a species j in a electrolyte at a constant temperature, a steady state is achieved, and D can be experimentally determined since $\partial C/\partial t = 0$; otherwise, a nonsteady state or transient diffusional process arises, and the concentration rate becomes $\partial C/\partial t > 0$ [22].

In general, methodological sources for determining the diffusivity can be found elsewhere [22, 23]. In order for the reader to have a better understanding on diffusion, simple and yet important diffusion processes can be generalized by connecting a system A and a system B together at the same temperature T and pressure P. This causes a mixture of species A and B until a uniform composition is achieved due to diffusion of A into B and vice versa. Then, the average composition of the mixture is $C = (C_A + C_B)/2$.

Below are some common diffusion cases for mixing species A and B:

- Diffusion of gases, such as helium and argon to form a mixture in a closed vessel at temperature T and pressure P. As a result, the mixing is a mass transport phenomenon that does not require bulk motion. For example, mixing concentrated ammonia and hydrochloric acid solutions forms solid ammonium chloride. Nonetheless, molecules diffuse from high- to low-concentration areas.
- Diffusion of liquids, such as water and alcohol, water and liquid soap, food coloring and water, and so forth. Recall that diffusion is a process through which a species j moves from high- to low-concentration areas.
- Diffusion in solids, such as a bar of copper (Cu) and a bar of nickel (Ni) clamped together in a furnace at 1050 °C. Eventually, Cu and Ni atoms exchange atomic positions to form an alloy composition at the interface. Carburization is also a solid-state diffusion during which carbon atoms diffuse into steel to an extent, hardening the steel surface.

6.2.4 Stationary Boundaries

The open-ended capillary technique is used for determining the diffusivity of self-diffusing ions in dilute solutions [23]. The capillary shown in Fig. 6.12 has stationary boundaries in the liquid medium and an open upper end where ion (solute) interactions take place due to diffusion.

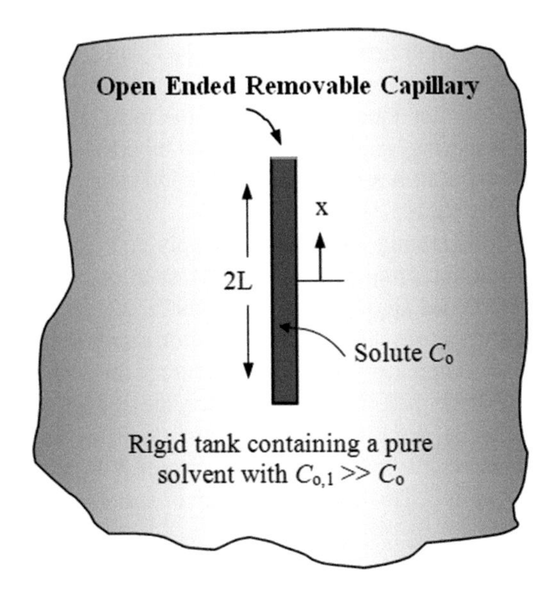

Fig. 6.12 Schematic sketch for open-ended capillary

The solution to this diffusion problem is based on the boundary conditions given below for nonsteady state diffusion, Eq. (6.29):

$$C = C_o \quad @\ x = 0,\ t \geq 0 \tag{a}$$

$$\partial C/\partial x = 0 \quad @\ x = 0,\ t \geq 0 \tag{b}$$

$$C = 0 \quad @\ x = L,\ t \geq 0 \tag{c}$$

$$C = C_x \quad @\ -L < x < L \tag{d}$$

Note that condition (b) dictates that diffusion must not occur across the central plane at $x = 0$. For nonsteady state diffusion of ions contained in a volume capillary with known initial concentration C_o, the general concept for an experimental setup is schematically shown in Fig. 6.10. The capillary is attached to other components for easy handling during immersion and withdrawing from the solution. The reader wishing to obtain design details of this capillary should consult the work of Kumar and Gupta [23] and Anderson and Saddington [24].

In this technique, the ions or species j in the capillary and in the solvent diffuse through the open-ended space at an infinite time. After time t has elapsed, the capillary is removed from the system for determining the final concentration and the concentration ratio as $C = f(x, t)$ and $C/C_o = f(x, t)$, respectively. The former is related to the diffusivity D.

The solution of Eq. (6.29) is based on methods such as Laplace transform and separation of variables. Using either method yields the solution of Eq. (6.29) as normalized concentration equation [15]:

$$\frac{C - C_o}{C_1 - C_o} = 1 - \frac{4}{\pi}\sum_{n=0}^{\infty}\frac{(-1)^n}{2n+1}\exp\left[-\frac{(2n+1)^2\pi^2 Dt}{4L^2}\right]\cos\left[\frac{(2n+1)\pi x}{2L}\right] \tag{6.56}$$

If $C_1 = 0$, then Eq. (6.56) reduces to

$$\frac{C}{C_o} = -\frac{4}{\pi}\sum_{n=0}^{\infty}\frac{(-1)^n}{2n+1}\exp\left[-\frac{(2n+1)^2\pi^2 Dt}{4L^2}\right]\cos\left[\frac{(2n+1)\pi x}{2L}\right] \tag{6.57}$$

The accuracy of Eq. (6.57) depends on the number of terms used to carry out the series. Assuming that one term in this series yields accurate enough results, then the diffusivity becomes

$$\frac{C}{C_o} = -\frac{4}{\pi}\exp\left(-\frac{\pi^2 Dt}{4L^2}\right)\cos\left(\frac{\pi x}{2L}\right) \quad \text{for } n = 0 \tag{6.58a}$$

$$D = \frac{4L^2}{\pi^2 t}\ln\left[\frac{\pi C}{4C_o}\sec\left(\frac{\pi x}{2L}\right)\right] \quad \text{for } C > C_o\ \&\ t > 0 \tag{6.58b}$$

Letting $x = L/2$ in Eq. (6.58b) yields $\sec(\pi/4) = \sqrt{2}$ and

$$D = \frac{4L^2}{\pi^2 t} \ln\left[\frac{\pi\sqrt{2}C}{4C_o}\right] \tag{6.59}$$

Consequently, $D = f(C_o/C)$ at $t > 0$ as previously assumed. For convenience, the concentration C can be defined from Eq. (6.59) as

$$C = \frac{4C_o}{\pi\sqrt{2}} \exp\left(\frac{\pi^2 Dt}{4L^2}\right) \tag{6.60}$$

Using

$$\exp(x) = \sum_{k=0} \frac{(-x)^k}{k!} \simeq 1 - x \tag{a}$$

on Eq. (6.60) yields

$$C \simeq \frac{4C_o}{\pi\sqrt{2}}\left(1 - \frac{\pi^2 Dt}{4L^2}\right) \tag{6.61}$$

$$C \simeq \frac{C_o}{\pi\sqrt{2}L^2}\left(4L^2 - \pi^2 Dt\right) \tag{6.62}$$

If $4L^2 >> \pi^2 Dt$, then

$$C \simeq \frac{4C_o}{\pi\sqrt{2}} = \frac{2\sqrt{2}}{\pi} C_o = 0.9C_o \tag{6.62a}$$

Example 6.4. *Plot Eq. (6.59) as $D/L^2 = f(C_o/C, t)$ in order to determine which variable has a strong effect on the diffusion coefficient D. The result should be a 3D surface. Also, plot $D/L^2 = f(t)$ for several C_o/C values to reveal the normalized diffusion coefficient behavior as a time-dependent variable.*

Solution. *The plot is given below without scales for convenience. It is clear that time t has a stronger effect on the normalized diffusion coefficient D/L^2 than the concentration ratio C_o/C. However, D/L^2 is strongly affected at short times since $D/L^2 \propto 1/t$. In general, this graphical representation in three dimensions (3D) is called a surface plot given by*

$$\frac{D}{L^2} = \frac{4}{\pi^2 t} \ln\left[\frac{\pi\sqrt{2}C}{4C_o}\right]$$

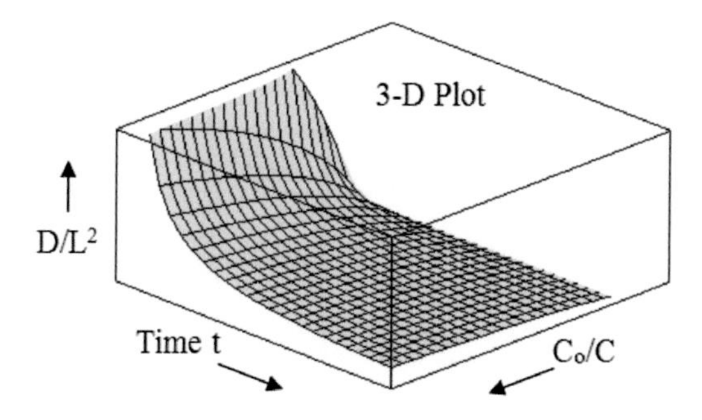

A two-dimensional plot along with $C_o/C \geq 1$ is also convenient. Thus,

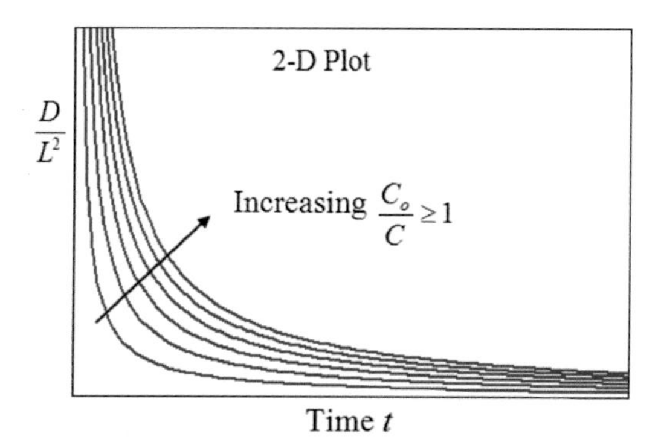

Notice that D/L^2 decreases very significantly at the initial state of the mass transport by diffusion and, subsequently, it exhibits a asymptotic behavior as a species j gets depleted in the solution. Therefore, diffusion of j is strongly significant at the early stage of mass transport.

6.3 Moving Boundary Diffusion

It is known that diffusion in liquids is usually faster than in solids of the same solute at the same temperature. However, the liquid geometry is not well understood because of the imperfect atomic arrangement. Despite of this problem, Einstein derived the diffusivity based on Brownian motion of fine particles in liquid water by assuming random direction and jump length of particles [25]. The main objective of a diffusion problem is to understand the problem and, subsequently, determine the proper boundary conditions needed for developing a suitable mathematical solution of Fick's second law equation. The boundary conditions, however, depend on the electrochemical system and the type of diffusion taking place.

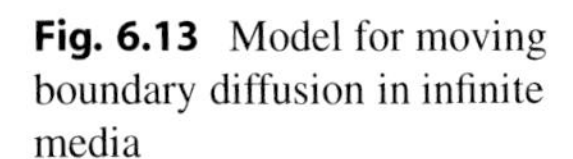
Fig. 6.13 Model for moving boundary diffusion in infinite media

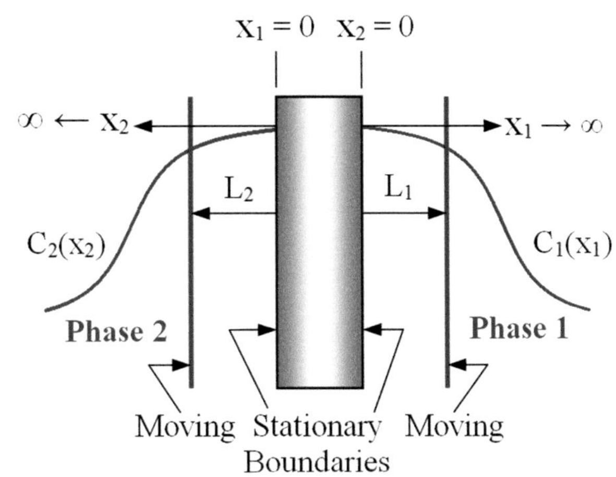

In this section, surface boundary motion is considered as an essential feature for solving such an equation. Thus, diffusion accompanied by an electrode thickness increment due to diffusion of metal cations (solute) and immobile species is considered to be a particular diffusion problem that resembles mass transfer diffusion toward the faces of cathode sheets in an electrowinning (EW) cell.

Figure 6.13 shows Crank's idealized model [15] for moving boundaries toward an aqueous bulk of solution and schematic concentration distribution. This diffusion problem requires that the medium be free of convective mass transfer and that boundary motion occurs along stationary x_1 and x_2 axes. In this case, the boundary motion occurs perpendicular to x_1 and x_2.

Metal diffusion can be simulated by letting a stationary electrode sheet be the cathode and be immersed in an electrolyte at temperature T and pressure P. A film of cationic solute M^{z+} in phase 1 (Fig. 6.13) deposits on the sheet by diffusion on the outer surfaces, and the film is constantly saturated with the electrolyte. Film thickness growth occurs in the x_1 and x_2 directions having a thickness L_1 in phase 1 and L_2 in phase 2, respectively. Thus, the spaces occupied by the deposited film on each side of the sheet are $0 \le x_1 \le L_1$ and $-L_2 \le x_2 \le 0$. Thus, the motion of one or two phases relative to the boundary is caused by the mass transfer diffusion across the interface. This diffusion problem resembles the tarnishing reactions forming an oxide film on a metal surface [15].

Assuming that Fick's second law is obeyed in the media along the interfaces of the moving film boundaries (Fig. 6.13) yields

$$\frac{\partial C_1}{\partial t} = D\frac{\partial^2 C_1}{\partial x^2} \quad (6.63)$$

$$\frac{\partial C_2}{\partial t} = D\frac{\partial^2 C_2}{\partial x^2} \quad (6.64)$$

According to the following boundary conditions

$$\underline{x_1\text{-direction}} \tag{6.65}$$

$$C_1 = C_{o,1} \quad \text{for } x_1 = 0, \ t = 0$$

$$C_1 = 0 \quad \text{for } x_1 = \infty, \ t > 0$$

$$\underline{x_2\text{-direction}} \tag{6.66}$$

$$C_2 = C_{o,2} \quad \text{for } x_2 = 0, \quad t = 0$$

$$C_2 = 0 \quad \text{for } x_2 = -\infty, \ t > 0$$

and assuming symmetry during diffusion on both sides of the moving boundaries so that $C_o = C_{o,1} = .C_{o,2}$, $x = x_1 = x_2$, and $D = D_1 = D_2$, the solution of Eqs. (6.63) and (6.64) for infinite media are easily obtainable using the analytical procedure included in Appendix A. Hence,

$$\frac{C}{C_o} = \operatorname{erf} c\left(\frac{x}{\sqrt{4Dt}}\right) \tag{6.67}$$

Thus,

$$C = -\frac{C_o}{\sqrt{\pi Dt}} \int_{\infty}^{y} \exp\left(-\frac{x^2}{4Dt}\right) dx \tag{6.68}$$

$$\frac{\partial C}{\partial x} = -\frac{C_o}{\sqrt{\pi Dt}} \exp\left(-\frac{x^2}{4Dt}\right) \tag{6.69}$$

where $y = x/\sqrt{4Dt}$.

Substituting Eq. (6.69) into (6.16) gives

$$J_x = \frac{DC_o}{\sqrt{\pi Dt}} \exp\left(-\frac{x^2}{4Dt}\right) \tag{6.70}$$

from which the rate of mass transfer is deduced as

$$\frac{dM}{dt} = A_s A_w J_x \tag{6.71a}$$

$$\frac{dM}{dt} = A_s A_w \frac{DC_o}{\sqrt{\pi Dt}} \exp\left(-\frac{x^2}{4Dt}\right) \tag{6.71b}$$

Example 6.5. *Use the data given in Example 6.3 and a diffusivity of* $10^{-5}\ cm^2/s$ *and* $0 \le x \le 0.14\,cm$ *and* $t = 1\,h$ *to determine the diffusion molar flux* (J_x) *profile as a function of distance from the electrode surface. Explain the resultant curve with respect to the effects of time and distance on the diffusion molar flux.*

Solution. *Given data:*

$$A_s = 53.10x10^4\,\text{cm}^2,\ A_w = 58.71\,\text{g/mol},\ D = 10^{-5}\,\text{cm}^2/\text{s}$$

$$C_o = 5.96 \times 10^{-4}\,\text{mol/cm}^3,\ t = 3600\,\text{s},\ z = 2,\ E = 2\,\text{V}$$

From Eq. (6.70), the flux is

$$J_x = C_o\sqrt{\frac{D}{\pi t}}\exp\left(-\frac{x^2}{4Dt}\right)$$

$$J_x = \left(5.96 \times 10^{-4}\,\text{mol/cm}^3\right)\sqrt{\frac{0^{-5}\,\text{cm}^2/\text{s}}{\pi t}}\exp\left[-\frac{x^2}{4\,(10^{-5}\,\text{cm}^2/\text{s})\,t}\right]$$

$$J_x = \left(1.0633 \times 10^{-6}\right)\sqrt{1/t}\exp\left[-25000x^2/t\right]$$

Substituting the given time periods in seconds and adjusting the y-axis scale give the following three equations for J_x in units of mol/cm^2 s. Thus,

$$J_x = (1.2532)\exp\left(-3.4722x^2\right)\quad \textit{for}\ t = 1.0\,\text{h}$$

$$J_x = (1.4470)\exp\left(-4.6296x^2\right)\quad \textit{for}\ t = 1.5\,\text{h}$$

$$J_x = (1.7722)\exp\left(-6.9444x^2\right)\quad \textit{for}\ t = 2.0\,\text{h}$$

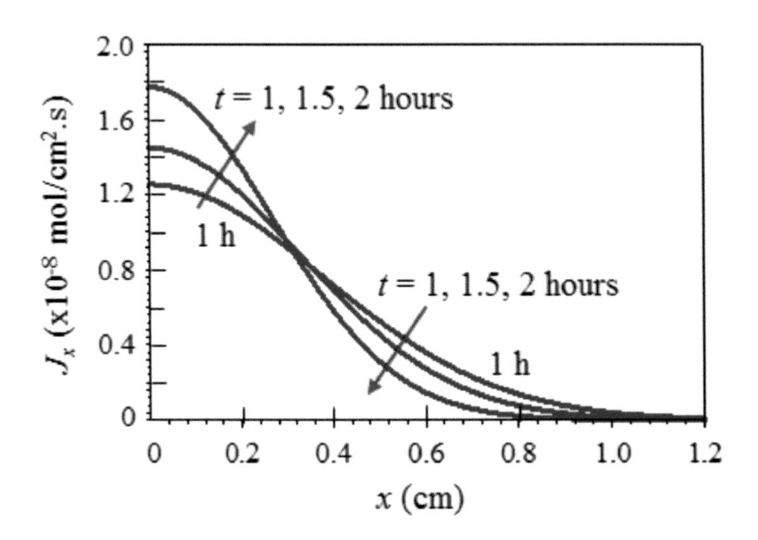

Therefore, the molar flux, J_x, diffusing perpendicularly through a unit cross-sectional area of electrolyte per unit time strongly depends on the diffusion distance x According to the scale being used for the above figure, the plots indicate that $J_x \to 0$ as $x \to 1$ cm.

6.4 Migration Molar Flux

Migrating of charged ions in an electric field strength accumulates at the electrode-electrolyte interface or deposit on the electrode surface as in electrodeposition processes, such as electroplating, electrowinning, and electrorefining. Specifically, mass transport in solid phases, such as polymers and carbon into steel during carburization, is much slower than in aqueous electrolytes, but it is a relevant mechanism used during charging and discharging of batteries.

Nonetheless, the movement of ions due to diffusion and migration mechanisms in electrolyte solutions between anodic and cathodic electrodes defines the ionic mass transport having certain properties. For instance, diffusion is due to a concentration gradient of all ions and migration to an electric field effects on charged species [3]. However, in principle, ion migration from low-concentration to high-concentration sites within the electrolyte is possible, provided that an external current supplies the driving force for this particular process, inducing migration into the bulk electrolyte solution [26].

In the presence of an electric field, the electric field strength, as defined in Chap. 2, is $E = \partial\phi/\partial x$, and assuming there is an initial concentration gradient of species j in the electrolyte, the total molar flux for a species j due to diffusion and migration mechanisms is defined as

$$J_{x,j} = -D_j \frac{\partial C_j}{\partial x} - \frac{z_j F D_j C_j}{RT} \frac{\partial \phi}{\partial x} \tag{6.72}$$

In principle, the driving forces for this coupled molar flux are the concentration gradient $\partial C_j/\partial x$ and the electric potential gradient $\partial\phi/\partial x$. A description of opposing the electric force E can be found elsewhere [3].

Mass transport in an electrochemical cell arises because of an electric field exerting electric forces on the mobile ions. In such a case, the migration molar flux representing the mass transport mode can be defined as [18]

$$J_m = -CB \frac{\partial \mu}{\partial x} \tag{6.73}$$

where B = ionic mobility $\left[\text{cm}^2/\text{V s} = \text{cm}^2\,\text{mol}/\,(\text{J s})\right]$

$\partial\mu/\partial x$ = chemical potential gradient [J/ (cm mol)]

Only the chemical potential gradient, $d\mu/dx$, is included in Eq. (6.73) since it strongly contributes to J_m. The influence of other contributing gradients to the total migration flux, such as concentration gradient (dC/dx), pressure gradient (dP/dx), and temperature gradient (dT/dx), is negligible in Eq. (6.73). Therefore, Eq. (6.73) represents an approximation mathematical model where the chemical potential gradient is given by

$$\frac{d\mu}{dx} = zq_e \frac{\partial \phi}{\partial x} \tag{6.74}$$

where $q_e = 1.6022x10^{-19}$ Coulomb (= J/V). Substituting Eq. (6.74) into (6.73) yields the migration molar flux in terms of electric potential gradient:

$$J_m = -zq_eCB\frac{\partial \phi}{\partial x} \tag{6.75}$$

The ionic mobility is defined as [16]

$$B = \frac{v}{F_x} \tag{6.76}$$

where v = kinematic velocity of the flowing species j (cm/s)
F_x = electric potential gradient acting on the species j (V/cm)

A more convenient approach to derive B is by combining Eqs. (6.4) and (6.75) along with Eq. (2.3), $F = q_eN_A$. Thus,

$$B = \frac{DN_A}{RT} = \frac{D}{\kappa T} \tag{6.77}$$

where T = absolute temperature
$R = 8.314$ J/mol K = gas constants
$N_A = 6.02213x10^{23}$ ions/mol = Avogadro's number
$\kappa = R/N_A = 1.38x10^{-23}$ J/K = Boltzmann constant

If $D \approx 10^{-5}$ cm^2/s for most ions [5], then the ionic mobility becomes $B \approx 2.43x10^{15}$ cm^2/J s at $T = 298$K. Solving Eq. (6.77) for D yields the **Nernst-Einstein equation** [7]:

$$D = B\kappa T \tag{6.78}$$

for a species j being treated as a spherical particle. Consider Einstein's assumption that spherical particles (ions) move through a continuous medium (solution) of viscosity η_v under **self-diffusion** conditions and that Stoke's drag or friction force F_x per unit length acts on these particles, and neglecting interatomic forces, then F_x is [7]

$$F_x = 6\pi R_i \eta_v v \tag{6.79}$$

then B and D become

$$B = \frac{1}{6\pi R_i \eta_v} \tag{6.80}$$

$$D = \frac{\kappa T}{6\pi R_i \eta_v} \tag{6.81}$$

where R_i = radius of spherical particles (cm)
η_v = viscosity $\left(\text{J s/cm}^3 \text{ or g s/cm}^2\right)$

6.5 Kinetics and Mass Transfer

This section describes the mass transfer to a planar and rotating electrodes by diffusion and migration mechanisms. Assume a coupled diffusion and migration mass transfer and neglect any reaction rate controlling the chemical process and that there is no convective motion of diluted electrolyte solutions. The coupled current density for this combined mass transfer is based on one-dimensional analysis under steady-state conditions.

In an electrochemical system, the Faradaic current density (i) that flows at any time is assumed to be a measure of the rate of the electrochemical reaction taking place at the electrode surface. Actually, the current density depends upon the rate mass transfer from the bulk solution and the rate of electron transfer across the electrode-solution interface (also known as the interfacial region).

It is important to recognize that the concentration of species, the concentration gradient, and the current density depend on position and time; that is,$C = f(x,t)$, $dC/dx = f(x,t)$ and $i = f(t)$. All these variables decay with time at a fixed distance from the electrode surface since they are proportional to the inverse square root of time. The previous analytical approach is the outcome of a potential-control process for which the current response is measured at a particular temperature. As of now, this analytical work has been done for determining the steady current density response as expressed by Eq. (6.19), which is the result of the concentration gradient that develops in the electrolyte due to kinetic mass transfer according to a simple reaction such as $M^{z+} + ze^- = M$, in which M^{z+} is reduced to metal M. Further analysis of the mass transfer in electrolytes subjected to potential-control process suggests that the current density response, Eq. (6.19), may solely be by natural diffusion. However, a potential-control electrochemical system generates an electric field, and it is apparent that both diffusion and migration molar fluxes are related to the concentration and potential gradients, respectively. The latter has a strong effect on the magnitude of the current density. Thus, the Nernst-Plank equation, Eq. (6.78), without the convective molar flux term may provide adequate results in most simplified models, which consider constant diffusion coefficients for particular ions. If the electric field is weak and the migration molar flux approaches zero ($J_m \to 0$), then the diffusion molar flux (J_d), as predicted by Fick's first law, is the controlling step for mass transfer under steady-state condition, in which the concentration rate is $\partial C/\partial t = 0$. On the other hand, if $\partial C/\partial t \neq 0$, it implies that diffusion molar flux is time-dependent, and it is described by Fick's second law of diffusion.

For a one-dimensional ionic transfer under electroneutrality condition, that is, $\sum z_j C_j = 0$, without the influence of an external current or potential source and convection effects, the total current density, Eq. (6.11), due to a coupled natural diffusion and natural migration mass transfer of a particular ion reduces to

$$i = zF\left[J_d + J_m\right] = zFD\frac{dC}{dx} + \frac{zFDC}{RT}\frac{\partial \phi}{\partial x} \tag{6.82}$$

and the concentration rate becomes [9]

$$\frac{\partial C}{\partial t} = D\left(\frac{\partial^2 C}{\partial x^2} + \frac{zF}{RT}\frac{\partial C}{\partial x}\frac{\partial \phi}{\partial x}\right) \quad \text{for nonporous media} \tag{6.83}$$

$$P\frac{\partial C}{\partial t} = D\left(\frac{\partial^2 C}{\partial x^2} + \frac{zF}{RT}\frac{\partial C}{\partial x}\frac{\partial \phi}{\partial x}\right) \quad \text{for porous media} \tag{6.84}$$

where $P =$ porosity

Actually, these expressions, in one form or another, have been used by Lorente [9] to optimize the ionic transfer by electrokinetic through porous media, such as reinforced concrete, and by Samson et al. [5] for determining the chloride ionic diffusion coefficient in reinforced concrete using migration tests.

If the interfacial potential gradient in Eq. (6.82) is $\partial\phi/\partial x = 0$, then natural (pure) diffusion controls the transport of ionic species due to an existing concentration gradient $\partial C/\partial x$. However, the effect of the interfacial potential (membrane potential) that arises from the action of ion transporters embedded in the electrolyte should not be neglected because this potential will always exist as long as there are anodic and cathodic sites.

Use Eq. (6.82) to take the derivative of i with respect to x, assume that ϕ depend on the coordinate x only, and set $di/dx = 0$ so that

$$\frac{d^2\phi}{dx^2} + \frac{1}{\sqrt{\pi D t}}\frac{d\phi}{dx} = 0 \tag{6.85}$$

Solving this ordinary differential equation gives a theoretical definition of the interfacial potential gradient as

$$\frac{d\phi}{dx} + \frac{\phi}{\sqrt{\pi D t}} = 0 \tag{6.86}$$

The solution of Eq. (6.86) is

$$\phi = \phi_o \exp(-\frac{x}{\sqrt{\pi D t}}) = \phi_o \exp(-x/\delta) \tag{6.87}$$

which is similar to Eq. (4.1) with $\phi_o = RT/zF$. Here, δ is the thickness of the electrical double layer as defined by Eq. (6.52). If $x = 0$, then $\phi = \phi_o > 0$ at the electrode surface.

Furthermore, if $dC/dx = 0$ in Eq. (6.82), then the current density due to migration mass transfer becomes

$$i_m = \frac{zFDC_o}{\delta} = \frac{zFDC_o}{\sqrt{\pi D t}} \quad \text{for } t > 0 \tag{6.88}$$

and that for diffusion mass transfer along with $d\phi/dx = 0$ is

$$i_d = \frac{zFDC_o}{\delta} = \frac{zFDC_o}{\sqrt{\pi Dt}} \qquad \text{for } t > 0 \tag{6.89}$$

Hence, for a coupled diffusion and migration without an external electrical system, the total current density becomes

$$i = i_m + i_d = \frac{2zFDC_o}{\sqrt{\pi Dt}} \qquad \text{for } t > 0 \tag{6.90}$$

This mathematical approach is an approximation scheme or an engineered mechanism since the diffusion coefficient (D) has been assumed to be constant. For instance, in chloride contaminated steel-reinforced concrete, the diffusion coefficient (D) of chloride ions has been reported to depend on time [10] and concentration gradient (dC/dx) and electric potential gradient ($d\phi/dx$).

For a reversible electrochemical cell, the overpotential for concentration polarization is defined by Nernst equation. Using Eq. (3.31) with the solubility product $K_{sp} = C_x/C_o$ and the cathodic overpotential $\eta_c = E - E_o$ yields the cathodic overpotential in terms of concentration ratio:

$$\eta_c = -\frac{RT}{zF} \ln\left[\frac{C_x}{C_o}\right] \tag{6.91}$$

which indicates that η_c becomes more negative as the concentration C_x increases. Here, $C_x = C(x) > 0$ and $C_x = C(\infty) = C_b$ as the limited concentration in the bulk.

Evaluating Eq. (6.87) at $x = \infty$ and $x = 0$ yields the overpotential as

$$\eta_c = \phi(\infty) - \phi(0) \tag{a}$$

$$\eta_c = \phi_o \exp(-\infty) - \phi_o \exp(-0) \tag{b}$$

$$\eta_c = -\phi \exp(x/\delta) \tag{6.92}$$

since $\exp(-\infty) = 0$ and $\exp(-0) = 1$. Recall that $\delta = \sqrt{\pi Dt}$.

Combining Eqs. (6.91) and (6.92) yields the interfacial potential as

$$\phi = \frac{RT}{zF} \ln\left[\frac{C_x}{C_o}\right] \cdot \exp(-x/\delta) \tag{6.93}$$

From Eq. (6.91), the concentration for polarization becomes

$$C_x = C_o \exp\left(\frac{zF\eta_c}{RT}\right) \qquad \text{(Conc. Polarization, } C_b > C_o\text{)} \tag{6.94}$$

and for activation (act.) polarization is

$$C_x = C_o \exp\left(-\frac{zF\eta_a}{RT}\right) \quad \text{(Act. Polarization, } C_b < C_o\text{)} \tag{6.95}$$

from which the activation overpotential is defined as

$$\eta_a = \frac{RT}{zF} \ln\left[\frac{C_o}{C_x}\right] \tag{6.96}$$

The concentration (C_x) behavior, Eqs. (6.94) and (6.95), is now given by exponential functions, instead of the error function as previously treated in a diffusion process, Eqs. (6.37) and (6.39).

6.6 Limiting Current Density

The limiting current density is important in designing electrochemical cells for electrodeposition and in electrodialysis of water (for water desalination).

Assume that the mass transfer is due to diffusion and that the diffusion molar flux (J_x) is the chemical rate of the mass transfer under steady-state condition. Subsequently, the diffusion process can be described by Fick's first law of diffusion. Thus, Eq. (6.16) for reducing a metallic species j becomes

$$J_x = -D\frac{\Delta C}{\Delta x} = -D\left(\frac{C_x - C_o}{\delta}\right) \tag{6.97}$$

where $\Delta x = x - x_o = \delta = \sqrt{\pi D t}$ since $x_o = 0$ at the electrode surface.

Combining Eqs. (6.9) and (6.97) yields the cathodic current density for a steady-state condition with $\Delta C < 0$

$$i_c = -\frac{zFD\Delta C}{\delta} \tag{6.98}$$

Recall that zF product is the number of Coulombs required to convert 1 mole of cathodic reactant, such as hydrogen evolution ($H^+ + e = \frac{1}{2}H_2$ with one mole of H^+ and $z = 2$) and oxygen reduction reaction ($O_2 + 2H_2O + 4e = 4OH^-$ with one mole of dissolved O_2 and $z = 4$). Therefore, there must exist a maximum or limiting current density (i_L) defined as $i_L = -i_c$. This is accomplished when the concentration on the electrode surface approaches zero, $C_x \to 0$.

In light of the above, Eq. (6.98) along with since $C_o = 0$ at $x = 0$ (electrode surface) becomes

$$i_L = \frac{zFDC_o}{\delta} \tag{6.99}$$

This theoretical approach leads to changes in the surface concentration, and as a result, the cell potential defined by the Nernst potential equation, Eq. (3.31), undergoes changes as well since it is a concentration-dependent potential (and temperature dependent).

This leads to changes in the surface concentration and as a result, the cell potential defined by the Nernst potential equation undergoes changes as well since it is a concentration-dependent potential (and temperature-dependent).

With respect to Eq. (3.31), the electric potential is $E = f\left(T, K_{sp}\right)$, but $K_{sp} = f\left(C_{\text{product}}/C_{\text{reactants}}\right)$, and therefore, the Nernst potential becomes $E = f\left(T, C_{product}/C_{reac\tan ts}\right)$.

On the other hand, most corrosion problems emerge when the transport of cathodic species (dissolved O_2, ferric cations Fe^{3+}, hydrogen cations H^+ in acid solutions, etc.) is rate determining, and consequently, $i_L = i_{corr}$ at the corrosion point as schematically illustrated in Fig. 5.2.

In addition, if $i_L < i_{crit}$, then metal corrosion will occur, and if $i_L > i_{crit}$, passivation will take place [19]. Here, i_{crit} is a critical current density for passivation which will be dealt with in a later chapter.

Further, dividing Eq. (6.98) by (6.99) yields the current density ratio as

$$\frac{i_c}{i_L} = \frac{\Delta C}{C} = \frac{C_x - C_o}{C_x} = 1 - \frac{C_o}{C_x} \tag{6.100}$$

Thus, the concentration ratio becomes

$$\frac{C_o}{C_x} = 1 - \frac{i_c}{i_L} \tag{6.101}$$

Substituting Eq. (6.101) into (6.96) gives the overpotential for concentration polarization in terms of the limiting current density:

$$\eta_{conc} = \frac{RT}{zF} \ln\left(1 - \frac{i_c}{i_L}\right) \quad \text{for } i_L > i_c \tag{6.102}$$

$$\eta_{conc} = \frac{2.303RT}{zF} \log\left(1 - \frac{i_c}{i_L}\right) \tag{6.103}$$

If $(1 - i_c/i_L) = x$ and $\ln(x) \simeq x - 1$, then Eq. (6.102) becomes a linear approximation:

$$\eta_{conc} = \frac{RT}{zF}\left(\frac{i_c}{i_L} - 1\right) \tag{6.104}$$

Recall that the overpotential for activation polarization is given by Eq. (5.38), but it is renumbered here for convenience. Thus,

$$\eta_{act} = \beta_c \log\left(\frac{i_c}{i_o}\right) = -\frac{2.303RT}{(1-\alpha)\,zF}\log\left(\frac{i_c}{i_{corr}}\right) \tag{6.105}$$

with $i_o = i_{corr}$. Now, the total cathodic overpotential for the assumed diffusion process is the sum of Eqs. (6.103) and (6.105):

$$\eta_c = \eta_{conc} + \eta_{act} \tag{6.106}$$

$$\eta_c = \frac{2.303RT}{zF}\left[\log\left(1 - \frac{i_c}{i_L}\right) - \frac{1}{(1-\alpha)}\log\left(\frac{i_c}{i_{corr}}\right)\right] \tag{6.107}$$

or

$$\eta_c = \frac{RT}{zF}\left[\ln\left(1 - \frac{i_c}{i_L}\right) - \frac{1}{(1-\alpha)}\ln\left(\frac{i_c}{i_{corr}}\right)\right] \tag{6.108}$$

But, $\eta_c = E_c - E_o$ where E_c is the applied cathodic potential and $E_o = E_{corr}$ is the corrosion potential. Thus, Eq. (6.108) becomes

$$E_c = E_o + \frac{2.303RT}{zF}\left[\log\left(1 - \frac{i_c}{i_L}\right) - \frac{1}{(1-\alpha)}\log\left(\frac{i_c}{i_{corr}}\right)\right] \tag{6.109}$$

This equation gives a cathodic curve from i_{corr} to i_L as schematically shown in Fig. 6.14. This curve is just part of a polarization diagram, which is used to extract kinetic parameters. Also depicted in this figure is the anodic curve indicating the anodic overpotential needed for anodically polarizing an electrode for corrosion studies. It should be pointed out that Fig. 6.14 only shows a portion of a practical polarization diagram since the aim in this section is to explain the significance of the corrosion current density and the limiting current density as the boundaries in cathodic polarization.

Notice from Fig. 6.14 that:

- A shift from (E_{corr}, i_{corr}) to point 1 gives the overpotential as $\eta = \eta_{a,1}$ and activation polarization controls the mass transfer.
- At point 2, $\eta = \eta_{a,2} + \eta_{c,2}$ and a mixed polarization exists, but $\eta_{a,2} > \eta_{c,2}$.
- At point 3, $\eta = \eta_{a,3} + \eta_{c,3}$ and a mixed polarization exists, but the current density at this point does not depend on potential and $i_3 = i_L$.
- According to Eq. (6.109), a change in temperature shifts the cathodic polarization curve.

In determining the oxidation or reduction state, which is just the valence, one needs to know the Tafel slopes and the symmetry factor (α) defined below Eq. (5.5). However, the valence z can be estimated using the relationship for concentration polarization given by Eq. (6.102) with $\eta_{conc} = E_c - E_{corr}$.

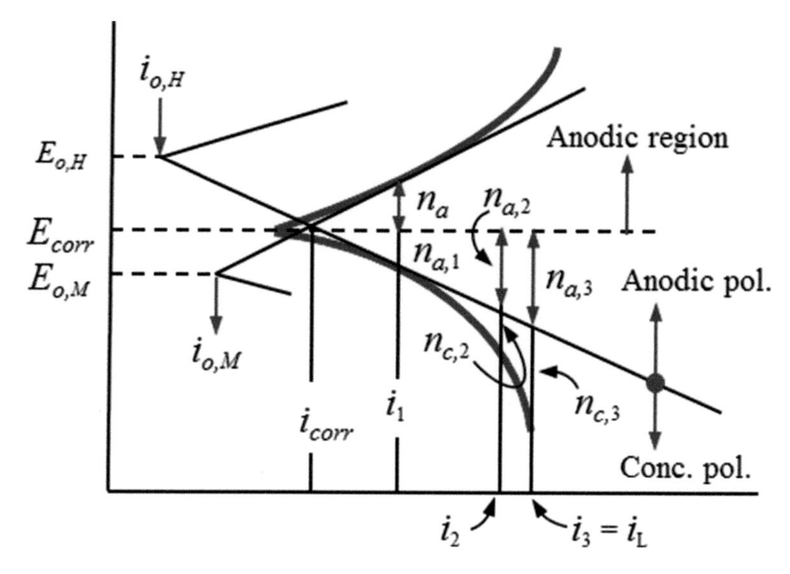

Fig. 6.14 Schematic curves illustrating the activation polarization (Act. Pol.) and the concentration polarization (Conc. Pol.) domains

Fig. 6.15 Linearized potential for determining the oxidation state (z) of a species j as per Eq. (6.110)

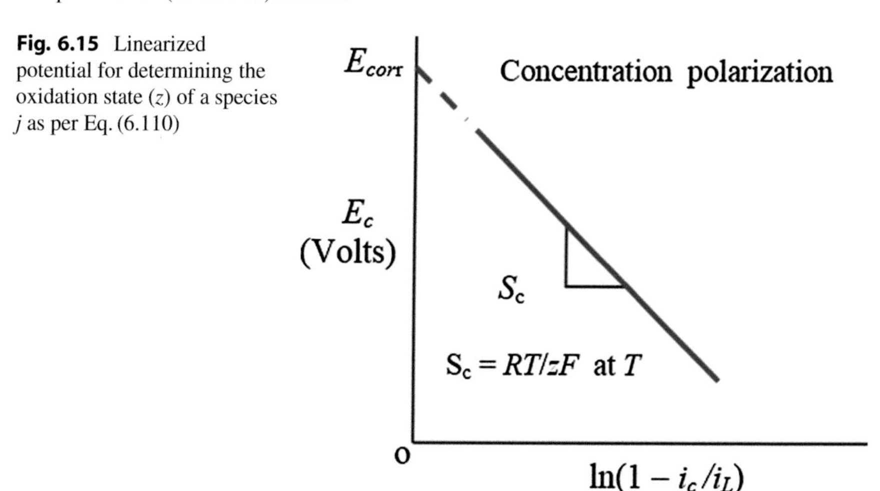

For a cathodic reaction, the electric potential becomes

$$E_c = E_{corr} - \frac{RT}{zF} \ln\left(1 - \frac{i_c}{i_L}\right) \tag{6.110}$$

Plotting E_c $vs.$ $\ln(1 - i_c/i_L)$ gives a straight line as schematically shown in Fig. 6.15. Now, extrapolating the straight line provides an intercept corresponding to E_{corr}.

The slope of the straight line is

$$S_c = \frac{\Delta E}{\Delta \ln (1 - i_c/i_L)} = \frac{RT}{zF} \tag{6.111}$$

Solving Eq. (6.111) for z yields an expression for approximating the oxidation state. Hence,

$$z = \frac{RT}{FS_c} \tag{6.112}$$

The slope can also be determined mathematically by taking the first derivative of Eq. (6.110) with respect to the cathodic current density. Thus,

$$\frac{dE_c}{di_c} = -\frac{RT}{zF(i_c - i_L)} \tag{6.113}$$

$$S_c = \frac{dE_c}{di_c} = \frac{\Delta E_c}{\Delta i_c} \tag{6.114}$$

Then, the oxidation number or valence is defined by

$$z = -\frac{RT}{S_c F(i_c - i_L)} \tag{6.115}$$

The preceding graphical approach is quite fundamental, but it is an essential element in electrochemistry. For instance, this linear approach is normally used in determining z for organic compounds in irreversible polarographic processes, which is a potential-control technique that requires a current response as a function of time for reducing species j [6]. The goal in this technique is to measure the limiting current $I_L = A_s i_L = f(t)$ where A_s is the surface area of the electrode. Hence, Eq. (6.51) becomes

$$I_L = \frac{zFA_s DC_b}{\sqrt{\pi Dt}} \tag{6.116}$$

Example 6.6. *Use the data given in Example 6.2, part (a), to calculate* **(a)** *the concentration gradient at* $x = o$, J_x, i_a *and* δ, *and* **(b)** η_a *and* i_L @ $x = 0.12\,mm$ *and* $t = 10\,min$. *The anodic reaction for copper is* $Cu \to Cu^{2+} + 2e^-$ *where* $z = 2$.

Data: $D = 1.5x10^{-6}\,cm^2/s$; $C_o = 10^{-2}\,mol/cm^3$; $C_b = 4x10^{-4}\,mol/cm^3$

$C_x = 7.89x10^{-3}\,mol/cm^3$ *at* $x = 0.12\,mm$ *and* $t = 10$ min $= 600\,s$

Solution. **(a)** *From Eq. (6.50),*

$$\frac{\partial C}{\partial x} = -\frac{(C_o - C_b)}{\sqrt{\pi Dt}} \exp\left(-\frac{x^2}{4Dt}\right) \quad for\ C_o > C_b\ at\ x = 0$$

$$\frac{\partial C}{\partial x} = -\frac{\left(10^{-2} - 4x10^{-4}\right)}{\sqrt{\pi\,(1.5x10^{-6})\,(10x60)}}$$

$$\frac{\partial C}{\partial x} = -0.18\,\text{mol/cm}^4$$

Then, from Eq. (6.16)

$$J_d = -D\partial C/\partial x = -\left(1.5x10^{-6}\text{cm}^2/\text{s}\right)\left(-0.176\,\text{mol/cm}^4\right)$$

$$J_d = 2.70 \times 10^{-7}\,\text{mol}/\left(\text{cm}^2\,\text{s}\right)$$

From Eq. (6.18),

$$i_a = zFJ_d = (2)\,(96,500\,)\left(2.70 \times 10^{-7}\,\text{mol}/\left(\text{cm}^2\,\text{s}\right)\right)$$

$$i_a = 0.052\,\text{A/cm}^2 = 52\,\text{mA/cm}^2$$

and from Eq. (6.52),

$$\delta = \sqrt{\pi Dt} = \sqrt{\pi\,(1.5x10^{-6})\,(10x60)} = 0.053\,\text{cm}$$

$$\delta = 0.53\,\text{mm}$$

(b) *Data: $x = 0.12\,mm$, $t = 600\,s$ and $T = 25\,°C = 298\,K$. From Eq. (6.96),*

$$\eta_a = \frac{RT}{zF}\ln\left[\frac{C_o}{C_x}\right]$$

$$\eta_a = \frac{(8.314)\,(298)}{(2)\,(96500)}\ln\frac{10^{-2}}{7.89x10^{-3}} = 0.003\,\text{V}$$

$$\eta_a = 3\,\text{mV}$$

From Eq. (6.99),

$$i_L = \frac{zFDC_x}{\delta}$$

$$i_L = (2)\,(96500)\left(1.5x10^{-6}\right)\left(7.89x10^{-3}\right)/\,(0.053)$$

$$i_L = 0.043\,\text{A/cm}^2 = 43\,\text{mV}$$

$$i_L < i_a$$

Example 6.7. *Plot Eq. (6.99) to determine a hypothetical limiting current I_L profile in a reduction metal M^{2+}/M process. The electrolyte contains a bulk concentration of $C_b = 10^{-4}\,mol/cm^3$ of cations M^{2+}, and the cathode electrode has a surface* *area of* $A_s = 25\,cm^2$. *Based on this information, explain the limiting current time dependency using the following two values of the diffusion coefficient of metal cations* M^{2+}: $D = 3x10^{-5}\,cm^2/s$ *and* $D = 8x10^{-5}\,cm^2/s$.

Solution. *From Eq. (6.99) and the given information,*

$$i_L = \frac{zFDC_x}{\delta} = \frac{zFDC_b}{\sqrt{\pi Dt}} = \frac{zFC_b\sqrt{D}}{\sqrt{\pi t}}$$

$$I_L = \frac{zFDC_bA_s}{\sqrt{\pi Dt}} = \frac{(2)\,(96500)\left(10^{-4}\right)(25)\sqrt{D}}{\sqrt{\pi t}}$$

$$I_L = (272.22\,\mathrm{A\,s/cm})\sqrt{D/t}$$

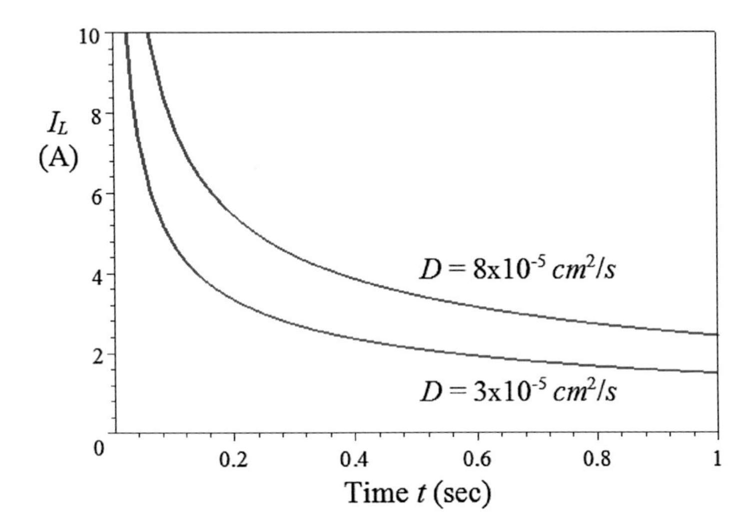

It is clear from these curves that the limiting current density I_L *is strongly dependent on a short time interval and on the diffusion coefficient D. As D increases, the* I_L *profile is shifted upward, implying that a higher current is needed for the reduction process.*

Furthermore, during cathodic processes, ions (M^{z+}) move toward the stationary cathode electrode because the bulk and surface concentrations are related as $C_b > C_s$; however, a limiting current density (i_L) arises due to ion depletion in a constant bulk solution volume. On the other hand, a rotating electrode (hydrodynamic method) increases i_L with increasing the velocity (v) and temperature (T) of the solution. Thus, the rate of mass transfer under laminar flow can be converted to liming current density i_L using Faraday's law of electrolysis. With respect to currents

($I = iA_s$), steady and transient currents are known as Faradaic and non-Faradaic currents, respectively. Henceforth, steady currents and also steady current densities are assumed throughout this textbook, unless stated otherwise.

The application of the classical **rotating-disk electrode** (hydrodynamic system) having a smooth surface is convenient for characterizing the effects of solution flow on corrosion reactions. For steady-state condition, I_L is defined by the Levich equation [27]:

$$I_L = 0.62\,(zFA_s)\left(\frac{DC_b}{\delta}\right)\left(\frac{K_v}{D}\right)^{1/3}\left(\frac{\delta^2\omega}{K_v}\right)^{1/2} \tag{6.117}$$

where r = rotating-disk radius (cm)
K_v = kinematic viscosity (cm^2/s)
ω = angular velocity (Hz= 1/s)

Modify Eq. (6.117) to get the Levich constant ε:

$$I_L = \varepsilon C_b \sqrt{\omega} \tag{6.118}$$

$$\varepsilon = 0.62\,(zFA_s)\left(\frac{D}{\delta}\right)\left(\frac{K_v}{D}\right)^{1/3}\left(\frac{\delta^2}{K_v}\right)^{1/2} \tag{6.119}$$

In addition, Eq. (6.118) suggests that the rotating smooth surface exposed to the solution is subjected to a transition from laminar to turbulent flow at a Reynolds number defined by [4]

$$\mathrm{Re} = \frac{\omega r_t^2}{K_v} = \frac{\omega r_t^2 \rho}{\eta_v} \tag{6.120}$$

where r_t = radius at which transition occurs (cm)
η_v = fluid viscosity (g/cm s)
ρ = density of particles (ions) (g/cm^3)

The transitional radius is smaller than the disk radius ($r_t < r$). Also, knowing the number of revolutions per unit time (N) of a rotating disk, the angular velocity is easily computed as $\omega = 2\pi N$. Commercial instrumentation is nowadays available for conducting electrochemical experiments using a rotating electrode systems at a wide range of electrode speeds. The reader wishing to obtain detailed information on commercial equipment should use a World Wide Web (www) search engine. In addition, a hydrodynamic system is further expanded in Chap. 9 for characterizing convection-diffusion mechanisms using empirical or semiempirical dimensionless numbers, such as Reynolds number, for mass transfer of a species j.

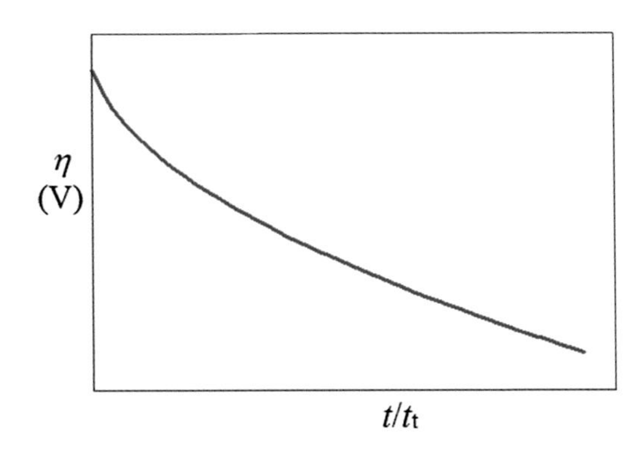

Fig. 6.16 Profile of the overpotential according to Eq. (6.124)

6.7 Galvanostatic Polarization

Using the galvanostatic technique gives rise to a current across the cathode electrode, and the potential response is measured prior to reaching steady state. According to Erdey-Gruz [28], the boundary condition used for solving Fick's second law can be deduced from Eq. (6.18). Hence,

$$D\frac{\partial C}{\partial x} = \frac{i}{zF} \quad \text{for } t > 0,\ \ x = 0,\ \ i = \text{constant} \tag{6.121}$$

Subsequently, the concentration equation as a function of time takes the following analytical form

$$C = C_b - \frac{2i}{zF}\sqrt{\frac{t}{\pi D}} \tag{6.122}$$

In addition, the transition diffusion time when $C = C_o = 0$ at $x = 0$ (electrode surface) is deduced from Eq. (6.122) as

$$t_t = \frac{\pi D}{4}\left(\frac{zFC_b}{i}\right)^2 \tag{6.123}$$

The diffusion overpotential as a function of time as well as temperature is given by [28]

$$\eta = \frac{RT}{zF}\ln\left(1 - \sqrt{t/t_t}\right) \tag{6.124}$$

The interpretation of Eq. (6.124) suggests that $\eta \rightarrow \infty$ when $t = t_t$ which is unrealistic. Instead, $\left(1 - \sqrt{t/t_t}\right) < 1$ so that $\eta < 0$. Also, another electrochemical process takes place when $t \rightarrow t_t$. The profile given by Eq. (6.124) is shown in Fig. 6.16.

6.8 Problems/Questions

6.1. Why convective mass transfer is independent of concentration gradient $\partial C/\partial x$ in Eq. (6.2)?

6.2. Use the information given in Example 6.3, part b, to determine the concentration rate in the x-direction at 1 mm from the column interface. Given data: $D = 10^{-5}\,\text{cm}^2/\text{s}$, $C_o = 10^{-4}\,\text{mol/cm}^3$, and $t = 10\,\text{s}$. [Solution: $1.96x10^{-16}\,\text{mol}/\left(\text{cm}^3\,\text{s}\right)$.]

6.3. What is the action of hydrogen on a steel blade exposed to an aqueous solution of hydrogen sulfide, $Fe + H_2S = FeS + 2H$? Assume that the steel blade gets damaged during service.

6.4. An aerated acid solution containing $10^{-2}\,\text{mol/l}$ of dissolved oxygen (O_2) moves at $2x10^{-4}\,\text{cm/s}$ in a stainless steel pipe when a critical current density of $10^3\,\mu\text{A/cm}^2$ passivates the pipe. Calculate **(a)** the thickness of the Helmholtz ionic structural layer (δ) and **(b)** the limiting current density (i_L) if the three modes of flux are equal in magnitude, and **(c)** Will the pipe corrode under the current conditions? Why or why not? Data:

$$O_2 + 2H_2O + 4e^- = 4OH^- \quad \text{(cathode)}$$

$D_{O_2} = 10^{-5}\,\text{cm}^2/\text{s}$	$T = 25\,°\text{C}$	$F = 96,500\,\text{A s/mol}$

[Solutions: **(a)** $\delta = 0.5\,\text{mm}$, **(b)** $i_L = 2.32x10^3\,\mu\text{A/cm}^2$, and **(c)** It will not].

6.5. **(a)** Determine the oxygen concentration for Problem 6.4 that will promote corrosion of the stainless steel pipe. **(b)** How long will it take for corrosion to occur? [Solutions: **(a)** $C_b = 1.30x10^{-2}\,\text{mol/l}$ and **(b)** $t = 1.34\,\text{min}$.]

6.6. Use the given data to calculate **(a)** the diffusivity of copper ions in a cathodic process at $25\,°C$ and **(b)** the valence z. The original concentration in an acidic solution is 60 g/l. [Solution: **(a)** $D = 10^{-5}\,\text{cm}^2/\text{s}$ and **(b)** $z = 2$.]

t (s)	i $(x10^{-2}\,\text{A/cm}^2)$	$\partial C/\partial x$ $(x10^{-3}\,\text{mol/cm}^4)$
0	0	0
5	0.1455	0.07531
10	0.1029	0.05325
15	0.0840	0.04350
20	0.0728	0.03766

6.7. It is desired to reduce copper, $Cu^{2+} + 2e^- = Cu$, at 35 °C from an electrolyte containing 60 g/l of Cu^{2+} ions. Calculate **(a)** the total molar flux that arises from equal amounts of diffusion and migration mass transfer under steady-state conditions and **(b)** the gradients $dc/dx, d\phi/dx$, and $d\mu/dx$, and **(c)** approximate the thickness of the diffusion layer at the cathode electrode surface. Operate the electrowinning cell for 10 min and let the diffusivity for Cu^{2+} ions be 10^{-5} cm^2/s. [Solutions: **(a)** $J = 6.88x10^{-8}$ mol/cm^2 s, **(b)** $\partial C/\partial x|_{x=0} = -3.44x10^{-3}$ mol/cm^4, $d\phi/dx = -4.83x10^{-2}$ V/cm, $d\mu/dx = -1.55x10^{-20}$ J/cm, and **(c)** $\delta = 1.37$ mm.]

6.8. Show that $J_{x=o} = DC_o/\delta$.

6.9. Show that $dC/dx = C_b/\sqrt{\pi D t}$.

6.10. Prove that the diffusivity D is constant in the Fick's second law, Eq. (6.17).

6.11. Assume that the total molar flux of a species j is due to diffusion and convection. The convective force acting on the species is $F_x = -(1/N_A)(d\Delta G/dx)$, where N_A is Avogadro's number and $d\Delta G/dx$ is the molar free energy gradient. Recall that the volume fraction is equal to the mole fraction divided by molar concentration; that is, $V = X/C$. Based on this information, show that

$$\frac{dC}{dx} = C\left[\frac{d\ln(C/C_o)}{dx}\right]$$

where $J_d = J_c$, $K = C/C_o < 1$, and $C_o = $ constant (mol/l) .

6.12. Use the information given in Problem 6.11 to show that

$$\frac{d\Delta G}{dx} = \frac{RT}{C}\frac{dC}{C}$$

6.13. Use the information given in Problem 6.12 to show that

$$v_x = -\frac{D}{C}\frac{dC}{dx}$$

6.14. If the migration flux is neglected in Eq. (6.2), approximate the total flux at low and high temperatures. Assume that $C_x >> C_o$ at a distance x from an electrode surface.

6.15. Consider the concentration plane shown below. This is a transient electrochemical system having finite dimensions. Derive a solution for Fick's second law, $\partial C/\partial t = D\partial^2 C/\partial x^2$ when $a \leq x \leq b$.

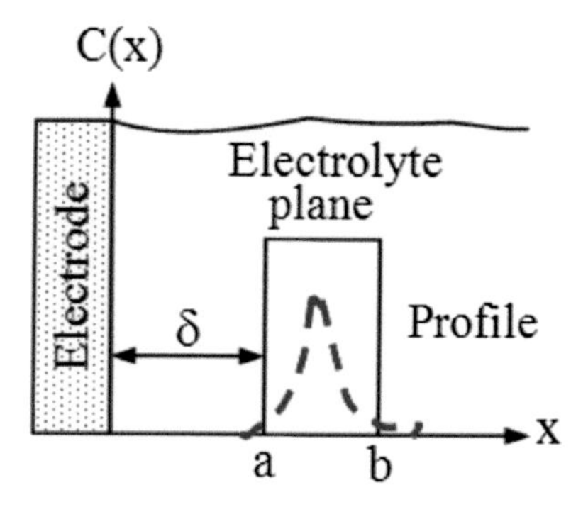

6.16. Derive Fick's second law if the volume element in Fig. 6.2 has a unit cross-sectional area and a thickness x.

6.17. What does Fick's first law mean in terms of atoms or ions of a single phase?

6.18. What will Fick's first law mean if D does not vary with x or C?

6.19. Derive Fick's second law if the volume element in Fig. 6.2 has a unit cross-sectional area and the diffusing plane is located between x and $x + dx$, where J_x is the entering molar flux at x and J_{x+dx} is that leaving at $x + dx$.

6.20. Find an expression for x when

$$\frac{C_x - C_b}{C_o - C_b} = 0.5205 = \text{erf}\left(\frac{x}{\sqrt{4Dt}}\right)$$

6.21. Derive Eq. (6.85).

6.22. Show that $d\phi/dx = -7\phi_o/\left(19\sqrt{\pi Dt}\right)$ in a coupled diffusion and migration mass transfer. let $x = \sqrt{\pi Dt}$.

6.23. For activation polarization, the concentration gradient is given by Eq. (6.50). If $C_o >> C_b$, then show that

$$\frac{C}{C_o} = \frac{1}{\sqrt{4\pi Dt}}\left(\frac{x^3}{12Dt} - x\right)$$

6.24. Derive Eq. (5.122) by converting Eq. (3.29) into chemical potential (μ) if only one ion participates in a reaction. The reaction constant is defined by $K_r = \gamma C$, where γ is the activity coefficient and C is the concentration. Let the molar diffusion and migration be equal and $J_c = 0$ in Eq. (6.10).

6.25. Follow the statement given in Example 6.3 for determining the average concentration as the starting point in this problem. Let $C_{o,1}$ and $C_{o,2}$ be the original concentrations of columns 1 and 2, respectively.

6.26. This problem requires the determination of the total molar flux and current density profiles due to diffusion, migration, and convection for the reduction of nickel from an electrolyte containing $8x10^{-4}\,\text{mol/cm}^3$ of Ni^{2+} cations. The electrochemical cell has a continuous and direct fluid flow system of $20\,\text{cm/s}$ at $35\,°\text{C}$. The diffusivity of Ni^{2+} cations is $D = 1.44 \times 10^{-5}\,\text{cm}^2/\text{s}$. **(a)** Use $0 \leq t \leq 100\,\text{s}$ for $J_x = f(t, x = 0)$ at the electrode surface and **(b)** $0 \leq x \leq 10\,\text{cm}$ for $J_x = f(x, t = 10)$, and **(c)** analyze the resultant profiles very succinctly and determine which molar flux dominates in this hypothetical reduction process for Ni^{2+} cations from solution. Can this process achieve a steady state?

6.27. **(a)** Derive an expression for the concentration in g/l as a function of temperature T and distance x from an oxidizing pure copper electrode in an electrolyte under an electrostatic field. Use the chemical potential gradient, $d\mu/dx = -(zF)\,d\phi/dx$ with an electric potential decay, $\phi_x = \phi_o \exp(-\lambda x)$. Let the free energy change be defined as $\Delta G = \Delta G_x$ @ x and $\Delta G = \Delta G_\infty$ @ $x = \infty$. **(b)** Plot the concentration C_x profile at two different temperatures. Let

$\phi_\infty = 0$	$\lambda = 0.2\,\text{cm}^{-1}$
$\phi_o = 8.314x10^{-3}\,\text{V}$	$0 \leq x \leq 3.8\,\text{cm}$
$T = 298\,\text{K}$	$C_\infty = 1.57x10^{-3}\,\text{mol/l} = 0.1\,\text{g/l}$
$T = 318\,\text{K}$	$F = 96,500\,\text{J/mol V}$

6.28. Derive an expression for $i_L = f(C_x, T, v_x)$ and explain its physical significance with respect to concentration polarization and metal reduction. Plot $\eta_c = f(i/i_L)$ when $i_L = 5 \times 10^{-3}\,\text{A}$ and $i_L = 10^{-2}\,\text{A}$ to support your explanation. When is the concentration polarization process discernible (apparent)?

References

1. M.G. Fontana, *Corrosion Engineering*, 3rd edn. (McGraw-Hill Book Company, New York, 1986)
2. T. Engel, P. Reid, *Physical Chemistry* (Pearson Education, Inc., Boston, 2013), p. 878
3. A.J. Bard, L.R. Faulkner, *Electrochemical Methods: Fundamentals and Applications*, Chap. 4, 2nd edn. (Wiley, New York, 2001), pp. 25–26
4. A.F. Mills, *Basic Heat and Mass Transport*, 2nd edn. (Prentice-Hall, Englewood Cliffs, 1999), pp. 330, 758, 820–823
5. E. Samsonl, J. Marchandl, K.A. Snyder, *Calculation of Ionic Diffusion Coefficients on the Basis of Migration Test Results*. Materials and Structures/Materiaux et Constructions, vol. 36 (April 2003), pp. 156–165
6. J. Wang, *Analytical Electrochemistry* (Wiley-VCH, New York, 1994), p. 32
7. G.H. Geiger, D.R. Poirier, *Transport Phenomena in Metallurgy* (Addison-Wesley Publishing Company, Readings, MA, 1973), pp. 449–459
8. J.R. Maloy, Factors affecting the shape of current-potential curves. J. Chem. Educ. **60**, 285–289 (1983)
9. S. Lorente, Constructal view of electrokinetic transfer through porous media. J. Phys. D: Appl. Phys. **40**, 2941–2947 (2007)
10. P. Begue, S. Lorente, Migration versus diffusion through porous media: time-dependent scale analysis. J. Porous Media **9**(7), 637–650 (2006)
11. U.R. Evans, *The Corrosion and Oxidation of Metals* (Arnold, London, 1961)
12. M.A. Biondi, S.C. Brown, Measurements of ambipolar diffusion in helium. Phys. Rev. **75**, 1700–1705 (1949)
13. B.A. Ruzicka, L.K. Werake, H. Samassekou, H. Zhao, Ambipolar diffusion of photoexcited carriers in bulk GaAs. Appl. Phys. Lett. **97**, 262119-1 (2010)
14. D. Nister, K. Keis, S.-E. Lindquist, A. Hagfeldt, A detailed analysis of ambipolar diffusion in nanostructured metal oxide films. Sol. Energy Mater. Sol. Cells **73**, 411–423 (2002)
15. J. Crank, *The Mathematics of Diffusion*, 2nd edn. (Oxford University Press, New York, 1975) [Reprinted 2001]
16. D.R. Gaskell, *An Introduction to Transport Phenomena in Materials Engineering* (Macmillan Publishing Company, New York, 1992), pp. 373–379, 483–89, 505
17. C.M.A. Brett, A.M.O. Brett, *Electrochemistry Principles, Methods and Applications*, chap. 5 (Oxford University Press Inc., New York, 1994)
18. J.W. Evans, L.C. De Jonghe, *The Production of Inorganic Materials* (Macmillan Publishing Company, New York, 1991), pp. 149, 186, 190
19. L.L. Shreir, Outline of electrochemistry, in *Corrosion Vol. 2, Corrosion Control*, ed. by L.L. Shreir, R.A. Jarman, G.T. Burstein (Butterworth-Heinemann, Boston, 1994)
20. T. Erdey-Cruz, *Kinetics of Electrode Processes* (Wiley-Interscience, Wiley, New York, 1972), p. 442
21. L.G. Twidwell, *Electrometallurgy*, Unit Process in Extractive Metallurgy, NSF Project SED 75-04821 (1978)
22. T. Rosenquist, *Principles of Extractive Metallurgy*, 2nd edn. (McGraw-Hill Book, New York, 1983), pp. 106–107
23. A. Kumar, R.K. Gupta, *Fundamentals of Polymers* (The McGraw-Hill Companies, Inc., New York, 1998), pp. 411–421
24. J.S. Anderson, K. Saddington, The use of radioactive isotopes in the study of the diffusion of ions in solution. J. Chem. Soc. S381–S386 (1949)
25. J.P. Broomfield, *Corrosion of Steel in Concrete* (E. & F.N. Spon, London, 1997)
26. J.L. Valdes, *Electrodeposition of colloidal particles*. J. Electrochem. Soc. **134**(4), 223C–225C (1987)
27. V.G. Levich, *Physicochemical Hydrodynamics* (Prentice-Hall, Englewood Cliffs, 1962)
28. T. Erdey-Gruz, *Kinetics of Electrode Processes* (Wiley-Interscience a Division of Wiley, New York, 1972), pp. 127–131

7 Corrosivity and Passivity

The electrochemical corrosion behavior of metals and alloys can be studied by generating polarization curves, which provide several electrochemical regions of diverse physical meaning. A material, specifically metallic solids, may exhibit some degree of corrosivity, that is, the tendency of an aggressive electrolyte (environment) to cause metal dissolution (corrosion through an oxidation reaction). However, the same material may undergo passivation in the same environment at relatively high potentials independent of the current. Thus, passivity may be manifested due to electrodeposition of an impervious metal-oxide film compound on the material surface, protecting it from deterioration. Further, increments in potential, the film usually breaks down leaving localized bare metal experiencing corrosion at relatively high potential.

This chapter is confined to analyze the complex aqueous corrosion phenomenon using the principles of mixed potential, which in turn is related to the mixed electrode electrochemical corrosion process. The resulting value of the potential of a given electrode with respect to a suitable reference electrode, such as the standard hydrogen electrode, is associated with a zero net current ($I_{net} = I_a + I_c = 0$). However, when electrodes are made of alloys, different redox couples may occur due to different species in an electrochemical system. Thus, the mixed potential is the potential of a working electrode against a reference electrode, despite that there may exist more than one coupled redox reaction taking place simultaneously. The principles and concepts introduced in this chapter represent a simplified approach for understanding the electrochemical behavior of corrosion (oxidation) due to coupled redox reactions.

Both Evans and Stern diagrams are included in order to compare and analyze simple and uncomplicated electrochemical systems. The concept of anodic control and cathodic control polarization is also introduced.

N. Perez, *Electrochemistry and Corrosion Science*,
DOI 10.1007/978-3-319-24847-9_7

7.1 Electrode Systems

7.1.1 Instrumentation

Electrochemical corrosion (*EC*) can be characterized using a proper electrochemical instrumentation. A custom design device is shown in Fig. 7.1 [1].

(WE) The power supply is a potentiostat for controlling the potential or a galvanostat for current control flow. The working electrode WE is embedded in epoxy resin along with a spot-weld wire. Actually, this type of WE-epoxy resin is just a metallographic sample, which is well polished prior to immersion in a working electrolyte. All parts immersed in the electrolyte must be inert to the electrolyte in order to avoid electrochemical complications.

Figure 7.2 shows a commercially available ASTM G-5 standard electrochemical cell designed by EG&G Princeton Applied Research as Model K47. These electrochemical cells (Figs. 7.1 and 7.2) produce similar polarization results, provided that electrochemical noise and ohmic effects are reduced significantly. See Appendix B for other cell arrangements.

This cell (Fig. 7.2) has two auxiliary graphite electrodes for providing a uniform current distribution to the WE surface, and it is known as a three-electrode cell.

Furthermore, corrosion and electrochemical studies can be performed using a well-calibrated electrochemical instrumentation equipment. Figure 7.3 illustrates an automated modern experimental instrumentation, which includes the commercially available device known as the Princeton Applied Research (PAR) *EG&G* Potentiostat/Galvanostat Model 273A and the electrochemical or polarization cell model K47. This particular cell is known as a three-electrode electrochemical cell, and it is used for characterizing the kinetics of the working electrode (WE) in a suitable environment at a desirable temperature.

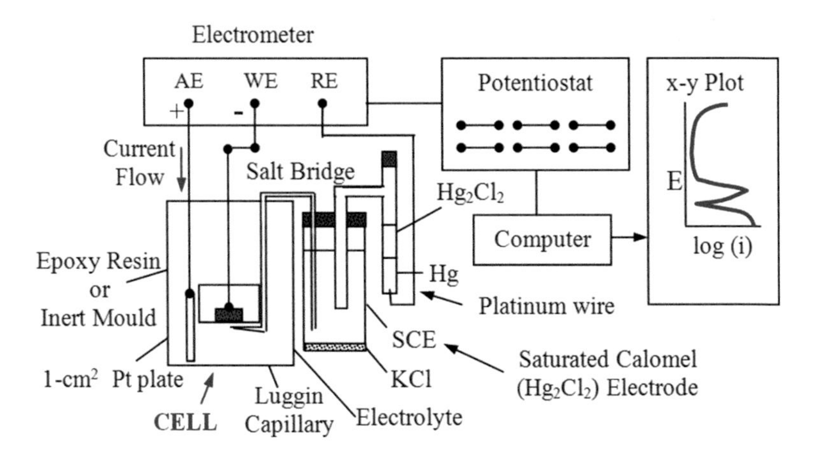

Fig. 7.1 Schematic laboratory electrochemical cell [1]

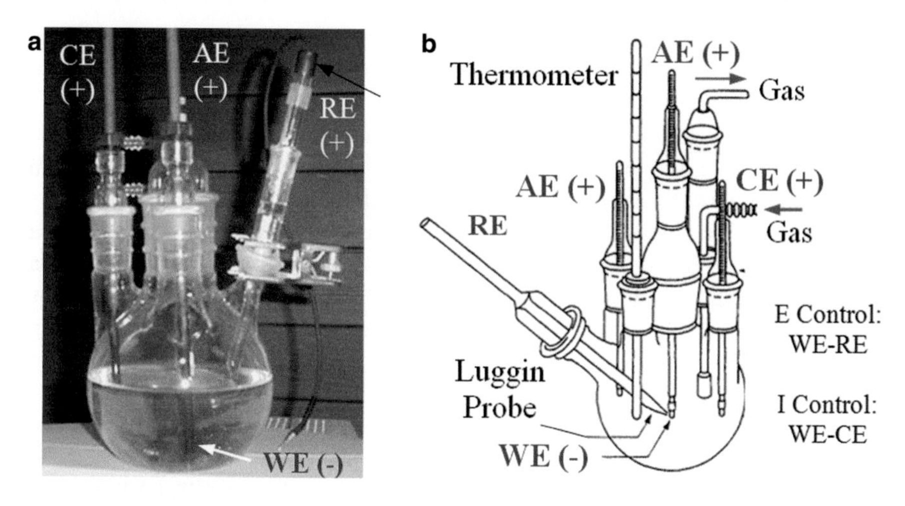

Fig. 7.2 Commercially EG&G Model K47 electrochemical cell. (**a**) Three-electrode cell. (**b**) Four-electrode cell

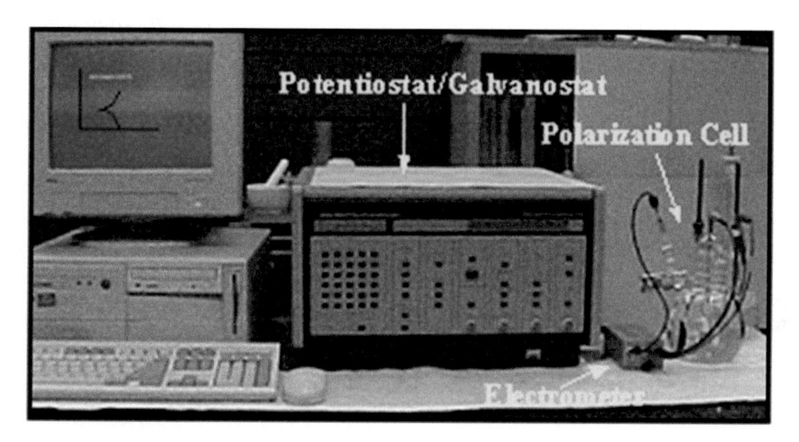

Fig. 7.3 Modern electrochemical instrumentation. Princeton Applied Research (PAR) EG&G Potentiostat Galvanostat Model 273A

Apparently, the most common electrochemical techniques are potentiostatic, potentiodynamic, galvanostatic, and galvanodynamic polarization. The 273A device can be used as potentiostat for measuring the current response due to an applied potential or as a galvanostat to measure the potential response at an applied current. Common corrosion studies using the above instrumentation include kinetic parameters.

The schematic electrical circuit for a basic potentiostat can be found elsewhere [2,3]. Nevertheless, the potentiostat uses R_s as the electrolyte resistance between the auxiliary electrode (AE) or counter electrode (CE) and the reference electrode (RE),

R_c as the resistance developed due to current flow between the working electrode (WE) and RE, E as an adjustable potential for keeping the WE potential at a constant value [2].

The working electrode WE potential is measured against the reference electrode RE, provided that an ohmic resistance gradient is reduced and that current flows only between the auxiliary electrode and the working electrode [4, 5].

The adjustable potential E is measured with a zero-resistance ammeter (ZRA) and an analog-to-digital converter (ADC). The main objective of the potentiostat is to control the potential difference between WE and RE by supplying a current flow through the AE.

7.1.2 Cell Components

The electrochemical cell components and their functions are described below.

Platinum (Pt) Auxiliary Electrode: It passes current to the working electrode (specimen) to be studied.

Luggin Capillary: It is a probe filled with an electrolyte to provide an ionic conductive path through the soluble ionic salt (KCl). The Luggin capillary and salt bridge connecting the cell and the reference electrode (RE) do not carry the polarizing current, and it serves the purpose of reducing the ohmic resistance gradient through the electrolyte between the WE and AE. In fact, some of the ohmic potential ($E = IR$) is included in the polarized potential [4].

Working Electrode (WE): A 1-cm^2 exposed surface area is desirable. The distance between the WE surface and the tip of the Luggin capillary should be in the range of 1 mm $\leq x \leq$ 2 mm. During an electrochemical corrosion experiment, the specimen immersion time must be constant in order to stabilize the electrode in the electrolyte prior to start the polarization test.

Potentiostat: A manually operated potentiostat is a stepwise instrument for measuring potential (E) and current density (i) and developing a E *vs.* $\log(i)$ plot as schematically shown in Fig. 7.4 along with the ideal parameters and polarization regions.

Nomenclature for Fig. 7.4:

i_p =passive current density
i_{corr} =corrosion current density
i_s =secondary current density
i_{crit} =critical current density
$i_{\max}$ =maximum current density

E_{pa} =Passive potential
E_{corr} =corrosion potential
E_{o_2} =oxygen evolution potential
E_{pp} =primary passive potential
E_p =pitting potential

A commercially programmable potentiostat, together with an electrometer, logarithmic converter, and data acquisition device, is an automated instrument (Fig. 7.3) that provides variability of continuous sweep (scan) over a desired potential range,

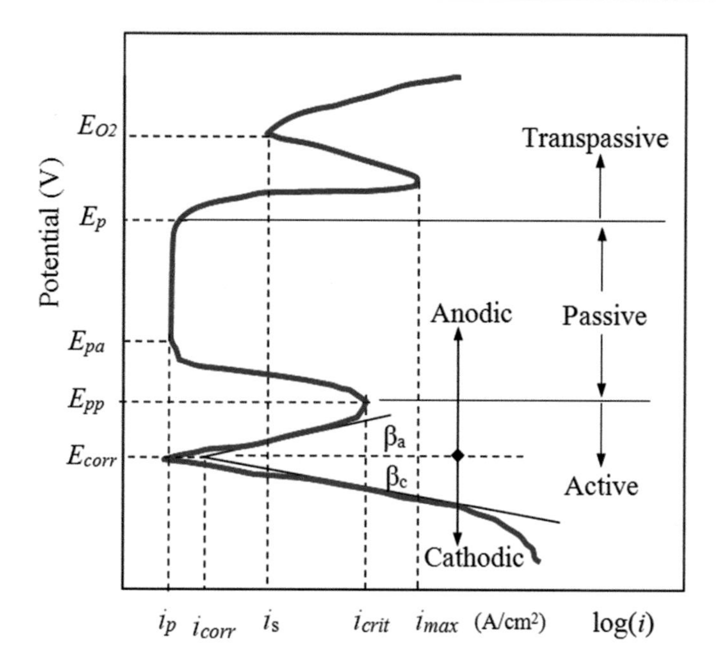

Fig. 7.4 Schematic ideal polarization curve

normally in the order of $-2\text{ V} < E < 2\text{ V}$ for obtaining an entire polarization curve (Fig. 7.4). This automated procedure is called potentiodynamic polarization technique.

Recall from Chap. 5 that the open-circuit potential is the corrosion potential (E_{corr}) in the absence of an externally applied potential and the corrosion current density (i_{corr}) cannot be measured or determined in this case. However, electrochemical techniques can be used for determining i_{corr} from a polarization curve generated by sweeping the WE surface with an applied potential and for measuring the response current or current density. This means that i_{corr} cannot be measured but graphically estimated. In addition, polarization is a measure of the overpotential, and the electrochemical process on the WE is basically a combination of kinetics and diffusion processes.

If the applied potential is at or very near E_{corr}, which can be defined as the electrochemical open-circuit potential in the presence of an applied potential for current to flow, the potential noise and the current noise can be monitored for evaluating the corrosion behavior of the WE. This implies that the potential difference becomes the overpotential (η), which represents the degree of polarization, and it is very small ($\eta \to 0$).

7.1.3 Electrochemical Noise

If heterogeneous (nonuniform) corrosion occurs when the anodic overpotential is $\eta_a \to 0$, the corrosion behavior can be characterized using an electrochemical noise (ECN) technique, in which measurements of the noisy signals quantify the discrete electrochemical events that perturb the steady state of the WE surface.

Electrochemical noise represents fluctuations (signal perturbations of low amplitude) in potential and current during an experiment. Thus, potential noise occurs between the WE and RE due to changes in the thermodynamic state of the WE, and the current noise arises between the WE and AE due to changes in the kinetic state of the corrosion process on the WE [6].

In general, potential (voltage) fluctuations induce capacitive background current as noise, causing difficulties in detecting the desirable Faradaic current. At times, the term charging current is used to designate noise as the amount of current needed to charge the electrical double layer between the working electrode and the electrolyte. In general, the electrochemical noise sources can be attributed to electrical connections, ohmic resistance effects, electrolyte composition, electrode mechanical surface defects, and microstructural features.

Currently, high resolution and automated instrumentation are commercially available for eliminating or significantly reducing electrochemical noise during measurements of electrochemical data. Hence, CorrElNoise® technique by ZAHNER is suitable for this purpose. In addition, instrumentation for electrochemical noise analysis can be found elsewhere [6–9].

7.2 Mixed Electrode Potential

Electrochemical corrosion systems can be characterized using the kinetic parameters previously described as Tafel slopes and exchange and limiting current densities. However, the mixed potential theory requires a mixed electrode system. This is shown in Fig. 7.5 for the classical pure zinc (Zn) electrode immersed in hydrochloric (HCl) acid solution [10–12]. This type of graphical representation of electrode potential and current density is known as Evans diagram for representing the electrode kinetics of pure zinc.

The Evans diagram requires anodic and cathodic straight lines and the corrosion potential (E_{corr}) and the corrosion current density (i_{corr}) point located where hydrogen reduction $\left(H^+/H_2\right)$ line and zinc oxidation $\left(Zn/Zn^{2+}\right)$ line converge. Furthermore, the exchange current densities (i_o) and the open-circuit potentials for hydrogen ($E_{o,H}$) and zinc ($E_{o,Zn}$) are necessary for completing the diagram as shown in Fig. 7.5 for a simple Zn-electrolyte electrochemical system. This type of potential-current density diagram can be developed for any solid material in contact with an electrolyte containing one or more oxidizing agents such as H^+, Fe^{2+}, etc.

In addition, connecting the i_o and i_{corr} provides the shown straight lines, which are accompany by their respective overpotential equation. With regard to the

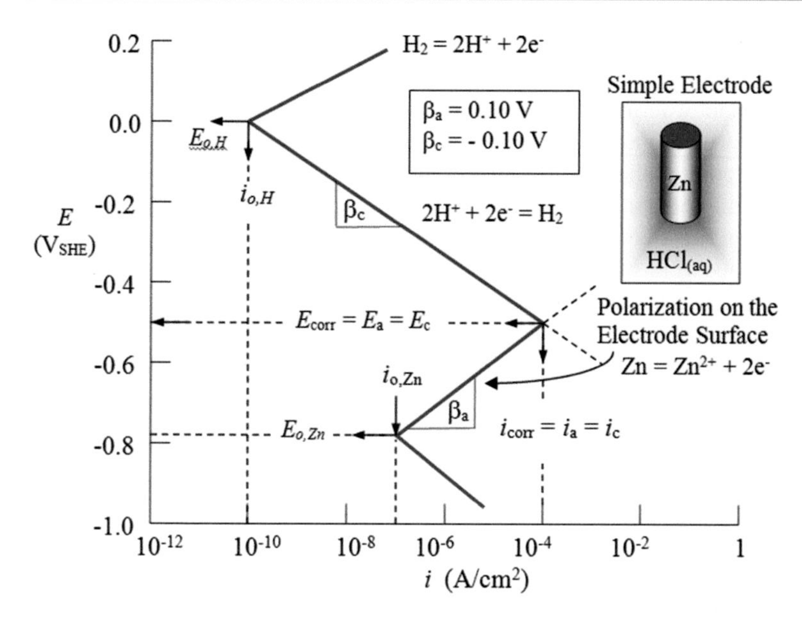

Fig. 7.5 Evans diagram for zinc in HCl acid solution

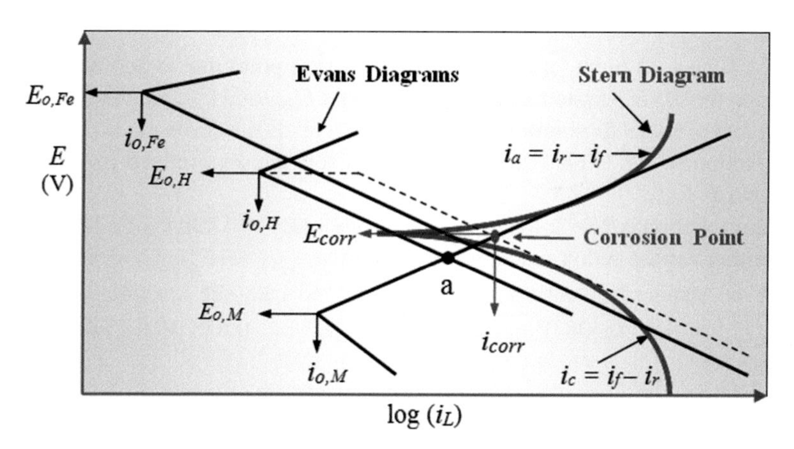

Fig. 7.6 Comparison of schematic polarization diagrams

E_{corr}-i_{corr} point in Fig. 7.5, the potentials and current densities are equal since this point represents electrochemical equilibrium. Conversely, Stern [13] used Eq. (5.22) to draw anodic and cathodic nonlinear polarization curves, which provide a more realistic representation of the electrode kinetics of an electrochemical system. Evans (straight lines) and Stern (solid curves) graphical methods are illustrated in Fig. 7.6 for a corroding hypothetical metal M immersed in a corrosive medium containing H^+ and F^{2+} oxidizers [14, 15].

7.2.1 Electrochemical Superposition

The addition of ferrous F^{2+} oxidizer to the electrolyte increases the corrosiveness of the electrolyte and requires an additional Evans diagram in order to determine the corrosion point of the metal M. In this particular case, the exchange current density, open-circuit potential, and the Tafel slopes for each oxidizing agent and the metal in question must be known prior to construct the Evans diagram. This, then, is a disadvantage of this technique for determining E_{corr} and i_{corr} for the metal M.

Furthermore, using the potentiostatic or potentiodynamic technique is very useful in determining the polarization curve near the corrosion point. However, the latter technique is a faster approach for obtaining even an entire polarization curve from a low potential as the limiting current density to a high potential for oxygen evolution. This dynamic approach is quite advantageous in obtaining a complete polarization curve in a few minutes.

Analysis of Fig. 7.6

Most corroding solutions contain more than one oxidizing agent, such as ferric $\left(Fe^{3+}\right)$ and ferrous $\left(Fe^{2+}\right)$ salts, and other ionic impurities due to contamination from corrosion products and existing oxidizers [4]. The electrochemical situation shown in Fig. 7.6 implies that:

- The (E_{corr}-i_{corr}) point "a" is shifted to corrosion point due to addition of Fe^{3+} ions to the M/H^{+} system. Consequently, both E_{corr} and i_{corr} are increased.
- The rate of metal dissolution is increased as well as the corrosion rate.
- Reduction of Fe^{3+}/Fe^{2+} and H^{+}/H_2 oxidizers causes a mixed potential shift known as E_{corr}.
- Redox reaction is $2Fe^{3+} + H_2 \rightarrow 2Fe^{2+} + 2H^{+}$ ($E^{o} = 0.438\ V_{SHE}$). The standard potential for the redox reaction is $E^{o} = E^{o}_{Fe^{3+}/Fe^{2+}} + E^{o}_{H_2/H^{+}} = 0.438\ V_{SHE}$, and the standard potentials for the anodic and cathodic reactions, $E^{o}_{Fe^{3+}/Fe^{2+}}$ ($E^{o}_{Fe^{3+}/Fe^{2+}} = 0.438\ V_{SHE}$) and $E^{o}_{H_2/H^{+}}$ ($E^{o} = 0.438\ V_{SHE}$), are given in Table 2.2.
- An addition of current density for reducing H^{+}/H_2 is needed so that H^{+}/H_2 follows the dashed line above the Fe^{3+}/Fe^{2+} line until it intersects the M/M^{z+} line.
- For one metal oxidation reaction, two reduction reactions are needed. Therefore, the addition of Fe^{3+} ions reduces the hydrogen evolution.
- The possible reactions that can occur due to the electrochemical system shown in Fig. 7.6 are

$$2Fe^{3+} + 2e^{-} \rightarrow 2Fe^{2+} \quad \text{(cathodic)} \tag{7.1}$$

$$2H^{+} + 2e^{-} \rightarrow H_2 \quad \text{(cathodic)} \tag{7.2}$$

$$2M \rightarrow 2M^{2+} + 4e^{-} \quad \text{(anodic)} \tag{7.3}$$

$$M + 2H^{+} + 2Fe^{3+} = M^{2+} + 2Fe^{3+} + 2H^{+} \quad \text{(redox)} \tag{7.4}$$

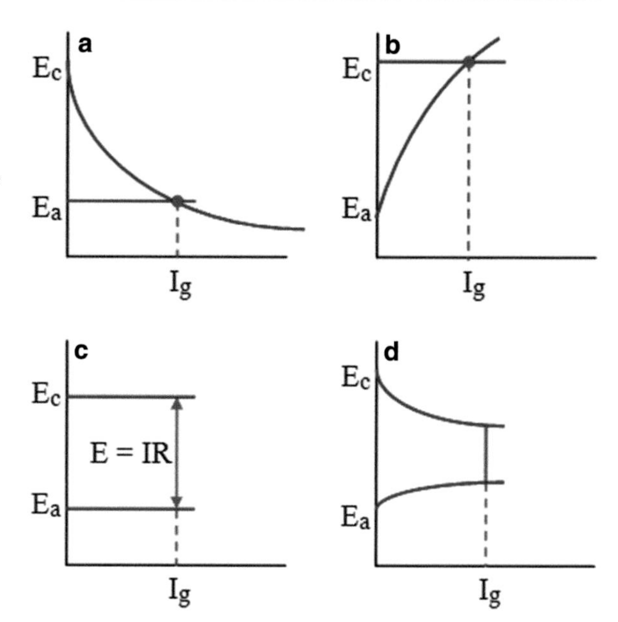

Fig. 7.7 Schematic polarization curves illustrating the polarization controlling modes. (**a**) Cathodic control, (**b**) anodic control, (**c**) resistance control, and (**d**) mixed control. After reference [16]

7.2.1.1 Polarization Controlling Modes

A polarization diagram predicts the type of half-cell controlling the electrochemical system. Figure 7.7 [16] depicts four schematic cases for determining the controlling mode of polarization. Here, I_g is the galvanic current, and $E = IR$ is Ohm's resistance potential.

7.2.2 Deviation from Equilibrium

Addition of an oxidizer to an electrolyte not necessarily increases the corrosion potential. It is the exchange current density that dictates such a situation. For instance, if $i_{o,Fe^{3+}/Fe^{2+}} << i_{o,M/M^{2+}}$, then Fe^{3+} ions do not have a strong effects on E_{corr} and i_{corr}. The opposite condition has an effect on E_{corr} and i_{corr}. Furthermore, if the applied potential is $E = 0$, then the current does not flow, and metal oxidation and oxidizer reduction occur simultaneously at the metal-electrolyte interface. Therefore, the oxidation and reduction current densities are equal and $i = i_{ox} - i_{red} = 0$. On the other hand, if $E \neq 0$, then polarization takes place due to a developed anodic or cathodic overpotential

$$\eta_a = (E - E_{corr}) > 0 \tag{7.5}$$

$$\eta_c = (E - E_{corr}) < 0 \tag{7.6}$$

The applied potential E is externally controlled, but it is the overpotential that should be analyzed because it is a kinetic parameter that indicates the degree of deviation from equilibrium. For corrosion studies, $\eta_a > 0$, and for reduction of a metal ion, $\eta_c < 0$. The latter electrochemical condition is a significant kinetic parameter for optimizing a large quantity of metal ion reduction from solution as in electrowinning of copper.

Anodic reactions represent metal deterioration since a corroding metal losses electrons. This is a surface electrochemical phenomenon that may cause drastic effects on metallic structures. Therefore, the driving force for corrosion is the overpotential (η) defined by

$$\eta = E_a - E_c \tag{7.7}$$

The subscripts "a" and "c" stand for anodic and cathodic. With regard to Fig. 7.5, assume that current does not flow, and then the local potentials become the open-circuit potentials of the corroding system. Thus,

$$E_o = \begin{cases} E_{o,c} = E_{H^+/H_2} \\ E_{o,a} = E_{Zn/Zn^{2+}} \end{cases} \tag{7.8}$$

On the other hand, if current flows, irreversible effects occur at the electrode surface due to electrochemical polarization effects. In this case, the corrosion potential and corrosion current density are compared with the cathodic and anodic terms. Thus,

$$E_c < E_{corr}, E_a \tag{7.9}$$

$$i_{corr} > i_{o,a} > i_{o,c} \tag{7.10}$$

$$i_{o,a} = i_{o,Zn/Zn^{2+}} \tag{7.11}$$

$$i_{o,c} = i_{o,H^+/H_2} \tag{7.12}$$

Additionally, electrochemical polarization is a measure of the overpotential and represents a deviation of the electrochemical state of half-cell electrodes induced by an applied external potential. Therefore, the driving force for electrochemical polarization is the overpotential.

Consequently, the electrochemical polarization is divided into two classifications, such as anodic polarization and cathodic polarization. The corresponding overpotentials are defined by Eqs. (5.58)a and (5.58)b as

$$\eta_a = \beta_a \log\left(\frac{i_a}{i_{corr}}\right) > 0 \tag{7.13}$$

$$\eta_c = -|\beta_c| \log\left(\frac{i_c}{i_{corr}}\right) < 0 \tag{7.14}$$

Recall that the overpotentials for the anodic and cathodic parts of a polarization curve represent deviation from equilibrium anodically or cathodically.

In general, the higher the current density, the higher the overpotential and the faster the electrochemical rate of reaction. Furthermore, other aspects of the polarization must be considered as a beneficial phenomenon for electrometallurgical operation, such as electrowinning, electrorefining, electroplating, and cathodic protection against corrosion or a detrimental process for material surface deterioration known as corrosion. Therefore, continuous polarization causes local potentials to change until a steady state is reached and the observed mixed potentials is the corrosion potential (E_{corr}) of the corroding electrochemical system. Thus far it has been assumed that the temperature and pressure are remain constant; otherwise, temperature and pressure gradients must included in the analysis of the electrochemical system.

7.2.2.1 Galvanic Coupling

Moreover, consider the case for polarizing a galvanic coupling, such as two different structural steels [17]. Particular hypothetical galvanic polarization curves for metals M_1 and M_2 are shown in Fig. 7.8.

These curves imply that metals M_1 and M_2 are connected (coupled) and electron flow occurs from M_1 to since M_2 since $i_{corr,M1} < i_{corr,M2}$. For convenience, assume that the metals have the same valence z. The main reactions that take place in the coupling are

$$M_1^{z+} + ze^- \rightarrow M_1 \qquad \text{(anode)} \tag{7.15}$$

$$M_2 \rightarrow M_2^{z+} + ze^- \qquad \text{(cathode)} \tag{7.16}$$

$$M_1^{z+} + M_2 \rightarrow M_1 + M_2^{z+} \qquad \text{(redox)} \tag{7.17}$$

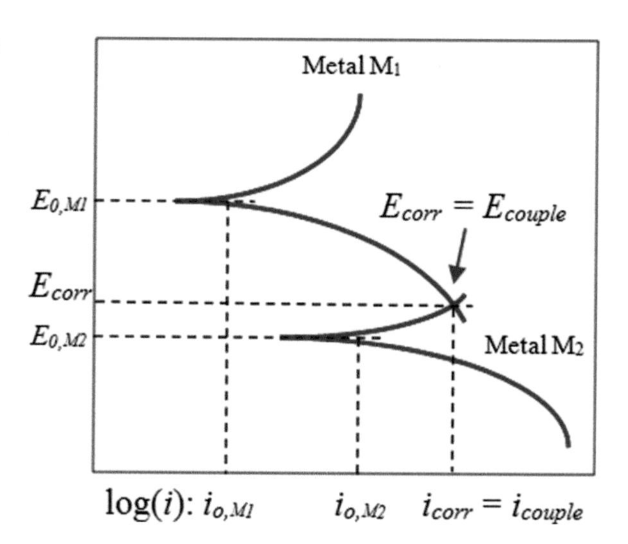

Fig. 7.8 Schematic galvanic couple polarization

The polarization behavior of metals M_1 and M_2 in a suitable electrolyte is normally characterized by using a polarization diagram, specially a potentiodynamic polarization diagram schematically depicted in Fig. 7.6 to study the hydrogen evolution mechanism on each these metals. Subsequently, assess their electrochemical behavior as a galvanic couple in the hydrogen-containing electrolyte. In practice, the kinetic parameters obtainable from a potentiodynamic polarization diagram serve as significant data for designing applications.

Thus, the coupling corrosion current density is depicted from Fig. 7.8 at the crossover (intersect) of the anodic and cathodic polarization curves. This couple current density has a value between the corrosion current density for metal M_1 and M_2, respectively:

$$i_{corr,M1} < i_{couple} = i_{corr} < i_{corr,M2} \tag{7.18}$$

When steady state is reached, the current densities and potentials become equal; that is,

$$i_{couple} = i_a = -i_c \qquad \text{@ equilibrium} \tag{7.19}$$

$$E_{couple} = E_{corr} \qquad \text{@ equilibrium} \tag{7.20}$$

It should be pointed out that the corrosion potential in the galvanic series (Table 2.3) and that in the emf series (Table 2.2) is not the same. The former may be measured as a coupled potential at different temperatures and ionic concentrations used in the latter series. Therefore, the galvanic series must be used with caution since it is temperature- and concentration-dependent kinetic parameter.

In addition, the surface area of the electrode in the corrosive medium is known as the relative area, and it influences the rate of galvanic corrosion. Therefore, the corrosion potential of a galvanic coupling is strongly dependent on the cathode-to-anode surface area ratio.

A galvanic couple is essentially a corrosion cell developed between two different metals in a common electrolyte. In this case, one metal is oxidized (anode), while the other is reduced (cathode). If the metal electrodes are in physical contact, then the less resistant metal oxidizes (corrodes) at the point of contact. Nonetheless, the corrosion potential (E_{corr}) and the corrosion current density (i_{corr}) do change (shift) significantly from their uncoupled values.

In a galvanic couple, the higher resistance metal turns cathodic, while the less resistant one becomes anodic. Typically, the cathodic material undergoes little or no corrosion at all in a galvanic couple.

7.3 Experimental Data

7.3.1 Potentiostatic Polarization

This is a stepwise technique in which the electric potential supplied by a potentiostat is increased from a prescribed potential in steps with current recorded at the end of a selected time [4]. Commonly, this technique is used for anodic polarization, but it can be used in a potential range to include cathodic measurements as shown in the potentiostatic polarization diagram in Fig. 7.9 for AISI 1080 carbon steel in deaerated 1 N sulfuric acid (H_2SO_4) solution at room temperature [18].

This polarization diagram (Fig. 7.9) is clearly divided into anodic and cathodic regions near the corrosion point along with kinetic parameters. From this diagram, $E_{corr} = -0.51\ V_{SCE}$ which is an electrochemical property based on the experimental conditions.

In principle, this potentiostatic polarization technique is similar to galvanostatic and potentiodynamic for characterizing anodic and cathodic behavior of a metal in an electrolytes. Generally, potentiostatic polarization measurements are also useful for assessing corrosion reactions on a metal surface, corrosion product identification, and alloy selection [4].

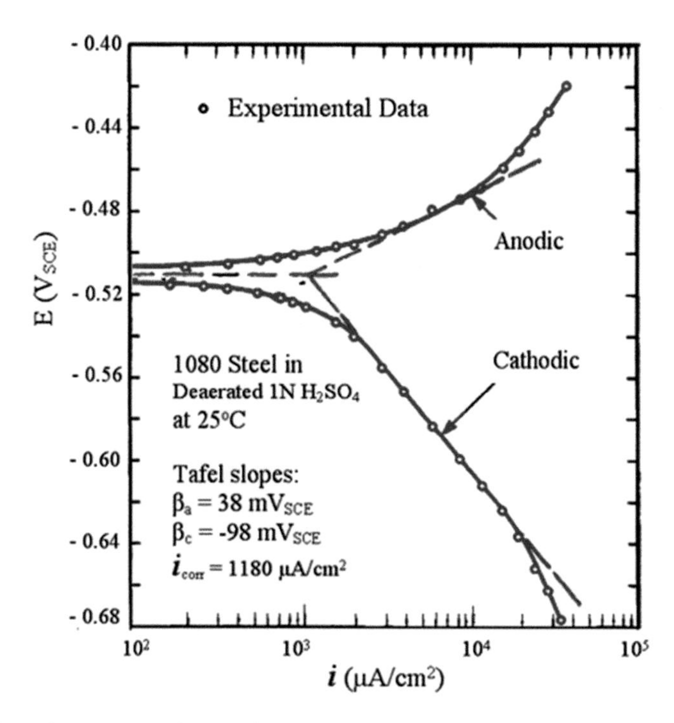

Fig. 7.9 Experimental stepwise polarization curve for 1080 eutectoid steel in deaerated 1 N sulfuric acid [18]

A polarization curve can provide evidence on whether or not a material is active, passive, or active–passive. Hence, potentiostatic testing method can give fairly reproducible results to determine the **passivity** of a metal and the **corrosivity** of an electrolyte.

7.3.2 Potentiodynamic Polarization

This is a technique that uses a programmed potentiostat to supply a continuously increasing potential (E) at a desired potential scan rate (dE/dt) along with a current logarithmic converter [4]. As a result, a potentiodynamic polarization diagram is recorded for characterizing the electrochemical behavior and extracting kinetic parameters for a working electrode. Incidentally, a polarization diagram using potentiostatic and potentiodynamic polarization techniques is equivalent. A potentiodynamic polarization is commonly used for assessing the anodic behavior of a metal, but an applied potential range can be used to include the cathodic curve.

7.3.2.1 Stainless Steels

Figure 7.10 shows the **effect of concentration** of aerated sulfuric acid (H_2SO_4) at room temperature on AISI 304 stainless steel. Clearly, potentiodynamic polarization revealed an active–passive diagram for this alloy. Among many design applications, this alloy is suitable for anodic protection.

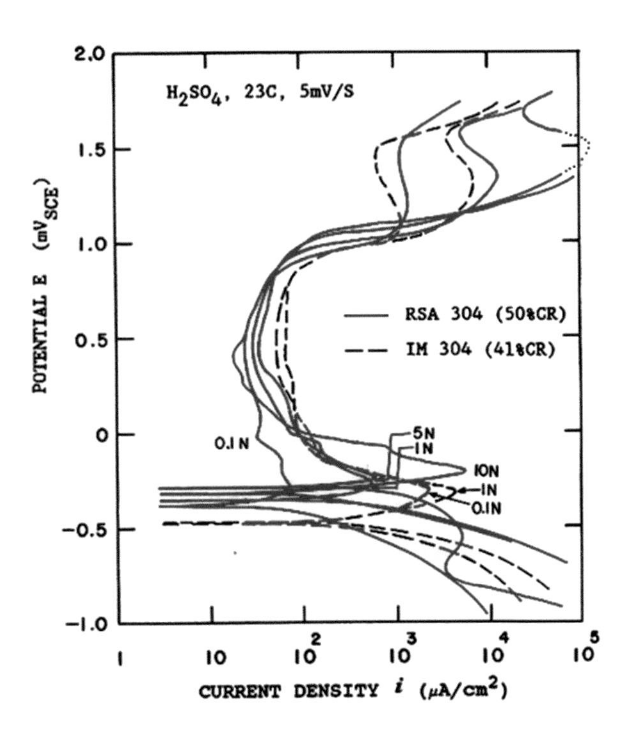

Fig. 7.10 Experimental potentiodynamic polarization curves for cold-rolled RSA 304 and IM 304 in sulfuric acid concentrations at room temperature [1]

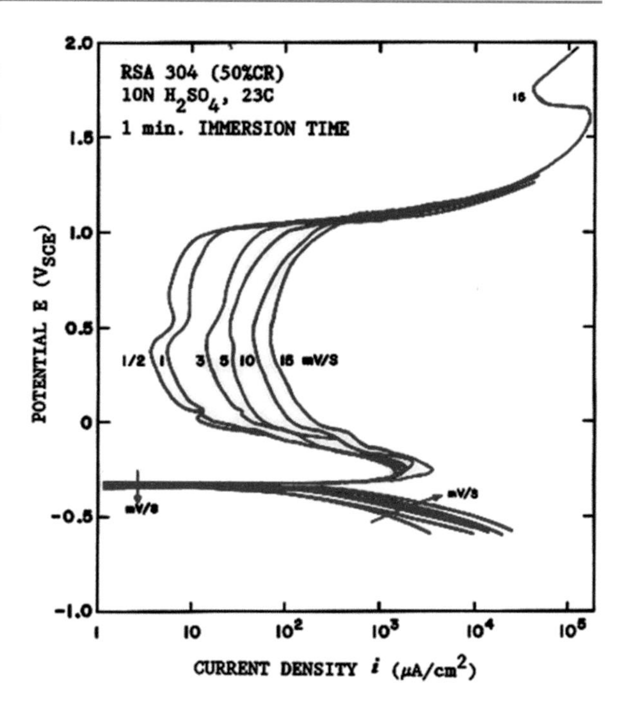

Fig. 7.11 Experimental potentiodynamic polarization curves for 50 %CR RSA 304 as functions scan rate in 10 N sulfuric acid solution at $pH = -1.2$ [1]

This stainless steel was prepared by rapid solidification and conventional ingot metallurgy solidification technologies. The former alloy is called RSA 304 and the latter IM 304. Initially, RSA 304 and IM 304 were 50 % and 41 % cold rolled (CR) at room temperature [1].

Figure 7.11 illustrates the **effect of scan rate** on the electrochemical behavior of RSA 304 stainless steel [1]. The electrochemical corrosion behavior of alloy is significantly affected by scan rate.

The importance of potentiodynamic polarization in experimentally revealing the electrochemical behavior of metals is clearly indicated in Fig. 7.11. Only the passive range is significantly affected by electric potential scan rate. The electric potentials, E_{corr}, E_{crit}, and E_p, approximately remained constant, which is a useful information for designing applications.

The passive current density i_p is not well defined, but the passive potential range is displaced to the right as scan rate increases. This passive potential range is approximately 1 V_{SCE}. In addition, the passive range is considered sufficiently wide for applications in the field of anodic protection. Other polarization parameters are not significantly affected by scan rate. For instance, the pitting (E_p) and critical (E_{crit}) potentials are approximately 1 V_{SCE} and -0.25 V_{SCE}, respectively, at all scan rates.

For comparison purpose, it is apparent from Figs. 7.10 and 7.11 that the polarization runs are adequate to evaluate in great details the electrochemical behavior of the stainless steels RSA 304 and IM 304. Conclusively, the RSA 304 is slightly

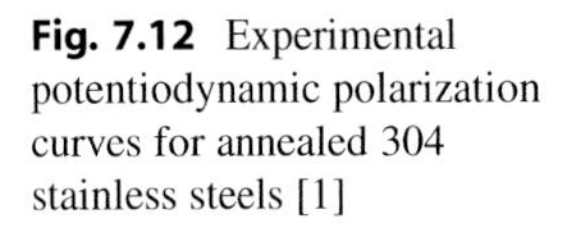

Fig. 7.12 Experimental potentiodynamic polarization curves for annealed 304 stainless steels [1]

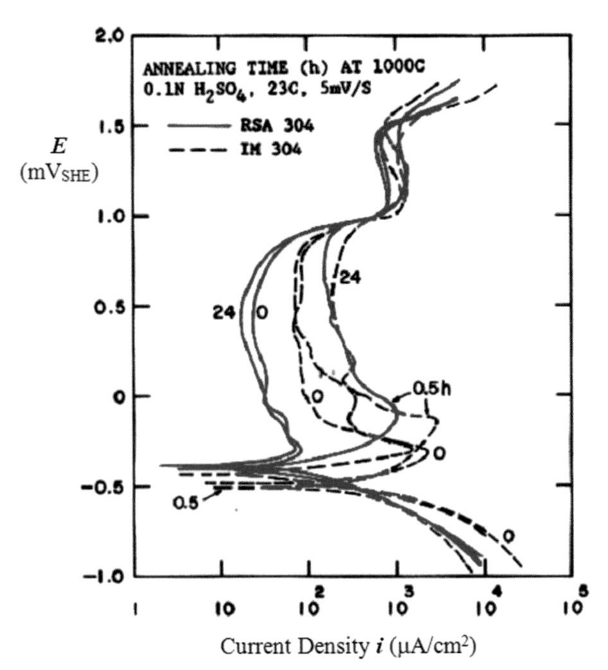

more corrosion resistant than its counterpart IM 304. The only significant different between these alloys is that foundry technique used to produce them. The RSA 304 was rapidly solidified from the melt, while IM 304 was produced using conventional solidification.

7.3.2.2 Ni-Mo-Based Alloy

Moreover, the **effect of heat treatment** (annealed microstructural conditions) on the electrochemical behavior of both stainless steels, RSA 304 and IM 304, is shown in Fig. 7.12 [1].

With respect to the corrosion rate (C_R) in familiar units, the RSA 304 heat-treated for 0.5 h at 1000 °C exhibited

$$C_R = 5.47\ \text{mm/year} \quad @ \quad i_c$$

$$C_R = 0.89\ \text{mm/year} \quad @ \quad i_p$$

$$C_R = 0.10\ \text{mm/year} \quad @ \quad i_{corr}$$

These results are very significant in design applications.

Basically, the RSA 304 is more corrosion resistant that IM 304 due to the attributable inherent properties induced by the rapid solidification processing (*RSP*) and to the formation of an oxide surface film of higher quality. It has been reported

[19] that the presence of chromium (Cr) in a H_2SO_4-containing solution is the main source for a Cr-rich oxide film and it is this surface oxide film responsible for maintaining passivity of stainless steel surface.

The most interesting case is the corrosion rate at the corrosion current density. Thus, $C_R = 0.098$ mm/year @ i_{corr} is very low in magnitude. In fact, both RSA 304 and IM 304 show high corrosion resistance in sulfuric acid solutions which can be attributed to the high-chromium (Cr) content for passivation due to the formation of a hydrated chromium oxyhydroxide, $CrO_x\,(OH)_{3-2x} \cdot nH_2O$, protective film [20].

Not all metals and alloys show active–passive behavior as 304 stainless steel in a particular environment. Aluminum and its alloys have the tendency to be active in most environments. Advanced alloys such as a $Ni - Mo$ base rapidly solidified alloy (RSA), $Ni_{53}Mo_{35}Fe_9B_2$, tested in a sulfuric acid solution passivates very briefly as indicated in Fig. 7.13 [1].

This rapidly solidified alloy was annealed for prolonged times at 1100 °C and characterized potentiodynamically in 0.10 N ($pH = 1.5$) sulfuric acid (H_2SO_4) solution at 25 °C. As a result, this RSA acts as an active material, despite that its corrosion current density is very low in the order of 11 $\mu A/cm^2$. The electrochemical behavior of this alloy in H_2SO_4 may be attributed to galvanic effects due to several boride particles embedded in the Ni-Mo matrix [21]. These borides were revealed by X-rays as

$$Mo_2NiB_2,\ Mo_2FeB_2 \text{ and } MoFe_2B_4$$

$$BFe_2 \text{ and } BFe_3$$

$$B_6Fe_{23}$$

In addition, **anodizing** is a process in which aluminum and aluminum alloys are readily oxidized by an adherent and protective oxide film on the surface and further oxidation of the film is by solid state diffusion. However, these materials are active in an electrochemical environment containing sodium chloride.

7.3.2.3 Al-Li Alloys

The high strength to weight ratio makes Al-Li alloys (2195, OX24, OX27) potential candidates for aerospace applications, but their electrochemical behavior render them as poor corrosion-resistant materials in certain environments. For instance, 2195 Al-Li alloy is susceptible to pitting corrosion in 3.5 % NaCl as shown in Fig. 1.11. This is attributed to aggressive chloride ions.

For instance, Fig. 7.14 shows experimental potentiodynamic polarization curves for 2195 Al-Li alloy being aged at 190 °C and tested in deaerated 3.5 %NaCl solution. All polarization curves show active electrochemical behavior since the formation of an oxide protective film does not occur on all aged conditions. Despite that this alloy passivates in air due to the anodizing process, it becomes active electrochemically since a critical current density does not develop in the tested solution. Therefore, the alloy does not passivate; instead, it oxides potentiodynamically when the applied potential is above the corrosion potential.

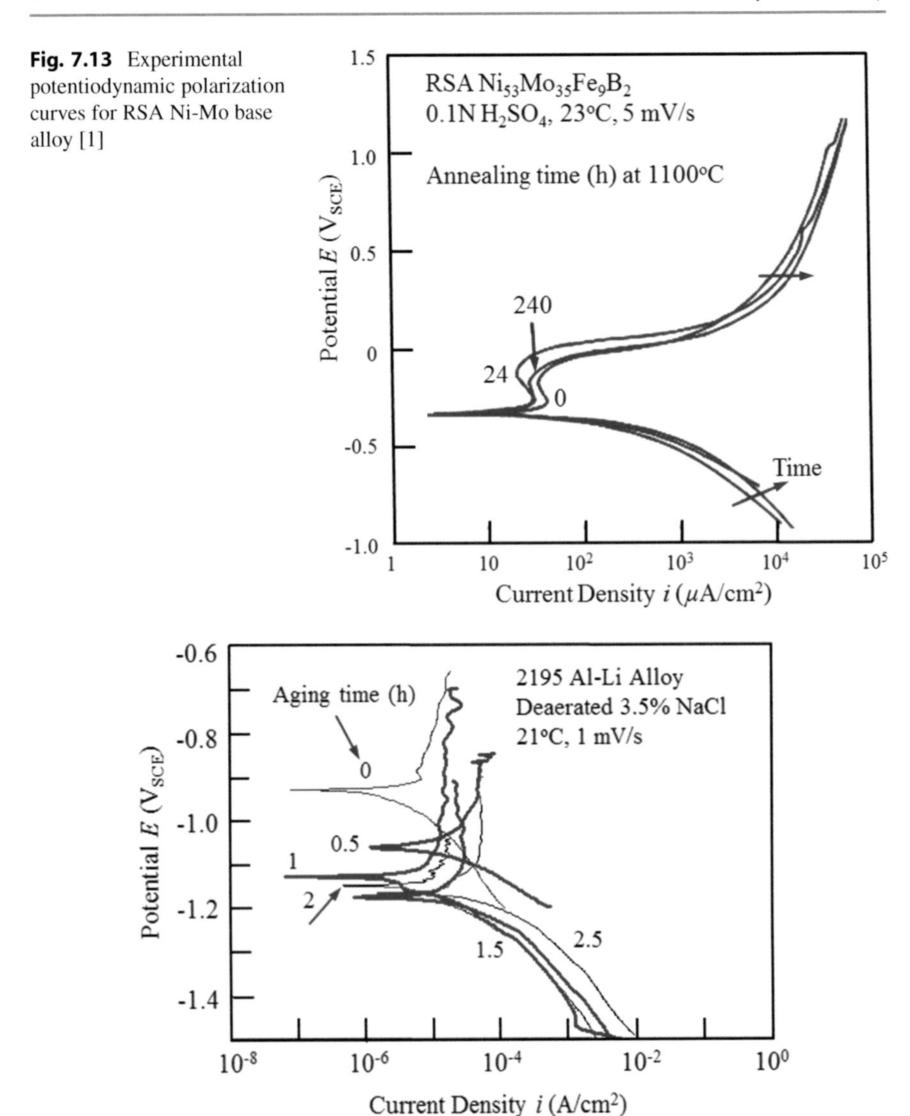

Fig. 7.13 Experimental potentiodynamic polarization curves for RSA Ni-Mo base alloy [1]

Fig. 7.14 Effect of aging time on polarization of 2195 Al-Li alloy in deaerated 3.5 % sodium chloride solution. Experimental data taken by Laura Baca (graduate student)

Furthermore, both corrosion potential and corrosion current density for the aged alloy are affected by the aging time (heat treatment time) due to the precipitation of secondary phases, which may form localized galvanic cells on the surface of the 2195 Al-Li alloy [22].

According to the experimental data in Fig. 7.14, heat treatment time improved E_{corr}, but I_{corr} slightly increased making the Al-Li alloy less corrosion resistant in 3.5 % NaCl. Therefore, potentiodynamic polarization is a fast technique for determining the electrochemical corrosion of the alloy.

7.4 Cyclic Polarization Curves

Essentially, a cyclic polarization curve is a representation of a mixed potential phenomenon in which the protective potential (E_{prot}) is determined by reversing the scan rate (reverse scan) until the forward scan curve is intercepted. This is schematically shown in Fig. 7.15 for active–passive and active curves. In general, E_{prot} is the potential point at which the reverse scan intersects the forward scan curve, and it is the potential for the repassivation of previously formed pits. Additionally, the smaller the hysteresis area, the more localized or pitting corrosion resistant a material ought to be in a particular corrosive environment [23], and pitting should not form or grow at $E < E_{prot}$ [17].

The active–passive cyclic curve shows the protective potential against pitting corrosion. This potential is below the pitting (breakdown) corrosion potential (E_p). The area of the hysteresis loop is virtually the power supplied to the electrode surface [24]. The more active both E_{prot} and E_p are, the more resistant the metal is to crevice corrosion [25]. Among all suitable applications of the cyclic polarization technique, the biological field can benefit since the pitting potential can be used as a measure of the resistance of metallic implants exposed to human body fluids. According to Fontana [10], a body fluid is an aerated physiological saline solution containing approximately 1 % $NaCl$, other salts and some organic compounds at 37 °C, and the corrosivity of the human body fluid is similar to that of aerated warm water. The Tyrode's [26] and Hank's [27] physiological saline solutions are nowadays commercially available.

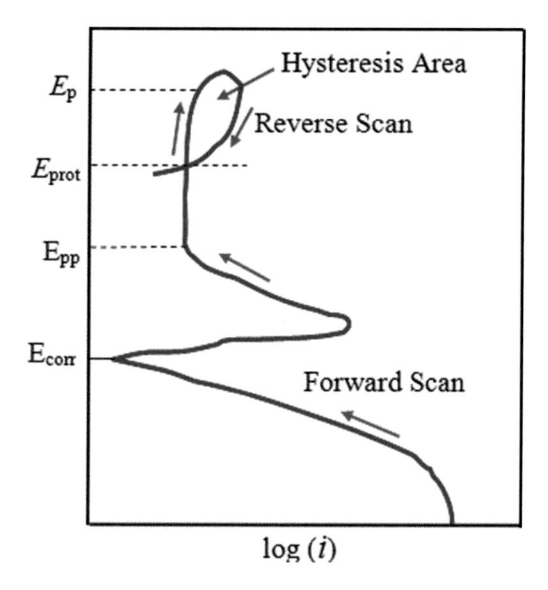

Fig. 7.15 Schematic cyclic polarization curves showing the protected potential E_{prot}

7.5 Metal-Oxide Film

Passivity is an electrochemical reduction mechanism that occurs on a metal-electrolyte interface. The resultant reduction product is a solid film of a metal-oxide compound having a stoichiometric reaction (a reaction that goes to completion) based on the corroding metal. Thus, passivity occurs on metals that are active–passive by **natural passivation** without the influence of external potentials or **artificial passivation** due an applied external anodic potential.

Hereafter, artificial passivation will be referred though out the context of this chapter, unless stated otherwise. The passive oxide film is a solid interfacial oxide compound that protects the metal against further oxidation and ranges from 1 to 10 nm in thickness [12].

Apparently, thin films in the order of 1–2 nm in thickness are of higher quality than thick films due to lesser atomic defects. A metal that exhibits passivity is thermodynamically unstable within a potential range independent or nearly independent of current or current density. This means that the metal is unstable in the passive state since a slight disturbance may increase the passive potential to or above the pitting potential causing film breakdown.

Furthermore, chemical passivity is a state related to cathodic reactions on the metal surface, while electrochemical passivity depends on an external anodic potential to force cathodic reactions to occur [28]. According to Fig. 7.4, the i_c-i_p region defines the active/passive transition. Passivation starts at i_c and ends at i_p. Thus, film thickness increases from the E_{pa} potential at i_p, but ionic conductance of the oxide film controls its thickness. Moreover, high ionic conductivity of metal cations promotes a thick film. The opposite occurs for a low ionic conductivity [29, 30].

7.5.1 Microstructural Defects

Most metals and alloys do not show a well-defined passive potential range as indicated schematically in Fig. 7.4. But, those that do, such as polycrystalline alloys, have numerous defects, such as voids, dislocations, grain boundaries, and the like, contributing to the passive behavior. This implies that the inherent formation of passive oxide films has a complex mechanism due to these defects. The passive oxide film and passive behavior are strongly dependent on the scan rate (dE/dt) as shown in Fig. 7.11.

Nonetheless, the right combination of an active–passive material and an electrolyte can induce the formation of a passive oxide film using the potentiodynamic polarization technique. Subsequent high magnification microscopic work provides the suitable images for revealing the oxide film and its surface characteristics, including defects. However, passivity depends on the metal-electrolyte system, in which the current density to maintain passivity is $i_p << i_c$ at high oxide film resistance. The stability of passivity can be achieved if the potential is kept low within the passivation region $E_{pp} \leq E < E_p$ (Fig. 7.4).

7.6 Mechanism of Oxide Growth

Figure 7.16a exhibits images of copper clusters (islands) taken after 1 second (1 s) exposure. Observe that the clusters in the ex situ SEM and the in situ STM microphotographs are almost uniform in size. The ex situ SEM image indicates that copper cluster formation was due to somewhat instantaneous nucleation kinetics [31].

On the other hand, Fig. 7.16b shows a STM image of copper clusters on $Zn(0001)$ after 41 min exposure. The symbol 0.25 ML stands for 0.25 milliliters of copper solution that was mixed with the electrolyte used in the experiments [31, 32].

This is a low-temperature oxidation process for the formation of a 3D metal-oxide film, such as the aluminum oxide Al_2O_3 film on pure aluminum. The metal-oxide film formation on a flat or an irregular metal surface occurs when cations (M^{z+}) and mainly oxygen anions (O^-) attract each through an electron balance process, which randomly produces metal-oxide islands (clusters) across the surface. The formation of these islands is an electrochemical ionic exchange event [32]. Island growth and subsequent new island formation take place through an activated process that is attributable to an image force between the metal and oxygen ions [33].

When the metal surface is covered by an initial film thickness, the film thickness increment due to ion movement into and through the metal oxide is attributable to a quantum mechanical process of electron tunneling through this initial thin metal-oxide film [34]. During this tunneling process, an electron can penetrate an energy barrier without thermal activation.

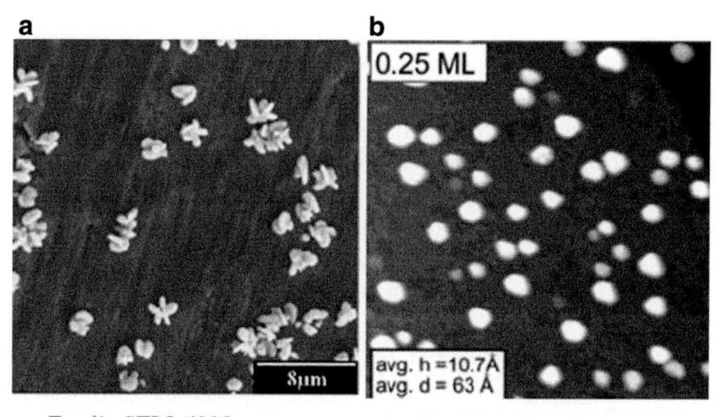

Fig. 7.16 (**a**) Copper clusters on Ti foil during copper plating for 1 s [31]. (**b**) Copper clusters (islands) (bright spots, 2-nm average diameter) on Zn(0001) surface after 41 min. The clusters height (h) and average diameter (d) are given in the image [32]

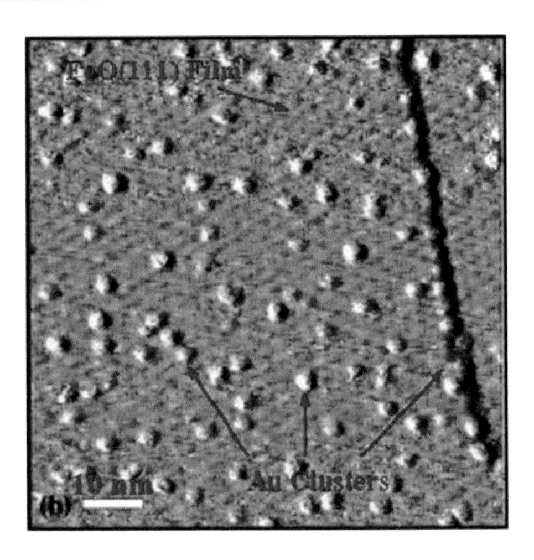

Fig. 7.17 In situ STM image of Au clusters (particles) on FeO(111) film taken at 2 mbar H_2 pressure and room temperature. This image is in Fure 4b of reference [37]

Despite that the rate of film thickness depends on the orientation of crystallographic planes on the film surface [35, 36], film thickness increases at the expense of a decrease in the electric field strength across the film, and therefore, the process ceases when the film reaches a characteristic thickness.

A full characterization of a metal-oxide film requires the determination of its thickness, morphology, chemical composition, degree of crystallinity, phase constitution, instability, electrochemical, and even mechanical properties at specific conditions.

The images in Fig. 7.16 clearly show random cluster formation on the substrates, and they represent the initiation of metal-oxide film formation. Eventually, the substrates surfaces are completely cover with a copper film, which thickens to an extent due to weakening of the electric strength field strength across the film.

Furthermore, the effects of relatively high pressure on metal-oxide films is important in understanding the behavior of a pure metal being deposited on it. One particular research finding is shown in Fig. 7.17. The in situ *STM* image exhibits gold (Au) clusters (particles) deposited on a thin $FeO(111)$ film at elevated pressures of CO, O_2, $CO + O_2$, and H_2 at room temperature [37].

It has been reported [37] that the relatively large (3 nm) Au clusters were quite stable at room temperature on the $FeO(111)$ films being exposed to above gaseous environments, except in CO gas. Apparently, gold clusters were somewhat mobile gold in CO ambient at room temperature.

7.7 Kinetics of Passivation

Assume a defect-free single crystal and a mechanism of oxide film growth by vacancy migration. Thus, the rate of film formation for a single crystal, related to Faraday's law of electrolysis, can be approximated as [28]

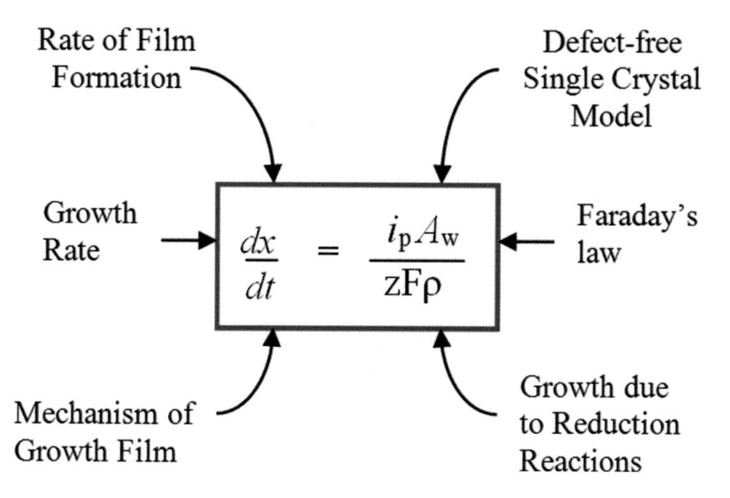

Fig. 7.18 Significance of the rate of film formation

$$\frac{dx}{dt} = \frac{i_p A_w}{zF\rho} \tag{7.21}$$

$$i_p = c_x \frac{dx}{dt} \tag{7.21a}$$

where x = film thickness (cm)
dx/dt = rate of film formation (cm/s)
i_p = passive current density (A/cm^2)
z = valence (oxidation state)
F = 96,500 C/mol (=A s/mol)
ρ = density of metal (g/cm^3)

Note that Eq. (7.21) mathematically resembles Eq. (5.90), but both have different meanings, and it is related to several factors shown in Fig. 7.18.

In addition, if the Arrhenius equation applies, then the rate of film growth is described by the forward/reverse current densities at equilibrium:

$$i_f = i_r = \alpha z F \exp\left(-\frac{Q}{RT}\right) \tag{7.22}$$

where α = rate constant (mol/s)
Q = activation energy (J/mol)
R = gas constant = 8.314 J/mol K
T = absolute temperature (K)

Now, apply an anodic overpotential for film growth to take place, and consequently, an electric field potential gradient (η/x) develops. If there exists a linear electric field across the film, then the forward (f) and reverse (r) current densities become [28], respectively,

$$i_f = \alpha zF \exp\left(\frac{-Q - zFL\eta/x}{RT}\right) \tag{7.23}$$

$$i_r = \alpha zF \exp\left(\frac{-Q + zFL\eta/x}{RT}\right) \tag{7.24}$$

Here, L is the distance from the electrode surface at which a potential drop exists in the electrolyte and $zFL\eta/x$ is the energy barrier due to an overpotential.

If the net current density at a distance x is $i_x = i_f - i_r$, then

$$i_x = i_o\left[\exp\left(\frac{B\eta}{x}\right) - \exp\left(-\frac{B\eta}{x}\right)\right] \tag{7.25}$$

$$i_x = 2i_o \sinh\left(\frac{B\eta}{x}\right) \tag{7.26}$$

where the exchange current density i_o and the constant B are

$$i_o = \alpha zF \exp\left(-\frac{Q}{RT}\right) \tag{7.27}$$

$$B = \frac{zFL}{RT} \tag{7.28}$$

If $B\eta/x \to \infty$, then Eq. (7.25) yields the high field equation

$$i_x = i_o \exp\left(\frac{B\eta}{x}\right) \tag{7.29}$$

If $B\eta/x \to 0$, then Eq. (7.26) yields

$$i_x = 2i_o\left(\frac{B\eta}{x}\right) = \frac{\eta}{R_x} \tag{7.30}$$

since the film resistance due to ohmic effect can be defined as

$$R_x = \frac{x}{2i_oB} \tag{7.31}$$

For film formation with $i_x < i_c$ (critical), Eq. (7.21) can be redefined as

$$\frac{dx}{dt} = \frac{i_xA_w}{zF\rho} \tag{7.32}$$

Substituting Eq. (7.26) into (7.32) yields

$$\frac{dx}{dt} = \frac{2i_o A_w}{zF\rho} \sinh\left(\frac{B\eta}{x}\right) \tag{7.33}$$

If

$$\lambda = \frac{2i_o A_w}{zF\rho} \tag{7.34}$$

$$\theta = \frac{B\eta}{x} \tag{7.35}$$

then Eq. (7.33) becomes

$$\frac{dx}{dt} = \lambda \sinh(\theta) \tag{7.36}$$

where $-1 \leq \theta \leq +1$ rad.

Since the overpotential $\eta > 0$ for anodic polarization, the profile for the rate of oxide film formation, Eq. (7.36), is theoretically shown in Fig. 7.19.

The thermodynamics of passivity together with the **Nernst equation** can be generalized in simple reaction steps for the formation of an oxide film (MO). Hence:

- Metal reduction

$$M^{2+} + 2e^- \rightarrow M \tag{7.37}$$

$$E_{M^{2+}/M} = E_M^o - \frac{RT}{zF} \ln \frac{[M]}{[M^{2+}]} \tag{7.38}$$

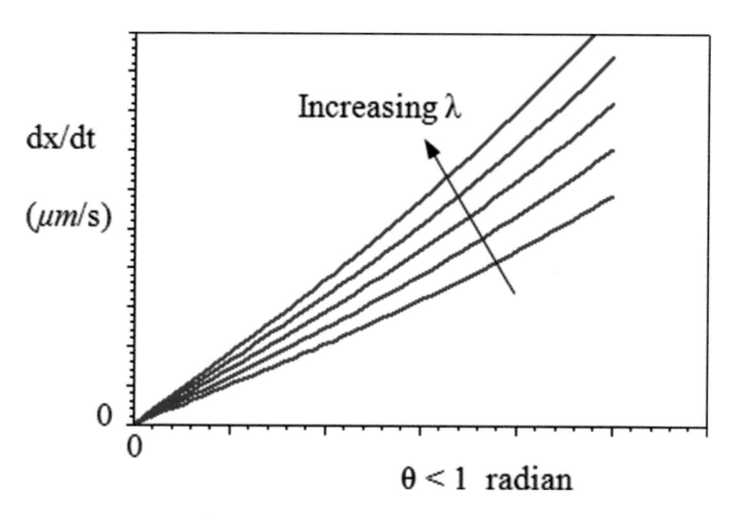

Fig. 7.19 Theoretical profile for the rate of oxide film formation

- Metal oxide reduction

$$MO + 2H^+ + 2e^- \rightarrow M + H_2O \tag{7.39}$$

$$E_{MO/M} = E^o_{MO/M} - \frac{RT}{zF} \ln \left[\frac{[M]\,[H_2O]}{[MO]\,[H^+]^2} \right] \tag{7.40}$$

$$E_{MO/M} = E^o_{MO/M} + \frac{2RT}{zF} \ln \left[H^+ \right] \tag{7.41}$$

since $[M] = [H_2O] = [MO] = 1$ mol/l.

- Formation of metal oxide

$$M^{2+} + H_2O \rightarrow MO + 2H^+ \tag{7.42}$$

$$E_{M^{2+}/MO} = E^o_{M^{2+}/MO} - \frac{RT}{zF} \ln \left[\frac{[MO]\,[H^+]^2}{[M^{2+}]\,[H_2O]} \right] \tag{7.43}$$

$$E_{M^{2+}/MO} = E^o_{M^{2+}/MO} - \frac{RT}{zF} \ln \left[\frac{[H^+]^2}{[M^{2+}]} \right] \tag{7.44}$$

$$E_{M^{2+}/MO} = E^o_{M^{2+}/MO} + \frac{RT}{2.3026zF} \left[2pH + \log \left[M^{2+} \right] \right] \tag{7.45}$$

where $[MO] = [H_2O] = 1$ mol/l and

$$pH = -\log \left[H^+ \right] \tag{7.46}$$

Passive oxide films are extremely difficult to characterize due to their thin thickness and strong adhesion on the metal surface. Some schematic passive oxide films are shown in Fig. 7.20 for some metals and alloys.

One particular application of passive films is through the electrogaining technique in which metal surface is roughen by pitting corrosion for printer's lithographic aluminum sheets in HCl or H_2NO_3 acid solution [38].

Apparently, metals and alloys that readily passivate may be due to rapid formation of the passive oxide film. If the film forms, then the metal is easily protected. This may be an indication of the higher quality of passive films, which induce higher pitting potentials and lower corrosion rates.

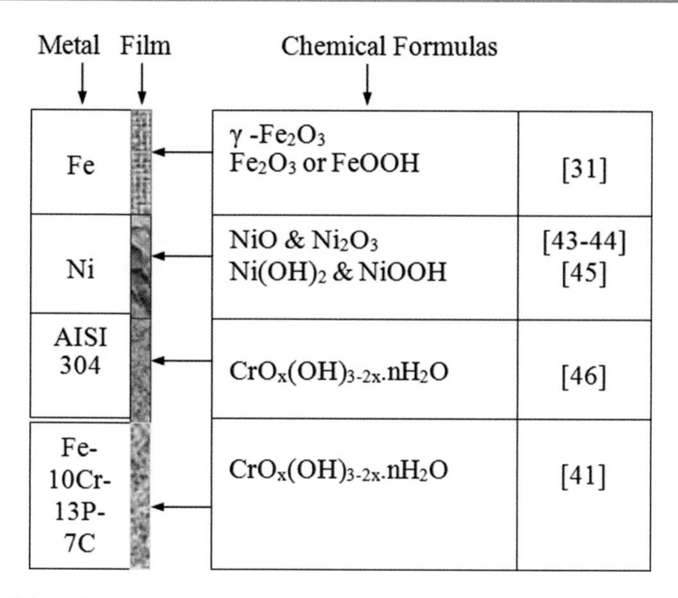

Fig. 7.20 Schematic oxide passive films

7.7.1 Ex Situ and In Situ Measurements

Performing experimental measurements require use of appropriate instruments. Each field of science has its own techniques which provide certain experimental results, such as data or simply images for further analysis. Some measurements are considered in situ when done on site and others are ex situ when performed after an event has occurred. This implies that electrochemical measurements can be in situ or ex situ.

Nonetheless, describing all available techniques for in situ and ex situ measurements is out of the scope of this section. Only a list of techniques are cited below as references.

In electrochemistry, the structure, composition, and thickness of a passive oxide film may be determined using ex situ (without potential-control) measurements. The advantages of ex situ measurements is that they generally require short time periods. Thus, ex situ (after an event) measurements can be performed using techniques like [39, 40]

- Electron spectroscopy (AES)—for electronic structures
- Ion mass spectroscopy (IMS)—for chemical compounds
- X-ray diffraction (XRD)—for crystallography studies
- X-ray photoelectron spectroscopy (XPS)—for elemental composition

Alternatively, in situ measurements are done on site and in the electrochemical environment. The advantages of in situ measurements are that they are directly implemented in the electrochemical environment and that are they precise in qualitative and quantitative analysis of oxide films. Nonetheless, further characterization of oxide films can be done in situ (under potential control) using techniques such as [39–41]

- Ellipsometry—for dielectric properties
- Scanning tunneling microscope (STM)—for imaging surfaces
- Scanning electrochemical microscopy (SECM)—for behavior of interfaces
- Atomic force microscopy (AFM)—for identification of atoms
- Frequency response analysis (FRA)—for output spectrum of systems
- Photocurrent spectroscopy (PS)—for photoelectric effects

Ellipsometry is an optical technique that uses polarized light, generally in an elliptic state, to probe the dielectric properties of thin films, such as film thickness, morphology, and chemical composition.

7.7.2 Formation of Oxide Films

A distinction between surface layers and passive oxide films should be made. If a metal is insoluble in an environment, it may form an insoluble oxide corrosion product on the metal surface as a crystalline and poorly adherent surface layer (nonpassive thick layer), such as the blue-green layer on corroded copper plumbing and well-known brown "rust" layer on iron and iron alloys.

On the other hand, if a metal is soluble in a solution, it oxidizes and dissolves in the solution as cations. These cations react with dissolved oxygen and are deposited back on the metal as oxides, forming a passive oxide film under the influence of an electric field. Also, the metal may be oxidized readily on the metal-solution interface forming a spontaneous passive oxide film, protecting the base metal from corrosion [28].

An example of reaction steps for a passive film formation is given below. According to Burstein [28], a passive nickel (Ni) oxide film may form through several reaction steps on the metal surface. One possible mechanism is as follows:

$$Ni + H_2O \rightarrow NiOH_{ads} + H^+ + e^- \quad \text{(insoluble monolayer)} \tag{7.47}$$

$$NiOH_{ads} + H^+ \rightarrow Ni^{2+} + H_2O + e^- \quad \text{(protonation)} \tag{7.48}$$

$$NiOH_{ads} \rightarrow NiO_{film} + H^+ + e^- \quad \text{(deprotonation)} \tag{7.49}$$

The subscript "ads" stands for adsorption, which means that the product $NiOH_{ads}$ is in contact with the metal surface, and it is said to be adsorbed, while $NiOH_{film}$ is a thin and coherent film on the metal surface forming a metallic couple, and it may be crystalline having its own atomic structure.

Despite the protective action of a passive oxide film, it is thermodynamically unstable in particular environments. However, the film may be amorphous in a specific environmental condition. Suppose that nickel oxide (NiO) film has a crystalline structure, and for comparison, $CrO_x\,(OH)_{3-2x}\cdot nH_2O$ is amorphous, then it can be deduced that the latter is of higher film quality than the former due to the lack of crystal defects, such as grain boundaries. This Cr oxide along with Fe oxide gives stainless steel its common characteristics of being a corrosion-resistant ferrous (Fe) alloy containing chromium (Cr).

In fact, it is apparent that passive oxide films are amorphous and insoluble, and they isolate the metal surface from solution. However, if a passive oxide film has defects, thin films from 1 to 2 nm in thickness are preferred over thicker ones due to fewer defects [28].

The passive film thickness (x) at time t can be predicted using Eq. (7.21), provided that the passive current density $\left(i_p\right)$ remains constant in the passive potential range, $E_{pp} \leq E \leq E_p$, as shown in Fig. 7.4. Hence, integrating Eq. (7.21) gives an expression for the thickness of a passive film:

$$\int_o^x dx = \int_o^t \frac{i_p A_w}{zF\rho} dt \tag{7.50}$$

$$x = \frac{i_p A_w t}{zF\rho} \tag{7.51}$$

Example 7.1. *A hypothetical stainless steel pipe is used to transport an aerated acid solution containing* $10^{-6}\,mol/cm^3$ *of dissolved oxygen at room temperature. The electrical double layer is* 0.4 *and* 0.8 *mm under static and flowing velocity, respectively. Assume that the critical current density (Fig. 7.4) for passivation and the diffusion coefficient of dissolved oxygen are* $300\,\mu A/cm^2$ *and* $10^{-5}\,cm^2/s$*, respectively.* ***(a)*** *Determine whether or not corrosion will occur under the given conditions, and* ***(b)*** *if passivation occurs, then calculate the passive film thickness at 2-min exposure time. Assume that the density of the film and the molecular weight of the stainless steel are* $7.8\,g/cm^3$ *and* $55.85\,g/mol$*, respectively, and that the predominant oxidation state is* 3.

Solution. *Given data:*

$F = 96{,}500\,A\,s/mol$ $\quad C_x = 10^{-6}\,mol/cm^3$
$z = 3$ $\quad A_w = 55.85\,g/mol$
$D = 10^{-5}\,cm^2/s$ $\quad \rho = 7.8\,g/cm^3$
$i_{crit} = 300\,\mu A/cm^2$ $\quad \delta = 0.04\,cm\ (static)$
$t = 2\,min = 120\,s$ $\quad \delta = 0.08\,cm\ (dynamic)$

The reduction reaction is

$$O_2 + 2H_2O + 4e = 4OH^- \quad with \quad E^o_{O_2} = 0.401\,V_{HSE}$$

(a) *From Eq. (6.99), the limiting current density for both conditions is*

$$i_L = \frac{zFDC_x}{\delta} = \frac{(3)\,(96500\,\text{A s/mol})\,(10^{-5}\,\text{cm}^2/\text{s})\left(10^{-6}\,\text{mol/cm}^3\right)}{0.07\,\text{cm}}$$

$$i_L \simeq 72.38\,\mu\text{A/cm}^2 < i_{crit}\ \textit{for corrosion under static solution}$$

$$i_L \simeq 36.19\,\mu\text{A/cm}^2 > i_{crit}\ \textit{for passivation under flowing solution}$$

(b) *The limiting current density for passivation under flowing solution becomes the passive current density, and then the passive film thickness can be estimated using Eq. (7.51) along with a density of 7.8 g/cm*^{3} *and a molecular weight of 55.85 g/mol. Hence,*

$$x = \frac{i_p A_w t}{zF\rho}$$

$$x = \frac{\left(482.50x10^{-6}\,\text{A/cm}^2\right)(55.85\,\text{g/mol})\,(120\,\text{s})}{(3)\,(96500\,\text{A s/mol})\left(7.8\,\text{g/cm}^3\right)}$$

$$x = 14.36\,\text{nm}$$

This is a reasonable result despite the crude approximation technique being used in this example problem. Also, the passive oxide film is assumed to be chemically stable, and it may have the formula $CrO_x(OH)_{3-2x} \cdot nH_2O$ *as per Fig. 7.20.*

7.8 Concrete Corrosion

In this section, concrete corrosion is referred to as corrosion of steel bars embedded in concrete. Concrete is predominantly composed of aggregates, hydraulic cement, and water. Other cementitious materials and chemical admixtures can be added to improve the basic concrete properties. Embedding carbon steel or stainless steel bars in concrete is a common technique for reinforcing the mechanical strength of the resultant cementitious composite known as "reinforced concrete." Further improvement of the reinforced concrete is achievable by adding chemical admixtures to the basic concrete mixture.

In general, failure of reinforced concrete induced by external forces and thermal cycling effects can be prevented by a maintenance and prevention plan. However, steel-reinforced concrete is used in severe environments like seawater and de-icing salts. The latter environment contains chloride ions (Cl^-) which diffuse into the concrete, and when it reaches the passive steel bars, it breaks down the passive film, and consequently, the steel bars corrode by general rust formation or by pitting mechanism.

Also, carbon dioxide (CO_2) gas from the atmosphere can diffuse into concrete defects and enhance the corrosion process [33, 42]. Therefore, corrosion mitigation strategies included materials selection, design approach, protective coatings, modification of environment, and cathodic protection.

7.8.1 Significance of Maintenance

In general, reinforced concrete structures are constantly exposed to diverse environmental conditions and can last decades without significant damage. However, concrete deterioration can be accelerated when subjected to negligence, abuse, and lack of maintenance in severe or aggressive environments. Thus, routine inspections help establish a long-term maintenance plan for the reinforced concrete structure. For instance, the common deficiencies that can be visually detected on a flat roof (slab) include pondings, splitting, blistering (swelling that feels spongy when pressed), fishmouthing (a raising of a portion of the roof caused by moisture buildup, and lifting of laps at the edges of bituminous roofs) and punctures.

Specifically, ponding (the formation of ponds of water on the roof) increases the rate of deterioration substantially because it creates a large reservoir of water that will further damage the interior (if a leak occurs), and it is usually the result of structural settlement and localized roof failure. On the other hand, splitting is the occurrence of long cracks in the roof membrane which can be attributed to physical stress, ponding, cracking of the substrate, or bad workmanship. Consequently, the penetration of chlorides, carbon dioxide (CO_2), water, and oxygen becomes a significant problem in due time since the reinforced steel bars will corrode, causing cracking, spalling, and rust staining of the ceiling (roof interior) [33, 42]. In such a case, aesthetic appearance becomes an issue. Nonetheless, installing an insulating layer on the roof corrects any deficiency and prevents concrete corrosion from taking place.

Since reinforced concrete structures are subjected to corrosion from airborne salts, moisture, and carbon dioxide (CO_2) from the atmosphere, the quality and pertinent maintenance of the porous concrete play a major role in the penetration (ingress) of these species and time to corrosion. In effect, CO_2 diffuses through the porous concrete (carbonation) and neutralizes the alkalinity of the concrete ($pH < 12$), leading to the destruction of the passive oxide film on the reinforcing steel, and doubtlessly it contributes to the onset of chloride-induced corrosion. Categorically, concrete structures exposed to hot tropical marine environments are especially susceptible to concrete deterioration since corrosion rates are greatly influenced by humidity, temperature cycles, and resistivity of the concrete. Therefore, a suitable maintenance plan must prevail at all times in order to avoid structural damage due to steel corrosion.

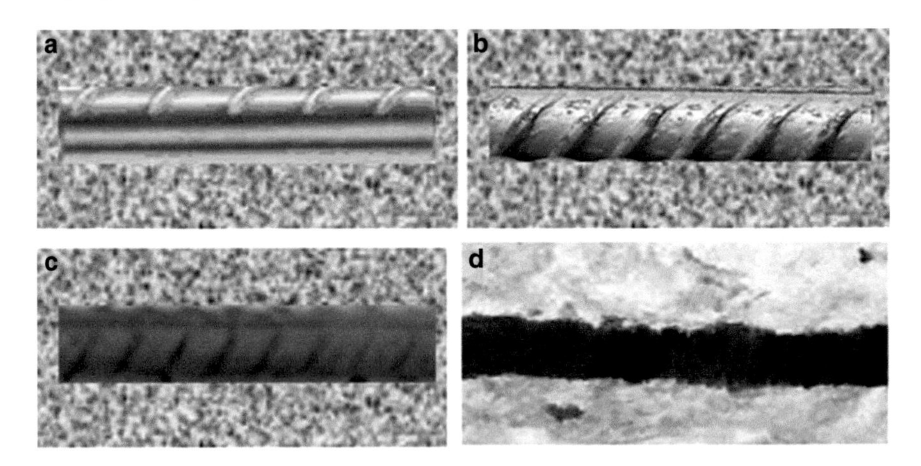

Fig. 7.21 Four stages of a hypothetical carbon steel bar embedded in concrete containing limited amount of water in pores. (**a**) Corrosion free bar. (**b**) Slightly passivated. (**c**) Passivated bar. (**d**) Heavily corroded

7.8.2 Corrosion of Steel in Concrete

Although the electrochemical corrosion process of steel in reinforced concrete containing chloride ions has been studied for decades [24, 40, 43, 44], the corrosion rate of steel due to the dynamic relationship between chloride Cl^- and ferrous Fe^{2+} ions is affected by chemical and physical processes that take place on the corroding steel surface. However, once corrosion starts, its rate may be accelerated, leading to severe structural damage.

Figure 7.21 illustrates a possible four-stage corrosion case for a steel bar being embedded in concrete.

The steel bar goes from sound (corrosion free) and clean to heavily corroded conditions. The steel bar in Fig. 7.21d is the only picture being taken from a damaged residence ceiling, which underwent spotty corrosion, cracking, and spalling because the hydrated ferric hydroxide or brown rust, $Fe_2O_3 \cdot 3H_2O$, increased its volume as corrosion propagated in a general form. In most cases, the surface appearance (aesthetic) of the spotty corrosion or general corrosion is not an acceptable state.

Furthermore, Fig. 7.22 shows pictures taken from residence ceilings (exposed to high humidity in the Caribbean) that were damaged by spotty corrosion induced by chloride ions.

It is event that corroded steel bars cracked the concrete, leading to spalling. The top left picture in Fig. 7.22 shows some steel bars being placed very close to the bottom of the concrete slab. This is a construction defect. The top right picture shows a rust stain and a branched wide crack, indicating that the embedded steel bars imparted very high tensile stresses in the concrete. The bottom left picture shows an

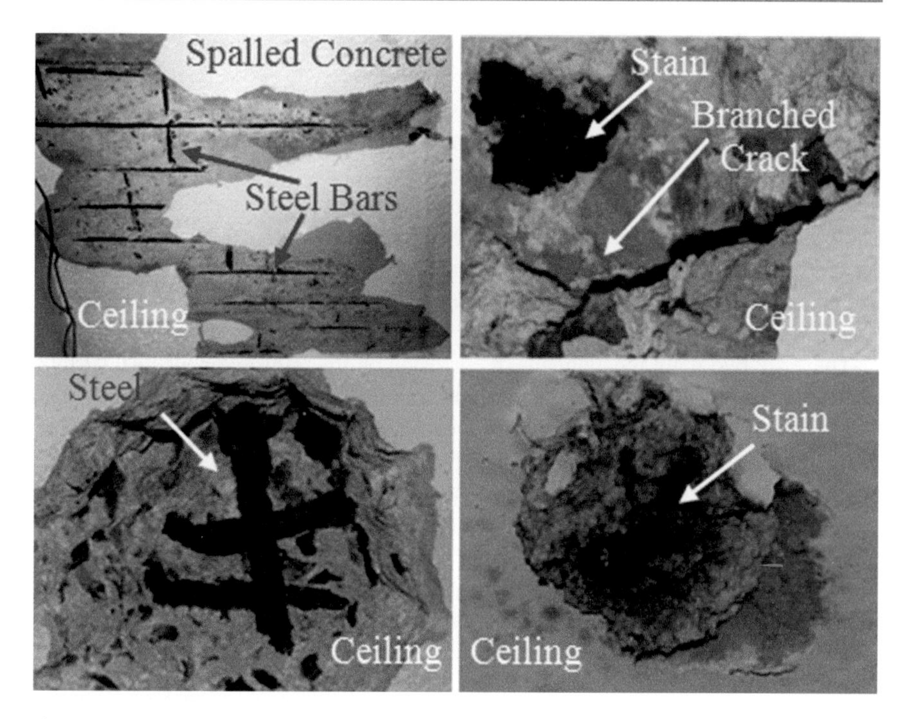

Fig. 7.22 Spotty corrosion in four different residences in the Caribbean. Cracking, spalling, and rust staining are evident

excavated away spot revealing heavily corroded steel bars responsible for concrete spalling. The bottom right picture only shows a rust stain and spalled concrete.

In general, these pictures represent typical damage of reinforced concrete in coastal areas, such as the tropical islands which are exposed to hot weather, high relative humidity, and salty air.

Generally, structural steel bars (reinforcements) in concrete are protected against corrosion by a passive (protective) oxide film, which is principally preserved by the high alkaline state of the concrete matrix in the absence or presence of insufficient chloride ions (Cl^-) and CO_2 gas in the concrete.

If these conditions prevail, then the steel is in a passive state because the passive film is not disrupted, and fortunately, the steel does not corrode. This ideal case, however, does not hold in some reinforced concrete structures exposed to de-icing salts, seawater, and high humidity because of penetration of these species from the moist atmosphere. Consequently, the reinforcement steel embedded in concrete normally corrodes, despite that corrosion may take years to cause structural damage. Thus, the corrosion of steel in moist atmosphere occurs due to differences in the electric potential between anodic and cathodic sites.

7.8.3 Mechanism of Corrosion in Concrete

It is known that iron on the steel surface oxidizes by liberating electrons and oxygen reduces by accepting these electrons. The resulting ions are transported through the moist pores in concrete as part of the electrolyte. In general, the extent of the reinforced concrete structural damage depends on the carbonation of concrete (which reduces pH), the content of chloride ions $[Cl^-]$, the properties of the concrete, the microstructural features of the steel, and sustained mechanical stresses and environment. Relevant theoretical details on this matter can be found elsewhere [32, 46, 47].

There are two main mechanisms of corrosion on the reinforcement steel surface embedded in concrete. Firstly, penetration (ingress) of carbon dioxide (CO_2) gas from the atmosphere through incipient capillaries and cracks in the reinforced concrete slab (roof top of buildings, bridges, parking lots, columns, etc.). Thus, the rate of penetration is assumed to be dependent of the concrete permeability.

As a result, CO_2 gas diffuses toward the steel bars and changes the alkalinity of the concrete from $pH > 13$ to $7 < pH \leq 9$ at the steel-concrete interface for neutrality. Therefore, any oxide passive film protecting the steel is disrupted (depassivated) at flaw sites by a process called concrete carbonation or carbonation-induced corrosion, which is controlled by diffusion of $\mathrm{CO_2}$ molecules through the concrete cover. Furthermore, carbonation is a reaction between atmospheric carbon dioxide and calcium hydroxide in the cement paste in the presence of water [33, 42].

It should be mentioned that the high alkalinity of concrete is due to sodium hydroxide, potassium hydroxide, and calcium hydroxide released during the various hydration reactions. Nonetheless, two particular reactions involved in carbonation are [33, 42]

$$Ca(OH)_2 + CO_2 \rightarrow CaCO_3 + H_2O \tag{7.52}$$

$$2NaOH + CO_2 \rightarrow Na_2CO_3 + H_2O \tag{7.53}$$

Secondly, diffusion of free chloride ions (Cl^-) through the concrete pores initiate pitting corrosion on the steel oxide passive film at inherent defects, causing a difference in electrochemical potential across the oxide film due to the formation of micro galvanic cells [35, 36].

This is a time-dependent localized attack known as chloride-induced corrosion (CIC) due to depassivation of the protective oxide film at discrete flaw sites. During this process, it is assumed that some of the chloride ions are chemically and physically bound to concrete (cement hydrates), cannot move, and do not participate in the depassivation process at a specific time.

On the other hand, some Cl^- ions are free to diffuse into flaws on the steel passive oxide film due to their mobility within the concrete pores and cracks. Consequently, spotty corrosion occurs at discrete sites on steel bars embedded in concrete.

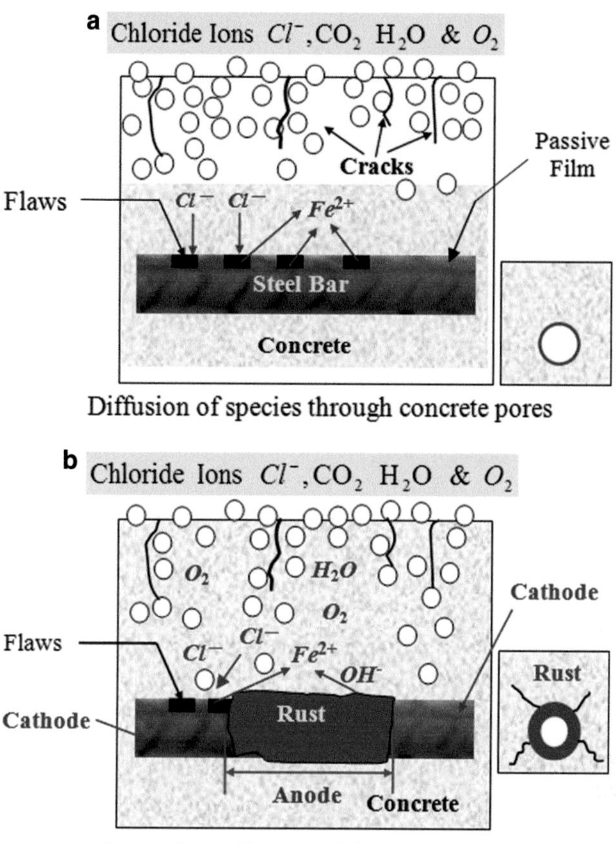

Fig. 7.23 (**a**) Corrosion model for diffusion of chloride salts, carbon dioxide, water, and oxygen molecules through the porous concrete. (**b**) Formation of rust. After references [40, 42]

Figure 7.23 illustrates a corrosion model for the formation of hydrated ferric oxide $Fe\,(OH)_2$ (rust) on the reinforcement steel [31, 32]. This compound is the corrosion product responsible for the generation of internal tensile stresses in concrete that causes concrete cracking and concrete spalling.

This model suggests that corrosion of concrete takes place when chloride in the form of salt, carbon dioxide (CO_2), water, and oxygen molecules penetrate and, subsequently, dissolve in the concrete pores. Therefore, corrosion is mainly induced by the presence of hydrogen and chlorine ions and to a lesser extent by carbon dioxide.

When CO_2 dissolves, carbonic acid is formed. This acid reacts with the alkali in the concrete to form carbonates and to lower the pH level of the concrete. In concrete construction, a high-quality concrete cover (well-consolidated, low permeability concrete) over the reinforcement steel bars (also referred to as rebars) is desirable in order to prevent Cl^- ions and CO_2 molecules from reaching the rebars prematurely. However, once the steel protective layer is disrupted, the corrosion of the rebars occurs. If the steel rebars are under load, then the local stresses increase due to a reduction in cross-sectional area. This may lead to failure.

In principle, the possible reactions related to rust formation in the presence of chloride ions, oxygen, and water are [37, 48, 49]

$$Fe^{2+} + 2Cl^- \rightarrow FeCl_2 \tag{7.54}$$

$$FeCl_2 + 2OH^- \rightarrow Fe(OH)_2 + 2Cl^- \tag{7.55}$$

$$2Fe(OH)_2 + \frac{1}{2}O_2 + H_2O \rightarrow 2Fe(OH)_3 \text{ (rust)} \tag{7.56}$$

Furthermore, corrosion takes place in the presence of water and oxygen because the steel bar becomes depassivated. As corrosion continues, more rust forms on the steel surface, expands in volume, and compresses the adjacent concrete at a high pressure. This increase in volume increases the tensile stresses in the concrete, and consequently, the concrete cracks, delaminates, and/or spalls. Just a little increment of rust volume, leading to a little reduction of the steel bar diameter, is needed to crack the concrete because it is a brittle refractory material. Therefore, corrosion is accelerated because new pathways are opened for water, oxygen, chlorides, and/or CO_2 to penetrate the concrete.

It is worth mentioning that the anodic-controlled passivation of steel is attributed to an electron-conducting oxide film. Subsequently, the steel is protected from corrosion since this film suppresses anodic reactions. And it is the inherent insolubility and the tightly adherence nature of this electron-conducting oxide film in concrete that makes the structural steel a very attractive reinforcing material.

Though, the sensitivity of the film coupled with low-pH concrete to chloride ions, it imposes a barrier to the maintainability, durability, and longevity of reinforced concrete structures. Moreover, the inherent defects of the film are potential sites for chloride-ion diffusion at relatively slow rate under time-dependent process. Inevitably, these ions act as oxidizers, making the film defects local anodic sites and forcing the steel surface undergo chloride-pitting corrosion (also known as chloride-induced corrosion).

These localized anodic sites can develop electrochemical potentials in the range of $-0.26\ V_{SHE} \leq E \leq -0.04\ V_{SHE}$ [50]. Hence, the potential equation below, Eq. (7.57), can be solved for the current density i, which can easily converted to corrosion rate (C_R) in units of mm/year using Faraday's law:

$$E = E_o + \beta_a \log(i/i_o) \tag{7.57}$$

This potential range seems low but sufficiently high for the chloride-induced corrosion process that takes place in concrete. This corrosion process is fundamentally electrochemical in nature because of the electron flow between anodic and cathodic sites on the rebars. A brief review on pitting corrosion in concrete can be found elsewhere [51].

Moreover, knowing the electric potential E, one can determine the Gibbs free energy for rust formation using Eq. (3.14). Thus,

$$\Delta G = -zFE \tag{7.58}$$

Recall that $\Delta G < 0$ is required for a reaction to proceed as written; otherwise, the direction of the reaction is reversed.

7.8.4 Galvanic Cells in Concrete

In addition, galvanic micro-cells and macro-cells are contributing factors for reinforced concrete corrosion. Thus, the concrete acts as the electrolyte, and the metallic conductor is provided by wire ties and the steel bars. Figure 7.24a illustrates how a galvanic macro-cell can develop from differences in chloride ion concentration and potential [52, 53].

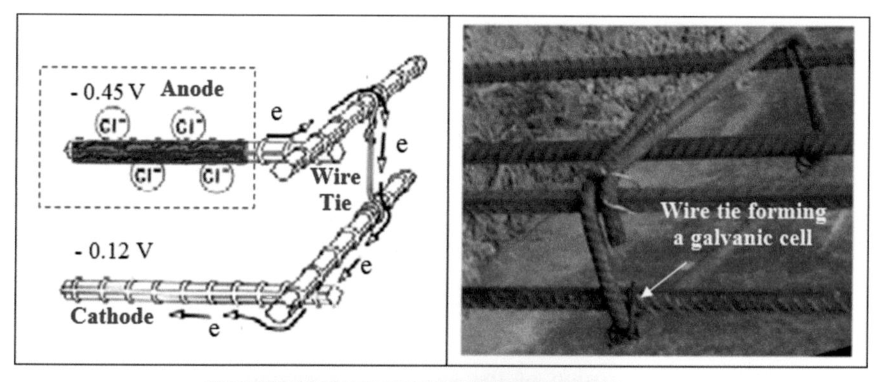

Fig. 7.24 Galvanic macro-cells in reinforced concrete. (**a**) Model of a galvanic macro-cell [49, 50]. (**b**) Galvanic macro-cell recently built for this section. (**c**) Corrosion due to galvanic macro-cells

Figure 7.24b shows an actual galvanic macro-cell built recently for this section, and Fig. 7.24c clearly elucidates a galvanic macro-cell found in a 17-year-old residence located in the Caribbean.

The obvious step for eliminating spotty corrosion sites on reinforced concrete structures is to prepare the damaged areas for patching with fresh concrete. However, another corrosion mechanism may develop due to the three-phase patched areas, which are (a) the old chloride-contaminated concrete having its own chloride concentration, (b) the cleaned steel bars, and (c) the fresh or new chloride-free concrete. Consequently, a new and strong galvanic macro-cell may develop on the steel surface due to a large chloride concentration.

7.8.4.1 Diffusion in Reinforced Concrete

The nonsteady state of chloride ions toward the embedded steel bars in concrete, as depicted in Fig. 7.23, can be explained, to an extent, using Fick's second law of diffusion for an isotropic and semi-infinite porous medium (heterogeneous). Thus, the steel-concrete interface is the low chloride concentration site, and Fick's second law of nonsteady diffusion, Eq. (6.29), is the generally accepted model to predict chloride-induced corrosion [26, 27, 40, 41, 54–59].

The solution of Fick's second law, Eq. (6.29), along with the boundary conditions $C(x,t) = 0$ for $x > 0$ and $t = 0$ and $C(x,t) = C_s$ for $x = 0$ and $t > 0$ is

$$C(x,t) = C_s\left[1 - \operatorname{erf}\left(\frac{x}{2\sqrt{Dt}}\right)\right] \tag{7.59}$$

where $C(x,t)$ = chloride concentration at x and $t > 0$
C_s = chloride concentration at the surface
D = diffusivity of chloride anions
x = concrete cover depth

The initiation of corrosion of the steel reinforcement bar occurs when the concentration of Cl^-, $C(x,t) = C_{th}$ at $t = t_i$, is known as the threshold (critical) chloride content. Thus, Eq. (7.59) becomes

$$C_{th} = C_s\left[1 - \operatorname{erf}\left(\frac{x}{2\sqrt{Dt_i}}\right)\right] \tag{7.60}$$

It has been reported [49] that a practical range of chloride content in kilograms of chloride ions per cubic meter of concrete is

$$0.60\ \text{kg/m}^3 \leq C_{th} \leq 0.83\ \text{kg/m}^3 \tag{7.61}$$

which activates the onset of corrosion at a prolonged time T_i. On the other hand, chloride threshold level for corrosion initiation in percentage by mass of cement is also used [3]. For bare structural steel in concrete,

$$0.02\,\% < C_{th} < 1\,\% \tag{7.62}$$

A considerable variation of C_{th} is expected because it depends on the type of steel, electrochemical environment in concrete, permeability, and testing method [59].

Furthermore, solving Eq. (7.60) for t_i yields the theoretical time to initiate corrosion:

$$t_i = \frac{x^2}{4D}\left[\mathrm{erf}^{-1}\left(1 - \frac{C_{th}}{C_s}\right)\right]^{-2} \tag{7.63}$$

According to Pfeifer [60], a mere 25 μm reduction (penetration) of the steel bar diameter (d) due to corrosion can be used to predicted the time for cracking and spalling the steel-reinforced concrete:

$$t_c = \frac{d_o - d}{C_R} = \frac{25\ \mu\mathrm{m}}{C_R} \tag{7.64}$$

where $d_o > d$ is the original bar diameter. Moreover, the time to repair the cracked concrete may be estimated as [61]

$$t_r = t_i + t_c \tag{7.65}$$

Conversely, t_i, t_c, and t_c are implicitly related to the formation of corrosion product (rust) and the corresponding molar flux J_x. In order to represent corrosion through the formation of rust, Eq. (7.66) is used as the main chemical reaction that forms ferric hydroxide:

$$2Fe(OH)_2 + \frac{1}{2}O_2 + H_2O \rightarrow 2Fe(OH)_3 \tag{7.66}$$

Now, one mole of ferrous hydroxide $Fe(OH)_2$ produces one mole of ferric hydroxide $Fe(OH)_3$, and subsequently, the rate (flux) of these compounds are equal. Thus, $J_{Fe(OH)_3} = J_{Fe(OH)_2}$. Dividing each molar flux by the corresponding molecular weight of the compounds yields [26, 27]

$$\frac{J_{Fe(OH)_3}}{J_{Fe(OH)_2}} = \left[\frac{M_{Fe(OH)_3}}{M_{Fe(OH)_2}}\right] \tag{7.67}$$

$$J_{Fe(OH)_3} = \left[\frac{M_{Fe(OH)_3}}{M_{Fe(OH)_2}}\right] J_{Fe(OH)_2} \tag{7.68}$$

where $M_{Fe(OH)_2} = 89.85\ \mathrm{g/mol}$
$M_{Fe(OH)_3} = 106.85\ \mathrm{g/mol}$

For $Fe(OH)_2$ production, the relationship between an anodic current density, $i_a > i_{corr}$, and the flux is

$$J_{Fe(OH)_2} = \frac{i_a}{zF} \tag{7.69}$$

and that for rust, $J_{Fe(OH)_3}$, becomes

$$J_{Fe(OH)_3} = \left[\frac{M_{Fe(OH)_3}}{M_{Fe(OH)_2}}\right]\frac{i_a}{zF} = \frac{1.19 i_a}{zF} \tag{7.70}$$

Solving for the anodic current density yields

$$i_a = zFJ_{Fe(OH)_3}\left[\frac{M_{Fe(OH)_2}}{M_{Fe(OH)_3}}\right] = 0.84zFJ_{Fe(OH)_3} \tag{7.71}$$

This expression suggests that current density variations are expected due to nonuniform and porous $Fe(OH)_3$ rust layer. In addition, $Fe(OH)_3$ is known as iron(III) oxide-hydroxide, but it may form under specific circumstances as hydrated iron(III) oxide, $Fe_2O_3 \cdot nH_2O$. If rust is not uniform and flaky, then it provides no protection to steel as in reinforced concrete applications.

The main stream of reinforced concrete corrosion, in general, is the chloride ingress through the concrete porous. This leads to corrosion of the reinforcing steel, and subsequently, the steel strength and the local half-cell potential are reduced to an extent. This means that the potential becomes more negative increasing the probability of corrosion. With regard to the corrosion potential, the ASTM C876 standard test method for half-cell potentials of reinforcing steel in concrete provides meaningful information on survey potential measurements as related to the probability of corrosion at a particular moment. The recommended reference electrode is a copper-copper sulfate ($Cu/CuSO_4$) half-cell circuitry, and the recommended potential range to be analyzed for corrosion is illustrated in Table 7.1.

The Nernst equation defines the equilibrium setup between the metal and its ions in the concrete electrolyte as a function of the ion concentration for the products and reactants [20].

For an anodic overpotential $\eta_a = (E_a - E_{corr}) > 0$, the anodic current density that activates corrosion is defined by Eq. (5.79). Thus,

$$i_a = i_{corr} \exp\left[\frac{2.303\,(E_a - E_{corr})}{\beta_a}\right] \tag{7.72}$$

Table 7.1 Probability of corrosion using $Cu/CuSO_4$ reference electrode [62]

Potential E (Volts = $V_{Cu/CuSO_4}$)	Probability of corrosion
$E > -0.20$ V	$P > 90\,\%$ of no corrosion
$-0.35\text{ V} \le E \le -0.20$ V	Uncertain corrosion activity
$E < -0.35$ V	$P > 90\,\%$ of active corrosion

Inserting Eq. (5.39) into (7.72) yields

$$i_a = i_{corr} \exp\left[\frac{\alpha zF\left(E_a - E_{corr}\right)}{RT}\right] \tag{7.73}$$

In addition, the rate of corrosion (C_R) of steel in concrete containing moisture and oxygen depends on the magnitude of the anodic current density (i_a), which is $i_a > i_{corr}$, where i_{corr} is the corrosion current density related to the corrosion potential (E_{corr}).

The current density also depends on the molar flux ($J_{Fe^{2+}}$) and on the ionic conductivity of the concrete electrolyte. In general, the concrete electrolyte is just a capillary-pore system filled with water and oxygen. Thus, the transport of corrosive species from an external environment into the reinforced concrete can be controlled by using a high-quality concrete cover [49].

The corrosion rate as per Faraday's law of electrolysis, Eq. (5.90), is

$$C_R = \frac{i_a A_w}{zF\rho} \tag{7.74}$$

$$C_R = \frac{J_{Fe^{2+}} A_w}{\rho} \tag{7.75}$$

Example 7.2. *For the reinforced concrete specimen (concrete slab) shown below,* ***(a)*** *derive two expressions for chloride ion penetration depth (x) as a function of time in a coupled diffusion and migration ionic transfer. One expression must be related to a natural condition and the other related to an external electrical source supplying a constant potential in the range* $2\,V \leq E \leq 3\,V$*. The specimen has an effective thickness of* 8 *cm(concrete cover), and the chloride ionic diffusion coefficient is* $8x10^{-5}\,cm^2/s$ *at room temperature.* ***(b)*** *Plot the expression and interpret the profiles. Assume a nonporous oxide film.*

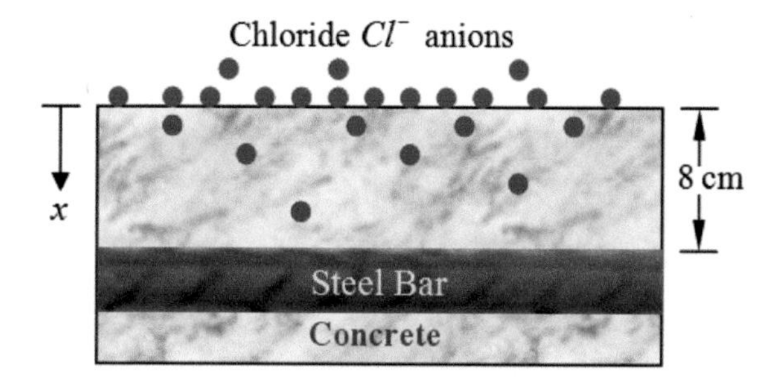

Solution. (a) *For mass transfer due to migration, Eq.* (6.83) *becomes*

$$\frac{\partial C}{\partial t} = \frac{zFD}{RT}\frac{\partial C}{\partial x}\frac{\partial \phi}{\partial x} \tag{a}$$

$$\frac{\Delta C}{t} \simeq \frac{zFD}{RT}\frac{\Delta C}{x}\frac{\partial \phi}{\partial x} \tag{b}$$

$$\frac{1}{t} \simeq \frac{zFD}{RT}\frac{1}{x}\frac{\partial \phi}{\partial x} \tag{c}$$

Solving for the penetration depth x along with $z = |-1| = 1$ yields

$$x \simeq \frac{zFDt}{RT}\frac{\partial \phi}{\partial x} \tag{d}$$

where $\partial \phi / \partial x$ is given by Eqs. (6.12) *and* (6.13)

$$\left(\frac{d\phi}{dx}\right)_{diffusion} = \frac{RT}{xF} \tag{6.12}$$

$$\left(\frac{d\phi}{dx}\right)_{electric} = \frac{E}{L} \tag{6.13}$$

Then, the penetration depth for diffusion and migration are, respectively,

$$x_d = \sqrt{zDt} = \left(8.9443 \times 10^{-3}\ \mathrm{cm/s^{1/2}}\right) t^{1/2} \tag{e}$$

$$x_m = \left(\frac{zFDE}{RTL}\right) t = \left(3.8949 \times 10^{-4}\ \frac{\mathrm{cm}}{\mathrm{V\,s}}\right) Et \tag{f}$$

(b) *The penetration depth profiles, Eqs.* (e) *and* (f), *for three potential values are given below:*

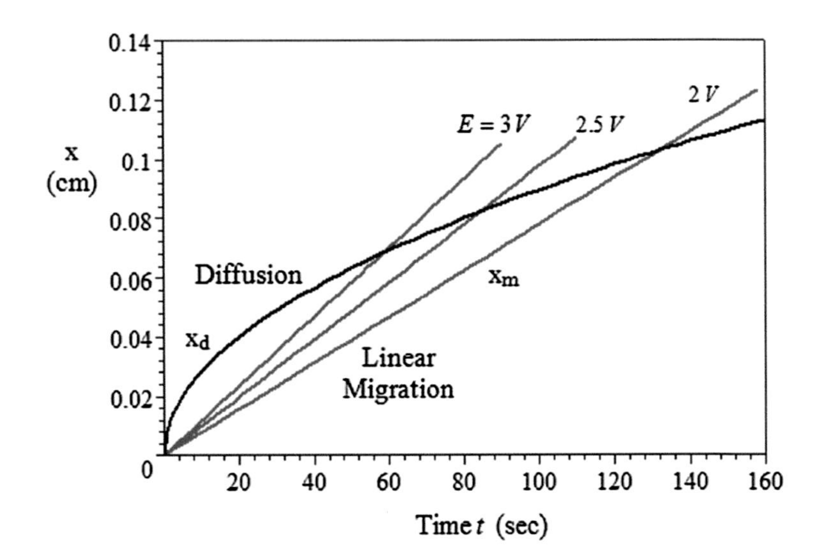

According to Lorente [63], the penetration due to diffusion is greater than the penetration due to electrical effects at the beginning of the ionic transport. The transition from diffusion to migration control is faster at higher potential.

Example 7.3. *A reinforced concrete slab has a concrete cover 50-mm deep. Let the threshold and the surface concentrations of chloride anions be, respectively,* $C_{th} = 0.60\ kg/m^3$ *and* $C_s = 18\ kg/m^3$ *and* $D = 31.61\ mm^2/year$ *be the diffusivity of chloride anions. Determine* ***(a)*** *the chloride molar flux,* ***(b)*** *the time* (T_i) *to initiate corrosion on an uncoated reinforcing steel surface (Fig. 7.21c). Explain.* ***(c)*** *Plot the concentration profile of chloride anions at the corrosion initiation time* T_i*. Calculate* ***(d)*** *the rate of iron hydroxide* $Fe(OH)_2$ *production (molar flux and mass flux) at the anodic regions for an anodic current density of* $1\ \mu A/cm^2$*,* ***(e)*** *the corrosion rate* (C_R) *of steel in mm/yr,* ***(f)*** *the time required for cracking and spalling due to the formation of a critical rust* $(Fe(OH)_3)$ *volume if the bar diameter is reduced only* $25\ \mu m$ *and* ***(g)*** *the time for repair and the apparent mass of rust per area.*

Solution. (a) *The chloride molar flux at* $t_i = 11$ *years can be calculated using Eq. (6.8). Thus,*

$$J_{Cl^-} = \frac{i}{zF} = \frac{1\ \mu\text{A/cm}^2}{(1)\,(96500\ \text{A s/mol})} \simeq 10^{-11}\ \frac{\text{mol}}{\text{cm}^2\ \text{s}}$$

(b) *The time to initiate corrosion, Eq. (7.63), is*

$$t_i = \frac{x^2}{4D}\left[\text{erf}^{-1}\left(1 - \frac{C_{th}}{C_s}\right)\right]^{-2}$$

$$t_i = \frac{(50\ \text{mm})^2}{4\,(31.61\ \text{mm}^2/\text{year})}\left[\text{erf}^{-1}\left(1 - \frac{0.6\ \text{kg/m}^3}{18\ \text{kg/m}^3}\right)\right]^{-2}$$

$$t_i \simeq 11\ \text{years}$$

(c) *The concentration profile of chloride ions,* $C(x, t_i)$ *at the approximated initiation time of* 11 *years, can be determined using the solution of Fick's second law of diffusion, Eq. (7.59).*

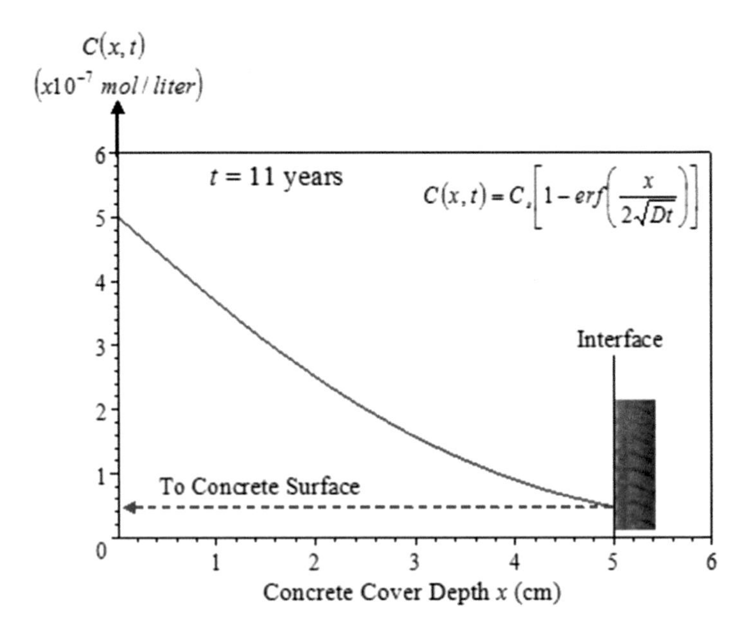

(d) *From Eq.* (7.69) *along with* $i_a = i_{corr}$

$$J_{Fe(OH)_2} = \frac{i_{corr}}{zF} = \frac{1x10^{-6}\ \mathrm{A/cm^2}}{(2)\,(96500\ \mathrm{A\,s/mol})}$$

$$J_{Fe(OH)_2} = 5.18 \times 10^{-12}\ \mathrm{mol/\left(cm^2\,s\right)}$$

Multiplying this result by the iron molar mass, 55.85 *g/mol, yields*

$$J_{Fe(OH)_2} = 2.893 \times 10^{-10}\ \mathrm{g/\left(cm^2\,s\right)} = 9.123 \times 10^{-3}\ \mathrm{g/\left(cm^2\,year\right)}$$

$$J_{Fe(OH)_2} \simeq 91.23\ \mathrm{g/\left(m^2\,year\right)} \simeq 0.091\ \mathrm{kg/\left(m^2\,year\right)}$$

Thus, the rate of rust [$Fe(OH)_3$] *production, Eq.* (7.68), *is*

$$J_{Fe(OH)_3} = \left[\frac{M_{Fe(OH)_3}}{M_{Fe(OH)_2}}\right] J_{Fe(OH)_2} \qquad (rust)$$

$$J_{Fe(OH)_3} = \left(\frac{106.85\ \mathrm{g/ml}}{89.85\ \mathrm{g/mol}}\right)\left[91.23\ \mathrm{g/\left(m^2\,year\right)}\right]$$

$$J_{Fe(OH)_3} \simeq 108.49\ g/\left(\mathrm{m^2\,year}\right) \qquad (rust)$$

(e) *The uniform corrosion rate* (C_R) *of ferrous ions* (Fe^{2+}), *Eq.* (7.74), *at the end of the 11th year is*

$$C_R = \frac{i_a A_w}{zF\rho} = \frac{\left(1 \times 10^{-6}\ \text{A/cm}^2\right)(55.85\ \text{g/mol})}{(2)(96500\ \text{A s/mol})\left(7.86\ \text{g/cm}^3\right)}$$

$$C_R = 3.6817 \times 10^{-11}\ \text{cm/s} \simeq 0.012\ \text{mm/year}$$

$$C_R = 12\ \mu\text{m/year}$$

Therefore, the corrosion rate, 12 $\mu m/year$, *is rather low, but sufficient to be noticed after* 11 *years. This corrosion rate corresponds to pitting formation at the 11th year; however, if the chloride ion supply continues, then the corrosion eventually propagates.*

(f) *If the reduction in diameter of the uncoated steel bar is* 25 μm, *then the time for some rust formation that causes cracking and spalling, Eq.* (7.64), *is*

$$t_c = \frac{25\ \mu\text{m}}{12\ \mu\text{m/year}} = 2.0833\ \text{year} = 25\ \text{months}$$

(g) *The repair time, Eq.* (7.65), *is*

$$t_r = t_i + t_c = 11\ \text{year} + 2.0833\ \text{year}$$

$$t_r = 13.0833\ \text{year}$$

Therefore, the first concrete repair is required at 13.0883 *years, that is, 13 years and 1 month. On the other hand, the apparent mass of rust per area is*

$$m_{rust} = J_{rust} . t_c = \left[108.49\ \text{g}/\left(\text{m}^2\ \text{year}\right)\right](2.0833\ \text{year})$$

$$m_{rust} = 226.02\ \text{g/m}^2 = 0.226\ \text{g/cm}^2$$

Therefore, this area density is considered low, but it is significant in corrosion assessment.

In the light of the above calculations, one can assume that the production of rust is an electrochemical process and that the roughness and porosity of the rust layer can be attributed to sufficiently high anodic current density. The risk of the reinforcing steel to lose its strength is unavoidable in most design applications. Hence, metal loss occurs at the anode surface sites through the electrochemical oxidation reaction, $Fe \rightarrow Fe^{2+} + 2e^{-}$. *These electrons diffuse to the cathode surface areas which remain protected against corrosion. Loss of metal, in general, means that the steel diameter decreases at the anodic sites being covered by porous rust layers.*

7.9 Voltammetry

7.9.1 Linear Sweep Voltammetry

Among several electrochemical methods found elsewhere [64–66], the linear sweep voltammetry (LSV) is briefly described from an analytical point of view so that the reader gets a clear understanding on how to interpret a voltammogram. This is a voltammetric method which requires a time-dependent linear potential sweep applied to an electrode for measuring the current response. The resultant experimental data is a current-potential plot under conditions that permit polarization of a working electrode. Basically, cyclic voltammetry is a type of potentiodynamic electrochemical technique since it associates potential (E) and current (I) of an electrochemical cell. Figure 7.25 schematically illustrates typical plots for the reversible $O + ze^- = R$ reaction [67].

Figure 7.25a shows several linear potential plots at different scan rates. The current response is clearly depicted in Fig. 7.25b where the plots have similar shapes, but the peak current increases as the potential scan rate (dE/dt) increases. The forward scan produces an apparent exponential current response before a current peak is achieved. Thus, a reduction layer (interfacial layer) of the species j begins to form on the electrode surface, but its growth is limited by depletion of the concentration of the species j (analyte), and as a result, a current peak is attained, and then it falls off as the concentration of the analyte is depleted close to the electrode surface. For reversible electron-transfer reactions, the peak current (I_p) occurs at the same potential (E_p) as shown in Fig. 7.25b. This is possible for oxidation reactions having the same rate constant (K_r). The interfacial layer, as it is known in the literature, is fundamentally a reduction layer produced by diffusion and migration mechanisms.

Denote that the peak current (I_p) increases as the scan rate increases, attributable to a slow growth of the diffusion layer. The faster the potential scan rate, the slower

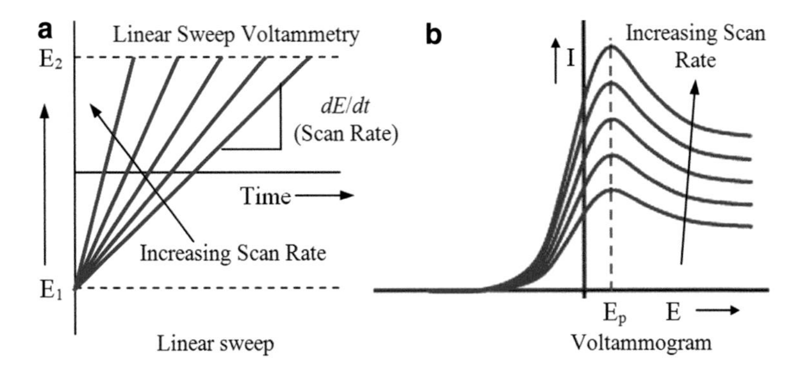

Fig. 7.25 Schematic linear sweep voltammetry (LSV) plots for reversible electron transfer. (**a**) Linear excitation signal and (**b**) current response at several potential scan rates [67]

the layer growth. At the peak, the diffusion layer has grown sufficiently on the electrode surface. Consequently, the flux of reactants (J) toward the electrode is not fast enough, and the current begins to decrease as the potential $E \to E_2$.

The basic electrochemical process of metallic corrosion (reduction) in aqueous solution is represented by the following reactions under charge transfer conditions across the metal-solution interface:

$$M \to M^{z+} + ze^- \quad \text{(oxidation)} \tag{7.76}$$

$$O + ze^- \to R \quad \text{(reduction)} \tag{7.77}$$

$$M + O \to M^{z+} + R \quad \text{(redox)} \tag{7.78}$$

where M is the metallic species, M^{z+} is the hydrated metal ion, O is an oxidant, and R is a reductant. The overall or redox corrosion reaction is given by Eq. (7.78). These reactions are strongly dependent on the electrode potential E, which is associated with the Fermi level E_F (energy level of electrons). This energy level can be approximated as

$$E_F = q_e E \tag{7.79}$$

where $q_e = 1.6x10^{-19}$ C and E has units of voltage (V). Thus, E_F has units of energy in joules (J) or electron volts (eV).

7.9.2 Cyclic Voltammetry

This is a potential sweep technique (sweep reversal method) used in kinetic studies and qualitative analysis. Normally CV requires stationary electrodes, but it can also use hydrodynamic electrodes [14, 67]. A three-electrode cell is normally used for forward and reverse potential scan rates. Figure 7.26a schematically shows a one-cycle potential scan.

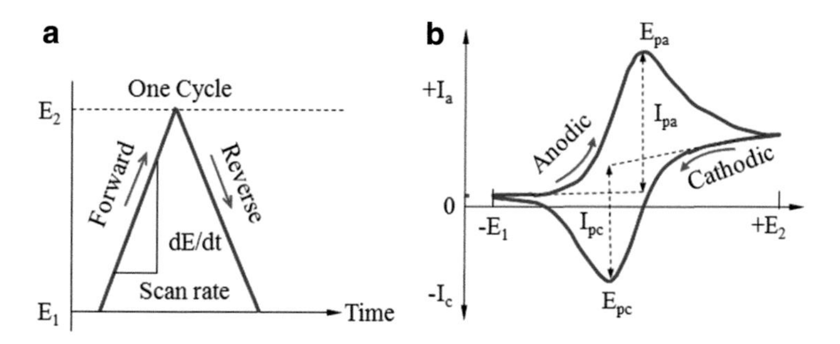

Fig. 7.26 Schematic (**a**) potential sweep for one-cycle and (**b**) one-cyclic voltammogram showing the positive anodic and negative cathodic current peaks for a simple redox couple. After [67]

This excitation signal provides forward and reverse linear scan rates within a suitable potential range. The resultant current response to this potential excitation is called cyclic voltammogram (voltammetric wave) as schematically depicted in Fig. 7.26b.

Reversible System For a reversible system with fast enough electron transfer, the reversible reaction is generalized as

$$R = O + ze^- \quad \text{(Oxidation)} \tag{7.80}$$

During the production of the diffusion layer (δ) constituted by the species "R," there is a continuous change in the surface concentration of the species "O" and δ increases in thickness, and consequently, the current begins to decrease as the steady-state diffusion flux, $J = -D\Delta C/\delta$, decreases [64]. On reverse scan rate, the reduced layer is reoxidized, and an inverse anodic peak current of reverse polarity from the forward potential scan is achieved (Fig. 7.26b). The I–E curves are symmetrical for reversible electron-transfer reactions; otherwise, they are antisymmetric for irreversible reactions.

For the forward scan rate, oxidation is represented by the anodic reaction given by Eq. (7.80), say, for iron

$$Fe \rightarrow Fe^{2+} + 2e^- \tag{7.81}$$

The redox potential for the above redox couple is

$$E^o = 0.5\left(E_{pa} + E_{pc}\right) \tag{7.82}$$

The separation peak potential difference can be used to determined the number of electrons transferred [64]. This potential difference is defined as

$$\Delta E_p = E_{pa} - E_{pc} = \frac{RT}{zF} \tag{7.83}$$

For a linear sweep, the potential scan rate (v) is done from E_1 to E_2, stopping at E_2 at a time $t > 0$. Initially, select a value for E_1 at which no oxidation reaction takes places. In principle, the scan direction can be positive (from E_1 to E_2) or negative (from E_2 to E_1). Here, E_1 is the starting potential, and on reaching $t > 0$, the potential sweep direction is inverted at, say, E_2 as shown in Fig. 7.26b for one cycle only. As illustrated in Fig. 7.26b, an anodic current peak (I_{pa}) appears between E_1 and E_2 and a cathodic current peak (I_{pc}) between E_2 and E_1 potential limits. This procedure can be repeated as desired for obtaining a series of voltammograms. Upon increasing the sweep rate (v), the shape of the voltammogram changes to a smaller size [67, 68].

A potential scan from E_1 to E_2 region is a potential perturbation which has a positive anodic current (I_a) response on the working electrode (substrate) surface. Conversely, a potential scan from E_2 to E_1 region causes a cathodic current density (I_c) response (decay). Recall that E_1 is the starting potential at which no current should flow and E_2 is the switching potential. Thus, anodic and cathodic current peaks (I_p) emerge in these regions followed by diffusion double-layer expansion. At the peaks, oxidation on the substrate occurs at (E_{pa}, I_{pa}) and reduction at (E_{pc}, I_{pc}), where E_{pa} and E_{pc} are the anodic and cathodic potentials and I_{pa} and I_{pc} are the anodic and cathodic currents, respectively.

The potentials for the plot in Fig. 7.26b are deduced as follows:

$$E = E_1 + vt \quad \text{(anodic)} \tag{7.84a}$$

$$E = E_2 - vt \quad \text{(cathodic)} \tag{7.84b}$$

where $v = dE/dt$ is the scan rate (V/s or mV/s) and t is the time (s).

Furthermore, oxidation ($M \rightarrow M^{+z} + ze^-$) starts between $E > E_1$ and $E < E_{pa}$ range, and completion is achieved at $E = E_{pa}$, where the anodic current is positive. Then, an electrical double layer expands, and the current drops as the potential increases further ($E_{pa} \leq E \leq E_2$). Reduction ($M^{+z} + ze^- \rightarrow M$) occurs between $E_{pc} \leq E \leq E_2$ range, and a negative current peak is achieved at $E = E_{pc}$. Subsequently, the current response on the reverse scanning increases when $E \rightarrow E_1$ and the M^{+z} ions are further reduced.

According to Fig. 7.26b, the electrochemical system is considered reversible when the electrode kinetics are much faster than the rate of diffusion, and for a small potential separation, $E_{pa} \simeq E_{pc}$.

Thus, the Nernst equation for $R \rightarrow O + ze^-$ and the anodic-to-cathodic concentration ratio at the working electrode surface in a reversible system are, respectively, [67]

$$E = E^o - \frac{RT}{zF} \ln \left[\frac{C_O(0,t)}{C_R(0,t)} \right] \tag{7.85}$$

$$\frac{C_O(0,t)}{C_O(0,t)} = \exp\left[-\frac{zF}{RT}(E - E^o) \right] \tag{7.86}$$

where E^o is the standard (formal) potential given by Eq. (7.82).

Assume that the ratio of anodic-to-cathodic peak currents in a reversible system is $I_{pa}/I_{pc} \simeq 1$. In this case, there is no electrodeposition, and the peak current (I_p) for a reversible linear potential sweep is given by the Randles-Sevcik equation (see Appendix 7A) [14, 67, 69–71]:

$$I_p = 0.4463 zFA_s C_{O,b} \left(\frac{zFD_O v}{RT} \right)^{1/2} \tag{7.87a}$$

$$I_p = kz^{3/2} A_s C_{O,b} D_O^{1/2} v^{1/2} \tag{7.87b}$$

where $k = 2.69x10^5$ A s/(V$^{1/2}$ mol) at 25 °C
z = valence
A_s = electrode surface area (cm^2)
D_O = diffusion coefficient (cm$^2/s$)
$C_{O,b}$ = concentration on the electrode surface (mol/cm^3)
$v = dE/dt$ = potential scan rate (V/s)
F = Faraday constant (C/mol)
R = gas constant [VC/(K mol)]
T = temperature (K)

Moreover, $I_p = f(v^{1/2})$ yields a linear plot with a slope defined as

$$Slope = \frac{dI_p}{d\left(v^{1/2}\right)} = kz^{3/2} A_s C_{O,b} D^{1/2} \tag{7.88}$$

If $C_{O,b}$ is known, then D for the species j is determined using Eq. (7.88).

Example 7.4. *Use the given cyclic voltammetry data [68] for determining the diffusion coefficient of ferrous Fe^{+3} ions in dimethylformamide (DMF) containing Fe^{+3} at room temperature. Assume that $D_{Fe^{+2}} = D_{Fe^{+3}}$.*

v (mV/s)	0	200	100	50	20
I_{pa} (μA)	0	12.8	8.80	5.90	3.70

Consider the cathodic reaction $Fe^{+3} + e^- \rightarrow Fe^{+2}$ where the electron is supplied by the electrode. The working electrode has a diameter of 2 mm immersed in the electrolyte containing $C_{O,b} = C_{Fe^{+3}} = 6 \times 10^{-7}$ mol/cm^3.

Solution. *The electrode exposed surface area is estimated as*

$$A_s = (\pi/4)\, d^2 = (\pi/4)\,(0.2 \text{ cm})^2 = 3.1416 \times 10^{-2} \text{ cm}^2$$

Plotting the given data and forcing the linear current response to start from zero yields the curve fitting equation as

$$I_{pa} = \left[27.50 \frac{\mu A}{(V/s)^{1/2}}\right] x$$

where $x = v^{1/2}$. The linear plot is

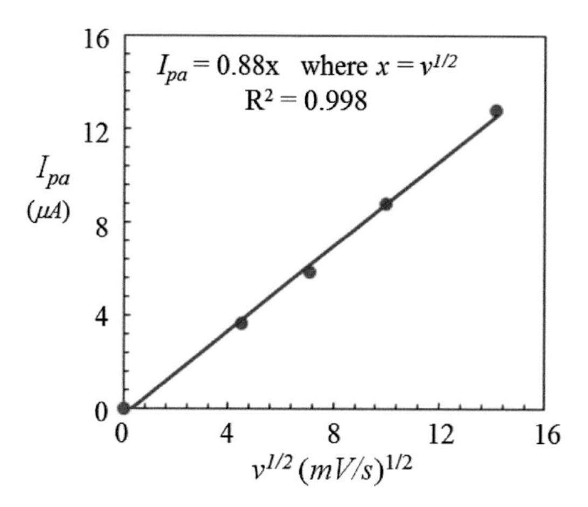

From Eq. (7.88), the slope of the linear plot along with $z = 1$ is

$$Slope = \frac{dI_p}{d\left(v^{1/2}\right)} = kz^{3/2}A_sC_{O,b}D^{1/2} = 27.50\ \mu\text{A V}^{1/2}/\text{s}^{1/2}$$

from which $D_{Fe^{+3}}$ is

$$D_{Fe^{+3}} = D = \left(\frac{Slope}{kz^{3/2}A_sC_{O,b}}\right)^2 = 2.94 \times 10^{-5}\ \text{cm}^2/\text{s}$$

Irreversible System If a cyclic voltammogram for one-cycle oxidation process does not show a cathodic peak on reverse potential scanning, then the system is **irreversible**. This is attributed to the electrode kinetics being slower than the rate of diffusion. In this case, the anodic current peak is normally reduced in size and widely separated, and no inverse cathodic peak emerges on changing the scan direction at E_2 (Fig. 7.27) [14, 67].

Quasi-Reversible System When the anodic and cathodic peaks have a relatively large potential separation ($E_{pa} >> E_{pc}$), the system is **quasi-reversible**. In this case, the current is controlled by both the charge transfer and mass transport by diffusion. This type of system is schematically depicted in Fig. 7.27 [67].

In addition, cyclic voltammetry has also been used [72] for studying the influence of cement factors on the corrosion of embedded iron and steel in hardened cement paste. Figure 7.28 depicts the cyclic voltammogram for corrosion of steel in a solution saturated with $Ca(OH)_2$ at 25 °C. It is clear that there are various voltammetric waves corresponding to different species or compounds.

This cyclic voltammogram exhibits several current peaks attributed to the formation of iron oxides during the forward sweep. The reverse scan shows current peaks for the dissolution of these oxide compounds. The highest and lowest current

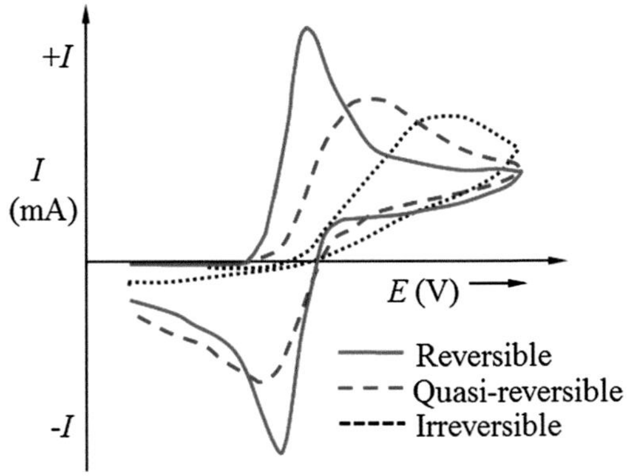

Fig. 7.27 Schematic cyclic voltammetric waves for reversible, quasi-reversible, and irreversible system [14, 67]

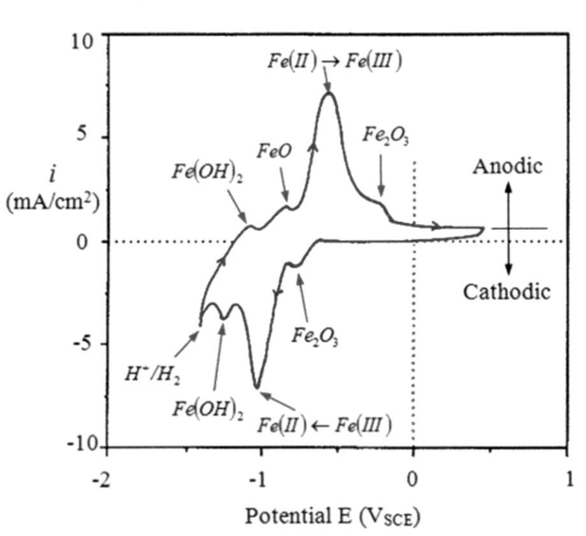

Fig. 7.28 Standard steady-state voltammogram for cement-covered steel wire electrode (0.08 cm^2 exposed area). Solution: saturated $Ca(OH)_2$ at 25 °C. Voltage limits of $-1.4\,V_{SCE}$ and $+0.4\,V_{SCE}$ at a linear sweep rate of 50 mV/s. Data obtained after 200 cycles using a rapid cyclic voltammetry (RCV) [72]

peaks are for $Fe\,(II)\,/Fe\,(III)$ and $Fe\,(III)\,/Fe\,(II)$. Here, $Fe\,(II) = FeO$ (wüstite) and $Fe\,(III) = Fe_2O_3$ (hematite) are compounds having peak currents at different potentials (Fig. 7.28).

Nevertheless, Fig. 7.28 is a typical voltammogram obtained using the rapid cyclic voltammetric (*RCV*) method on cement paste electrodes having only 3-day cures, and it provides information on the corrosiveness of the ordinary portland cement (*OPC*) paste in the absence of chloride solution surrounding the embedded steel bars [72]. Denote in Fig. 7.28 that the cyclic voltammogram is not as symmetrical as being schematically shown in Fig. 7.26b, but it contains actual qualitative information for corrosion studies of steel in reinforced concrete structures.

Assuming a steady-state diffusion process, the current density for oxidation and reduction of a species can be predicted using Eqs. (6.8) and (6.9), respectively. For instance, the cathodic current density, Eq. (6.8), is defined as

$$i = zFJ = -zFD\left(\frac{\partial C}{\partial x}\right)_{x=o} \tag{7.89}$$

$$\left(\frac{\partial C}{\partial x}\right)_{x=o} = \frac{C_x - C_o}{\Delta x} \tag{7.90}$$

Here, C_x is the concentration of a species at $x > 0$, C_o is the concentration of a species at $x = 0$, x is the distance from the electrode surface, and $i \propto \partial C/\partial x$ and $\Delta x = \delta$ are the thickness of the diffusion layer. Using the generalized electrochemical reaction, $R \to O + ze^-$, the cell potential is predicted by the Nernst equation, Eq. (7.85).

For electrodeposition, the generalized reversible reaction $M^{+z} + ze^- = M$implies that at least a thin metallic film is adhered on the cathode electrode when a reversible potential is reached. Thus, the peak current becomes known as Berzins–Delahay equation [73]:

$$I_p = 0.6105\,(zF)^{3/2}\,A_s C_{O,b}\left(\frac{Dv}{RT}\right)^{1/2} \tag{7.91}$$

where $C_{O,b}$ is the bulk concentration in Eq. (7.91). Denote that Eqs. (7.87a) and (7.91) give slight different values of I_p.

7.10 Appendix 7A: Laplace Transform

Laplace Transform [74] In general, the Laplace transform is a method for solving differential equations and corresponding initial and boundary conditions. Let the general function $f(t)$ be defined for all time $t \geq 0$, multiply it by $\exp(-st)$, and integrate so that the Laplace transform, $L[f(t)]$, becomes

$$L[f(t)] = F(s) = \int_0^\infty f(\tau)\exp(-st)\,d\tau \tag{7A.1}$$

This implies that the function of time $f(t)$ is transformed into a new function called frequency function $F(s)$. Also, the notation $L[f(t)]$ is used to indicate the Laplace transform of $f(t)$ and $s = \sigma + jw$, where σ and w are real and imaginary parameters, respectively.

The Laplace inverse transform is also defined as

$$f(t) = L^{-1}[F(s)] \tag{7A.2}$$

and the Laplace of the derivatives of $f(t)$ are

$$L\left[f'(t)\right] = sL[f(t)] - f(0) \tag{7A.3}$$

$$L\left[f''(t)\right] = sL\left[f'(t)\right] - f'(0) = s\{sL[f(t)] - f(0)\} - f'(0) \tag{a}$$

$$L\left[f''(t)\right] = s^2L[f(t)] - sf(0) - f'(0) \tag{7A.4}$$

In general,

$$L\left[f^{(n)}(t)\right] = s^nL[f(t)] - s^{n-1}f(0) - s^{n-2}f'(0) - \cdots \cdot f^{(n-1)}(0) \tag{7A.5}$$

Example 7A.1. *Use Fick's second law of diffusion (which is a partial differential equation) and corresponding initial condition to apply the Laplace transform to obtain an ordinary differential equation:*

$$\frac{\partial C(x,t)}{\partial t} = D\frac{\partial^2 C(x,t)}{\partial^2} \quad \text{with } C(x,0) = C_b$$

Solution. *The partial differential equation can be rearranged as*

$$\frac{\partial C(x,t)}{\partial t} - D\frac{\partial^2 C(x,t)}{\partial^2} = 0$$

Laplace transform gives

$$L\left[\frac{\partial C(x,t)}{\partial t}\right] - L\left[D\frac{\partial^2 C(x,t)}{\partial x^2}\right] = L[0] = 0$$

According to the Laplace transform definition, Eq. (7A.3),

$$L\left[f'(t)\right] = sL[f(t)] - f(0)$$

$$L\left[\frac{\partial C(x,t)}{\partial t}\right] = sC(x,s) - C(x,0)$$

$$L\left[\frac{\partial C(x,t)}{\partial t}\right] = sC(x,s) - C_b \tag{a}$$

and from Eq. (7A.1),

$$L[f(t)] = \int_0^\infty f(\tau)\exp(-st)\,d\tau$$

$$L\left[D\frac{\partial^2 C(x,t)}{\partial x^2}\right] = \int_0^\infty D\frac{\partial^2 C(x,t)}{\partial x^2}\exp(-st)\,d\tau$$

$$= D\frac{\partial^2}{\partial x^2}\int_0^\infty C(x,t)\exp(-st)\,d\tau$$

$$= D\frac{d^2 C(x,s)}{dx^2} \tag{b}$$

Then,

$$L\left[\frac{\partial C(x,t)}{\partial t}\right] - L\left[D\frac{\partial^2 C(x,t)}{\partial x^2}\right] = 0$$

$$sC(x,s) - C_b - D\frac{d^2 C(x,s)}{dx^2} = 0 \tag{c}$$

$$\frac{d^2 C(x,s)}{dx^2} - \frac{s}{D}C(x,s) + \frac{C_b}{D} = 0 \tag{d}$$

This ordinary differential equation (ODE) has a solution of the form

$$C(x,s) = \frac{C_b}{s} + A(s)\exp\left(-x\sqrt{s/D}\right) \tag{e}$$

In the literature, $C(x,s) = \overline{C}(x,s)$ *where* $\overline{C}(x,s)$ *is the transformed concentration [14]. In addition, using the definition of the diffusion flux as per Fick's first law and eq. (e) gives the transform of the current,* $\overline{I}(s)$, *as*

$$\overline{I}(s) = zFA_sD\left[\frac{\partial \overline{C}(x,s)}{\partial x}\right]_{x=0} \tag{f}$$

where the transform of the concentration gradient is

$$\frac{\partial \overline{C}(x,s)}{\partial x} = -A(s)\sqrt{s/D}\exp\left(-x\sqrt{s/D}\right) \tag{g}$$

Convolution Theorem [74] This is a property of the Laplace transform defined in standard notation as

$$h(t) = (f * g)(t) = \int_o^t f(\tau)\,g(t-\tau)\,d\tau \tag{7A.6}$$

$$H(t) = L^{-1}[h(t)] = L^{-1}[(f * g)(t)] \tag{7A.7}$$

where $h(t)$ is the Laplace transform of the $H(t)$, $*$ is the convolution notation in $(f * g)(t)$, and $f(t)$ and $g(t)$ are Laplace functions. The inverse $h(t)$ is written $(f * g)(t)$, which is called convolution of $f(t)$ and $g(t)$.

Voltammetry Normally, convolution voltammetry follows Laplace transform, and it has comprehensive applications in chemistry and electrochemistry. It is intended in this section to elucidate the use of this technique for deriving the well-known Randles-Sevcik equation for a reversible reaction, $O + ze^- \rightarrow R$.

For a reversible electrochemical system, the time-dependent potential, $E(t)$, can be defined by the Nernst equation and the linear scanning voltammetry (LSV).

In fact, the $E(t)$ equations given below are the initial expressions for deriving numerically the cyclic voltammetry peak current known as the Randles-Sevcik equation [14, 71]. For a reversible reaction, $O + ze^- \rightarrow R$, with only O initially present in solution, the time-dependent potential expressions are

$$E(t) = E^o + \frac{zF}{RT} \ln \frac{C_O(0,t)}{C_R(0,t)} \quad \text{(Nernst equation)} \tag{7A.8}$$

$$E(t) = E_i - vt \quad \text{for } 0 < t \leq t_\lambda \ \text{(LSV)} \tag{7A.9}$$

where t_λ is the time for potential inversion in a *LSV* experiment. From Eq. (7A.8), the concentration ratio is

$$\frac{C_O}{C_R} = \frac{C_O(0,t)}{C_R(0,t)} = \exp\left[\frac{zF}{RT}(E_i - vt - E^o)\right] \tag{7A.10}$$

which is defined as a boundary condition redefined as [14, 71]

$$\frac{C_O}{C_R} = \theta S(t) \tag{7A.11}$$

where

$$C_O = \theta S(t) C_R \tag{7A.12}$$

$$\theta = \exp\frac{zF}{RT}(E_i - E^o) \tag{7A.13}$$

$$S(t) = \exp(-at) \quad \text{for } t \leq t_\lambda \tag{7A.14}$$

$$S(t) = \exp(at - 2at_\lambda) \quad \text{for } t \geq t_\lambda \tag{7A.15}$$

$$a = \left(\frac{zF}{RT}\right) v \tag{7A.16}$$

$$at = \left(\frac{zF}{RT}\right) vt = \left(\frac{F}{RT}\right)(E_i - E) z \tag{7A.17}$$

Considering a stationary electrode and a semi-infinite diffusion case leads to a direct use of Fick's second law of diffusion for the reversible reaction $O + ze^- \rightarrow R$ which takes place at a plane electrode surface and gives the rate of concentration for the oxidized (O) and reduced (R) ions, respectively:

$$\frac{\partial C_O}{\partial t} = D_O \frac{\partial^2 C_O}{\partial x^2} \tag{7A.18}$$

$$\frac{\partial C_R}{\partial t} = D_R \frac{\partial^2 C_R}{\partial x^2} \tag{7A.19}$$

For a reduction reaction $O + ze^- \rightarrow R$ and an initial negative potential sweep direction, the boundary conditions are [14, 67, 71]

$$C_O = C_{b,O} \text{ and } C_{b,R} = 0 \ \text{ for } t = 0,\ x = 0 \tag{7A.20}$$

$$C_O \rightarrow C_{b,O} \text{ and } C_{b,R} \rightarrow 0 \ \text{ for } t > 0,\ x \rightarrow \infty \tag{7A.21}$$

$$D_O\left(\frac{\partial^2 C_O}{\partial x^2}\right) = -D_R\left(\frac{\partial^2 C_R}{\partial x^2}\right) \ \text{ for } t > 0,\ x = 0 \tag{7A.22}$$

$$E = E_i - vt \quad \text{for } 0 < t \le t_\lambda \tag{7A.23}$$

$$E = E_i - vt - 2vt_\lambda \quad \text{for } t > t_\lambda \tag{7A.24}$$

where $C_{b,O}$ and $C_{b,R}$ are bulk concentrations. On the other hand, for an oxidation reaction $R \rightarrow O + ze^-$ and an initial positive potential sweep direction the boundary conditions are [14, 67, 71]

$$C_R = C_{r,O} \text{ and } C_{b,o} = 0 \ \text{ for } t = 0,\ x = 0 \tag{7A.25}$$

$$C_R \rightarrow C_{b,R} \text{ and } C_{b,o} \rightarrow 0 \ \text{ for } t > 0,\ x \rightarrow \infty \tag{7A.26}$$

$$D_O\left(\frac{\partial^2 C_O}{\partial x^2}\right) = -D_R\left(\frac{\partial^2 C_R}{\partial x^2}\right) \ \text{ for } t > 0,\ x = 0 \tag{7A.27}$$

$$E = E_i - vt \quad \text{for } 0 < t \le t_\lambda \tag{7A.28}$$

$$E = E_i + vt + 2vt_\lambda \quad \text{for } t > t_\lambda \tag{7A.29}$$

For $O + ze^- \rightarrow R$, take the Laplace transform of Eqs. (7A.18)–(7A.21), solve for the transformed concentrations, and apply the convolution theorem to get [71]

$$C_O = C_{b,O} - \frac{1}{\sqrt{\pi D_O}} \int_0^t \frac{f(\tau)\, d\tau}{\sqrt{t-\tau}} \tag{7A.30}$$

$$C_R = \frac{1}{\sqrt{\pi D_R}} \int_0^t \frac{f(\tau)\, d\tau}{\sqrt{t-\tau}} \tag{7A.31}$$

Using Eq. (6.121) gives the diffusion flux, $J_x = f(t)$, and the current (I)

$$f(t) = D_O\left(\frac{\partial C_O}{\partial x}\right)_{x=0} = \frac{I}{zFA_s} \tag{7A.32}$$

Substituting Eq. (7A.11) into (7A.30) yields

$$\theta S(t)\, C_R = C_{b,O} - \frac{1}{\sqrt{\pi D_O}} \int_0^t \frac{f(\tau)\, d\tau}{\sqrt{t-\tau}} \tag{7A.33}$$

Inserting Eq. (7A.31) into (7A.33) gives

$$\frac{\theta S(t)}{\sqrt{\pi D_R}} \int_0^t \frac{f(\tau)\, d\tau}{\sqrt{t-\tau}} = C_{b,O} - \frac{1}{\sqrt{\pi D_O}} \int_0^t \frac{f(\tau)\, d\tau}{\sqrt{t-\tau}} \tag{7A.34}$$

Rearrange Eq. (7A.34) so that

$$\int_0^t \frac{f(\tau)\, d\tau}{\sqrt{t-\tau}} = \frac{C_{b,O}\sqrt{\pi D_O}}{1+\theta S(t)\sqrt{D_O/D_R}} \tag{7A.35}$$

$$\int_0^t \frac{f(\tau)\, d\tau}{\sqrt{t-\tau}} = \frac{C_{b,O}\sqrt{\pi D_O}}{1+\gamma\theta S(t)} \tag{7A.36}$$

where

$$\gamma = \sqrt{D_O/D_R} \tag{7A.37}$$

Let $\tau = z/a$ and $f(t) = g(at)$ so that $d\tau = dz/a$ and Eq. (7A.36) takes the form [14, 71]

$$\int_0^{at} \frac{g(z)\, dz}{\sqrt{a}\sqrt{at-z}} = \frac{C_{b,O}\sqrt{\pi D_O}}{1+\gamma\theta S(at)} \tag{7A.38}$$

If [71]

$$g(at) = C_{b,O}\left(\pi a D_O\right)^{1/2} \chi(at) \tag{7A.39}$$

then Eq. (7A.38) becomes the final integral of the form

$$\int_0^{at} \frac{\chi(z)\, dz}{\sqrt{at-z}} = \frac{1}{1+\gamma\theta S(at)} \tag{7A.40}$$

where

$$\chi(z) = \frac{I}{zFA_s C_O\left(\pi a D_O\right)^{1/2}} \tag{7A.41}$$

Manipulating Eqs. (7A.10) and (7A.11) along with $E_{1/2} \simeq E^o$ yields [71]

$$\left(E - E_{1/2}\right) z = \frac{RT}{F}\left[\ln(\gamma\theta) + \ln S(at)\right] \tag{7A.42}$$

$$\gamma\theta S(at) = \exp\left[\frac{F}{RT}\left(E - E_{1/2}\right) z\right] \tag{7A.43}$$

Table 7A.1 Values of $\sqrt{\pi}\chi(at)$ and $\phi(at)$ at 25 °C [14, 67, 71]

$(E - E_{1/2})z$	$\sqrt{\pi}\chi(at)$	$\phi(at)$	$(E - E_{1/2})z$	$\sqrt{\pi}\chi(at)$	$\phi(at)$
40	0.160	0.173	−5	0.400	0.548
35	0.185	0.208	−10	0.418	0.596
30	0.211	0.236	−15	0.432	0.641
25	0.240	0.273	−20	0.441	0.685
20	0.269	0.314	−25	0.445	0.725
15	0.298	0.357	−28.50	0.4463	0.7516
10	0.328	0.403	−30	0.446	0.763
2	0.355	0.451	−35	0.443	0.796
0	0.380	0.499	−40	0.438	0.826

Inserting Eq. (7A.43) into (7A.40) gives

$$\int_0^{at} \frac{\chi(z)\,dz}{\sqrt{at - z}} = \frac{1}{1 + \exp\left[\frac{F}{RT}\left(E - E_{1/2}\right)z\right]} \tag{7A.44}$$

This integral is solved numerically for $\chi(at)$ when values of $\left(E - E_{1/2}\right)z$ are known. Then, $\left(E - E_{1/2}\right)z$ vs. $\pi^{1/2}\chi(at)$ for planar electrodes and $\phi(at)$ for spherical correction results are given in Table 7A.1 [14, 67, 71]. The spherical correction dimensionless function $\phi(at)$ solution can be found elsewhere [14, 71].

The current equation for a planar electrode at any time on the *LSV* plot is given by Eq. (7A.23):

$$I = \left[(\pi)^{1/2}\chi(at)\right] zFA_sC_{b,O}\left(aD_O\right)^{1/2} \tag{7A.45}$$

From Table 7A.1, the maximum numerical result for the dimensionless function $\pi^{1/2}\chi(z)$ is 0.4463 when $\left(E - E_{1/2}\right)z = -28.50$ mV at 25 °C [14, 67, 71]. Hence, Eq. (7A.45) yields the peak current as

$$I_p = 0.4463zFA_sC_{b,O}\left(aD_O\right)^{1/2} \tag{7A.46}$$

Substituting Eq. (7A.16) into (7A.46) yields the usual Randles-Sevcik equation used in cyclic voltammetry (sweep voltammetry) for predicting the peak current in a voltammogram [14, 71]:

$$I_p = 0.4463zFA_sC_{b,O}\left(\frac{zFD_O}{RT}\right)^{1/2} v^{1/2} \tag{7A.47}$$

This mathematical expression is exactly equal to Eq. (7.87a), and it is the well-known Randles-Sevcik equation which depends on $C_{b,o}$ and D_O of the species "*O*," testing temperature (*T*), and the applied potential scan rate $v = dE(t)/dt$.

Example 7A.2. *Use Fick's second law of diffusion and corresponding initial and boundary conditions to apply the Laplace transform to obtain the concentration equations for O and R as per reaction $O + ze^- \rightarrow R$. Assume a semi-infinite linear diffusion ($x \rightarrow \infty$) at a planar electrode. Subsequently, apply the convolution theorem to derive the current equation given by Eq.* (7A.47). *The concentration rate equation for the O and R are given by Eqs.* (7A.18) *and* (7A.19)*:*

$$\frac{\partial C_O(x,t)}{\partial t} = D_O \frac{\partial^2 C_O(x,t)}{\partial x^2} \tag{7A.18}$$

$$\frac{\partial C_R(x,t)}{\partial t} = D_R \frac{\partial^2 C_R(x,t)}{\partial x^2} \tag{7A.19}$$

The initial and boundary conditions are

$$C_O(x,t) = C_{O,b} \text{ and } C_{R,b} = 0 \text{ for } t = 0,\ x = 0 \tag{7A.20}$$

$$C_O(x,t) \rightarrow C_{O,b} \text{ and } C_{R,b} \rightarrow 0 \text{ for } t > 0,\ x \rightarrow \infty \tag{7A.21}$$

$$D_O \left[\frac{\partial^2 C_O(x,t)}{\partial x^2} \right] = -D_R \left(\frac{\partial^2 C_R(x,t)}{\partial x^2} \right) \text{ for } t > 0,\ x = 0 \tag{7A.22}$$

Solution. *The Laplace transform of Eqs.* (7A.18) *and* (7A.19) *are [75]*

$$\overline{C}_O(x,s)_{x=0} = \frac{C_{O,b}}{s} - \frac{\overline{I}}{zFA_s} \frac{1}{\sqrt{sD_O}} \tag{a}$$

$$\overline{C}_R(x,s)_{x=0} = -\frac{\overline{I}}{zFA_s} \frac{1}{\sqrt{sD_R}} \tag{b}$$

For convolution, separate the product $\left[\overline{I}/(zFA_s)\right]\left[1/\sqrt{sD_O}\right]$ *as shown below :*

$$L^{-1}\left[\frac{\overline{I}}{zFA_s}\right] = \frac{I}{zFA_s} \tag{c}$$

$$L^{-1}\left[\frac{1}{\sqrt{sD_O}}\right] = \frac{1}{\sqrt{\pi t D_O}} \tag{d}$$

Similarly,

$$L^{-1}\left[\frac{1}{\sqrt{sD_{OR}}}\right] = \frac{1}{\sqrt{\pi t D_R}}$$

Thus, Eqs. (a) *and* (b) *become*

$$C_O(x,t) = \frac{C_{b,O}}{s} - \frac{1}{\sqrt{\pi D_O}} \int_0^t \frac{\overline{I}(\tau)}{zFA_s} \cdot \frac{1}{\sqrt{t-\tau}} d\tau \tag{e}$$

$$C_R(x,t) = \frac{1}{\sqrt{\pi D_R}} \int_0^t \frac{\overline{I}(\tau)}{zFA_s} \cdot \frac{1}{\sqrt{t-\tau}} d\tau \tag{f}$$

Get $C_O(x,t)/C_R(x,t)$ *ratio from these two equations and equate the resultant expression to Eq.* (7A.10) *so that*

$$\int_0^t \frac{\overline{I}(\tau)}{zFA_s} \cdot \frac{1}{\sqrt{t-\tau}} d\tau = \frac{C_{b,O}\sqrt{\pi D_O}}{1 + \sqrt{D_O/D_R}\exp\left[(zF/RT)(E_i - vt - E^o)\right]} \tag{g}$$

The solution of Eq. (g) *is given by Nicholson–Shain [71] as*

$$I_p = zFA_s C_{b,O}\left(\frac{zFD_O}{RT}\right)^{1/2} v^{1/2}\left(\pi^{1/2}\chi\right) \tag{h}$$

$$I_p = 0.4463 zFA_s C_{b,O}\left(\frac{zFD_O}{RT}\right)^{1/2} v^{1/2} \tag{7A.47}$$

7.11 Problems

7.1. Determine **(a)** the electric potential E in millivolts, **(b)** the time in seconds, and **(c)** the growth rate in μm/s for electroplating a 3-μm-thick *Cr* film on a *Ni*-undercoated steel part, provided that the electrolyte contains 10^{-4} mol/l of Cr^{3+} cations at 25 °C, the *Ni*-steel part has a 10-cm^2 surface area, and the cell operates at 50 % current efficiency and at a passive current of 0.8 A. Assume that the oxide passive film has a density of 7.19 g/cm^3. [Solutions: (a) $E > -0.823$ V, (b) $t = 5$ min, $dx/dt = 0.01$ μm/s.]

7.2. If chromium oxide film flow rate is 0.03 μm/s, **(a)** how long will it take to grow a 5 μm film on a Ni substrate. Assume that the cell operates at 50 % current efficiency. Calculate **(b)** the current density. [Solutions: (a) $t = 100$ s, (b) $i = 0.25$ A/cm^2.]

7.3. Briefly, explain why the corrosion rate and the corrosion potential increase on the surface of some metallic materials in contact with an oxygen-containing acid solution. In this situation, oxygen acts as an oxidizer. Use a schematic polarization diagram to support your explanation. Let the exchange current densities and the open-circuit potentials be $i_{o,O_2} < i_{o,H_2} < i_{o,M}$ and $E_{o,O_2} > i_{o,H_2} > i_{o,M}$. What effect would have oxygen on the metal dissolution if $i_{o,O_2} << i_{o,H_2} < i_{o,M}$ and $E_{o,O_2} > i_{o,H_2} > i_{o,M}$.

7.4. Assume that a stainless steel pipe is used to transport an aerated acid solution containing $1.2x10^{-6}\,\mathrm{mol/cm^3}$ of dissolved oxygen at room temperature and that the electrical double layer is 0.7 and 0.8 mm under static and flowing velocity, respectively. If the critical current density for passivation and the diffusion coefficient of dissolved oxygen are $300\,\mu\mathrm{A/cm^2}$ and $10^{-5}\,\mathrm{cm^2/s}$, then determine **(a)** whether or not corrosion will occur under the given conditions and **(b)** the passive film thickness at 5-min exposure time if $i_p = 50\,\mu\mathrm{A/cm^2}$. Also assume that the density of the film and the molecular weight of the stainless steel are 7.8 $\mathrm{g/cm^3}$ and 55.85 g/mol, respectively, and that the predominant oxidation state is 3. Given data:

$$
\begin{array}{ll}
F = 96{,}500\,\mathrm{A\,s/mol} & C_x = 1.2x10^{-6}\,\mathrm{mol/cm^3} \\
z = 3 & \delta(static) = 0.07\,\mathrm{cm} \\
D = 10^{-5}\,\mathrm{cm^2/s} & \delta(flowing) = 0.008\,\mathrm{cm} \\
i_{crit} = 300\,\mu\mathrm{A/cm^2} & \rho = 7.8\,\mathrm{g/cm^3}
\end{array}
$$

7.5. For the reinforced concrete specimen (concrete slab) shown in Example 7.2, calculate **(a)** the chloride-ion penetration depth (x) when the potential gradient due to diffusion and migration is equal and **(b)** the potential gradients. [Solution: (a) $x = 1.03\,\mathrm{mm}$.]

7.6. For a reinforced concrete slab having a concrete cover 50 mm deep, the threshold and the surface concentrations of chloride anions are $C_{th} = 0.60\,\mathrm{kg/m^3}$ and $C_s = 19\,\mathrm{kg/m^3}$. Also, $D = 32\,\mathrm{mm^2/year}$ is the diffusivity of chloride anions. Determine **(a)** the chloride molar flux and **(b)** the time (T_i) to initiate corrosion on an uncoated reinforcing steel surface (Fig. 7.21c). Explain. **(c)** Plot the concentration profile of chloride anions at the corrosion initiation time T_i. Calculate **(d)** the rate of iron hydroxide $Fe(OH)_2$ production (molar flux and mass flux) at the anodic regions for an anodic current density of $1.5\,\mu\mathrm{A/cm^2}$, **(e)** the corrosion rate (C_R) of steel in mm/year, **(f)** the time required for cracking and spalling due to the formation of a critical rust ($Fe(OH)_3$) volume if the bar diameter is reduced only 25 μm, and **(g)** the time for repair and the apparent mass of rust per area. [Solution: (b) $t_i = 13.43$ years.]

7.7. Assume that all electrochemical reactions are governed by the Nernst equation, which specifies the relationship between the potential of an electrode and the concentrations of the two species Fe^{3+} and Fe^{+2} involved in the reversible redox reaction at the working electrode. Assume a hydrogen-rich electrolyte. **(a)** Write down the electrochemical reaction for Fe^{3+} and Fe^{2+} and the potential equation. **(b)** What is the driving force when the concentrations of the species at the working electrode surface are equal? **(c)** What is the driving force at the working electrode surface when $C_{Fe^{3+}}(\infty, t) > 2C_{Fe^{3+}}(0, t)$ and $C_{Fe^{2+}}(0, t) = 0.5C_{Fe^{3+}}(0, t)$, where $t > 0$ at 25 °C? **(d)** Derive the current equation for the case described in part (c). Recall that the current is simply the flow rate of electrons. **(e)** Draw schematic concentration profiles for $C_O = C_{Fe^{3+}}$ when $I = 0$, $I > 0$ for reduction (Fe^{3+} cation moves to the electrode surface) and $I < 0$ for oxidation so that the Fe^{3+} cation moves away from the electrode surface.

7.8. Consider the reversible reaction $O + ze^- = R$ (as in $Fe^{3+} + e^- = Fe^{2+}$) for deriving the half-wave potential ($E_{1/2}$) using $I = 0.5I_L$, where I_L is the limiting current for a voltammogram. Assume that the initial solution contains only the species O and that diffusion is the form of mass transport so that the current in a voltammetric cell is

$$I = K_O\left([O]_b - [O]_{x=0}\right) \quad \text{(a)}$$

$$I = K_R\left([R]_{x=0} - [R]_b\right) = K_R\,[R]_{x=0} \quad \text{(b)}$$

where $K_O = zFA_sD_O/\delta$ and $K_R = zFA_sD_R/\delta$, $[R]_b = 0$ in the bulk solution initially. Use the Nernst equation for defining the voltammetric cell potential. Recall that the applied potential reduces O to R and that the current depends on the rate at which O diffuses through the diffusion layer (δ).

References

1. N. Perez, *Evaluation of Thermally Degraded Rapidly Solidified Fe-Base and Ni-Base Alloys*. Ph.D. Dissertation, University of Idaho, Moscow (1989)
2. J. Wang, *Analytical Electrochemistry* (Wiley-VCH, New York, 1994), p. 32
3. P.S. Mangat, M.C. Limbachiya, Effect of initial curing on chloride diffusion in concrete repair materials. Cem. Concr. Res. **29**, 1475–1485 (1999)
4. D.A. Jones, *Principles and Prevention of Corrosion* (Macmillan Publishing Company, New York, 1992)
5. T. Engel, P. Reid, *Physical Chemistry* (Pearson Education, Boston, 2013), pp. 123, 878
6. D.W. Shoesmith, Kinetics of aqueous corrosion, in *Corrosion, ASM Handbook*, vol. 13, 9th edn. (American Society for Metals International, Metals Park, 1987)
7. P.J. Boden, Effect of concentration, velocity and temperature, in *Corrosion: Metal/Environment Reactions*, 3rd edn., ed. by L.L. Shreir, R.A. Jarman, G.T. Burstein (Butterworth-Heinemann, Oxford, 1998), pp. 2:3–2:30
8. T. Erdey-Gruz, *Kinetics of Electrode Processes* (Wiley-Interscience a Division of Wiley, New York, 1972), pp. 127–131
9. L.O. Timblin Jr., T.E. Backstrom, Study of depassivation of steel in concrete, in *25th Conference, National Association of Corrosion Engineering*, Houston, TX (1970), pp. 88–97
10. M.G. Fontana, *Corrosion Engineering*, 3rd edn. (McGraw-Hill Book Company, New York, 1986)
11. U.R. Evans, The distribution and velocity of the corrosion of metals. J. Frankl. Inst. **208**, 45–58 (1929)
12. U.R. Evans, *The Corrosion and Oxidation of Metals* (Arnold, London, 1961)
13. M. Stern, A.L. Geary, Electrochemical polarization. J. Electrochem. Soc. **104**(1), 56–63 (1957)
14. A.J. Bard, L.R. Faulkner, *Electrochemical Methods*, Chap. 6 (Wiley, New York, 2001)
15. J.R. Scully, *Electrochemical Methods of Corrosion Testing*, 9th edn. Corrosion, vol. 13. Laboratory Testing (ASM International, Materials Park, 1987), p. 213
16. X.G. Zhang, Galvanic corrosion, in *Corrosion, Vol. 1, Metal/Environment Reactions*, Chap. 8, ed. by L.L. Shreir, R.A. Jarman, G.T. Burstein (Butterworth-Heinemann, Boston, 1994)

17. H.H. Uhlig, R.W. Revie, *Corrosion and Corrosion Control*, 3rd edn., Chaps. 5 and 7 (Wiley, New York, 1985)
18. U.R. Evans, *The Corrosion and Oxidation of Metals* (Arnold, London, 1961)
19. A.A. Hermas, Polarisation of low phosphorus AISI 304 stainless steel in sulphuric acid containing arsenites. Br. Corros. J. **34**, 132–138 (1999)
20. J.R. Maloy, Factors affecting the shape of current-potential curves. J. Chem. Educ. **60**, 285–289 (1983)
21. A.F. Mills, *Basic Heat and Mass Transport*, 2nd edn. (Prentice-Hall, Englewood Cliffs, 1999), pp. 330, 758, 820–823
22. K.L. Moore, J.M. Sykes, S.C. Hogg, P.S. Grant, Pitting corrosion of spray formed Al–Li–Mg alloys. Corros. Sci. **50**, 3221–3226 (2008)
23. G. Balabanic, N. Bicanic, A. Durekovic, Mathematical modeling of electrochemical steel corrosion in concrete. J. Eng. Mech. 1113–1122 (1996)
24. R.E. West, W.G. Hime, Chloride profiles in salty concrete. Mater. Perform. **24**(7), 29–36 (1985)
25. C.E. Locke, C. Dehghanian, Embeddable reference electrodes for chloride contaminated concrete. Mater. Perform. **18**(2), 70–73 (1979)
26. Z.P. Bazant, Physical model for steel corrosion in concrete sea structures EM dash theory. ASCE J. Struct. Div. **105**(6), 1137–1153 (1979)
27. Z.P. Bazant, Physical model for steel corrosion in concrete sea structures Em dash application. ASCE J. Struct. Div. **105**(6), 1155–1166 (1979)
28. M. Stern, Evidence for a logarithmic oxidation for stainless steel in aqueous systems. J. Electrochem. Soc. **106**(4), 376–381 (1959)
29. E.M. Purcell, *Electricity and Magnetism*. Berkeley Physics Course, vol. 2 (McGraw-Hill Book Company, New York, 1965), p. 420
30. L.L. Shreir, Outline of electrochemistry, in *Corrosion Vol. 2, Corrosion Control*, ed. by L.L. Shreir, R.A. Jarman, G.T. Burstein (Butterworth-Heinemann, Boston, 1994)
31. T.S. Butalia, Corrosion in concrete and the role of fly ash in its mitigation, in *Energeia*, vol. 15, no. 4 (CAER University of Kentucky, Center for Applied Energy Research, 2004)
32. A.M. Neville, *Properties of Concrete*, 4th edn. (Wiley, New York, 1996)
33. L.J. Parrott, A review of carbonation in reinforced concrete. BRE/C&CA Report C/1-0987 (July 1987), 369
34. J.A. Gonzalez, S. Algaba, C. Andrade Corrosion of reinforcing bars in carbonated concrete. Br. Corros. J. **3**, 135–139 (1980)
35. K.E. Curtis, K. Mehta, A critical review of deterioration of concrete due to corrosion of reinforcing steel, in *Durability of Concrete, Proceedings Fourth CANMET/ACI International Conference*, Sydney, 1997
36. A. Rosenberg, C.M. Hansson, C. Andrade, Mechanisms of corrosion of steel in concrete, in *Materials Science of Corrosion I* (The American Ceramic Society, Westerville, OH, 1989)
37. D.W. Whitmore, Impressed current and galvanic discrete anode cathodic protection for corrosion protection of concrete structures. Paper 02263, Corrosion 2002
38. J. Crank, *The Mathematics of Diffusion*, 2nd edn. (Oxford University Press, Oxford, 1975)
39. S. Lorente, D. Voinitchi, P. Bgu-Escaffit, X. Bourbon, The single-valued diffusion coefficient for ionic diffusion through porous media. J. Appl. Phys. **101**(2), 024907 (2007)
40. T. Luping, J. Gulikers, On the mathematics of time-dependent apparent chloride diffusion coefficient in concrete. Cem. Concr. Res. **37**(4), 589–595 (2007)
41. J.Z. Zhang, I.M. McLaughlin, N.R. Buenfeld, Modelling of chloride diffusion into surface-treated concrete. Cem. Concr. Compos. **20**, 253–261 (1998)
42. J.A. Gonzalez, S. Algaba, C. Andrade, Corrosion of reinforcing bars in carbonated concrete. Br. Corros. J. **3**, 135–139 (1980)
43. S.W. Bishara, Rapid, accurate method for determining water-soluble chloride in concrete, cement, mortar, and aggregate. Application to quantitative study of chloride ion distribution in aged concrete. ACI Mater. J. **88**(3), 265–270 (1991)
44. G. Balabanic, N. Bicanic, A. Durekovic, Mathematical modeling of electrochemical steel corrosion in concrete. J. Eng. Mech. 1113–1122 (1996)

45. J.P. Broomfield, *Corrosion of Steel in Concrete* (E. & F.N. Spon, London, 1997)
46. J. Ozbolt, F. Orsanic, G. Balabanic, Modeling corrosion-induced damage of reinforced concrete elements with multiple-arranged reinforcement bars. Mater. Corrosion **67**(5), 433–562 (2016)
47. H.H. Uhlig, R.W. Revie, *Corrosion and Corrosion Control*, 3rd edn. (Wiley, New York, 1985)
48. K. Suda, S. Misra, K. Motohashi, Corrosion products of reinforcing bars embedded in concrete. Corros. Sci. **35**, 1543–1549 (1993)
49. Y. Liu, Modeling the time-to-corrosion cracking of the cover concrete in chloride contaminated reinforced concrete structures. Ph.D. dissertation, 1996
50. H. Arup, The mechanisms of the protection of steel by concrete, in *Corrosion of Reinforcement in Concrete Construction*, ed. by A.P. Crane (Ellis-Horwood Limited, West Sussex, 1983), pp. 151–157
51. A. Kuter, P. Moller, M.R. Geiker, Corrosion of steel in concrete - thermodynamical aspects, in *Proc. NKM* 13, Reykjavik, 18–20 April 2004, pp. 1–13
52. G.G. Clemena, M.B. Pritchett, C.S. Napier, Application of cathodic prevention in a new concrete bridge deck in Virginia. Final Report VTRC 03-R11, Virginia Transportation Research Council in Cooperation with the U.S. Department of Transportation Federal Highway Administration, Charlottesville, VA, Copyright 2003 by the Commonwealth of Virginia (February 2003)
53. S.F. Daily, Understanding corrosion and cathodic protection of reinforced concrete structures. CP of concrete structures, Corrpro Companies, Inc. (2007). http://www.cathodicprotection.com/essays4.htm
54. O. Poupard, A. Ait-Mokhtar, P. Dumargue, Corrosion by chlorides in reinforced concrete: determination of chloride concentration threshold by impedance spectroscopy. Cem. Concr. Res. **34**, 991–1000 (2004)
55. N. Gowripalana, V. Sirivivatnanonb, C.C. Lima, Chloride diffusivity of concrete cracked in flexure. Cem. Concr. Res. **30**, 725–730 (2000)
56. J. Zhang, Z. Lounis, Sensitivity analysis of simplified diffusion-based corrosion initiation model of concrete structures exposed to chlorides. Cem. Concr. Res. **36**(7), 1312–1323 (2006)
57. Y. Xi, Z.P. Bazant, Modeling chloride penetration in saturated concrete. J. Mat. Civ. Eng. **11**(1), 58–65 (1999)
58. O.B. Isgor, A.G. Razaqpur, Advanced modelling of concrete deterioration due to reinforcement corrosion. Can. J. Civ. Eng. **33**, 707–718 (2006)
59. C. Alonso, C. Andrade, M. Castello, P. Castro, Chloride threshold values to depassivate reinforcing bars embedded in a standardized OPC mortar. Cem. Concr. Res. **30**, 1047–1055 (2000)
60. D.W. Pfeifer, High performance concrete and reinforcing steel with 100-year service life. PCI J. **45**(3), 46–54 (2000)
61. P. Zia, High corrosion resistance MMFX microcomposite reinforcing steels. Appraisal Report, Concrete Innovations Appraisal Service (CIAS), CIAS Report 03-2 (31 May 2003), 1–42
62. Standard test method for half cell potentials of reinforcing steel in concrete, ASTM C 876-91. Annual Book of ASTM Standards, volume 04.02 (1991)
63. S. Lorente, Constructal view of electrokinetic transfer through porous media. J. Phys. D: Appl. Phys. **40**, 2941–2947 (2007)
64. J. Wang, *Analytical Electrochemistry*, 3rd edn. (Wiley-VCH, New York, 2006)
65. A.J. Bard, L.R. Faulkner, *Electrochemical Methods: Fundamentals and Applications*, 2nd edn. (Wiley, New York, 2000)
66. P.M.S. Monk, *Fundamentals of Electro-Analytical Chemistry* (Wiley, New York, 2001)
67. C.M.A. Brett, A.M. Oliveira Brett, *Electrochemistry Principles, Methods and Applications*, Chap. 9 (Oxford University Press, New York, 1994)
68. P. Kanatharana, M.S. Spritzer, Cyclic voltammetry of the Iron(II)-Iron(III) couple in dimethylformamide. Anal. Chem. **46**(7), 958–959 (1974)
69. P. Zanello, *Inorganic Electrochemistry: Theory, Practice and Application* (The Royal Society of Chemistry, Cambridge, 2003)

70. J.E.B. Randies, Trans. Faraday Soc. **44**, 327 (1948); A. Sevcik, Collect. Czech. Chem. Commun. 349 (1948) . References cited in [67]
71. R.S. Nicholson, I. Shain, Theory of stationary electrode polarography: single scan and cyclic methods applied to reversible, irreversible, and kinetic systems. Anal. Chem. **36**(4), 706–723 (1964)
72. F.R. Foulkes, P. McGrath, A rapid cyclic voltammetric method for studying cement factors affecting the corrosion of reinforced concrete. Cem. Concr. Res. **29**, 873–883 (1999)
73. P. Delahay, *New Instrumental Methods in Electrochemistry* (Interscience Publishers, New York, 1954), p. 123
74. E. Kreyszig, *Advanced Engineering Mathematics*, Chap. 5 (Wiley, New York, 1999)
75. D.J.Rose, Analysis of antioxidant behaviour in lubricating oils. Dissertation, The University of Leeds, School of Chemistry, March 1991

8 Design Against Corrosion

The mechanism of corrosion involves metal dissolution due to an electrochemical phenomenon. Thus, corrosion is associated with current flow over finite distances from the corroding metal, and the amount of corrosion that can be accounted for is quantitatively determined by the amount of current passing through the metal. The electrochemical phenomenon occurs because of differences in potential between areas of the corroding metal surface. Therefore, the **driving force of corrosion** is the decrease in free energy associated with the formation of corrosion product on the metal surface. In contrast, preventing corrosion leads to cathodic protection (CP) under steady-state conditions.

Simple engineering structures, such as a spherical steel tank for storing water in hospitals, need to be protected against corrosion. A thin coating of paint can protect the tank. However, complex structures, such as buried steel pipelines near steel tanks, steel bridges, multiple storage tanks in a refinery, offshore oil-drilling rigs, seagoing ships, and many more metallic structures, have to be protected against corrosion.

Electrochemical methods are useful for designing against corrosion, provided that the electrochemical polarization diagram is available. The main objective in protecting a metallic structure is to eliminate or reduce corrosion rate by supplying an electron flow to a structure to reduce or eliminate metal dissolution (oxidation). This implies that the anodic reaction is suppressed on the surface of the structure. This can be accomplished using secondary materials and appropriate instrumentation to supply electrons to the structure.

Cathodic protection is an electrochemical technique in which a cathodic potential is applied to a structure in order to prevent corrosion from taking place. This implies that Ohm's law, $E = IR_x$, can be used to control the potential so that $E < E_{corr}$ and implicitly the current must be $I < I_{corrr}$.

N. Perez, *Electrochemistry and Corrosion Science*,
DOI 10.1007/978-3-319-24847-9_8

8.1 Cathodic Protection

8.1.1 Electrochemical Principles

Cathodic protection is an electrochemical technique in which a cathodic (protective) potential is applied to an engineering structure in order to prevent corrosion from taking place. This implies that Ohm's law, $E = IR$, can be used to control the potential, as well as the current. Hence, metal oxidation is prevented since the potential must be below the corrosion potential ($E < E_{corr}$). This is the main reason for this potential-control technique. In principle, all structures can be protected cathodically, but structural steels being the most common ferrous materials used to build large structures are cathodically protected by an external potential (impressed potential).

Figure 8.1 shows the general diagram for protecting a structure from corrosion. The useful techniques for protecting metallic structures can be solely applied or coupled with other techniques, such as coating supplemented with impressed-current cathodic protection (ICCP).

According to the electrochemical principles, the formation of a solid oxide corrosion product on a metal M immersed in an electrolyte depends on the constituents in solution. For instance, consider the presence and absence of oxygen in solution. The following reactions for a hypothetical oxidizing metal M are used to classify the type of solution:

$$M + \frac{1}{2}O_2 + H_2O \rightarrow M(OH)_2 \quad \text{(Aerated)} \tag{8.1}$$

$$M^{2+} + 2H_2O \rightarrow M(OH)_2 + 2H^+ \quad \text{(Deaerated)} \tag{8.2}$$

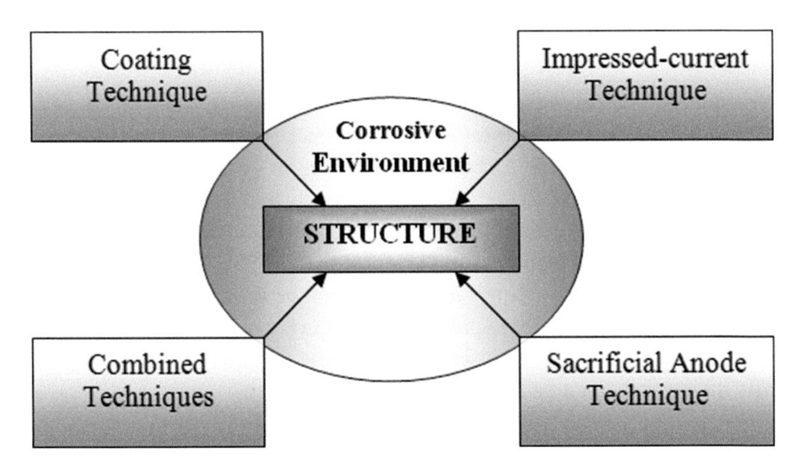

Fig. 8.1 Diagram illustrating protection techniques against corrosion

Thus, an aerated solution contains oxygen, and its deaerated counterpart lacks of dissolved oxygen. The former is the most common reaction encountered in industrial schemes for the formation of a metallic hydroxide surface layer, $M\,(OH)_2$. For corroding iron or steel, the coupled anodic and cathodic reactions in the presence of oxygen generate an overall (redox) reaction similar to Eq. (8.1) [see Eq. (1.14)]. A possible reaction sequence can be outlined as follows:

$$2Fe \rightarrow 2Fe^{2+} + 4e^{-} \qquad \text{(anode)} \tag{8.3}$$

$$O_2 + 2H_2O + 4e^{-} \rightarrow 4OH^{-} \qquad \text{(cathode)} \tag{8.4}$$

$$2Fe + O_2 + 2H_2O \rightarrow 2Fe^{2+} + 4OH^{-} \qquad \text{(redox)} \tag{8.5}$$

Then, Fe^{2+} and OH^{-} form ferrous hydroxide, $Fe\,(OH)_2$, but $Fe\,(OH)_2$ in the presence of oxygen transforms to ferric hydroxide, $Fe\,(OH)_3$, known as rust. The reactions for the formation of ferric hydroxide are [1]

$$2Fe^{2+} + 4OH^{-} \rightarrow 2Fe\,(OH)_2 \downarrow \qquad \text{(unstable)} \tag{8.6}$$

$$2Fe\,(OH)_2 + \frac{1}{2}O_2 + H_2O \rightarrow 2Fe\,(OH)_3 \downarrow \equiv Fe_2O_30.3H_2O \downarrow \tag{8.7}$$

In principle, this overall reaction represents the corrosion product, and the polarization diagram in Fig. 8.2 illustrates the corrosion behavior for iron (Fe) or carbon steel.

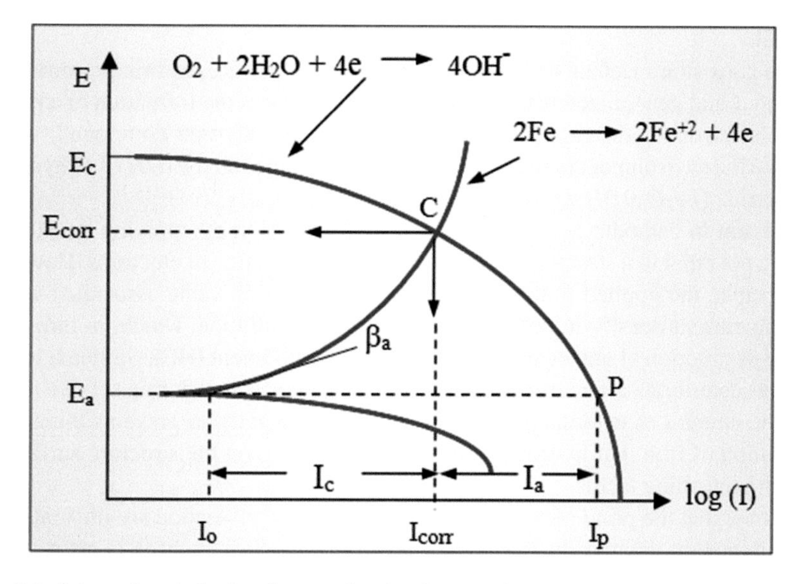

Fig. 8.2 Schematic polarization diagram showing the corrosion point "C" and the protective point "P" for a hypothetical metal immersed in an oxygen-containing electrolyte

Figure 8.2 schematically shows a polarization diagram for iron or carbon steel exposed to the atmosphere. The interpretation of this polarization diagram is as follows:

- In Fig. 8.2, point "C" is the corrosion point of the system, and "P" is the ideal protective point, where $E = E_a$ and $I_a = |I_c| = I_p$. Hence, E is the open-circuit potential of the metal when the electrochemical cell is in equilibrium (no current flows at this potential).
- However, if $E < E_a$, the structure is also protected at a higher value, but hydrogen evolution evolves leading to hydrogen embrittlement of steel under a level of elastic stress or internal pressure.
- In the case of coated steel pipelines, the coating can undergo localized damage called holiday. Also, shown in Fig. 8.2 is the anodic Tafel slope (β_a), which is related to the anodic overpotential (η_a) defined by Eq. (5.39).
- The schematic electrochemical polarization curve suggests that metal dissolution occurs along the E_a-C curve, provided that both the applied potential and the current response are $E > E_a$ and $I > I_a$, respectively.
- If $E_a < E < E_{corr}$ and $I_a < I < I_{corr}$, oxidation takes place which is a qualitative measure of corrosion. However, knowing the value of I and the exposed surface area of the specimen or metal part, the corrosion rate (C_R) can be estimated using Faraday's law of electrolysis.

Theoretically, cathodic protection can be applied to any structure susceptible to corrode in an electrochemical environment, but this method is commonly used to protect carbon steel structures in diluted or alkaline electrolytes, such as seawater and soil.

The corrosion mechanism on iron or carbon steel surface, as briefly introduced in Chap. 1 and generalized hereafter by Eq. (8.7), includes the formation of a partial unstable ferrous hydroxide $[Fe\,(OH)_2]$ as the initial corrosion compound, which reacts in the environment to form ferric hydroxide compound $[Fe\,(OH)_3]$ or hydrated ferric oxide ($Fe_2O_3 0.3H_2O$) known as "rust."

The aim in cathodic protection is to prevent corrosion by applying an external electric potential to a structure along with a DC as the source of electrons. However, in principle, the applied potential has a practical limiting value associated with a limiting current density in order to avoid hydrogen evolution, which, in turn, may induce destruction of any coating or hydrogen embrittlement (HE). Suffice it to say, coatings deteriorate after structures are cathodically protected during service [2]. In fact, the amount of external electrons reduces significantly or prevents the rate of dissolution of iron, but hydroxyl ions (OH^-) still form on the structure surface as the pH-controlling factor.

Despite that the principles of corrosion and cathodic protection are illustrated by the polarization diagram in Fig. 8.2 for iron (Fe), kinetics analysis is apart of the cathodic protection scheme since corrosion may occur at a slow or fast rate in an aerated electrolyte. But once iron atoms release electrons, the reduction of water takes place to form hydroxyl ions (OH^-) in the electrolyte, and as a result, the electrolyte pH changes.

Moreover, assume that oxygen reduction occurs and no other oxidizing agents are present. This, then, becomes an ideal case for protecting buried or underground structures. Recall that oxygen is an uncharged species which implies that its mass transfer depends on its concentration gradient [3].

8.1.2 Cathodic Protection Techniques

8.1.2.1 Coating Technique

This is a cost-effective technique, providing the most the protection, but it is susceptible to have surface defects called **holidays**, which promote current drainage points causing localized corrosion sites. Therefore, a combination of coating and cathodic protection (through an impressed-current) techniques is adequate to protect a metallic structure, commonly made out of steel. For instance, coal tar pitch (black resin) is used for coating underground steel structures. Anyway, coating defects develop due to chemical reactions and mechanical damage, leading to an increase in defect density and current requirements for protection. An increase in the protective current due to low resistance defects can be determined using Ohm's law, $I = E/R_x$.

8.1.2.2 Sacrificial Anode Technique

If the coating technique is insufficient to cathodically protect a structure from corrosion, then a combination of coating and an external current suffices for the protection. Fundamentally, CP dictates that the structure is treated as the cathode ($-$) and an external electrode as the anode ($+$) by connecting them to a voltmeter (conductor) to complete an electrical circuit as a galvanic cell or galvanic couple. Thus, the principle of electrochemistry assures that the CP technique impresses an external current on the cathode. This galvanic couple is referred to as a sacrificial anode technique (SAT) where the anode is active and the cathode is noble [2, 4]. This principle is schematically shown in Fig. 8.3 under ideal conditions.

Hereafter, the sacrificial anode is just a galvanic anode (+) that used a counter electrode (CE), and the structure is the cathode ($-$). Field measurements of corrosion potential require a reference electrode connected to the positive terminal of a voltmeter.

Assume that the structure is made of steel (Fe) and the sacrificial anode is a magnesium (Mg) alloy. The superimposed polarization curves intercept at (I_{xorr}, E_{corr}), where the element Mg oxidizes very fast and the CP design life reduces significantly without the proper protection, say, in soil. However, an applied potential E_p produces an external current response $I_x < I_{corr}$, where $I_x = I_c - I_a$. Thus, point "P" on the structure cathodic curve becomes the ideal protective point for CP using a sacrificial anode since $I_{corr} \rightarrow I_a$. Suitable sacrificial anodes are commonly made out of Al, Mg, and Zn alloys.

Sacrificial nodes can also be used to protect uncoated (bare) structures, seagoing vessels, offshore tanks, offshore platforms, water heater, and the like. The anodes

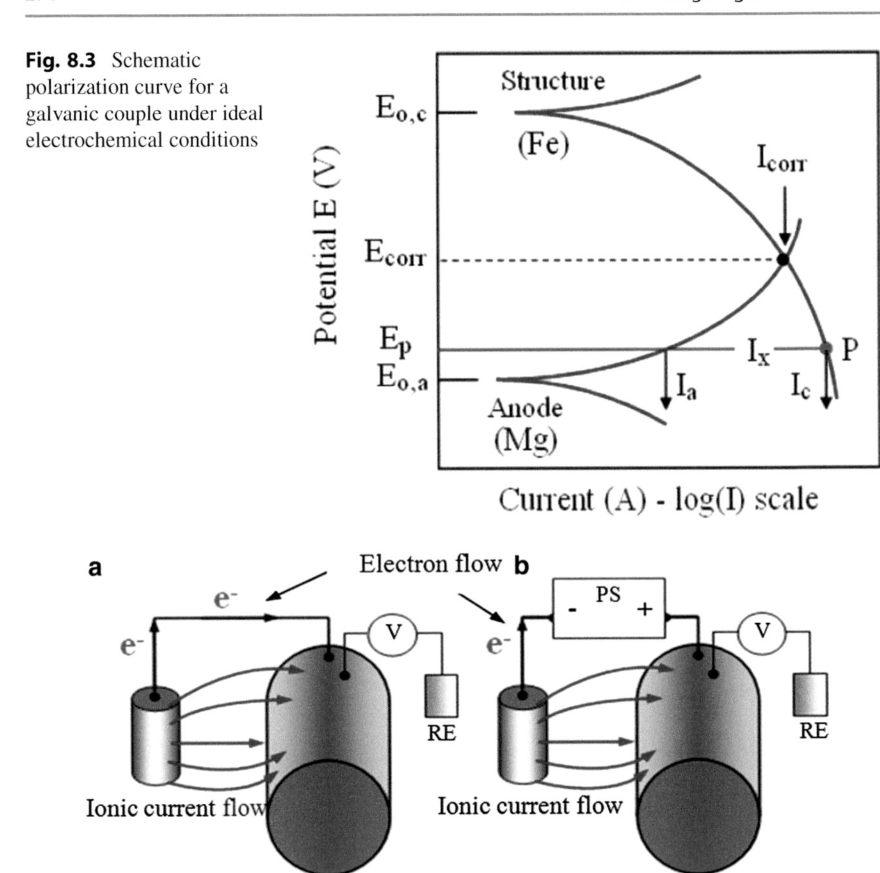

Fig. 8.3 Schematic polarization curve for a galvanic couple under ideal electrochemical conditions

Fig. 8.4 Schematic galvanic cell layout for cathodic protection of a structure with one anode

must be monitored frequently because they dissolve sacrificially (purposely) and become incapable of delivering the necessary current to the structure. However, if the SAT design does not deliver a design current for polarizing the galvanic couple, a power supply (PS) is connected between the structure and the anodes for such a purpose.

Figure 8.4 schematically shows [5] the general layouts for **sacrificial anode technique** (SAT) and **impressed-current technique** (ICT) for cathodic protection (CP) of a structure. The former technique requires several sacrificial anodes (SAs) for supplying electrons to large cathodic structure, while the latter usually requires less anodes than the former one because of the rectifier Thus, a protective potential-current (E, I) is determined as explained above.

Figure 8.4b illustrates an impressed-current cathodic protection (ICCP) technique for providing the electrons to large coated and bare structures (pipelines, gasoline tanks, etc.) using a DC power supply (PS), commonly known as a rectifier. The initial polarization for ICCP is provided by an applied $(E, I)_{initial}$ pair, and once CP is in progress, the structure's potential (E) against a suitable reference electrode (RE) can be measured using a voltmeter (V), which is drawn in Fig. 8.4 for both SAT and impressed current (IC).

Buried structures in direct contact with soil, the soluble soil salts, and moisture content constitute the electrolyte of the corrosion cell. In such a case, CP against corrosion seems suitable to form galvanic cells as illustrated in Fig. 8.4 so that the anode is sacrificed naturally or forcedly.

8.1.3 Sketches of Industrial Cases

The impressed-current technique is a simple and, yet, significant form of cathodic protection of underground steel pipelines as shown in Fig. 8.5a [1]. For comparison, Fig. 8.5b illustrates the same galvanic couple without the rectifier. The former technique is most common for underground large structures.

A buried pipeline is connected to the negative terminal of a rectifier (power supply) and the anode to the positive terminal. Both terminals must be well insulated; otherwise, current leakage (stray current) occurs, and the structure may not be protected adequately [6].

In both cases, the current flows from the rectifier to the inert anode or sacrificial anode (graphite) through the soil (electrolyte) to the cathode. The purpose of the rectifier is to convert alternating current (AC) to uniform direct current (DC).

Other cathodic protection systems are schematically shown in Fig. 8.6 [7]. For instance, Fig. 8.6a illustrates the three-electrode technique to conduct measurements of (a) the potential difference between the working electrode (WE) and the reference electrode (RE) using a voltmeter (V) and (b) the current flowing to the anodic auxiliary electrode (AE) using an ammeter (A).

Figure 8.6b illustrates another CP system encountered at industrial sites using embedded anodes [7], where the length of the anode-backfill column, known as active anode bed, must not be chosen arbitrarily because it is related to the current density entering the anode bed. Hence, the backfill current density (i_b) is simply defined by the ground current divided by the column surface area:

$$i_b = \frac{I}{\pi D h} \tag{8.8}$$

where I = ground current

D = diameter of the column
h = length of the column

Example 8.1. *Suppose that a buried steel structure is to be cathodically protected using a rectifier capable of delivering 1 V and 5 A through the wiring system. Assume*

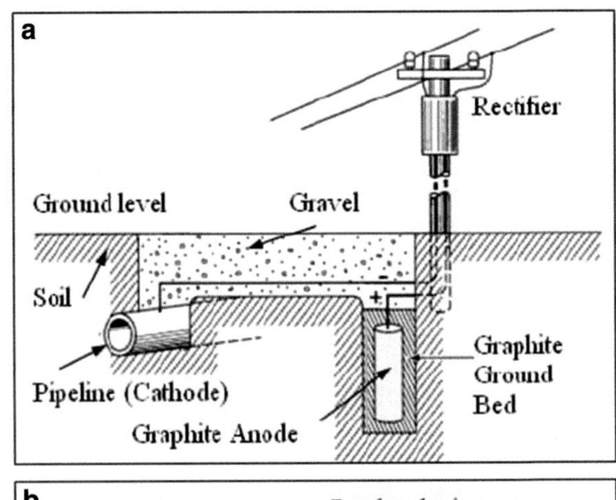

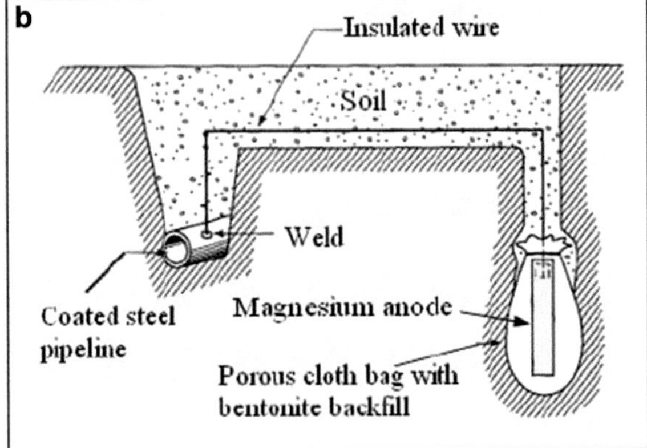

Fig. 8.5 Schematic cathodic protection technique [1]. (**a**) Impressed current. (**b**) Sacrificial anode

that the soil resistance is 80 % of the external resistance. Calculate **(a)** *the soil resistance when the anode bed is* 4 *m long and* 0.10 *m in diameter and* **(b)** *the current density* i_b.

Solution. (a) *Using Ohm's law yields the external electric resistance as*

$$R_x = \frac{E}{I} = \frac{1\,\text{V}}{5\,\text{A}} = 0.20\,\Omega$$

Thus, the soil electric resistance is

$$R_s = 0.80 R_x = 0.16\,\Omega$$

a

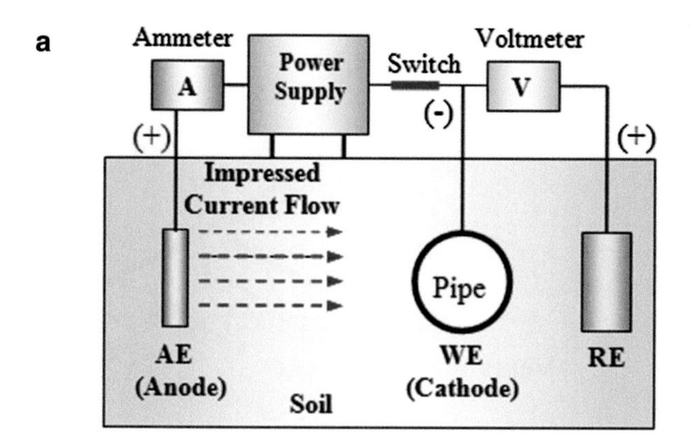

b

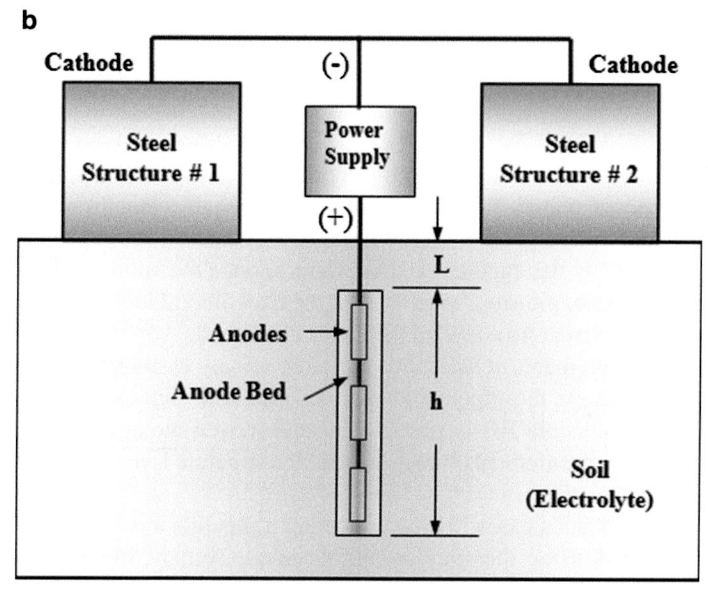

Fig. 8.6 Cathodic protection systems of two steel structures. After [7]. (**a**) Three-electrode technique; (**b**) Embedded anodes

(b) *From Eq. (8.8),*

$$i_b = \frac{I}{\pi D h} = \frac{5\,\text{A}}{\pi\,(0.10\,\text{m})\,(4\,\text{m})}$$

$$i_b \approx 4\,\text{A/m}^2 \quad @\ h = 4\,\text{m}$$

Figure 8.7 shows a modern CP system (http://www.corrpro.co.uk/pdf/TANK-PAK-impressed_current_system.pdf) known as a gasoline station, gas station, or service station.

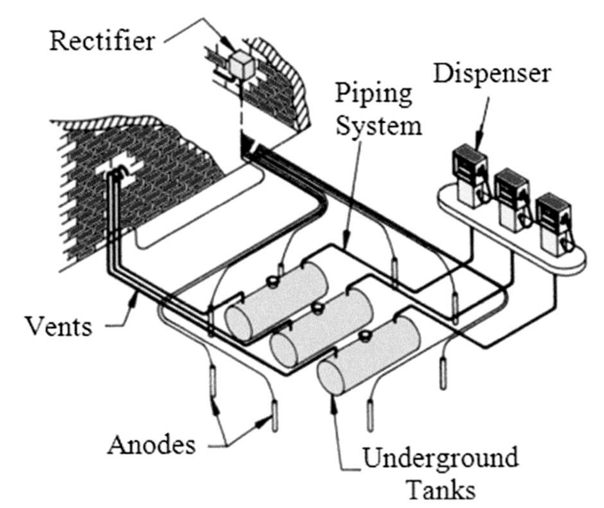

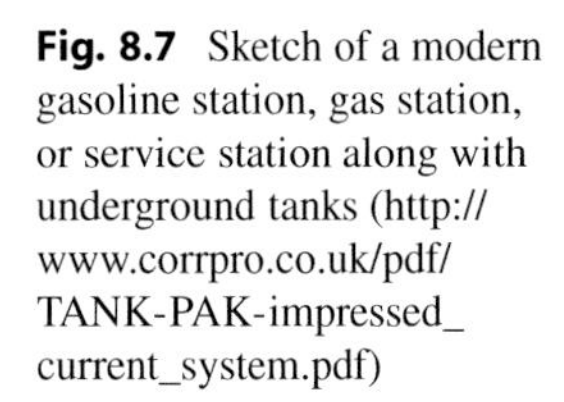
Fig. 8.7 Sketch of a modern gasoline station, gas station, or service station along with underground tanks (http://www.corrpro.co.uk/pdf/TANK-PAK-impressed_current_system.pdf)

The design and installation method for a buried structure using magnesium or any other suitable anode material as the source of current requires that the wiring system be well insulated with a good electric resistance material; otherwise, the insulation may be destroyed by the anodic reaction products once the anode is energized by a power supply. Also, moisture must exist at the backfill-soil interface to provide a path for uniform current flow toward the active anode bed.

Potential measurements of structures buried in soil are carried out along the soil surface directly above the pipeline at discrete intervals. Similarly, for immersed structures, a submersible RE is placed near and moved along the pipeline. The outcome of these measurements is to determine if a structure is maintained polarized at the design potential.

Let us use Fontana's cases [5] for protecting a pipeline against corrosion near a buried steel tank using the stray-current technique. Firstly, Fig. 8.8a shows the detrimental effect of stray current (leakage current), which has the following path (dashed lines) in the cathodic circuit:

$$\text{Power supply} \rightarrow \text{anode} \rightarrow \left\{ \begin{array}{c} \text{soil} \\ \text{pipeline} \end{array} \right. \rightarrow \text{steel tank} \rightarrow \text{power supply} \tag{8.9}$$

As a result, the pipeline corrodes near the steel tank. Secondly, the solution to this problem relies on placing and connecting another anode to the circuit as shown in Fig. 8.8b. Thus, current flows from both anodes toward the tank and pipeline, and consequently, both structures are cathodically protected by this uniform stray-current flow.

In the above cathodic protection design, the current being discharged from the anodes is picked up by the tank and pipeline in the directions shown as dashed lines. In treating the tank and the pipeline as a whole structure, the total cathodic potential

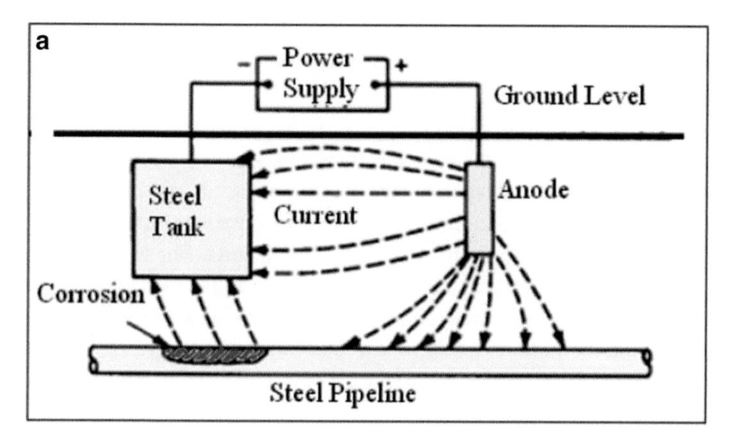

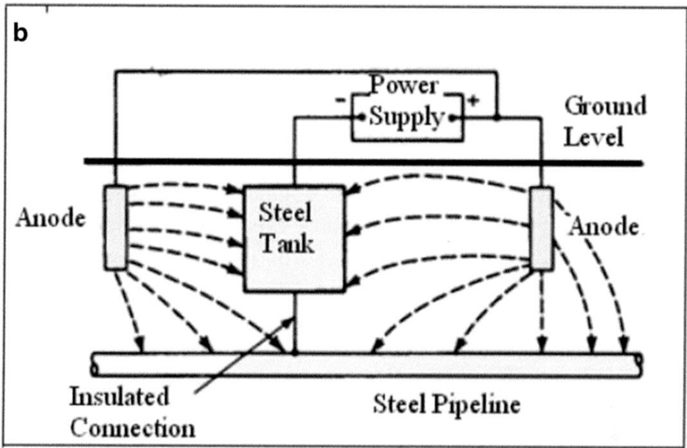

Fig. 8.8 Stray-current technique for cathodic protection [5]. (**a**) Stray-current corrosion and (**b**) stray-current protection

(E) can be measured with respect to a reference electrode (RE), excluding any ohmic potential drop (IR_x). This can be accomplished by using the recommended switch-off technique, in which the rectifier is switched off for interrupting the current flow and eliminating IR_x, and measuring the potential immediately with a high-impedance voltmeter. Consequently, the ohmic potential drop is dissipated immediately, while polarization of the structure decays at a low rate [3]. With regard to Fig. 8.8a, the stray currents flow through the soil to the tank and to the pipeline, and corrosion occurs where the stray currents leave the pipeline in a confined area. This is the reason for the shown localized corrosion. Stray-current corrosion is an unusual metal deterioration of buried or submerged structures.

The source of stray currents is an electric generator, such as an electric welding machine, a grounded DC power supply, a cathodic protection system, and an electroplating cell [6]. Some structures are more vulnerable than others due to localization, design, and surrounding electric generators. Nevertheless, the principal remedy for stray-current corrosion is to eliminate the current leakage source whenever possible. Otherwise, an undesirable metal deterioration or even catastrophic failure may occur.

Stray currents flow through paths other than the designed circuit, and the effect of this flow can render designs very ineffectively. In fact, stray-current corrosion is different from natural corrosion since it is induced by an external electric current. Corrosion may be accelerated when other corrosion mechanisms induced by environmental factors are coupled with stray currents. For instance, both stray-current corrosion and galvanic corrosion have cathodic and anodic sites on a structure, but the former may vary with time, and metal dissolution is induced by the current flow, while the latter is a continuous electrochemical process independent of an external current and proceeds at a constant rate. These sites may be meters apart, and corrosion damage occurs on the metal surface since the structure has a higher electrical conductivity than a particular internal fluid. However, if the stray-current flow reaches the internal fluid, corrosion damage may occur on the inside of the pipeline, making matters worse because undetected corrosion may be ignored and, consequently, a disastrous and regrettable problem may occur.

Pipelines and other steel structures may be cathodically protected using aluminum, zinc, and magnesium sacrificial anodes. These buried sacrificial anodes are usually backfilled with a high conductive material, such as granulated coke or graphite flakes, for preventing anode consumption and spreading the current in a relatively large area [2, 7]. Other anodes can be placed directly on the structure, such as offshore and inland tanks.

In summary, the impressed-current technique requires the following:

- A power supply (rectifier) that converts AC to DC in order to cathodically polarize the structure.
- In principle, impressed-current anodes can be made out of materials having a variety of shapes, but they have to be capable of providing electrons to the cathodes.
- The applied potential and current must be $E << E_{corr}$ and $I << I_{corr}$, respectively.
- Despite that $I << I_{corr}$ is required, this current should not be large in magnitude; otherwise, hydrogen evolution may cause embrittlement of the structure. This is a well-known fact being studied and reported in the literature for decades.

Table 8.1 illustrates a list of impressed-current anodes used in known corrosive environments. These anodes are high corrosion-resistant solid materials. For instance, platinum (Pt) is used as the primary anode material due to its excellent corrosion resistance leading to low consumption.

Graphite anodes are made by mixing calcined petroleum coke particles and coal tar pitch binder. The desired shape is heated at 2800 °C in order to convert amorphous carbon to crystalline carbon (graphite), which is immune to chloride solutions [9, 10]. These anodes are normally impregnated with linseed oil or synthetic resins for reducing porosity and spalling. In addition, the anodes are buried

Table 8.1 Impressed-current anodes [7,8]

Anode	Application
Platinized Ag	Chloride-containing tanks
Platinized Ti	Surface pipelines
Platinized Ta	Pipelines
Ti_4O_7 and Ti_5O_9	Steel-reinforced concrete, buried steel structures
Base ceramics	Chloride-containing steel tanks,
	electrowinning electrodes, batteries
Polymers	Steel-reinforced concrete roof,
	bridge decks, parking garages,
	steel-reinforced bars
$Fe - 14Si - 4Cr$	Offshore structures
Graphite	Buried pipelines
	Onshore pipelines

in the soil and backfilled with coke breeze for a uniform distribution of the electrical conductivity. The coke breeze surrounding the graphite anodes increases the surface area and disperses the anode reactions forming CO_2 and O_2 gases, which are vented through the soil porosity [8].

For cathodic protection, the National Association of Corrosion Engineers (NACE) recommends a potential of $E = -0.85\,\text{V}$ vs. $Cu/CuSO_4$ reference electrode for steel and iron buried in the soil [8] and $E = -0.81\,\text{V}$ vs. $Ag/AgCl$ reference electrode for Mg-alloy anodes buried in the soil and in domestic water heater [11]. For submerged structures, the polarizing potential is normally within the range $-1.05\,\text{V} < E < -0.80\,V_{Ag/AgCl}$ as reported in the literature using dual anodes made out of Al-Zn-Hg alloys having an outer layer of Mg [12–14]. Recent papers describing experimental procedures using laboratory cells for testing specimens in seawater can be found elsewhere [12–16].

With regard to sacrificial anode in cathodic protection, this technique uses the natural potential difference between the cathodic structure and the anode as the driving force for protection against corrosion of the structure. Therefore, the anode is sacrificed for a long period of time since it becomes the source of electron transfer.

Table 8.2 lists relevant data for common sacrificial anode materials used for protecting steel structures. The anode capacity C_a is mathematically defined in later section. In fact, a sacrificial anode has to be more electronegative than the structure and be sufficiently corrosion resistant; otherwise, anode dissolution would be too fast, and the design lifetime would be shorten significantly.

Essentially, Al, Zn, and Mg alloys are commonly used in order to comply with these requirements. Also, the electrons released by the anodes must be transferred to the steel structure to support the cathodic reactions on the structure and maintain cathodic polarization [17].

Table 8.2 Sacrificial anode materials and operating parameters for protecting steel structures [3, 17]

Alloy	Potential (V) vs. $Ag/AgCl$ in seawater	Bare steel potential (V)	Anode capacity C_a (A h/Kg)
Al-Zn-In	-0.95 to -1.10	0.15–0.3	1300–2650
Al-Zn-Hg	-1.00 to -1.05	0.20–0.25	2600–2850
Al-Zn-Sn	-1.00 to -1.05	0.20–0.25	925–2600
Zn-Al-Cd	-1.05	0.20	780
Mg-Al-Zn	-1.50	0.70	1230
Mg-Mn	-1.70	0.90	1230
Zn	-0.95 to -1.03	0.15–0.23	750–780

Among many requirements, impressed-current anodes must be good electrical conductor, have low corrosion rate and low cost, and be capable of withstanding stresses during installation and service.

8.1.4 Cathodic Protection Criteria

The electrochemical foundation for cathodic protection is based on the mathematical model described by Eq. (5.32) or (5.43), where $i_{corr} = i_a = -i_c$ at E_{corr} or $i_{corr} = i_a + i_c = 0$. This model states that the corrosion rate on a metal surface is zero when the forward and reverse current densities become exactly the same at equilibrium ($E = E_{corr}$). This criterion was recognized by Mears and Brown [18] in 1938, and it has become a common practice for protecting steel structures.

However, designing a cathodic protection system requires an initial current greater than the maintenance value for a specific design life, and fortunately, the structure can be cathodically polarized during the lifetime. A particular design using an anode material containing a layer of magnesium (Mg) is very attractive in the early stages of cathodic protection. For soil structures, Mg strips can be used [19], and those in marine environments, Al alloy plus a layer of Mg have been reported [20] to be effective in polarizing the structures by allowing Mg dissolution at the initial current density prior to steady-state conditions at the maintenance current density. Therefore, Mg is the starter material that is consumed in the early stages of protection, and eventually, the polarized structure reaches steady state.

Half-Cell Potential Criterion For steel structures, cathodic protection is achieved when polarized at the iron (Fe) equilibrium half-cell potential [1]. In neutral environments (soil and seawater), the half-cell potential is based on the following reactions, and it is determined by the Nernst equation.

For pure iron,

$$Fe^{2+} + 2e = Fe \text{ with } K_{Fe} = \frac{[Fe]}{[Fe^{2+}]} \text{ and } z = 2 \tag{8.10}$$

$$E_{Fe} = E^o_{Fe} - \frac{RT}{zF} \ln(K_{Fe}) \tag{8.11}$$

$$E_{Fe} = -0.44 V_{SHE} + \frac{2.303 RT}{zF} \log\left[Fe^{2+}\right] \tag{8.12}$$

For ferrous oxide,

$$Fe(OH)_2 = Fe^{2+} + 2(OH^-) \text{ with } K_{Fe(OH)_2} = 1.8x10^{-15} \tag{8.13}$$

$$E_{Fe(OH)_2} = E^o_{Fe(OH)_2} - \frac{RT}{zF} \ln\left(K_{Fe(OH)_2}\right) \tag{8.14}$$

The activity of ferrous Fe^{2+} and hydroxyl OH^- ions in Eq. (8.13) is related by

$$\left[Fe^{2+}\right] = 2\left[OH^-\right] \tag{8.15}$$

Consequently, the solubility constant of $K_{Fe(OH)_2}$ and $\left[Fe^{2+}\right]$ becomes

$$K_{Fe(OH)_2} = \left[Fe^{2+}\right]\left[OH^-\right]^2 = \frac{1}{4}\left[Fe^{2+}\right]^3 \tag{8.16}$$

and $\left[Fe^{2+}\right] = 1.931x10^{-5}$ mol/l. Hence, the theoretical potential for polarizing iron and carbon steel, say, at $T = 25\,°C = 298\,K$ is

$$E = E_{Fe} \approx -0.60 V_{SHE} = -0.92 V_{Cu/CuSO_4} \tag{8.17}$$

which is close to the National Association of Corrosion Engineers (NACE) recommended potential of $-0.85\ V_{Cu/CuSO_4}$. Consult Table 2.5 for conversions. In fact, a potential lower than $-0.92\ V_{Cu/CuSO_4}$ enhances hydrogen evolution.

For accurate and correct application of NACE standards, refer to:

NACE TM0497—Measurement Techniques Related to Criteria for Cathodic Protection on Underground or Submerged Metallic Piping Systems

NACE SP0108—Corrosion Control of Offshore Structures by Protective Coatings

pH Criterion The above theoretical potential for polarizing a steel structure can be determined using a different approach. If pH is defined as given in Table 3.2, then

$$pH = 14 + \log a(OH^-) = 14 + \frac{1}{2.303} \ln\left[OH^-\right] \tag{8.18}$$

Thus, the hydroxyl activity becomes

$$[OH^-] = \exp[2.303\,(pH - 14)] \tag{8.19}$$

Combining Eq. (8.19) and $K_{Fe(OH)2} = \left[Fe^{2+}\right][OH^-]^2 = 2\,[OH^-]^3$ yields

$$\ln\left(K_{Fe(OH)_2}\right) \approx 7pH - 96 \tag{8.20}$$

Once more, the Nernst equation gives the theoretical potential for polarizing iron and steel:

$$E = E_{corr} + \frac{RT}{zF}\ln\left(K_{Fe(OH)_2}\right) \tag{8.21}$$

Ideally, cathodic protection is achieved when $E_{corr} = 0$. Thus, Eq. (8.21) becomes

$$E \approx \frac{RT}{zF}\ln\left(K_{Fe(OH)_2}\right) \tag{8.22}$$

Substituting Eq. (8.20) into (8.22) yields the potential as a function of both temperature T and pH:

$$E \approx \frac{RT}{zF}(7pH - 96) \tag{8.23}$$

For iron or steel in a neutral solution at $pH = 7$ and $T = 25\,°C = 298\,K$, Eq. (8.23) gives the same result as predicted by Eq. (8.17). Cathodic protection is unusually performed in acid environments, but it is possible, at least, theoretically. Normally, protection of a metal in acid solutions is done anodically, provided that the metal is active–passive and shows a significant passive region. Thus, the anodic protection technique would prevail in such a case.

Tafel Criterion According to Jones [2, 21], this is a criterion that requires known values of the Tafel anodic slope β_a (constant) of the metal to be protected cathodically and an applied potential (E) to reduce corrosion current density (i_{corr}) to a fixed polarization current density (i_p). Theoretically, the required potential is defined by Eq. (5.76) as

$$\eta_a = \beta_a \log\left(i_p/i_{corr}\right) \tag{8.24}$$

but $\eta_a = E - E_{corr}$, and Eq. (8.24a) becomes

$$E = [E_{corr} - \beta_a \log(i_{corr})] + \beta_a \log\left(i_p\right) \tag{8.25}$$

where $E_{corr} - \beta_a \log(i_{corr})$ is the intercept if $E = f\left(\log i_p\right)$.

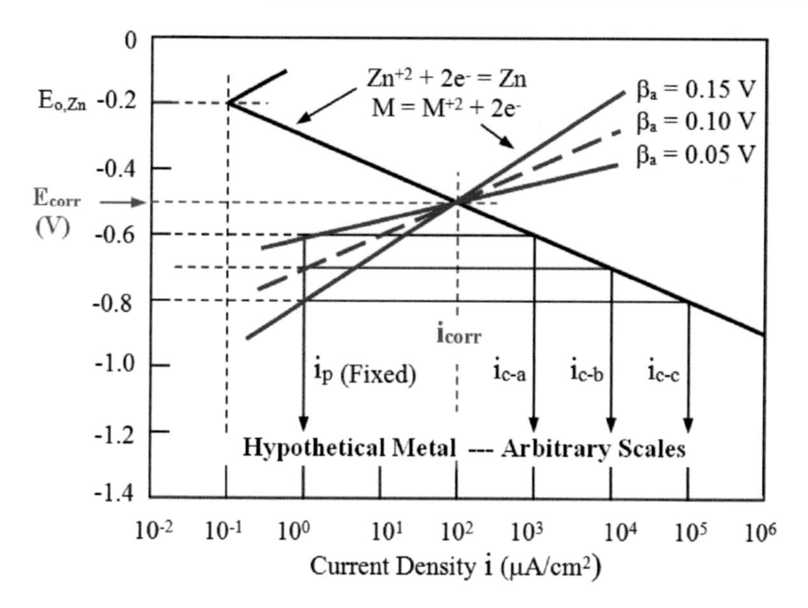

Fig. 8.9 Effect of Tafel anodic constant on overpotential for cathodic polarization [2, 21]

For $E_{corr} = -0.5\,\text{V}$, $i_{corr} = 100\,\mu\text{A/cm}^2$, and $i_p = 1\,\mu\text{A/cm}^2$, Eq. (8.24a) provides the anodic β_a plots depicted in Fig. 8.9 [2] for selected values of the anodic Tafel constant β_a.

For a fixed or protective anodic current density $i = i_p = 1\,\mu\text{A/cm}^2$, the applied potential for the three cases shown in Fig. 8.9 is calculated using Eq. (8.25). Hence,

$$
\begin{aligned}
E &= -0.5 - \beta_a \log i_{corr} + \beta_a \log(1) = -0.6\,\text{V for } \beta_a = 0.05\,\text{V} \\
E &= -0.5 - \beta_a \log i_{corr} + \beta_a \log(1) = -0.7\,\text{V for } \beta_a = 0.10\,\text{V} \\
E &= -0.5 - \beta_a \log i_{corr} + \beta_a \log(1) = -0.8\,\text{V for } \beta_a = 0.15\,\text{V}
\end{aligned} \tag{8.26a}
$$

for which the intercepts are

$$
\begin{aligned}
\text{Intercept} &= -0.5 - 0.05 \log(100) = -0.73\,\text{V for } \beta_a = 0.05\,\text{V} \\
\text{Intercept} &= -0.5 - 0.10 \log(100) = -0.96\,\text{V for } \beta_a = 0.10\,\text{V} \\
\text{Intercept} &= -0.5 - 0.15 \log(100) = -1.19\,\text{V for } \beta_a = 0.15\,\text{V}
\end{aligned} \tag{8.26b}
$$

8.1.5 Potential Attenuation

Consider the schematic buried pipeline in soil shown in Fig. 8.10 containing a current drain point at a holiday (coating defect). The pipeline is buried in the soil

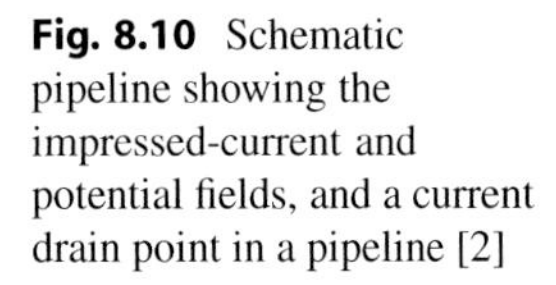

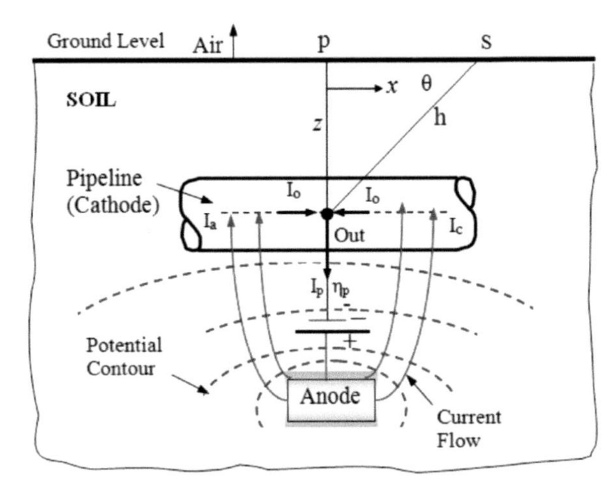

Fig. 8.10 Schematic pipeline showing the impressed-current and potential fields, and a current drain point in a pipeline [2]

at a remote distance from an sacrificial anode. This setup creates an impressed-current field induced by the anode in the soil of constant electrical resistivity. As Ashworth pointed out [3], this is just an approximation technique used for deriving an expression defining the drainage current or electric potential at infinite pipeline.

Additionally, stray-current corrosion is common on seagoing vessels. For example, powered battery chargers generate DC stray currents, which may flow indefinitely if proper precautions are ignored or overlooked. As a result, stray-current corrosion occurs, and it can manifest as pitting in confined areas, metal surface discoloration, rust formation on steel parts, or weakening of batteries. The magnitude of the current leaving or entering a buried pipeline in the soil may be determined by measuring the potential difference between two points on the soil surface as illustrated in Fig. 8.10.

The potential attenuation (fluctuation of potential) in an impressed-current system has a detrimental effect on corrosion prevention of a metallic structure containing defects. In this case, a coated pipeline buried in soil is subjected to experience current leakage due to coating defects known as holidays.

The following concrete assumptions are considered:

1. There is a uniform soil resistivity.
2. There is a uniform defect density in the pipeline coating. Detachment of the coating and crevice formation must be considered.
3. There is a positive current field for inducing an electric field potential gradient $(d\phi/dx)$ in the soil, specifically between the anode and the cathodic structure.
4. The current and potential fields are uniform around the buried pipeline.

Charged Disk Model Assume that the holiday of radius a is a uniform charged flat disk shown in Fig. 8.11a and that point A is located at a distance from the drainage point in the soil.

Fig. 8.11 Holiday as a current drainage point on a pipeline. (**a**) Model for finding the potential along the y-direction and (**b**) electric field. After [22]

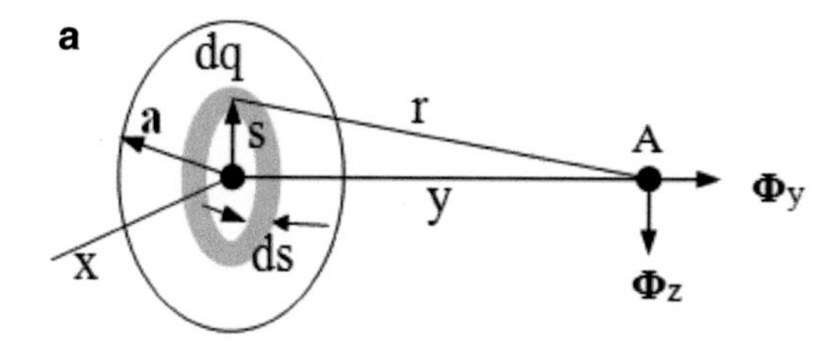

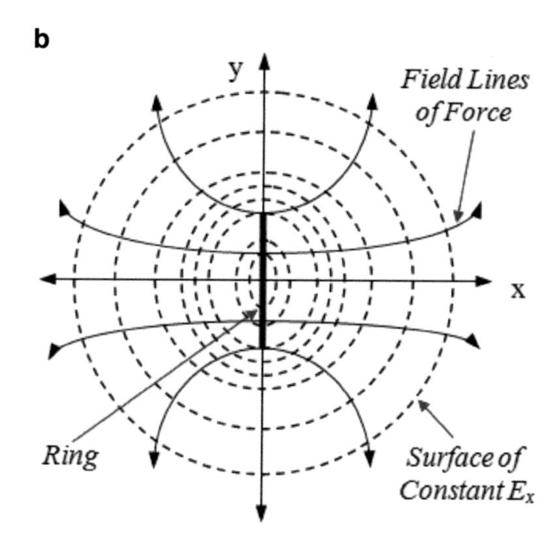

Also, assume that all charge elements in the ring-shaped annular segment of the disk lie at the same distance from point A and that the amount of charge on the disk is $dp = \sigma dA_s$, where σ is the charge density, $dA_s = 2\pi s ds$ is the area of the segment, s is the radius of the segment, and ds is the width of the segment. Figure 8.11b shows the electric field for a holiday standing alone. This is a typical model used in physics [22].

Thus, the potential along the x-direction (perpendicular to the disk flat surface) is defined as

$$\phi = \int \frac{dp}{r} = \int_o^a \frac{2\pi\sigma s ds}{\sqrt{s^2 + x^2}} \tag{8.27}$$

$$\phi = 2\pi\sigma\left[\sqrt{a^2 + x^2} \mp x\right] \tag{8.28}$$

The electric field around the disk is shown in Fig. 8.11b. The electric field in the y-direction can be defined as

$$E_x = -\frac{\partial\phi}{\partial x} = 2\pi\sigma\left[1 - \frac{x}{\sqrt{a^2 + x^2}}\right] \quad \text{For } x > 0 \tag{8.29}$$

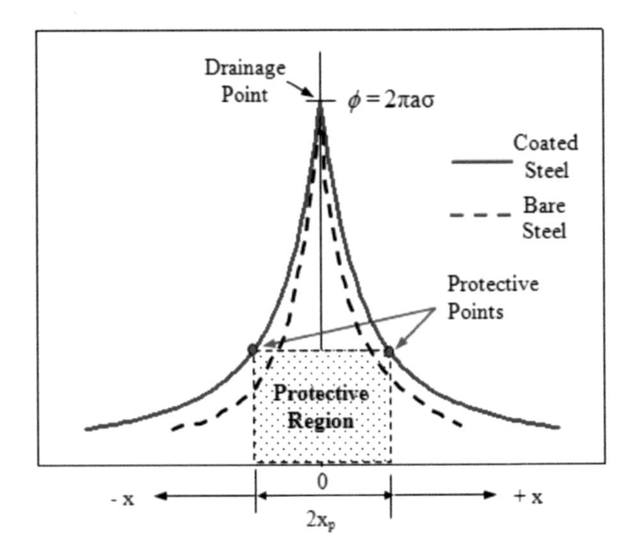

Fig. 8.12 Schematic potential attenuation (potential profile) along the x-direction for one drainage point

The potential attenuation (profile) based on this model is shown in Fig. 8.12 along with the ideal protective area for cathodic protection.

The protective points are located at a distance $2x_p$ between the anode and the pipeline structure. This potential attenuation indicates that the potential falls off from the center of the disk implying that the electric field has an outward component in the plane of the disk. The solid line is for the coated steel pipeline, while the dashed line is for possible nonlinear potential distribution of a bare steel pipeline.

Uhlig's Model [1] It is instructive to include Uhlig's model for deriving the potential change along a cathodically protected infinite pipeline. Thus, the potential and current attenuations of a buried pipeline can be approximated by using the free-body diagram depicted in Fig. 8.13.

In pursuing the mathematical details peculiar to Uhlig's model when the current is defined as $I_x = A_s i_x$, while the cross-sectional area of the pipe is $A_s = 2\pi rx$, the current gradient in the x-direction is

$$\frac{dI_x}{dx} = -2\pi r i_x \tag{8.30}$$

The overpotential at a distance x is $\eta_x = E - E_{corr}$, where E is the applied cathodic potential and the corrosion potential, E_{corr}, is zero in the cathodic protection system [3]. Thus, the potential gradient (dE_x/dx) and the pipe resistance per unit length (R_x) during a current (I_x) flow in the x-direction are related as

$$\frac{d\eta_x}{dx} = -R_x I_x \tag{8.30a}$$

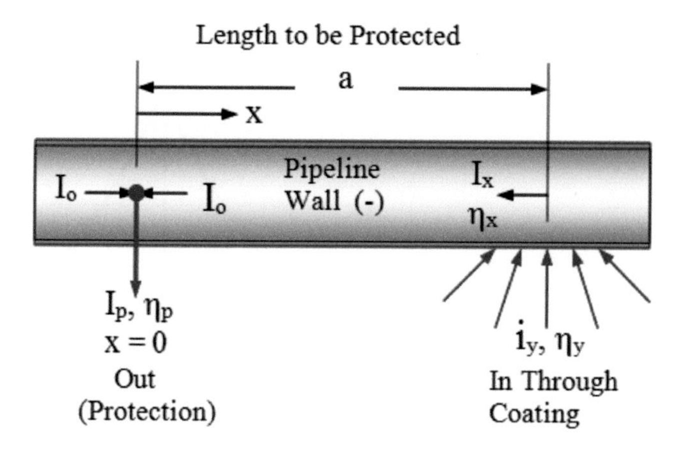

Fig. 8.13 Free-body diagram of an infinite pipeline wall showing the drainage point at a holiday or anode-wire-pipe connecting point. After [1]

From Eqs. (8.28) and (8.29),

$$\frac{d_x^2\eta}{dx^2} = (2\pi r i_x)\, R_x \tag{8.30b}$$

At small values of the current density (i_x), polarization of the pipeline is assumed to be linear so that [1]

$$\eta_x = \lambda R_L i_x \tag{8.31}$$

where R_L = leakage resistance of the pipe (Ω)
λ = constant

Combining Eqs. (8.30b) and (8.31) to eliminate the current density i_x yields the second-order partial differential equation for the overpotential at a distance x. Thus,

$$\frac{d^2\eta_x}{dx^2} - \left(\frac{2\pi r R_c}{\lambda R_L}\right)\eta_x = 0 \tag{8.32}$$

$$\frac{d^2\eta_x}{dx^2} - \left(\frac{R_s}{R_L}\right)\eta_x = 0 \tag{8.33}$$

Using the following boundary conditions for an infinite pipeline

$$\eta_x = 0 \quad \text{at} \quad x = \infty \tag{8.34a}$$

$$\eta_x = \eta_o \quad \text{at} \quad x = 0 \tag{8.34b}$$

the solution of Eq. (8.33) is

Fig. 8.14 Comparing potential attenuation model trends

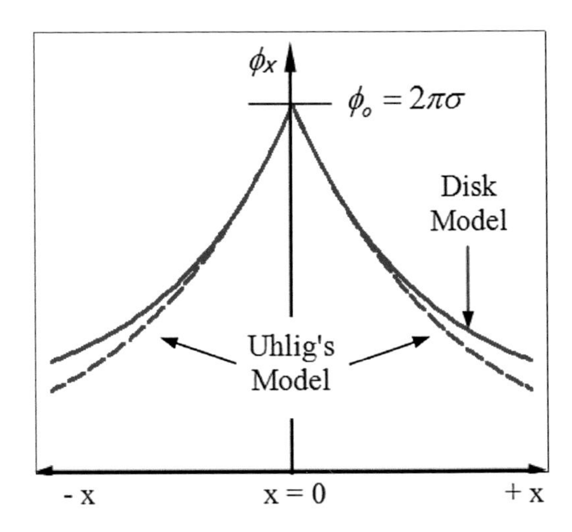

$$\eta_x = \eta_o \exp(-\alpha x) \tag{8.35}$$

where $\alpha = (R_s/R_L)^{1/2} = R_s/R_k =$ attenuation coefficient $\left(\text{cm}^{-1}\right)$

$\eta_o =$ overpotential coefficient or drainage overpotential (V)
$R_s =$ longitudinal pipe resistance per unit length (Ω/cm)
$R_k =$ characteristic resistance between the structure and earth $\left(\Omega/\text{cm}^{1/2}\right)$

The overpotentials are approximated as $\eta_x = \phi_x - E_{corr} \approx \phi_x$ and $\eta_o = \phi_o - E_{corr} \approx \phi_o$ due to $E_{corr} \approx 0$ since the pipeline is designed against corrosion. Then, Eq. (8.35) becomes

$$\phi_x = \phi_o \exp(-\alpha x) \tag{8.36}$$

Figure 8.14 depicts the potential plots using Eqs. (8.28) and (8.36). Denote that Uhlig's and disk models give similar potential trends near the drainage point where $\phi_x \to 0$ as $x \to \infty$.

On the other hand, the current and current density attenuations are [1]

$$I_x = I_o \exp(-\alpha x) \tag{8.37}$$

$$i_x = \frac{I_x}{\pi d L} \tag{8.38}$$

where $I_o = \alpha \eta_o / R_L =$ drainage current (A)

$d =$ pipe diameter (cm)
$L =$ pipe length at a distance x from a drainage point (cm)
$\pi d =$ circumference of the pipe

The protective parameters are $I_x = I_o$ and $E_x = \phi_x = \phi_o$ when $x = 0$ [1]. According to Ohm's law, the leakage resistance is equal to the potential drop from

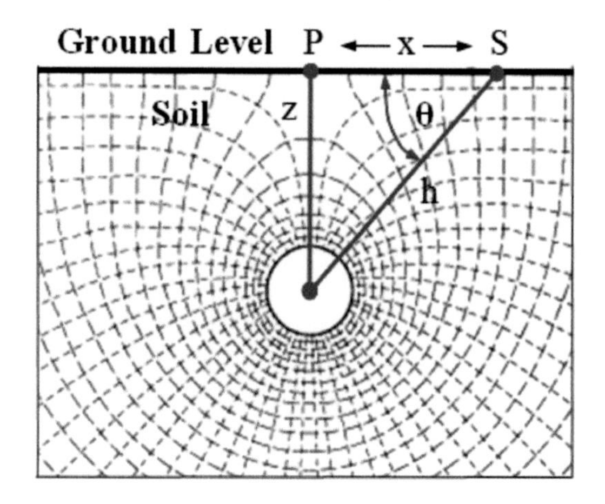

Fig. 8.15 Schematic buried pipeline in an electric field [23]

the leakage point divided by the polarization current difference at the coating-metal interface. Thus,

$$R_L = \frac{\phi_x - \phi_o}{I_x - I_o} \tag{8.39}$$

The preceding analytical approach for determining the potential and current distributions of a simple pipeline has been based on uniform current attenuations and uniform soil electric resistance. However, real structures being cathodically protected require more complex analyses due to either inherently imperfect coatings, nonuniform structure potential, complexity of the structure, and the like.

Consider the schematic side view of the pipeline shown in Fig. 8.15 illustrating two locations on the soil surface denoted as point P just above the pipeline and point S at a distance x from point P. Thus, the configuration forming a triangle is used to determine the potential difference. The dashed curves in Fig. 8.15 are current and potential profiles around a buried pipeline [23, 24].

According to the triangle in Fig. 8.15, the component of the current density along the x-direction is defined as [1]

$$i_x = \frac{2I}{2\pi rh}\cos\theta = \frac{I}{\pi rh}\frac{x}{h} = \frac{xI}{\pi rh^2} \tag{8.40}$$

where $2\pi rh$ is the surface area.

Using Ohm's law the potential gradient along the soil surface in the x-direction is

$$\frac{d\phi}{dx} = \rho_s i_x = \frac{x\rho_s I}{\pi rh^2} \tag{8.41}$$

$$\frac{d\phi}{dx} = \frac{\rho_s I}{\pi r}\frac{x}{x^2 + z^2} \tag{8.42}$$

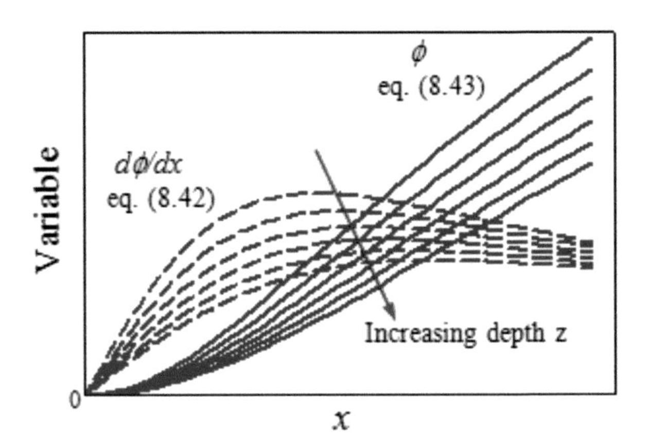

Fig. 8.16 Potential profile along the x-direction and depth z

Integrating Eq. (8.42) yields the potential along the x-direction at a depth z:

$$\phi = \frac{\rho_s I}{2\pi r} \ln \left[\frac{x^2 + z^2}{z^2} \right] \tag{8.43}$$

where ρ_s = soil resistivity (Ω cm)

z = depth directly above the pipeline (cm)
r = radius of the pipeline (cm)

Figure 8.16 shows the potential profile given by Eq. (8.43) as a function of both x and varying depth z.

Observe that the potential on the soil surface along the x-direction approaches zero as the depth approaches infinity. Thus, the boundary conditions for x and z become $0 < x < \infty$ and $0 < z < \infty$. If $x \to 0$, then $\phi \to 0$. The current (I) in Eq. (8.43) is either leaving or entering the buried pipeline.

Denote that the potential is not strongly dependent on depth z at a short distance x. The opposite trend is remarkably indicated in Fig. 8.16 at large distance x. The potential gradient, Eq. (8.42), is included in this figure for comparison purposes. Observe that this gradient increases with increasing x and has a maximum value. Subsequently, $d\phi/dx$ begins to decrease as the potential increases.

Example 8.2. *A pipeline buried in humid soil has an outer diameter of* 30.48 *cm (12 in.). If the longitudinal electric resistance* (R_s) *per unit length and the leakage electric resistance* (R_L) *of the structure are* $1.31x10^{-7}\,\Omega/cm$ *and* $3048\,\Omega$, *respectively, calculate* ***(a)*** *the characteristic electric resistance and* ***(b)*** *attenuation coefficient, and* ***(c)*** *determine the overpotential at* 0.1 *mm and 1 Km from a drainage point on a pipeline of external diameter d. Explain.*

Solution. **(a)** *The characteristic electric resistance:*

$$R_k = \sqrt{R_s R_L} = \sqrt{(1.31x10^{-7}\,\Omega/\text{cm})\,(3048\,\Omega)}$$

$$R_k = 0.02\,\Omega/\text{cm}^{1/2}$$

(b) *The attenuation coefficient:*

$$\alpha = R_s/R_k = \left(1.31x10^{-7}\,\Omega/\text{cm}\right)\left(0.02\,\Omega/\text{cm}^{1/2}\right) = 6.55x10^{-6}\,\text{cm}^{-1}$$

(c) *From Eq. (8.35),*

$$\eta_x/\eta_o = \exp\left(-\alpha x\right) = \exp\left[\left(-6.55x10^{-6}\,\text{cm}^{-1}\right)\left(10^5\,\text{cm}\right)\right]$$

$$\eta_x/\eta_o = 0.52 or \eta_o \simeq 2\eta_x$$

$$d\eta_x/dx = -\left(\eta_o/\alpha\right)\exp\left(-\alpha x\right)$$

$$d\eta_x/dx = -218.34\eta_o \simeq -436.68\eta_x$$

Integrating yields

$$\ln\left(\eta_x\right) = -436.68x$$

$$\eta_x = \exp\left(-436.68x\right) = \exp\left(-436.68 * 0.01\right) = 0.013\,\text{V}$$

$$\eta_x = \exp\left(-436.68x\right) = \exp\left(-436.68 * 10^5\right) = 0$$

Therefore, the overpotential on the structure vanishes at large distances from the drainage point.

8.1.6 Mass Transfer in a Crevice

Underground steel pipelines are normally protected to a large extent by external coatings, which are susceptible to mechanical or chemical damage in the form of holidays (coating flaws). A holiday is known as a pinhole with a radius of a few millimeters (*mm*). If the coating is partially detached beneath the holiday, the base metal undergoes crevice corrosion in an aggressive electrochemical environment. Figure 8.17 depicts a schematic delamination of a holiday [25]. Crevice corrosion beneath holidays is common on buried pipelines in low conductivity soils. In order to prevent crevice corrosion, the cathodic current has to flow to the pipeline surface.

The molar flux in this type of corrosion is mainly due to the mass transfer of Na^+, Cl^-, and OH^- ions and dissolved molecular oxygen (O_2) in the soil. According to the stoichiometry in Eq. (8.4), the hydroxyl ions (OH^-) form in the presence of water and oxygen and react with ferrous ions $\left(Fe^{2+}\right)$ to form ferrous hydroxide, $Fe\,(OH)_2$, as the crevice corrosion product, Eq. (8.6). However, crevice corrosion can be avoided by cathodic protection of holidays by controlling current leakage and related local potentials.

The mass transfer phenomena and current distribution can be modeled using an ideal circular crevice beneath a holiday as indicated in Fig. 8.17. Some relevant parameters are in the order of $r_h = 0.80\,\text{mm}$, $6\,\text{mm} < r_o < 80\,\text{mm}$, and $\delta = 0.8\,\text{mm}$ [25].

This model has been successfully used [25] for crevice cathodic protection using numerical analysis based on the dilute solution theory and reduction reaction of dissolved oxygen and Na^+, Cl^-, and OH^- ions at the crevice surface. Hence, the Nernst–Plank equation, Eq. (6.2), can be generalized as a differentiable and continuous scalar diffusion molar flux function:

$$\nabla J_j = -D_j \nabla \cdot C_j - z_j \left(\frac{F}{RT}\right) D_j \nabla \cdot \phi + C_j V_j \tag{8.44}$$

where $j = 1, 2, 3$, and 4 for the species Na^+, Cl^-, OH^-, and O_2, respectively

∇ = LaPlacian differential operator
r = radius of the circular crevice
z_j = valence of the species Na^+, Cl^-, OH^-, and O_2
C_j = concentration of the species Na^+, Cl^-, OH^-, and O_2
D_j = diffusivity of species Na^+, Cl^-, OH^-, and O_2

The ferrous Fe^{2+} ions are not included in the analysis since the crevice cathodic protection is for preventing the formation of this type of ion. For a coupled diffusion and migration molar flux under steady-state conditions, $\partial C_j / \partial dt = 0$, the molar flux becomes the continuity equation for mass transfer under steady state conditions:

$$\nabla J_j = 0 \tag{8.45}$$

Thus, Eq. (8.44) yields

$$\nabla^2 C_j + z_j \left(\frac{F}{RT}\right) \nabla \cdot \left[C_j \nabla \phi\right] = 0 \tag{8.46}$$

where

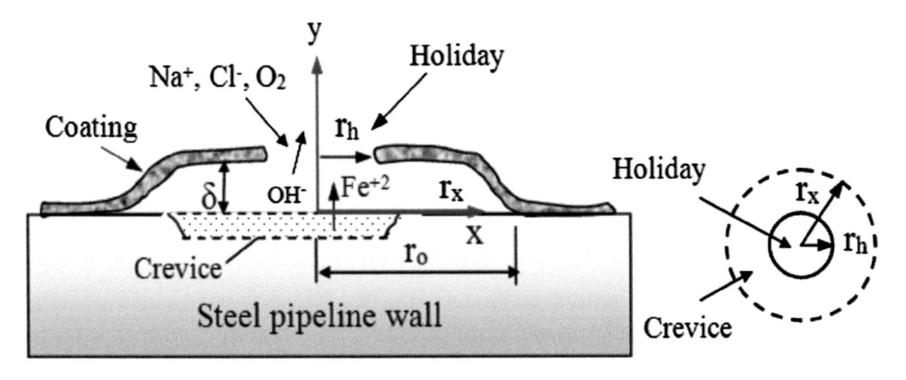

Fig. 8.17 Schematic circular crevice and holiday [25]

$$\nabla^2 = \frac{\partial^2}{\partial r^2} + \frac{1}{r}\frac{\partial}{\partial r} + \frac{\partial^2}{\partial y^2} \tag{8.47}$$

Expanding Eq. (8.46) into two-dimensional cylindrical coordinates gives a generalized expression that is solved numerically for particular species involved in the electrochemical process of crevice cathodic protection:

$$0 = \left[\frac{\partial C_j^2}{\partial r^2} + \frac{1}{r}\frac{\partial C_j}{\partial r} + \frac{\partial C_j^2}{\partial y^2}\right] + z_j\left(\frac{F}{RT}\right)\left[\frac{C_j}{r}\frac{\partial \phi}{\partial r} + C_j\frac{\partial \phi^2}{\partial r^2}\right. \tag{8.48}$$
$$\left. + \frac{\partial C_j}{\partial r}\frac{\partial \phi}{\partial r} + C_j\frac{\partial \phi^2}{\partial y^2} + \frac{\partial C_j}{\partial y}\frac{\partial \phi}{\partial y}\right]$$

For electric neutrality of the electrolyte dictates that $\nabla\phi = 0$ and

$$\sum_{j=1}^{N} z_j C_j = 0 \tag{8.49}$$

Also, the concentration of OH^- ions is related to pH as

$$pH = 14 + \log(C_3) \tag{8.50}$$

Furthermore, the electrolyte conductivity, Eq. (5.118), ionic mobility, Eq. (5.119), and the current density, Eq. (6.8), expressions are generalized as follows:

$$K_c = z_j F B_j C_j \tag{8.51}$$

$$B_j = \frac{D_j}{\kappa T} \tag{8.52}$$

$$i = z_j F J_j \tag{8.53}$$

Thus, Eqs. (8.48) through (8.53) are useful expressions in modeling the crevice cathodic protection. The model shown in Fig. 8.17 includes Cl^- ions as the principal species that drives the oxidation reaction $Fe \rightarrow Fe^{2+} + 2e^-$. Ideally, crevice CP ought to prevent this reaction from taking place by applying a potential $E < E_{corr}$.

On the other hand, the potential can be simulated using the following boundary conditions [25]:

$$\frac{\partial \phi}{\partial r} = 0 \text{ and } \frac{\partial C_j}{\partial r} = 0 \quad \text{for } y \geq 0,\ r = 0 \tag{8.54a}$$

$$\frac{\partial \phi}{\partial y} = 0 \text{ and } \frac{\partial C_j}{\partial y} = 0 \quad \text{for } y = \delta,\ r_h < r < r_o \tag{8.54b}$$

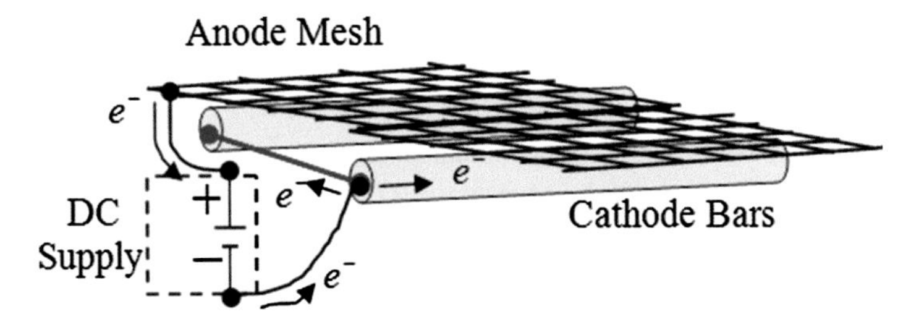

Fig. 8.18 Schematic impressed-current cathodic protection (CP) cell. After [29, 30]

Furthermore, there are empirical models for determining the pit growth rate [26–28], but henceforth the inward crevice growth rate (dy/dt) may be predicted using Faraday's law of electrolysis.

Dividing Eq. (5.1) by the metal density, R_F/ρ, and letting $dy/dt = R_F/\rho$ yield

$$\frac{dy}{dt} = \frac{IA_w}{zF\rho A_c} = \lambda I \tag{8.55}$$

where λ = rate constant (cm/A s)

$I = iA_c$ = current (A)
z = valence
F = 96,500 A s/mol
ρ = density of steel $\left(\text{g/cm}^3\right)$
$A_c = \pi r^2$ = crevice area $\left(\text{cm}^2\right)$
r = radius of crevice

8.1.7 Reinforced Concrete

Figure 8.18 illustrates a schematic impressed-current cathodic protection (CP) cell for designing against corrosion of steel bars embedded in concrete [29, 30]. Currently, an anodic titanium (*Ti*) mesh is used in the construction field for providing electrons to the cathodic steel bars through an external electrical source with a current density of the order of 1 $\mu\text{A/cm}^2$ [31]. It has been reported [30] that the *Ti* mesh has a life expectancy of at least 40 years when used in old concrete structures and over 100 years when employed in new reinforced concrete structures.

The current density range for this type of CP of bare steel bars is [30]

$$0.02\frac{\mu\text{A}}{\text{cm}^2} \le i \le 0.2\frac{\mu\text{A}}{\text{cm}^2} \quad \text{for new structures} \tag{8.56}$$

$$0.2\frac{\mu\text{A}}{\text{cm}^2} \le i \le 2\frac{\mu\text{A}}{\text{cm}^2} \quad \text{for old structures} \tag{8.57}$$

The main purpose of this type of CP is to force the DC to flow from the anode mesh to the steel cathode. As a result, the concrete high alkalinity is maintained at the steel-concrete interface, and theoretically, corrosion of steel bars is significantly reduced or eliminated. Thus, CP is an effective technique to protect the reinforcing steel in concrete bridges and other structures.

The essential components of an impressed-current cathodic protection system for a concrete structure are an external DC power source as indicated in Fig. 8.18, an anodic electrode (anode mesh), wiring, reference electrodes, etc. These components can be embedded in concrete, provided that the concrete is conductive enough or the anode mesh can be installed on the concrete structure.

Among all chemical and electrochemical reactions that may take place during CP, it can be hypothesized that chloride ions may react with the electrons flowing from the anode to form chloride gas. Thus, $Cl^- + e^- \rightarrow Cl_2$ gas may diffuse out of the porous concrete. If this is the case, then the oxide passive film on steel bars remains intact, and corrosion of the steel bars does not occur using a CP system.

8.1.8 Design Formulae

Sacrificial anodes are characterized by their capacity to electrochemically dissolve in a particular environment. Thus, the anode capacity (C_a) is used for this purpose since it is a measure of the electric energy per mass which can be determined using Faraday's law. Hence,

$$C_a = \frac{\epsilon z F}{A_w} \tag{8.58}$$

where ϵ = anode current efficiency: $\epsilon_{Mg} = 0.50,\ \epsilon_{Al} = 0.60,\ \epsilon_{Zn} = 0.90$ [32]

A_w = atomic weight (g/mol)
z = valence
$F = 96{,}500\ \text{A s/mol} = 26.81\ \text{A h/mol}$
C_a = anode capacity (A h/Kg)

Sacrificial magnesium (Mg) anodes are widely used in buried pipelines and domestic or industrial water heater applications. It seems that a Mg anode may protect as much as 8 Km of a coated pipeline buried in the soil [1].

Example 8.3. *Plot Eq. (8.58) for some Al, Ti, Ni, Cu, Nb, Cd and Pt metals, and explain the result.*

Solution. *The plot below shows an equilateral hyperbola profile for the anode capacity C_a of selected metals. This hyperbola behavior implies that the anode capacity being a measure of the electric energy per mass decreases with increasing atomic mass. Accordingly, Alhas the highest anode capacity and the lowest atomic mass among the selected metals. This, then, makes Al the obvious choice for a* *sacrificial anode.*

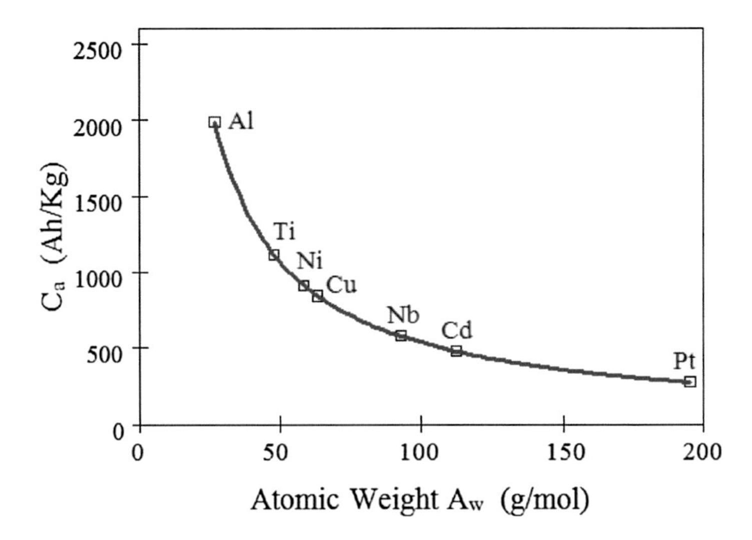

The given design guidelines included in this section are for cathodic protection of onshore, offshore, underground, and hybrid structures. The practical application of cathodic protection is inherently complex since an optimum design requires a theoretical knowledge of electrochemical principles, an imaginary engineering mind for proper judgment on engineering decision, and a sense of artistic approach. Thus, a cathodic protection designer must consider the following:

- The size and lifetime of the structure
- Environmental diversity and economical impact
- Impressed current, sacrificial anode, or hybrid systems
- Coating material and coating decay that leads to current drainage
- The applied current, potential, electrolyte-to-anode electric resistance, and wiring resistance
- The instrumentation capacity
- Experience

For cathodically protecting an offshore structure, use the current densities given in Table 8.3. Valuable formulae are given in Table 8.4 for coatings and anodes. The reader should consult the cited references for additional details.

The coating damages during installation and service are important factors to be considered in calculating the average applied current. For instance, the following coating damage is suggested in order to adjust the coated and uncoated surface areas per structural zones [7]:

$$\text{Zone} \rightarrow \text{installation}$$

Table 8.3 Current density for steel surfaces in seawater [7]

↘	Current Density i ($\mu A/cm^2$)			
	Uncoated Steel		Coated Steel	
	To Polarize	After Polarization	To Polarize	After Polarization
Moving Seawater	32 − 37	7 − 11	3 − 5	1 − 2
Stagnant Seawater	16 − 27	4 − 7	1 − 3	0.5 − 1
Soil Zone	4 − 5	1 − 2	0.5 − 1	0.1 − 0.5

Table 8.4 Formulae for coating and anode design life [7, 17]

Parameter	Formula		
Total required	$A_T = \sum \alpha' A_s \quad A_s = \pi NdL$		
Surface area	Coating life	α'	
	L_c (years)	Initial	Final
	10	0.02	0.10
	20	0.02	0.30
	30	0.02	0.60
	40	0.02	0.90
Total mass of anodes	$M = (iA_T L_d) / (F_e F_u C_a) = (IL_d) / (F_e F_u C_a)$ $M = (3\pi IL_d) / (7C_a)$ (average)		
Anode life	$L_d = (MF_e F_u C_a) / (I)$ $L_d = (7MC_a) / (3\pi I)$ (average)		
Number of anodes	$N = Ai/I$ $Nm \geq M$		

Where α' = coating breakdown factor
m = mass of a single anode
F_e = efficiency factor = 0.85–0.95
F_e = 0.90 (average)
F_u = utilization factor = 0.75–0.85
F_u = 0.80 (average)
i = current density $\left(A/cm^2\right)$
I = current $\left(A/cm^2\right)$

$$\begin{aligned} \text{Tidal zone } (TZ) &\rightarrow 10\,\%\text{ coating damage} \\ \text{Immersed zone } (IZ) &\rightarrow 15\,\%\text{ coating damage} \\ \text{Soil zone } (SZ) &\rightarrow 50\,\%\text{ coating damage} \end{aligned} \tag{8.59}$$

Although the CP clearly represents an electrochemical method for protecting large structures against corrosion, it requires a broad spectrum approach to comprehensively maintain the structure cathodically polarized within a time frame based on the electrochemical environment. For a successful CP design, a maintenance plan must include guidelines to adjust the current due to coating damage and electric potential fluctuations.

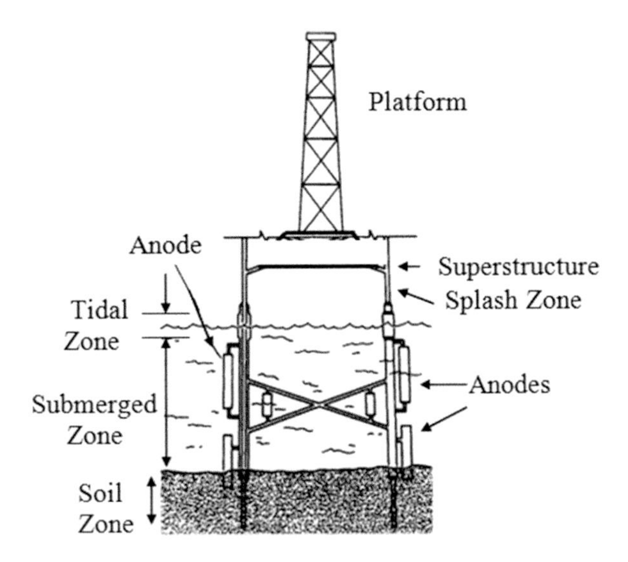

Fig. 8.19 Schematic submerged structure [7]

Figure 8.19 shows a complex offshore steel structure containing particular zones and suitable locations of anodes [9]. It is suggested that $\epsilon = 2\,\%$ coating damage per year is appropriate for a particular prolonged life (L_d). Hence, the applied current (I_p) for polarization and the final current (I_f) (protective current) can be defined as [7]

$$I_p = \sum I_{zone} \tag{8.60}$$

$$I_f = (1 - \lambda_i - \lambda_c L_d)\,(Ai)_{zone} \tag{8.61}$$

where λ_i = fraction of installation damage

λ_c = fraction of coating damage per year
A_{zone} = surface area per zone per uncoated or coated steel
i_{zone} = current density from Table 8.3 or predicted (μA/cm)

The required current for maintaining the structure cathodically protected can be defined as

$$I = \frac{I_p + I_f}{2} \tag{8.62}$$

Anode Electric Resistance to Soil The impressed-current system permits variability of the current out and monitor-control instrumentation. However, interactions with nearby structures may cause detrimental results. On the other hand, sacrificial anode systems do not require a power source and are easily installed, but large numbers of anodes are needed to protect structures against corrosion. For instance, offshore structures are cathodically protected by hybrid systems [33].

Theoretically, the potential distribution across the surface of an individual electrode in a galvanic couple can be obtained by solving Laplace's equation in a domain Ω. Assume that the concentration inducing oxidation is uniform, and the potential distribution in the electrochemical environment is governed by Laplace's equation in rectangular coordinates [34]. Thus,

$$\nabla^2 E\,(x, y, z) = \frac{\partial^2 E}{\partial x^2} + \frac{\partial^2 E}{\partial y^2} + \frac{\partial^2 E}{\partial z^2} = 0 \tag{8.63}$$

For steady-state conditions, the net current takes the form

$$\nabla I\,(x, y, z) = \frac{\partial I}{\partial x} + \frac{\partial I}{\partial y} + \frac{\partial I}{\partial z} = 0 \tag{8.64}$$

The solution of these equations can be achieved using the boundary element method (BEM), which has some advantages over other numerical methods like finite element methods (FEMs) and finite differences. Nonetheless, the approximate solution of this boundary value problem is an exact solution of the differential equation in the domain Ω, and only the boundary of the domain needs to be analyzed. Mathematical modeling of cathodic protection using BEM can be found elsewhere [34–36]. In this section, the classical mathematical approach prevails as a simple and comprehensive methodology.

In principle, numerical analysis gives comprehensive mathematical procedures for deriving potential and current distributions for an optimum cathodic protection of pipelines having coating holidays. If the exposed area within a coating holiday is taken as the sole corroding region, then corrosion is mainly attributed to the presence of Fe^{2+}, Cl^{-}, and OH^{-} ions in the holiday environment, which can be modeled as a bare pipeline.

Figure 8.20 schematically shows an anode rod being buried in soil. This model indicates that there are radial and longitudinal uniform current flow contours through earth cylindrical shells [37]. This simple model can be used to derive an expression for the vertical anode electric resistance (R_v) to ground.

Firstly, the current density in the x-direction

$$i_x = \frac{I}{A_s} = \frac{I}{2\pi x L} \tag{8.65}$$

where A_s = surface area of a cylindrical shell

L = length or depth of the cylindrical anode
x = distance from the anode rod surface

From Ohm's law in dimensional analysis, the magnitude of the electric field strength (electric field strength gradient) E_x is equals the product of the soil resistivity ρ_s and the current density [37]. Thus,

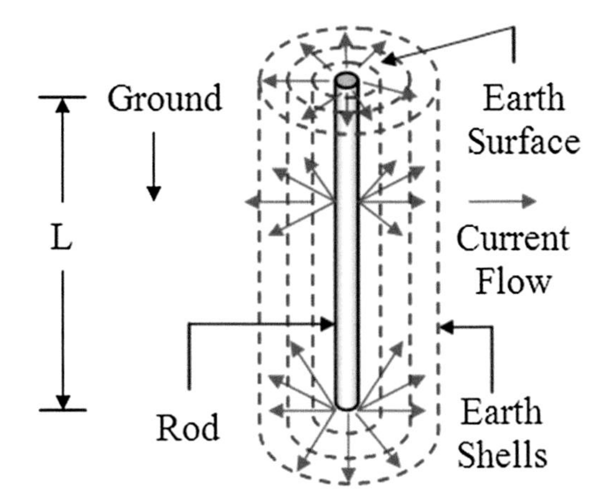

Fig. 8.20 Schematic anode rod and earth cylindrical shells. After [37]

$$E_x = \rho_s i_x = \frac{\rho_s I}{2\pi x L} \tag{8.66}$$

Integrating Eq. (8.66) yields the electric potential

$$\phi_x = \int_r^x E_x dx = \frac{\rho_s I}{2\pi L} \ln\left(\frac{x}{r}\right) \tag{8.67}$$

Now, the vertical anode electric resistance becomes

$$R_v = \frac{\phi_x}{I} = \frac{\rho_s}{2\pi L} \ln\left(\frac{x}{r}\right) \tag{8.68}$$

Secondly, assume that 95 % of the current dissipates so that $x = 4L$ [37]. Hence,

$$R_v = \frac{\rho_s}{2\pi L} \ln\left(\frac{4L}{r}\right) \tag{8.69}$$

which resembles the widely used Dwight's equation [38] for a buried straight wire of length L:

$$R_v = \frac{\rho_s}{2\pi L}\left[\ln\left(\frac{4L}{r}\right) - 1\right] \tag{8.70}$$

In addition, the classical and analytical procedures for deriving rather simple anode resistance equations can be found in Dwight's paper [38], but some common anode resistance expressions, known as Dwight's equations, are given in Table 8.5 as references.

Example 8.4. *This example deals with the design of a galvanic anode cathodic protection system, which considers several factors.* ***(a)*** *Determine the power supply*

Table 8.5 Resistance formulae for underground anodes [33, 38, 39]

Anode	Equations
Single horizontal rod or pipe	$R_h = \left[\rho_s/(2\pi L)\right][\ln(4L/d) - 1]$
Single vertical rod, $L \geq 4r$	$R_v = \left[\rho_s/(2\pi L)\right][\ln(4L/r) - 1]$
Single vertical rod, $L < 4r$	$R_v = \left[\rho_s/(2\pi L)\right]\left\{\begin{array}{l}\ln\left[(2L/r)\left(1+\sqrt{1+[r/2L]^2}\right)\right]\\ +r/2L - \sqrt{1+[r/2L]^2}\end{array}\right\}$
N vertical rods	$R_v = \left[\rho_s/(2\pi NL)\right]\{[\ln(8L/d) - 1]$
N parallel rods	$+ (2L/S)\ln(2N/\pi)\}$
Flush-mounted plate or bracelet	$R_f = \pi\rho_s/\left(10\sqrt{A_a}\right)$

where R_h, R_v, R_f = resistances (Ω)
ρ_s = soil resistivity (Ω cm)
L = length of anode (cm)
N = number of anodes
S = anode spacing (cm)
r = radius of anode rods
d = diameter of anode rods (cm)
$d = \sqrt{A_a/\pi}$ = for non-cylindrical shapes
A_a = anode surface area $\left(\text{cm}^2\right)$

output for cathodically protecting several onshore steel storage tanks using 15 *sacrificial anodes. Use* (4cm) x (4cm) x (150cm) *Al-Zn-Sn cast anodes, which are to be installed vertically in coke breeze backfill for a uniform current distribution. The anode spacing is* 500 *cm and the soil resistivity is 5000* Ω *cm. In addition, the anode-backfill couples are buried in the soil, and the backfill bags are large enough. The initial applied current is* 4 *A. Calculate* **(b)** *the amount of anode material (anode capacity) that will be required to protect the structure, the total mass of the anodes (M) for a lifetime of* 20 *years, and the individual anode mass (m).*

Solution. *Given data:*

$$N = 15\ anodes$$
$$I = 4\,A$$
$$L = 150\,cm$$
$$d = \sqrt{A_a/\pi} = \sqrt{(16\,\text{cm}^2)/\pi}$$
$$d = 2.26\,cm$$
$$\rho_s = 5000\,\Omega\,cm$$
$$S = 500\,cm$$
$$A_a = (4\,\text{cm})(4\,\text{cm}) = 16\,\text{cm}^2$$

(a) *From Table 8.5,*

$$R_v = \rho_s / (2\pi NL) \{[\ln(8L/d) - 1] + (2L/S) \ln(2N/\pi)\}$$

$$\rho_s / (2\pi NL) = \frac{(5000\,\Omega\,\text{cm})}{2\pi\,(15)\,(150\,\text{cm})} = 0.35368\,\Omega$$

$$\ln(8L/d) - 1 = \ln\left(\frac{(8)\,(150\,\text{cm})}{2.26\,\text{cm}}\right) - 1 = 5.2747$$

$$(2L/S)\ln(2N/\pi) = \frac{2\,(150\,\text{cm})}{(500\,\text{cm})} \ln\left[\frac{2\,(15)}{\pi}\right] = 1.3539$$

$$R_v = (0.35368\,\Omega)\,(5.2747 + 1.3539) = 2.34\,\Omega$$

From Ohm's law,

$$E = IR_v = 9.36\,\text{V}$$

The power supply should have a capacity greater than 9.36 V. That is, $E_{ext} >$ *9.36 V. (If 4-cm diameter rods are used, then* $R_v = 2.14\,\Omega$*, and* $E = 8.56$ *V with a difference of 8.5 %).*

(b) *From Table 8.4, the total exposed area is*

$$A_T = \pi NdL = 35,979.09\,\text{cm}^2 = 3.60\,\text{m}^2$$

$$i = I/A_T = 1.11x10^{-4}\,\text{A/cm}^2 \quad \textit{(initial current density)}$$

From Table 8.2, the average Al-Zn-Sn anode capacity is

$$C_a = (925 + 2600)/2 = 1762.50\,\text{A h/Kg}$$

From Table 8.4, the total average and individual masses of the anodes are

$$M = (3\pi IL_d) / (7C_a)$$

$$M = [(3\pi)\,(4\,\text{A})\,(20x365x24\,\text{h})] / [(7)\,(1762.50\,\text{A h/Kg})]$$

$$M = \simeq 535\,\text{Kg} \qquad \textit{(initial mass)}$$

$$m = M/N \simeq 36\,\text{Kg} \qquad \textit{(initial mass of one anode)}$$

So far the tank coating is intact since the tanks are coated after installation. Assume a 2 % ($\lambda_c = 0.02$*) deterioration of the coating per year so that the total damage in 20 years is 40 %. From Eq. (8.61), the final current along with zero installation damage (*$\lambda_i = 0$*) is*

$$I_f = (1 - 0.40)\,(A_T)\,(i) = 2.40\,\text{A}$$

From Table 8.3, the current density for polarizing the steel is $i_p = 1\,\mu\text{A/cm}^2$. *Then, the polarizing and maintenance (average) currents are, respectively,*

$$I_p = i_p A_T = \left(10^{-6}\,\text{A/cm}^2\right)\left(35,979.09\,\text{cm}^2\right) \simeq 0.04\,\text{A}$$

$$I = \left(I_f + I_p\right)/2 = 1.22A \quad (\textit{for maintenance})$$

Now, the estimated cathodic potential and the rectifier or power supply potential are, respectively,

$$E = IR_v = (1.22\,\text{A})\,(2.34\,\Omega) = 2.51\,\text{V}$$

$$E_{rect} > 2.51V,\ \textit{say}\ 5\,\text{V}$$

Thus, the final and individual masses of the anodes in 20 years are

$$M = (3\pi I L_d)\,/\,(7C_a)$$

$$M = [(3\pi)\,(1.22\,\text{A})\,(20x365x24\,\text{h})]\,/\,[(7)\,(1762.50\text{A}\,\text{h/Kg})]$$

$$M \simeq 163\text{Kg} \qquad (\textit{final total mass})$$

$$m = WM/N \simeq 11\,\text{Kg} \quad (\textit{final mass of one anode})$$

These results represent $(100\,\%)\,(36\,\text{Kg} - 11\,\text{Kg})\,/\,(36\,\text{Kg}) \simeq 69\,\%$ *reduction in anode mass at the end of the 20-year design life.*

The rate of mass reduction can be calculated approximately as

$$M_R = \frac{M}{L_d} = \frac{3\pi I}{7C_a} = \frac{3\pi\,(1.22\,\text{A})}{7\,(1762.50\,\text{A}\,\text{h/Kg})} = 9.32 \times 10^{-4}\,\text{Kg/h}$$

$$M_R \simeq 0.00036\,\text{g/s}$$

If the density of the anode alloy is $5\,\text{g/cm}^3$, *then corrosion rate becomes*

$$C_R = \frac{M_R}{\rho A_a} = \frac{0.00036\,\text{g/s}}{\left(5\,\text{g/cm}^3\right)\left(16\,\text{cm}^2\right)} = 4.50 \times 10^{-6}\,\text{cm/s} = 45 \times 10^{-6}\,\text{mm/s}$$

Example 8.5. *Cathodically protect an offshore steel platform in moving seawater. The platform has to be polarized using Al-Zn-In anodes. Use the data given below to determine the applied potential and applied current for maintenance. The procedure below can be found elsewhere [40].*

Solution. *The total current in the water and in the soil zones is needed in order to calculate the number of anodes for protecting the steel structure. Thus,*

Design life	$L_d = 20$ years
Single anode volume	$V = (0.25\,\text{m})\,x\,(0.25\,\text{m})\,x\,(2.50\,\text{m})$
Anode density	$\rho = 2.315\,\text{g/cm}^3 = 2315\,\text{Kg/m}^3$
Anode cross-sectional area	$A_a = 0.0625\,\text{m}^2 = 625\,\text{cm}^2$
Platform surface area	$A_w = 50\,\text{m}x60\,\text{m} = 3.0x10^7\,\text{cm}^2$ *(Water)* $A_s = (55\,\text{m})\,x\,(80\,\text{m}) = 4400\,\text{m}^2$ *(Soil)* $A_T = A_w + A_s = 7700\,\text{m}^2$
Water resistivity	$\rho_x = 25\,\Omega\,\text{cm}$
Polarized current density	$i_p = 5\,\mu\text{A/cm}^2$ *(from Table 8.3)*
Anode capacity	$C_a = 2650\,\text{A h/Kg}$ *(average, Table 8.2)*
Steel potential	$E_s = 0.25\,\text{V}$ *(from Table 8.2)*
Anode potential	$E_a = -1.00\,V_{Ag/AgCl}$ *(from Table 8.2)*

$$I_w = i_p A_w = \left(5\,\mu\text{A/cm}^2\right)\left(3.0x10^7\,\text{cm}^2\right) = 150\,\text{A}\ \textit{(water)}$$

$$I_s = i_p A_s = \left(5\,\mu\text{A/cm}^2\right)\left(4.4x10^7\,\text{cm}^2\right) = 220\,\text{A}\ \textit{(soil)}$$

$$I = i_p A_T = \left(5\,\mu\text{A/cm}^2\right)\left(7.7x10^7\,\text{cm}^2\right) = 385\,\text{A}\ \textit{(total)}$$

From Table 8.4, the total mass of the anode in water is

$$M = \frac{3\pi I L_d}{7C_a} = (3\pi/7)\,(385\,\text{A})\,(20\,\text{year})\,/\,(2650\,\text{A h/Kg})$$
$$M = 34,270.63\,\text{Kg}$$

The mass in water becomes

$$M_w = (3\pi I_w L_d)\,/\,(7C_a) = (3\pi/7)\,(150\,\text{A})\,(20\,\text{year})\,/\,(2650\,\text{A h/Kg})$$
$$M_w = 13,352.19\,\text{Kg}$$

and that in the soil is

$$M_s = M - M_w = 34,270.63\,\text{Kg} - 13,352.19\,\text{Kg}$$
$$M_s = 20,918.44\,\text{Kg}$$

The mass of each individual anode becomes

$$m = \rho V = \left(2315\,\text{Kg/m}^3\right)\left(0.15625\,\text{m}^3\right) = 361.72\,\text{Kg}$$

Now, the number of anodes is

$$N = \frac{M}{m} = \frac{34,270.63\,\text{Kg}}{361.72\,\text{Kg}} = 95$$

The number of anodes in the water and in the soil:

$$N_w = \frac{M_w}{m} = \frac{13,352.19\,\text{Kg}}{361.72\,\text{Kg}} = 37 \textit{ anodes in the water zone}$$

$$N_s = N - N_w = 95 - 37 = 58 \textit{ anodes in the soil zone}$$

Consider non-cylindrical anodes so that the characteristic anode diameter becomes

$$d = \sqrt{\frac{A_a}{\pi}} = \sqrt{\frac{625\,\text{cm}^2}{\pi}} = 14.10\,\text{cm}$$

From Table 8.5, the vertical anode resistance is

$$R_v = \frac{\rho_s}{2\pi L}\,[\ln(8L/d) - 1]$$

$$R_v = \frac{(25\,\Omega\,\text{cm})}{(2\pi)\,(2.5x10^2\,\text{cm})}\left\{\ln\left[\frac{(8)\,(2.5x10^2\,\text{cm})}{14.10\,\text{cm}}\right] - 1\right\}$$

$$R_v = 0.06294\,\Omega$$

From Ohm's law, the initial current and the rectifier output are

$$I_i = \frac{\Delta E}{R_v} = \frac{(E_a - E_s)}{R_v} = \frac{(1.00\,\text{V} - 0.25\,\text{V})}{0.06294\,\text{V/A}}$$

$$I_i = 11.92\,\text{A}$$

$$i = \frac{NI_i}{A_T} = \frac{(95)\,(11.92\,\text{A})}{625\,\text{cm}^2} = 1.81\,\text{A/cm}^2$$

$$E_{rect} > E = E_s - E_a = 1.00\,\text{V} - 0.25\,\text{V} = 0.75\,\text{V}$$

Maintenance current and potential:

$$I_m = \frac{I_i}{N} = \frac{385\,\text{A}}{95} = 4.05\,\text{A}$$

$$E_m = I_m R_v = (4.05\,\text{A})\,(0.06294\text{V/A}) = 0.25\,\text{V}$$

Therefore, the current $I_m = 4.05$ A is adequate for the design life because the calculated design life must be $L_{new} \geq L_d$. Thus, $L_{new} \geq (7MC_a)\,/\,(3\pi I_m) = 20$ years.

The above assumptions and calculations represent an approximation for cathodic protection of an offshore steel structure [41]. In fact, an offshore structure has diverse dimensional components exposed to different severity of corrosion susceptibility, and each structural section must be considered individually for protection against corrosion.

Example 8.6. *Cathodically protect a buried steel pipeline as shown in the sketch below. For fluid flow in a smooth pipe of uniform circular cross section, the characteristic length is the inner diameter (D) of the pipe, and the transition from laminar to turbulent fluid flow begins at a Reynolds number (R_e) of* 2100. *The pipeline has to be polarized using magnesium (Mg)-based anodes. Use the data given below to determine the applied potential and applied current for maintenance:*

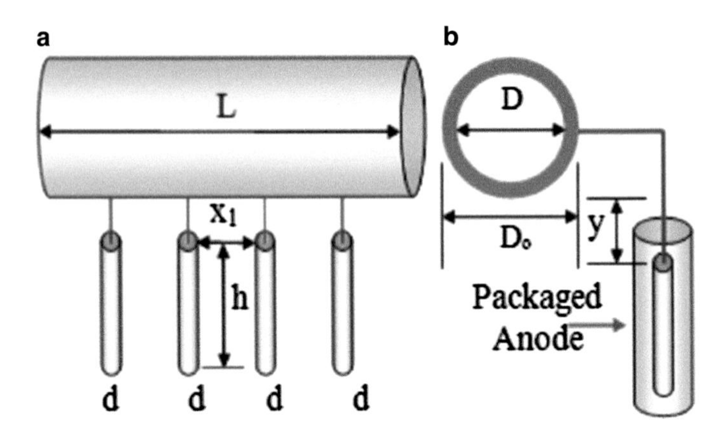

Data:

Production flow rate: $P_R = 60,000$ bpd (barrels per day)
Conversion: 1 *barrel of oil* = 1591 = $0.159\,\text{m}^3$
Density of crude oil: $\rho = 0.8207\,\text{Kg/m}^3$
Relative density of crude oil: $\rho_r = 0.8207$
Kinematic viscosity of crude oil: $K_v = 2.04\,\text{m}^2/\text{s}$
Dynamic (fluid) *viscosity of crude oil*: $\eta_v = 1.67x10^{-3}\,\text{Kg/(m s)}$
Pipeline surface roughness: $S_r = 0.025\,\text{mm}$
Temperature of flowing crude oil: $T = 40\,°\text{C}$
Pipe inner diameter: $D = 0.254\,\text{mm} = 0.254x10^{-3}\,\text{m}$
Pipe wall thickness: $t_w = 7.62\,\text{mm} = 7.62x10^{-3}\,\text{m}$
Length of the pipeline: $L = 12\,\text{Km}$
Design pipeline life: $t_d = 25$ years
Design current density: $i_d = 0.1\,\text{mA/m}^2$
Design potential: $E_d = -0.67\,\text{V}$
Design pressure: $P = 1100$ psig (76 bars)
Flow velocity: $v = 0.7 - 2.43\,\text{m/s}$
Corrosion rate of the pipeline: $C_R = 0.1\,\text{mm/yr}$
Depth of the ground cover: $y = 0.90\,\text{m}$

Ambient temperature: $T = 18\,°\text{C}–35\,°\text{C}$
Water in soil: $C_w = 30\,\%$
Mg-alloy anode capacity: $C_{a,Mg} = 0.125\,\text{A yr/kg}$ at 50 % efficiency
Mass of one anode: $M_{a,Mg} = 14.5\,\text{Kg}$
Mg-alloy anode size: $(0.5\,\text{m})\,x\,(13\,\text{cm})$ (cylinder)
Mg-alloy potential: $E_{Mg} = -1.75\,\text{V}$
Soil resistivity: $\rho_s = 500\,\Omega\,\text{cm}$

The procedure below is found elsewhere for crude-oil flow [42]. Use the upper limit of a variable whenever suitable to determine the short anode life (t_a) in years.

Solution. (a) *Number of anodes and anode separation at 0.90 m beneath the pipeline. The pipeline total surface area is the product of the circumference times length. Thus,*

$$A_s = \pi D_o L = 2\pi\,(D + 2t)\,L$$

$$A_s = \pi\left(0.254\,\text{m} + 2 * 7.62x10^{-3}\,\text{m}\right)\left(12x10^{3}\,\text{m}\right)$$

$$A_s = 10{,}150\,\text{m}^2$$

The protected current is

$$I_{prot} = A_s i_d = \left(10{,}150\,\text{m}^2\right)\left(0.1\,\text{mA/m}^2\right) = 1015\,\text{mA}$$

$$I_{prot} = 1.015\,\text{A}$$

For the magnesium alloy,

$$E_{avg} = 0.5\left(E_{Mg} + E_d\right) = 0.5\,(-1.75 - 0.67)\,V = -0.67\,\text{V}$$

Net initial driving potential is

$$E = E_{Mg} - E_{avg} = -1.75 - (-0.67) = -1.08\,\text{V}$$

From Table 8.5, the anode resistance in the vertical position along with the anode length $L_a = 0.5\,m$ is

$$R_v = \rho_s/\,(2\pi L_a)\,\{[\ln\,(8L_a/d) - 1] = 386.19\,\Omega$$

$$R_v = \frac{\rho_s}{2\pi L_a}\left[\ln\left(\frac{8L_a}{d}\right) - 1\right]$$

$$R_v = \frac{500x10^{-2}\,\Omega\,\text{m}}{2\pi\,(0.5\,\text{m})}\left[\ln\left(\frac{8\,(0.5\,\text{m})}{0.13\,\text{m}}\right) - 1\right]$$

$$R_v = 3.8619\,\Omega$$

According to Ohm's law, the maximum cathodic current is

$$I_{\max} = \frac{E}{R_v} = -\frac{1.08\,\text{V}}{3.8619\,\text{V/A}} = -0.28\,\text{A}$$

Then the number of anodes is

$$N = \frac{I_{prot}}{I_{\max}} = \frac{1.015}{0.28} \simeq 4$$

Now the anode distance becomes

$$x_1 = \frac{L}{N} = \frac{12\,\text{Km}}{4} = 3\,\text{Km}$$

(b) *Polarization of the pipeline:*
Polarize the pipeline with $E\,(NACE) = -0.85\,\text{V}$, *and get the net potential as the driving force for the anodes:*

$$E_{net} = E_{Mg} - E\,(NACE) = -1.75 - (-0.85) = -0.90\,\text{V}$$

Then the corresponding current becomes

$$I_{net} = \frac{E_{net}}{R_a} = -\frac{0.90\,\text{V}}{3.867\,\text{V/A}} = -0.23\,\text{A}$$

The anodes operate at 50 % efficiency. The short design life of one anode is determined using an expression given in Table 8.4. Hence,

$$t_a = \frac{7 M_{a,Mg} C_a}{3\pi I_{net}} \quad (average)$$

$$t_a = \frac{7\,(14.5\,\text{Kg})\,(0.125\,\text{A yr/kg})}{3\pi\,(0.23\,\text{A})} \simeq 6\ \text{years}$$

Caution: *External maintenance and nondestructive testing (NDT) should follow at least once a year. However, internal corrosion may occur due to the biological bacteria in the crude oil within the pipeline. Therefore, microbial corrosion mechanism leads to cathodic depolarization. A bacteria called sulfate-reducing bacteria (SRB) produces hydrogen sulfide gas* (H_2S), *which oxidizes in the presence of moisture to produce sulfuric acid* (H_2SO_4) *[5]. Perhaps, a better term for this type of corrosion may be bacteria-induced corrosion or biocorrosion because the bacteria waste product induces or aids anodic and cathodic reactions on steel and concrete. Therefore, bacteria-induced corrosion affects the electrochemical behavior of steel and the chemical behavior of concrete.*

One possible chemical reaction for the conversion of hydrogen sulfide to sulfuric acid is

$$H_2S + 2O_2 + 2H_2O \rightarrow H_2SO_4 + 2H^+ + 2OH^-$$

It is the bacteria waste product that reacts with oxygen and water to produce sulfuric acid, which, in turn, attacks or induces concrete corrosion and steel corrosion and lowers the pH of the environment.

For steel (iron) corrosion to take place, the possible reactions that may explain the bacteria-induced corrosion are

$$Fe + H_2SO_4 \rightarrow Fe^{2+} + 2H^+ + SO_4^-$$

$$Fe^{2+} + \frac{1}{2}O_2 + H_2O \rightarrow Fe\,(OH)_2$$

Therefore, corrosion of steel is due to the formation of ferrous hydroxide $Fe\,(OH)_2$*, which may react with the bacteria waste product for further surface deterioration of the steel surface.*

8.1.9 Pressure Vessels

The objective of this section is to provide the reader with some insights on pressure vessels used nowadays in our daily life and subsequently derive expressions for predicting the current and potential developed during cathodic protection of pressure vessels using sacrificial anodes. Among many types of pressure vessel designs, a domestic electric water heater schematically depicted in Fig. 8.21 is of interest in this section [5].

There are also gas water heaters available in the market. According to the *ASME* Code Sections IV and VIII, electric water heaters operate at 125 or 150 psi working pressure. The pressure vessel is a cylinder normally made of a carbon steel, which is commonly coated internally with vitreous silica compound known as porcelain enamel linings (glass coating) for protecting the steel walls against the corrosive action of portable water and possibly dissolved minerals, which in turn may enhance the electrical conductivity of the water. The thickness of the glass coating is in the order of 254 μm (10 mils), which serves as the primary protection against corrosion. During the manufacturing process of the vessel, the glass coating is accomplished by spraying a uniform layer of slurry silicate on the steel inner surface, and subsequently, the layer is fired at approximately 900 °C. As a result, the glass coating protects the steel due to its highly corrosion resistance. With respect to the electric water heater components, the *TPR* valve stands for temperature-pressure relief valve, which is a safety device that releases pressure when the temperature or pressure reaches an unsafe level.

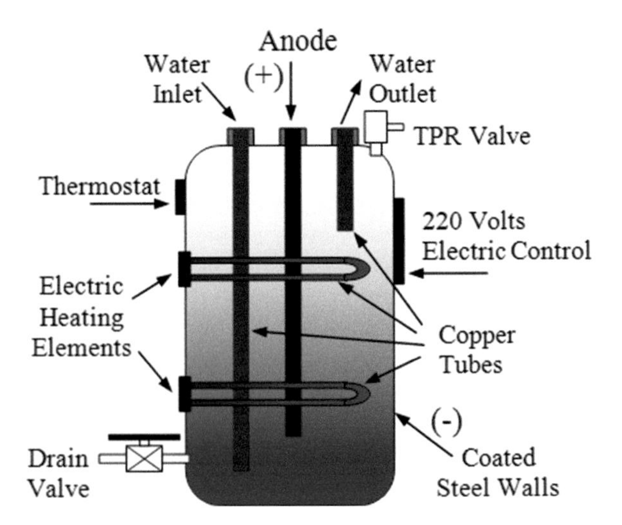

Fig. 8.21 Schematic electric water heater

Table 8.6 Chemical composition (%) of anode rods and average anode capacity (C_a) of 1125 A h/Kg [43]

Alloy	*Mg*	*Al*	*Mn*	Zn	*Si*	*Cu*	*Fe*	*Ni*
					(Maximum)			
Galvorod	Bal.	2.5–3.5	$\geq$0.20	0.7–1.3	0.05	0.01	0.002	0.001
Galvomag	Bal.	$\leq$ 0.01	0.5–1.3	–	–	0.02	0.03	0.001

Further protection of the steel vessel against corrosion is achieved by placing one or two magnesium (*Mg*) or *Mg*-alloy anode rods in a vertical position as shown in Fig. 8.21. This, then, is a secondary or supplemental corrosion protection to mitigate localized galvanic cells in the interior of the vessel. For convenience, the chemical composition and the average anode capacity (C_a) of two commercially available extruded *Mg*-alloy anode rods are given in Table 8.6. According to the electromotive force (emf) series given in Table 2.2, *Mg* is anodic to *Fe*, and consequently, the *Mg* anode is sacrificed during cathodic protection. This means that the electron flow due to the electrochemical reaction $Mg \rightarrow Mg^{2+} + 2e$ is a measure of current flow inside the pressure vessel. The electron flow is directed to bare steel inner areas, such as sharp corners and fittings. In addition, the oxidation potential ranges for Galvorod and Galvomag are 1.4–1.5 V and 1.6–1.7 V vs. $Cu/CuSO_4$ reference electrode, respectively [43].

Despite that the glass coating and the sacrificial anode cathodically protect the pressure vessel, a through thickness crack, which is an imperfection, in the glass coating is detrimental to steel because it would eventually corrode in hot water. In fact, galvanic corrosion may occur in the interior of the pressure vessel due to the presence of impurities in the water, oxygen, and dissimilar metallic materials, such as brass (*Cu-Zn* alloy) electric heating elements, brass fittings and drain, and inlet and outlet nipples. This implies that localized electrochemical galvanic cells may

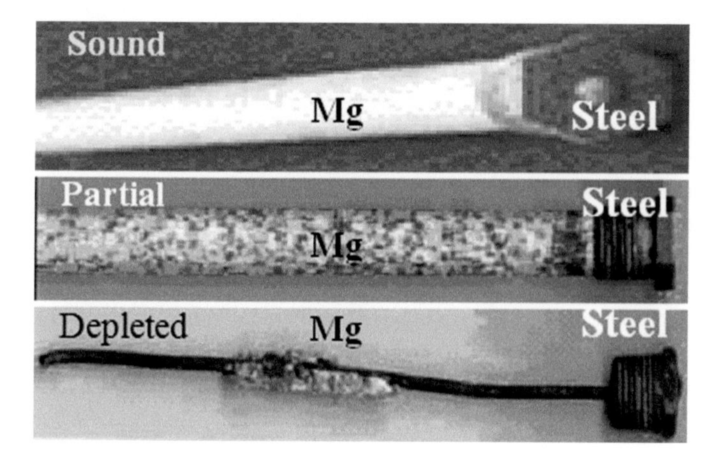

Fig. 8.22 Photos of Mg anode rods. The photo for the sound rod was taken from [40] and the other two photos were taken from [44]

Fig. 8.23 Corroded gas water heater and sediment (www.waterheaterrescue.com)

develop and proceed readily at relatively high temperatures, and consequently, the vessel lifetime is reduced due to high corrosion rates. For instance, the usefulness of the sacrificial anode for corrosion protection is clearly shown in Fig. 8.22, which depicts sound, partially sacrificed, and depleted *Mg* anode rods [44].

This figure also indicates the galvanic corrosion of the *Mg* anode rods. In fact, each anode has a steel core wire placed lengthwise through its center. The threaded upper part is for easy installation or replacement. In addition, the corrosion rate of the depleted *Mg* anode rod is usually low, and therefore, its lifetime is long enough in terms of years, provided that the water heater works efficiently.

Eventually, the cathodic protection may fail, and consequently, the inner steel surface corrodes. One particular case is shown in Fig. 8.23 for an old gas water heater and its pertinent sediment (www.waterheaterrescue.com).

The lifetime of the anode depends on the water temperature, the glass coating quality, and the water chemistry. Therefore, inspection of the anode must be done according to the manufacturer recommendations or experience in order to avoid corrosion in the interior of the vessel.

Apparently, a 5-year period after anode installation may be a good practice, but it depends on the water heater efficiency. If a water heater operates very efficiently, then the anode can last 10 or more years.

Corrosion due to a small leak over a long period of time is a slow process. The remedy is based on maintenance.

The second part of this section is based on Faraday's law of electrolysis, which states that the rate of thickness reduction is a measure of corrosion rate, which, in turn, is a direct representation of structural damage. Let us start with the rate of mass loss equation given by Eq. (6.6):

$$\frac{m}{t} = \frac{iA_s A_w}{zF} \tag{8.71}$$

Divide Eq. (8.71) by the product $A\rho = Am/V = m/B$, where A is the vessel exposed area and ρ is the density of the metal. Solving for the thickness B yields

$$B = \frac{iA_w t}{zF\rho} \tag{8.72}$$

For a cylindrical pressure vessel, the thickness B in Eq. (8.72) must satisfied the average hoop stress (average tangential stress) equation defined by

$$\sigma = \frac{PD}{2B} \tag{8.73}$$

where P = internal pressure (MPa)

D = internal diameter

Crudely, combine Eqs. (8.72) and (8.73) to get the current density:

$$i = \frac{zF\rho PD}{2\sigma t A_w} \tag{8.74}$$

Now, the corrosion rate, Eq. (5.90), becomes

$$C_R = \frac{iA_w}{zF\rho} = \frac{PD}{2\sigma t} \tag{8.75}$$

For coated structures, wisely assume or determine the coating efficiency in order to compute the bare surface and the corresponding current that must protect the bare surface from corrosion.

The bare area and the current are easily determined as

$$A_s = A_o (1 - \lambda_s) \tag{8.76}$$

$$I = i_{corr} A_o (1 - \lambda_s) \tag{8.77}$$

where λ_s = fraction of the protected surface area

A_o = total surface area

In addition, cathodic protection based on sacrificial anodes requires a cell potential as the driving force for a self-imposed spontaneous protective current imparted by sacrificial anodes liberating electrons at a specific rate [32].

The driving force for this current is the potential difference (overpotential) between the sacrificial anode and cathode; that is, $E = E_c + E_a$.In fact, the resultant electrochemical system is a naturally polarized galvanic cell, in which the anode is oxidized or sacrificed to provide electrons to the cathodic structure.

In general, an alloy, instead of a pure element, with known standard potential and being anodic to the a structure is normally used as a sacrificial anode (Table 8.6). Hence, a galvanic cell is created for protecting the structure against electrochemical corrosion.

For carbon steel structures, the total applied potential is defined as the cell potential given by

$$E = E_c + E_a \tag{8.77a}$$

where $E_c = -0.85\, V_{Cu/CuSO_4}$ is the protective or cathodic potential recommended by NACE and E_a is the anodic potential. Use Table 2.2 (the emf series) to select an element anodic to iron. Then, change the sign of E_a and convert V_{SHE} to $V_{Cu/CuSO_4}$. The total potential is then used to determine the cell free energy change: $\Delta G = -zFE$.

Example 8.7. *A vertical cylindrical steel pressure vessel containing oxygenerated water is subjected to 15-MPa internal pressure, and it is to be cathodically protected using one Mg-alloy (Galvorod) anode rod and the NACE recommended potential of* $-0.85\ V_{Cu/CuSO_4}$*. The hoop stress is limited to half the yield strength of the steel; that is,* $\sigma = 300\, MPa$*. The tank inside diameter is* $D = 40\, cm$*. If the design life* (L_d) *is 10 years and if the bare inner surface area is 98 % coating protected, calculate* ***(a)*** *the corrosion rate as a measure of the rate of thickness reduction of the vessel and its thickness at the end of its lifetime. Will the tank burst at the end of 10 years? Now, cathodically protect the vessel. Determine* ***(b)*** *the cell potential;* ***(c)*** *the effective mass of the Mg-alloy anode rod, which has 50 % current efficiency due to hydrogen evolution;* ***(d)*** *the system current density, the anode, and the system electric resistances; and* ***(e)*** *the net current density and the reduction in anode diameter. Data for steel:* $i_{corr} = 65\, \mu A/cm^2$ *(uniform),* $A_w = 55.85\, g/mol$, $\rho = 7.70\, g/cm^3$, $B = 1.30\, cm$, *and* $H = 80\, cm$ *(height of tank). For the sacrificial Mg-alloy anode rod:* $L = 30\, cm$, $d = 12\, cm$, $A_w = 24.31\, g/mol$, *and* $\rho = 1.74\, g/cm^3$*. The water resistivity is* $\rho_w = 1000\, \Omega\, cm$*:*

Solution. (a) Without Cathodic Protection: *Corrosion of the steel tank is manifested through the anodic reaction*

$$Fe \rightarrow Fe^{2+} + 2e$$

Thus, Eq. (8.75) gives the uniform corrosion rate of the bare steel vessel without cathodic protection:

$$C_R = \frac{i_{corr} A_w}{zF\rho}$$

$$C_R = \frac{\left(6.50x10^{-5}\,\text{A/cm}^2\right)(55.85\,\text{g/mol})}{(2)\,(96,500\,\text{A s/mol})\left(7.70\,\text{g/cm}^3\right)} = 0.77\,\text{mm/year}$$

The thickness reduction of the steel vessel based on C_R is

$$B_x = tC_R = (10\,\text{year})\,(0.77\,\text{mm/year})$$

$$B_x \simeq 7.70\,\text{mm} \quad \textit{(lost thickness)}$$

In this part of the example, the minimum vessel thickness (B_m) is calculated using the theory of thin-walled vessels and the average hoop stress (average tangential stress). Thus,

$$B_m = \frac{PD}{2\sigma}$$

$$B_m = \frac{(15\,\text{MPa})\,(40\,\text{cm})}{(2)\,(300\,\text{MPa})}$$

$$B_m = 1\,\text{cm} = 10\,\text{mm}$$

Now, the final thickness $\left(B_f\right)$ due to corrosion at the end of 10 years is

$$B_f = B - B_x$$

$$B_f = (13\,\text{mm} - 7.7\,\text{mm}) = 5.30\,\text{mm}$$

Therefore, the vessel will burst because $B_f < B_m$. This particular design can be enhanced by using a thicker steel vessel so that

$$B > B_m + B_x = (10\,\text{mm} + 7.7\,\text{mm}) > 17.70\,\text{mm}$$

The actual design thickness is subjected to the designer's experience or a particular design code, such as the ASME code for pressure vessels.

(b) Cathodic Protection: *The above corrosion rate is to be eliminated by sacrificing one Mg-alloy anode. Thus, the main electrochemical reactions during the cathodic protection process are*

$$2Mg \to 2Mg^{2+} + 4e \qquad \text{(anode)}$$
$$2H^+ + 2e \to H_2 \qquad \text{(cathode)}$$
$$2Mg + 2H^+ \to 2Mg^{2+} + H_2 \qquad \text{(overall)}$$

The average anodic potential for the chosen Mg-alloy anode rod (Table 8.6) is $E_a = 1.45\,V_{Cu/CuSO_4}$*, and that recommended by NACE is* $E_c = -0.85\,V_{Cu/CuSO_4}$*. Thus, the cell potential is*

$$E = E_c + E_a = -0.85\,\text{V} + 1.45\,\text{V}$$
$$E = 0.60\,\text{V}$$

(c) *The anode volume and its useful mass based on 50 % anode current efficiency* ($\epsilon = 0.50$) *are, respectively,*

$$V = \frac{\pi d^2 h}{4} = \frac{\pi\,(12\,\text{cm})^2\,(30\,\text{cm})}{4} = 3392.92\,\text{cm}^3$$
$$m = \epsilon\rho V = (0.5)\left(1.74\,\text{g/cm}^3\right)\left(3392.92\,\text{cm}^3\right)$$
$$m = 2.95\,\text{Kg}$$

(d) *The inner bare area of the vessel is*

$$A_s = A_o\,(1 - \lambda_s) = \pi DH\,(1 - \lambda_s)$$
$$A_s = \pi\,(40\,\text{cm})\,(80\,\text{cm})\,(1 - 0.98)$$
$$A_s = 201.06\,\text{cm}^2$$

This is the area to be protected by the anode for 10 years. Thus, the average self-imposed spontaneous protective current (Table 8.4) and current density imparted by the sacrificial Mg-alloy anode are, respectively,

$$I = \frac{7mC_a}{3\pi L_d} = \frac{(7)\,(2.95\,\text{Kg})\,(1125\,\text{A h/Kg})}{(3\pi)\,(87{,}600\,\text{h})}$$
$$I = 28.14\,\text{mA}$$

$$i = \frac{I}{A_s} = \frac{28.14\,\text{mA}}{201.06\,\text{cm}^2} = 0.14\,\text{mA/cm}^2$$

Then, the anode and the system resistances are, respectively,

$$R_a = \frac{E_a}{I} = \frac{1.45\,\text{V}}{28.14x10^{-3}\,\text{A}} = 51.53\Omega$$

$$R_s = \frac{E}{I} = \frac{0.60\,\text{V}}{28.14x10^{-3}\,\text{A}} = 21.32\,\Omega$$

(e) *The net current density is controlled by the cathodic current density of the steel structure. That is,*

$$i_{net} = i_c - i_a \simeq i_{corr}$$

The corrosion rate of the Mg alloy is

$$C_R = \frac{iA_w}{zF\rho}$$

$$C_R = \frac{\left(0.14 \times 10^{-3}\,\text{A/cm}^2\right)(24.31\,\text{g/mol})}{(2)\,(96,500\,\text{A s/mol})\left(1.74\,\text{g/cm}^3\right)}$$

$$C_R = 3.20\,\text{mm/year} \quad \textit{(anode)}$$

Assuming a uniform anode corrosion, one can predict the diameter reduction as

$$d_x = L_d C_R = (10\,\text{year})\,(3.20\,\text{mm/year}) = 32\,\text{mm}$$

Thus,

$$\Delta d = d - d_x = (120 - 32)\,\text{mm} = 88\,\text{mm}$$

which represents 73 % *reduction in anode diameter in* 10 *years. Now, the cell free energy change is then calculated as*

$$\Delta G = -zFE = (2)\,(96,500\,\text{J/mol V})\,(0.60\,\text{V}) = -1.158 \times 10^5\,\text{J/mol}$$

$$\Delta G = -0.1158\,\text{MJ/mol}$$

Therefore, the electrochemical cell operates as described since the electrochemical reactions proceed as written because $\Delta G < 0$. *In addition, the electrochemical cell is also an energy-producing device.*

8.2 Anodic Protection

The use of anodic protection (AP) can be considered when coating and cathodic protection techniques are not suitable for protecting a structure against corrosion. The AP technique is normally used very successfully in aggressive environments, such as sulfuric acid (H_2SO_4) and mild solutions [45–48]. In general, the paper- and fertilizer-making industries use anodic protection to prevent corrosion on the inner surfaces of tanks and containers. The main requirement for AP is that the material to be protected must exhibit an active–passive polarization behavior for the formation of a passive film, but the passive potential range must be wide enough and the passive current density (i_p) be stable and sufficiently lower than the corrosion current density (i_{corr}). Thus, a potentiostat is used to supply a protective potential (E_x) in the passive region of a polarization diagram, but potential monitoring is necessary; otherwise, a potential deviation can make the structural material become active, and the anodic protection is lost.

8.2.1 Design Criterion

Anodic protection (AP), like cathodic protection (CP), is an electrochemical technique in which the metal or structure is polarized from the critical current density (i_c) as schematically shown in Fig. 8.24, which schematically illustrates the Pourbaix and polarization diagrams.

In order to anodically protect a structure, such as sulfuric acid (H_2SO_4) storage steel tanks, heat exchangers, and transportation vessels [49], the critical current density i_c should be high in the active state, but the goal is to force the active state change to a passive state at a potential E_x as shown in Fig. 8.24. Thus, the protective parameters are E_x and i_x, where

$$E_{pa} < E_x < E_p \tag{8.78}$$

$$i_x = i_p < i_{corr} \tag{8.79}$$

Nonetheless, i_c is the current density required to obtain passivity, and i_p is the current density to maintain passivity. Also included in Fig. 8.24 is an schematic Pourbaix diagram (potential-pH diagram) showing the electrochemical regions needed for characterizing the electrochemical behavior of a metal immersed in an electrolyte. This diagram is related to the polarization diagram (Fig. 8.24b). Observe the correspondence of potential in both diagrams. As indicated in Chap. 3, the Pourbaix diagram does not give an indication of the rate of reaction, but the polarization curve does at a pH value indicated by a downward arrow in Fig. 8.24a.

Hence, a structure is anodically protected if the metal is active–passive and shows a sufficiently large passive potential range, Eq. (8.78), due to the formation of a dynamic oxide film. This implies that the current density depends on time and, therefore, the power supply must provide the required potential E_x so that

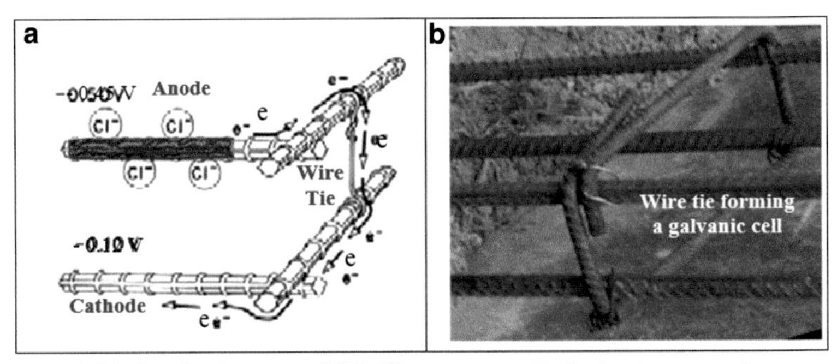

Model of a galvanic macro-cell [62-63]

Real galvanic macro-cell

Corrosion due to galvanic macro-cells

Fig. 8.24 Schematic Pourbaix diagram and active–passive polarization curve showing the passive region for anodic protection at E_x. (**a**) Model of a galvanic macro-cell [62-63]. (**b**) Real galvanic macro-cell. (**c**) Corrosion due to galvanic macro-cells

$i_x < i_{corr}$ [2]. Thus, anodic polarization results due to the formation of an insoluble oxide film of a few μm in thickness.

The effectiveness of anodic protection depends on the quality of the oxide film in a particular environment and the applied potential. For instance, if the applied potential is $E_x \geq E_p$, then the film corrodes by pitting, which is a localized electrochemical process. On the other hand, if $E_x \leq E_{pa}$, the metal corrodes by general and uniform process.

Passivation may be accomplished due to the accumulation of reacting ions forming an oxide film (metallic coating) on the anode surface. Consequently, the current flowing from the anode to the electrolyte reduces since the metallic coating has a high electric resistance. This can be defined by Ohm's law:

$$E_x = I_x R_x \tag{8.80}$$

In anodic protection, the structure to be protected is the anode by connecting it to the positive terminal of an electrochemical circuit, whereas the negative cathode is made of steel or graphite. On the other hand, in cathodic protection, the structure to be protected is made of the cathode (negative).

The potential E_x is assumed to be constant since anodic protection (*AP*) is a potential-control technique, in which the power supply is a potentiostat capable of supplying a constant potential; otherwise, a significant change in potential causes a change in the current, and the structure may become unprotected due to film breakdown.

Figure 8.25 shows a schematic setup for *AP* of a steel tank for storing an acid solution. Normally, sufficiently large cathodes are needed in order to compensate for the large structure surface area and the highly conductive electrolyte (acid). This indicates that *AP* is achieved if the circuit resistance (R_x) due to ohmic potential drop is controlled and maintained. This phenomenon is known as the ohmic effect.

Since the anodic protection technique is restricted to materials that exhibit active–passive transitions, it is important to determine the effects of solution-containing contaminants, such as chloride ions, and temperature on the polarization behavior of metallic materials. Figure 8.26 schematically illustrates the possible polarization trends based on the above variables.

For instance, the protective potential E_x being a measure of anodic protection and the passive potential range, $\Delta E = E_p - E_{pa}$, decreases, and the protective current density i_x increases as either chloride concentration in solution or temperature increase.

In fact, anodic protection of a material such as stainless steel becomes a complex task if the magnitude of the passive potential range is not sufficiently large. Furthermore, E_x is inversely proportional to i_x as the metallic coating resistance decreases due to the effects of chloride concentration and temperature. Nevertheless, a high critical current density is required to cause passivation, and a low current density is needed for maintaining passivation [2].

Clearly, i_c is strongly affected by temperature. Denote that i_c increase very rapidly at temperatures greater than 60 °C. The critical current density trend

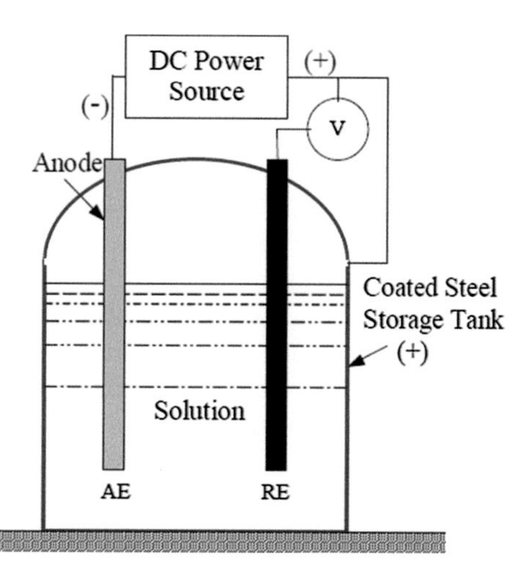

Fig. 8.25 Schematic anodic protection system after [10]

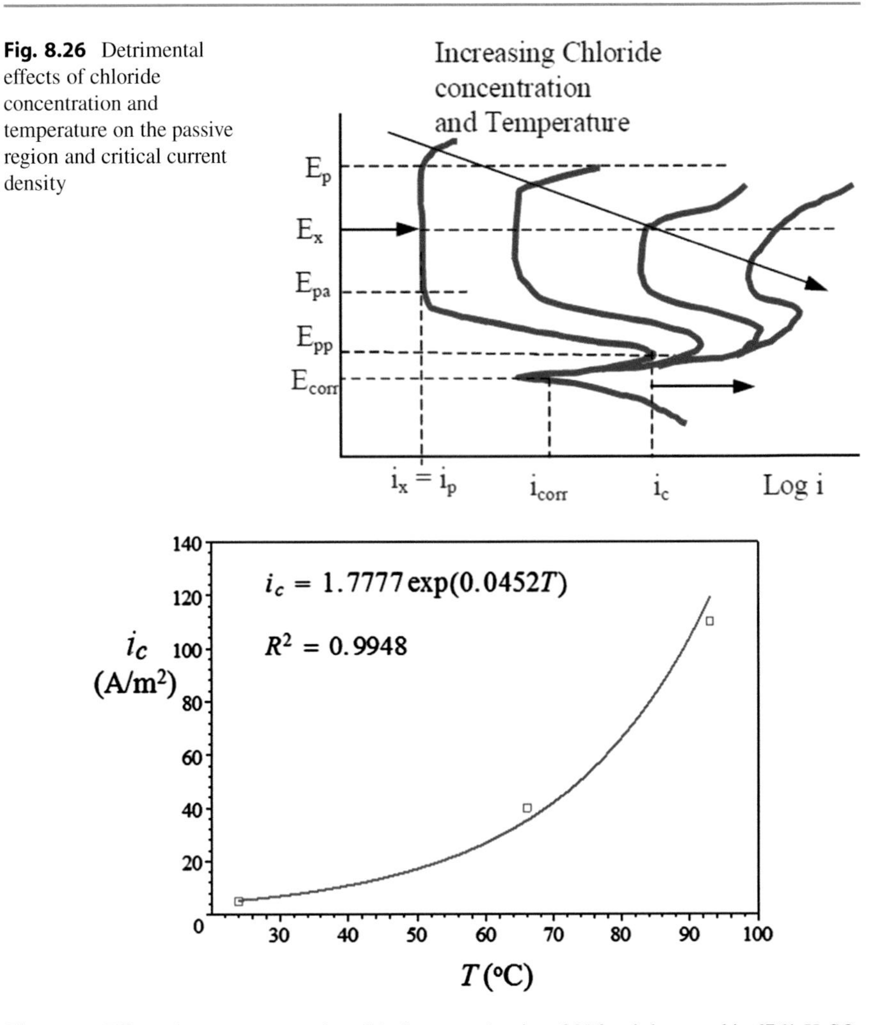

Fig. 8.26 Detrimental effects of chloride concentration and temperature on the passive region and critical current density

Fig. 8.27 Effect of temperature on the critical current density of 316 stainless steel in 67 % H_2SO_4 solution [50]

illustrated in Fig. 8.27 can be considered very unique for 316 stainless steel in 67 % sulfuric acid (H_2SO_4) solution since it exhibits a significant critical current density (i_c) profile [50]. Thus, it follows an exponential behavior in a large temperature range despite the three data points being available. This is supported by the nonlinear least-squares fitting, providing rather good correlation coefficients (R^2).

Moreover, the plot in Fig. 8.27 indicates that the rate of corrosion is very high as high temperatures. Computing the atomic weight (A_w), determining the bulk density (ρ), and determining the oxidation state (z) of the 316 stainless steel in 67 % H_2SO_4 solution, one can use Eq. (8.75) and the curve fitting equation in Fig. 8.27 to estimate the corrosion rate as

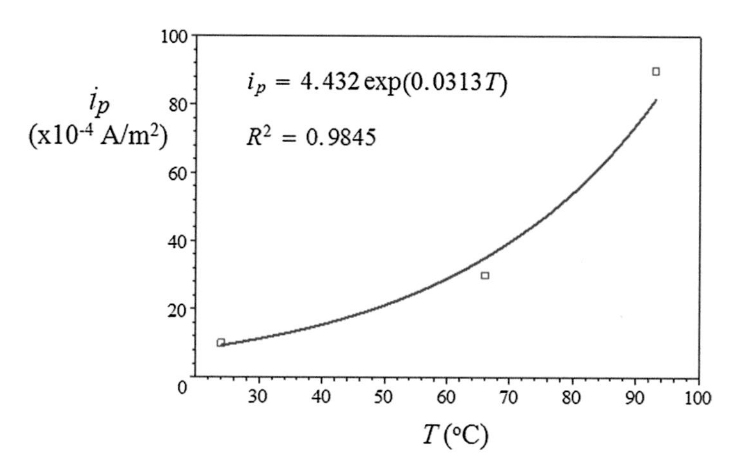

Fig. 8.28 Effect of temperature on the passive current density of 316 stainless steel in 67 % H_2SO_4 solution [50]

Table 8.7 Reference electrodes for anodic protection [46]

Electrode	Environment
Calomel	Sulfuric acid
Silicon cast iron	Sulfuric acid
Silver–silver chloride	Sulfuric acid, Kraft digester liquid
Copper	Hydroxylamine sulfate
Stainless steel	Liquid fertilizers (nitrate solutions)
Nickel-plate steel	Chemical nickel plating solutions
Hastelloy	Liquid fertilizers, sulfuric acid

$$C_R = \frac{i_c A_w}{zF\rho} = \frac{A_w}{zF\rho}\left[1.7777\exp\left(0.0452T\right)\right] \tag{8.81}$$

Figure 8.28 depicts the passive current density (i_p) for 316 stainless steel under the above condition. Despite that $i_p = f(T)$ has lower values than $i_c = f(T)$, it also shows an exponential behavior. A successful anodic protection design not only requires a controlled potential/current, but a high-quality passive film which must be insoluble in the aggressive solution. In fact, the very low anodic corrosion rate in terms of the passive current density i_p (Fig. 8.28) is apparently due to the limited ionic mobility in the passive film [2].

Furthermore, Table 8.7 lists some reference electrodes and corrosive media that have been used for anodic protection [46]. The reader should consult reference [46] for significant details on this topic.

Despite that a reference electrode is used in anodic protection in order to control the potential of the protected structural wall, it has to be stable as a function of time, temperature, and electrolyte composition. This implies that the reference electrode must ideally have a constant electrical potential. The anodic structure, the cathodic electrode, the reference electrode (RE), and the solution are the components of

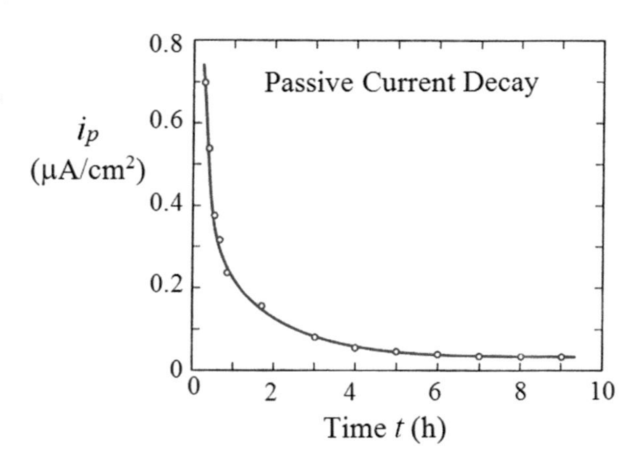

Fig. 8.29 Potentiostatic passive current density decay at room temperature for some polycrystalline material [2]

the electric circuit that must deliver a DC between the anode inner wall and the cathode [46].

8.2.2 Kinetics of Passive Oxide Growth

This section is devoted to metallic passivity related to Pourbaix and potential-current anodic diagrams. A metal oxide film (several nanometers thick) can be produced by (1) anodic polarization in which the metal oxidizes and then passivates in an electrolyte containing dissolved oxygen and (2) exposing the electrode to an electrolyte containing the metal ion and dissolved oxygen in an electrochemical cell. Conversely, a metal oxide produced at high-temperature oxidation is known as an oxide scale, which is porous and thick. This topic is dealt with in Chap. 10.

During anodic polarization the passive film begins to form at a potential E_{pp} (Fig. 8.24). When $E > E_{pp}$, the slowly growing passive film thickens, and consequently, the potentiostatic passive current density i_p decreases with increasing time. This is depicted in Fig. 8.29 as a decay behavior for some polycrystalline material [2].

The i_p decay shown in Fig. 8.29 has the following assumed mathematical definition:

$$i_p = \frac{c_p}{t} \tag{8.82}$$

Here, c_p is a constant. In addition, i_p is also assumed to be related to the rate of oxide film growth or oxide film thickening, dx/dt, as indicated below [51]:

$$i_p = c_x \frac{dx}{dt} \tag{8.83}$$

where c_x is also a constant. This expression is also defined by Eq. (7.21a).

At any rate, the passive oxide film quality is strongly influenced by the type of surface preparation, the ionic impurities in solution, and the reaction steps during film formation in an electrochemical field. Basically, electrochemistry is an aqueous engineering approach for controlling metallic corrosion, metallic oxide formation, or metallic electrodeposition. However, characterizing metal oxide formation is vital for structural integrity in a multidimensional space due to the effects of concentration, potential-pH value, temperature, and even pressure that enhance or deteriorate the electrochemical state of a metallic structural component.

The analysis of passive films on metal surfaces under anodic polarization conditions is a deterministic approach toward the quality and protectiveness of the oxide films. Fundamentally, the electrochemical electric potential (E) as a function of pH, known as Pourbaix diagrams, is essential from a thermodynamic standpoint. If fact, a particular $E - pH$ data can be located on a known Pourbaix diagram for a pure metal to determine either the metal ion due to oxidation or the oxide product due to passivation. The latter serves as a guideline in design applications against corrosion, but the kinetics associated with oxide formation is essential in determining the optimum electrochemical cell setup. Therefore, the Pourbaix and polarization diagrams provide the thermodynamical and electrochemical maps for a better understanding of the oxide film formation and related properties. Obviously, microscopy should follow for a complete overview of the passive oxide film morphology and surface defects.

The Pourbaix diagram lacks information of kinetics, and the anodic polarization diagram lacks information on pH, but they illustrate the potential range for passivity. Hence, these diagrams are useful in predicting and characterizing passive oxide films from a graphical point of view. Most Pourbaix diagrams are available for pure metals in water at standard conditions, and conversely, some polarization diagrams are available for pure metals and their alloys.

8.3 Problems/Questions

8.1. Why cathodic protection is not generally recommended for stress corrosion problems on high-strength ferritic steels?

8.2. A steel structure exposed to seawater is to be cathodically protected using the sacrificial anode technique. Calculate **(a)** the number of *Zn* anodes (N) that will be consumed in a year if the *Zn* anode capacity is 770 A h/Kg (average value as per Table 8.2) and the current is 0.70 A and **(b)** the individual weight (M) of anodes so that $NM \geq W$, where W is the total theoretical weight of the anodes. [Solutions: (a) $N \simeq 1$ Zn anode and (b) $m = 7.48$ Kg.]

8.3. Cathodically protect a vertical pressurized steel tank ($\sigma_{ys} = 414\,\text{MPa}$ and $P = 12\,\text{MPa}$) using one flush-mounted *Mg-Al-Zn* bracelet ($B_i = 0.9525\,\text{cm}$). Assume 2 % tank coating damage per year, and neglect any installation damage if the tank is coated after installation. Use a bracelet with dimensions equal to $(0.50\,\text{cm})\,x\,(3\,\text{cm})\,x\,(6\,\text{cm})$. Calculate **(a)** the maintenance current and current density if the initial current density and the design lifetime are $0.43\,\mu\mathrm{A/cm^2}$ and 25 years. Will it be convenient to protect the tank having a thickness of 1.8 cm? The tank height and the internal diameter are 65 and 15.24 cm, respectively. **(b)** Will the tank explode at the end of the anode lifetime? Assume that the anode *Mg* alloy and $\rho_x = 5000\,\Omega$ cm for the soil. Given data for the *Mg*-alloy anode: $\rho = 1.74\,\mathrm{g/cm^3}$, $A_w = 24.31\,\mathrm{g/mol}$, and $C_a = 1230$ A h/Kg for the *Mg-Al-Zn* anode (Table 8.2). [Solution: $I_i = 7.74\,\mu$A, $W_i = 1.86$ g of anode, $E_i = 2.87$ mV, $\bar{I} = 6.64\,\mu$A, and $E_f = 2.38$ mV.]

8.4. A 2-m diameter steel tank containing water is pressurized at 200 kPa. The hoop stress, thickness, and height are 420 MPa, 2 cm, and 8 cm, respectively. If the measured corrosion rate is 0.051 mm/year, determine **(a)** the tank life and **(b)** the developed current. Data: $\rho = 7.86\,\mathrm{g/cm^3}$ and $A_w = 55.86\,\mathrm{g/mol}$. [Solution: **(a)** $t = 9.34$ years and **(b)** $I = 92.49$ A.]

8.5. **(a)** Derive Eq. (8.58), and **(b)** calculate the theoretical anode capacity (C_a) for magnesium (*Mg*) and zinc (*Zn*). [Solution: **(b)** $C_{a,Mg} = 2969$ A h/Kg and $C_{a,Zn} = 1104$ A h/Kg.]

8.6. Show that $\alpha = \sqrt{R_s/R_L}$.

8.7. Use an annealed (@ 792 °C) 1040 steel plate of $B = 0.635$-cm thickness to design a cylindrical pressurized vessel using a safety factor (S_F) of 2. The annealed steel has a yield strength of 355 MPa. Calculate **(a)** the average allowable pressure according to the theory for thin-walled vessels. Will the vessel fail when tested at 2 MPa for 10 min? If it does not fail, then proceed to cathodically protect the cylindrical pressure vessel as a domestic or home water heater having a cylindrical magnesium (*Mg*) anode rod. All dimensions are shown below. Calculate **(b)** the applied hoop stress, the current, the potential for polarizing the heater, and the resistance of the system, **(c)** the theoretical anode capacity and the anode lifetime of the anode in years, and **(d)** how much electric energy will deliver the *Mg* anode rod. The steel vessel has a 98 % internal glass coating (vitreous porcelain enamel lining) for protecting the steel inner wall against the corrosive action of portable water. Assume a water column of 76.20 cm and an internal pressure of 1 MPa. Assume that the exterior surface of the steel vessel is coated with an appropriate paint for protecting it against atmospheric corrosion and that the anode delivers $1.5\,V_{Cu/CuSO_4}$. The NACE recommended potential for steels is $-0.85\,V_{Cu/CuSO_4}$. [Solution: $E = 0.65$ V.]

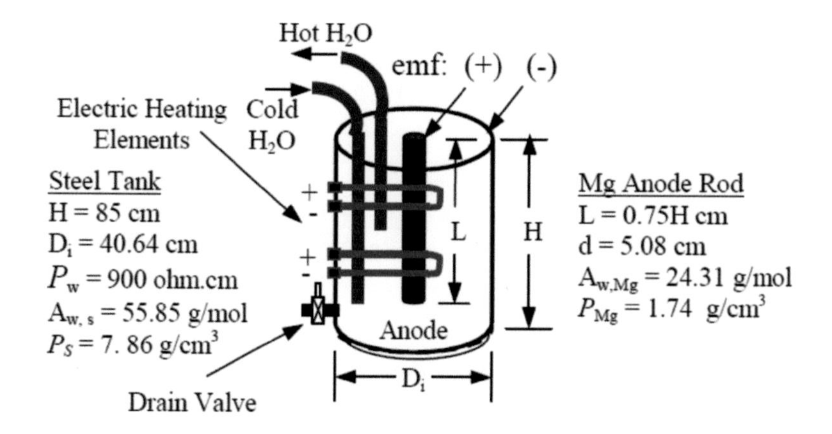

8.8. For anodic protection, the interface potential of the structure is increased to a passive domain, provided that the metal exhibits active–passive behavior. Thus, prevention of corrosion is through an impressed anodic current. Explain including the mathematical definition of the applied anodic current density.

8.9. What are the implications of cathodic protection of steels in acid solutions?

8.10. Why is passivation important for anodic protection? What is the mechanism that explains the electron motion for reactions to take place?

8.11. What are the passive film-preferred breakdown sites?

8.12. Calculate (a) the potential for the two titanium reactions given below. These reactions represent the passivation process for titanium-oxide film formation. Thus, anodic protection is natural because the TiO_2 film acts as a semiconductor barrier to protect the underlying metallic Ti from corrosion, $Ti \rightarrow Ti^{2+} + 2e^-$ at $E^o = 1.63\ V_{SHE}$. (b) Determine the pH. [Solution: $pH = 16.78$.]

$$Ti + O_2 = TiO_2 \qquad \Delta G^o = -852.70\,\text{kJ/mol}$$

$$Ti + 2H_2O = TiO_2 + 4H^+ + 4e^- \quad \Delta G^o = -346.94\,\text{kJ/mol}$$

8.13. The effect of sulfuric acid (H_2SO_4) concentration on polarization behavior of a metal at 25 °C can be assessed obtaining experimental potentiodynamic polarization curves. Below is a data set for the critical current density (i_c) of a hypothetical metal:

$C(\%)$	0	20	30	40	50	60	70
i_c (A/cm^2)	46	30	23	17	12	8	5

Plot $i_c = f(C)$, and explain whether or not passivation is affected by the concentration of sulfuric acid.

8.14. Below is a data set taken from Table 12.7 in reference [10] for determining the effect of concentration of sulfuric acid (H_2SO_4) at 24 °C on the critical current density of 316 (UNS S31600) stainless steel at 24 °C:

$\%C_{H_2SO_4}$	0	40	45	55	65	75
$i_c\ (\text{mA/cm}^2)$	4.7	1.6	1.4	1.0	0.7	0.4
$C_{R.i_{corr}}$ (mm/year)	0	2.2	5.6	8.9	7.8	6.7

(a) Do nonlinear curve fitting on i_c versus $\%C_{H_2SO_4}$ and plot $i_c = f\ (\%C_{H_2SO_4})$. **(b)** Assume that Fe, Cr, Mo, and Mn oxides simultaneously go into solution. Now calculate C_R for these four elements using the given experimental i_c values; do curve fitting on the new C_R data set and the experimental $C_{R.i_{corr}}$ values versus $\%C_{H_2SO_4}$. Plot the C_R functions for Fe, Cr, Mo, and Mn, and plot the experimental $C_{R.i_{corr}}$ data and its linear curve fit equation in order to compare the corrosion behavior of the stainless steel at i_{corr} and these elements i_c. Explain.

8.15. This problem deals with the nature of the anodic oxide film that forms on titanium in 0.9 % $NaCl$. This media is isotonic with human blood for surgical implants made out of titanium and its alloys. Normally, the passive oxide film on titanium mainly consists of TiO_2. The table given below contains data for the passive current density (i_p) as a function of sweep rate (de/dt in mV/s) [52] on titanium in 0.9 % $NaCl$ containing electrolyte. **(a)** Convert the i_p data to corrosion rate C_R in mm/year. Then plot both i_p and C_R as functions of sweep rate $v = dE/dt$. (b) Explain if this titanium material would be suitable for a human transplant for a prolonged time:

$v = dE/dt$ (mV/s)	10	50	100	300	500
$i_p\ (\text{mA/cm}^2)$	0.13	0.25	0.50	1.58	2.10

References

1. H.H. Uhlig, R.W. Revie, *Corrosion and Corrosion Control* (Wiley, New York, 1985), pp. 217–223
2. D.A. Jones, *Principles and Prevention of Corrosion* (Macmillan, New York, 1992), pp. 132–139, 456
3. V. Ashworth, Principles of cathodic protection, in *Corrosion Control*, ed. by L.L. Shreir, R.A. Jarman, G.T. Burstein, 3rd edn. (Butterworth-Heinemann, Oxford, 1994), pp. 10:24
4. E. Bardal, *Corrosion and Protection* (Springer, London, 2004), pp. 266–281
5. M.G. Fontana, *Corrosion Engineering*, Chapter 6, 3rd edn. (McGraw-Hill, New York, 1986)
6. S.L. Pohlman, Stray-current corrosion, in *Corrosion*, vol. 13, 9th edn. ASM Handbook (ASM International, Novelty, OH, 1987), p. 87

7. R.H. Heidersbach, Cathodic protection, in *Corrosion*, vol. 13. ASM Handbook (ASM International, Novelty, OH, 1987), pp. 466–477
8. J.W.L.F. Brand, P. Lydon, Impressed-current anodes, in *Corrosion Control*, vol. 2, 3rd edn., ed. by L.L. Sheir, R.A. Jarman, G.T. Burstein (Butterworth-Heinemann, Boston, 1994), pp. 10:56–10:87
9. G.D. Brady, Graphite anodes for impressed current cathodic protection: a practical approach. Mater. Perform. **10**(10), 20–22 (1971)
10. P.R. Roberge, *Handbook of Corrosion Engineering*, Chapters 11 & 12 (McGraw-Hill, New York, 2000)
11. W. Wang, W.H. Hartt, S. Chen, Sacrificial anode cathodic polarization of steel in seawater: Part 1. Corrosion control on steel, fixed offshore platforms associated with petroleum production. NACE EP-01-76 (1976)
12. W.H. Hartt, S. Chen, Path dependence of the potential-current density state for cathodically polarized steel in seawater. Corrosion **56**(1), 3–11 (2000)
13. S. Rossi, P.L. Bonora, R. Pasinetti, L. Benedetti, M. Draghetti, E. Sacco, Laboratory and field characterization of a new sacrificial anode for cathodic protection of offshore structures. Corrosion **54**(12), 1018–1025 (1998)
14. J.D. Burk, Dual anode field performance evaluation-cathodic protection for offshore structures. Corrosion/91, Paper No. 309, (NACE, Houston, TX, 1991)
15. W. Wang, W.H. Hartt, S. Chen, Sacrificial anode cathodic polarization of steel in seawater: Part 1 - A novel experimental and analysis methodology. Corrosion **52**(6), 419–427 (1996)
16. W.H. Hartt, S. Chen, D.W. Townley, Sacrificial anode cathodic polarization of steel in seawater: Part 2 - Design and data analysis. Corrosion **54**(4), 317–322 (1998)
17. L. Sherwood, Sacrificial anodes, in *Corrosion Control*, vol. 2, 3rd edn., ed. by L.L. Sheir, R.A. Jarman, G.T. Burstein (Butterworth-Heinemann, Boston, 1994), pp. 10:29–10:54
18. R.B. Mears, R.H. Brown, A theory of cathodic protection. Trans. Electrochem. Soc. **74**(1), 519–531 (1938)
19. R.P. Howell, Potential measurements in cathodic protection designs. Corrosion **8**(9), 300–304 (1952)
20. P. Pierson, K.P. Bethude, W.H. Hartt, P. Anathakrishnan, A new equation for potential attenuation and anode current output projection for cathodically polarized marine pipelines and risers. Corrosion **56**(4), 350–360 (2000)
21. D.A. Jones, The application of electrode kinetics to the theory and practice of cathodic protection. Corros. Sci. **11**(6), 439–451 (1971)
22. E.M. Purcell, *Electricity and Magnetism*. Berkeley Physics Course, vol. 2 (McGraw-Hill, New York, 1965), pp. 43–48
23. M.E. Parker, *Pipeline Corrosion and Cathodic Protection* (Gulf Publication, Houston, TX, 1962)
24. Z.D. Jastrzebski, *The Nature and Properties of Engineering Materials* (Wiley, New York, 1987), pp. 600–603
25. D.T. Chin, G.M. Sabde, Modeling transport process and current distribution in a cathodically protected crevice. Corrosion **56**(8), 783–793 (2000)
26. D.D. McDonald, C. Liu, M. Urquidi-Macdonald, G.H. Stickford, B. Hindin, A.K. Agrawal, K. Krist, Prediction and measurement of pitting damage functions for condensing heat exchangers. Corrosion **50**(10), 761–780 (1994)
27. D.D. McDonald et al., SRI Inter. Report to Gas Research Inst. (GRI) No. 5090–260-1969 (January 1992)
28. C. Liu, D.D. McDonald, M. Urquidi-Macdonald, Pennsylvania State University Report to Gas Institute, GRI Contract No. 5092–260-2353, NTIS GRI 93/0365, CAM 9311 (September 1993)
29. G.G. Clemeña, M.B. Pritchett, C.S. Napier, Application of cathodic prevention in a new concrete bridge deck in Virginia. Final Report VTRC 03-R11, Virginia Transportation Research Council in cooperation with the U.S. Department of Transportation Federal Highway Administration, Charlottesville, Virginia, Copyright 2003 by the Commonwealth of Virginia (February 2003)

30. S.F. Daily, *Understanding Corrosion and Cathodic Protection of Reinforced Concrete Structures*, CP of Concrete Structures. (Corrpro Companies, Medina, OH 2007). http://www.cathodicprotection.com/essays4.htm
31. S.R. Sharp, G.G. Clemena Y.P. Virmani, G.E. Stoner, R.G. Kelly, Electrochemical chloride extraction: influence of concrete surface on treatment. Report submitted to Office of Infrastructure Research and Development Federal Highway Administration, 6300 Georgetown Pike McLean, Virginia, pp. 22102–2296, (09 September 2002)
32. D.L. Piron, The electrochemistry of corrosion. NACE **231**, 237 (1991)
33. G.W. Currer, J.S. Gerrard, Practical application of cathodic protection, in *Corrosion Control*, vol. 2, 3rd edn., ed. by L.L. Sheir, R.A. Jarman, G.T. Burstein (Butterworth-Heinemann, Boston, 1994), pp. 10:93
34. M.E. Orazem, J.M. Esteban, K.J. Kennelley, R.M. Degerstedt, Mathematical models for cathodic protection of an underground pipeline with coating holidays: part 1 - theoretical development. Corrosion **53**(4), 264–272 (1997)
35. J.-F. Yan, S.N.R. Pakalapati, T.V. Nguyen, R.E. White, Mathematical modeling of cathodic protection using the boundary element method with a nonlinear polarization curve. J. Electrochem. Soc. **139**(7), 1932–1936 (1992)
36. K.J. Kennelley, L. Bone, M.E. Orazem, Current and potential distribution on a coated pipeline with holidays. Corrosion **49**, 199 (1993)
37. J.M. Tobias, Surface ground device. Report number CECOM-TR-94-9, US Army CECOM, ATTN: AMSEL-SF-SEP, Fort Monmouth, NJ (June 1994)
38. H.B. Dwight, Calculation of resistances to ground. Electr. Eng. **55**(7), 1319–1328 (1936)
39. W.T. Bryan (ed.) *Designing Impressed-Current Cathodic Protection Systems with Durco Anodes*, 2nd edn. (The Duriron Company, Dayton, OH, 1970)
40. R.H. Heidersbach, R. Baxter, J.S. Smart III, M. Haroun, Cathodic protection, in *Corrosion*, vol. 13. ASM International Handbook (ASM International, Novelty, OH, 1987), pp. 919–924
41. R. Singh, *Corrosion Control for Offshore Structures: Cathodic Protection and High-Efficiency Coating*, 1st edn. (Elsevier, Boston, 2014), pp. 63
42. M.T. Lilly, S.C. Ihekwoaba, S.O.T. Ogaji, S.D. Probert, Prolonging the lives of buried crude-oil and natural-gas pipelines by cathodic protection. Appl. Energy **84**(9), 958–970 (2007)
43. www.farwst.com (2003)
44. www.rheem.com (2003)
45. C. Edeleanu, Corrosion control by anodic protection. Platin. Met. Rev. **4**(3), 86–91 (1960)
46. C.E. Locke, Anodic protection, in *Corrosion*, vol. 13, 9th edn. Metals Handbook (ASM International, Novelty, OH, 1987), p. 463
47. W. Zhao, Y. Zou, D.X. Xia, Z.D. Zou, Effects of anodic protection on scc behavior of X80 pipeline steel in high-pH carbonate-bicarbonate solution. Metall. Mater. **60**(2A), 1009–1013 (2015)
48. R. Walker, Anodic protection, in *Corrosion*, vol. 2. Corrosion Control, ed. by L.L. Shreir, R.A. Jarman, G.T. Burstein (Butterworth-Heinemann, Boston, 1994), p. 10:160
49. O.L. Riggs, C.E. Locke, *Anodic Protection: Theory and Practice in the Prevention of Corrosion* (Plenum, New York, 1981)
50. R. Walker, A. Ward, Metallurgical Review No. 137. Met. Mater. **3**(9), (1969) 143
51. M. Stern, Evidence for a logarithmic oxidation process for stainless steel in aqueous systems. J. Electrochem. Soc. **106**(5), 376–381 (1959)
52. Huang, Y.Z., Blackwood, D.J.: Characterisation of titanium oxide film grown in 0.9 % NaCl at different sweep rates. Electrochim. Acta **51**, 1099–1107 (2005)

9 Electrodeposition

Electrodeposition is an electrochemical deposition of metallic ions (cations) from electrolyte solutions containing at least one type of metal, such as copper. This process is based on electrochemical reduction reactions on substrates called cathodes. The reactions proceed through an electrochemical nucleation and growth mechanism for the formation of atom cluster on solid substrate surfaces during cathodic polarization at potentials $E < E_{corr}$ and current densities $i < i_L$.

From a nanotechnology point of view, details on electrochemical cells relative to atom cluster formation and related implications, as introduced in Chap. 4, are obtainable by using techniques like scanning tunneling microscopy (STM) and electrochemical scanning tunneling microscopy (ESTM). These techniques are very useful for observing and characterizing the shape and size of atom clusters (known as islands) and their spatial distributions on the atomic scale. Hence, electrodeposition at a nanoscale provides an extraordinary approach for nanofabrication of electronic devices [34].

In the engineering field, details on electrochemical cells are conventionally analyzed relative to mass production of electrodeposited metals. The production of pure metals and related implications are detailed in this chapter by using electrodeposition techniques at a macroscale. These techniques are known as electroplating, electrowinning, and electrorefining. Obviously, electrodeposition of a metal on a foreign substrate starts at a nanoscale and proceeds to reach a macroscale. This means that a high island density is initially needed for a successful atomic scaling process.

The engineering side of electrochemistry dictates that not only pure metal electrodeposition, as described by conventional methods, dominates the macroscale electrodeposition, but alloy electrodeposition is possible as a codeposition mechanism. The outcome of alloy electrodeposition is related to engineering applications, such as corrosion protection and mechanical deformation resistance under a loading mode. Nonetheless, electrodeposited alloys have superior mechanical properties

N. Perez, *Electrochemistry and Corrosion Science*,
DOI 10.1007/978-3-319-24847-9_9

than pure metals, and their usage at a large scale depends on engineering problems. The electrodeposition mechanism of alloys is anticipated to be a complex ionic process due to the interaction of ions on electrodes.

9.1 Disciplines in Metal Extraction

The objective of this section is to illustrate the application of electrochemical principles to electrolytically produce or win pure metals or refine electrodeposited metals containing impurities, which codeposit during production. Before describing details related to electrowinning and electrorefining a metal, it is adequate to schematically show in Fig. 9.1 a simplified block diagram of engineering fields involved in metal production:

- Geology is the science that studies the earth rocks, composition, and processes by which they change.

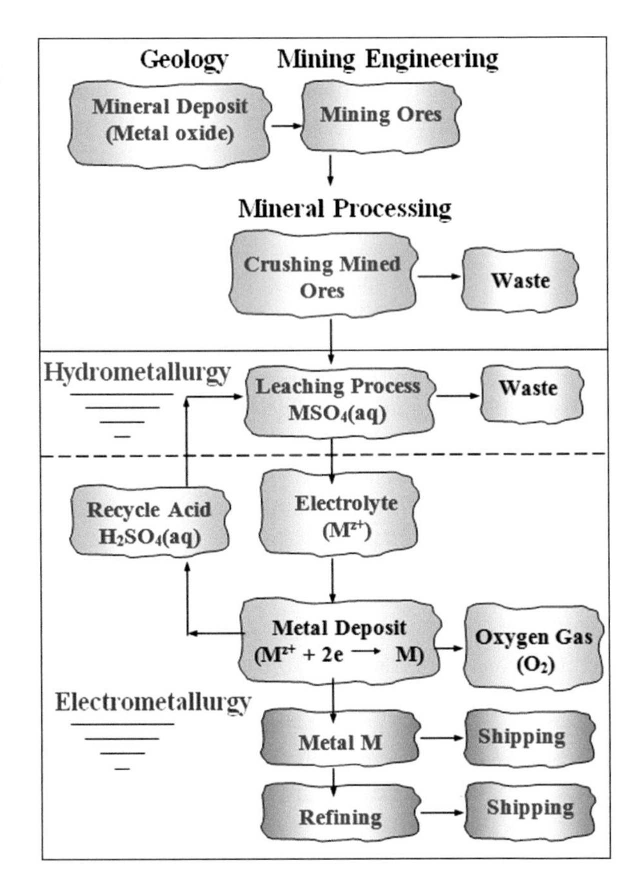

Fig. 9.1 Block diagram for electrowinning of a metal M

- Mining engineering is a discipline that extracts the earth minerals from their geological deposits or mineral reserve.
- Mineral processing is a discipline that crushes mineral rocks and processes the separation of minerals (ores). This discipline is also known as mineral dressing.

9.2 Metal Extraction

Extractive metallurgy is one of the branches of metallurgical engineering that uses natural mineral deposits known as ores. An ore is a naturally occurring mineral rock containing a low or high concentration of a metal oxide. The method of metal extraction strongly depends on the metal reactivity to chemical dissolution. For instance, iron (Fe) may be extracted by electrochemical reduction from electrolyte solutions, while aluminum (Al) is extracted by electrolysis. Common ores for extracting their principal metals are bauxite (Al_2O_3), rutile (TiO_3), hematite (Fe_2O_3), chalcopyrite ($CuFeS_2$), and so on.

Extractive metallurgy is divided into different areas as shown in Fig. 9.2 and described below [3, 9, 18, 36, 40, 49, 52].

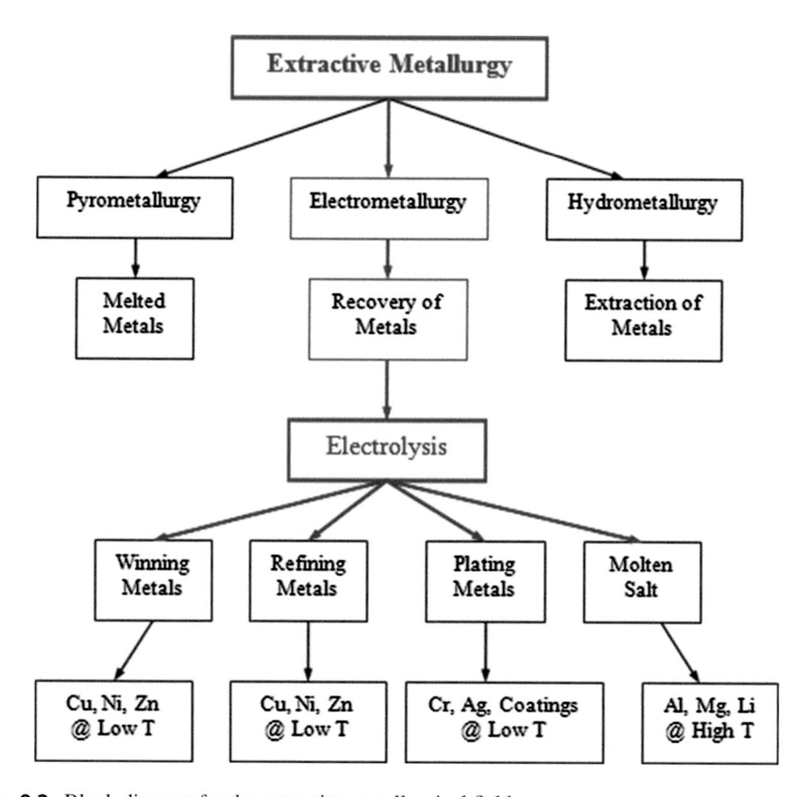

Fig. 9.2 Block diagram for the extractive metallurgical field

Pyrometallurgy is a technique used for melting metals at high temperatures. The melted metals are cast into several shapes using slow solidification or into ribbons or powders by rapid solidification. This technique deals with chemical reactions between different solids, liquids, and gases at high temperatures for extracting metals by converting metal sulfides into metal oxides and metal oxides into metals. The Ellingham diagram (see Chap. 10) is used to predict high-temperature chemical reactions through calcining and roasting operations.

Hydrometallurgy is an engineering discipline for extracting metals from their ores using solvent extraction (*SX*) and ion exchange (*IE*). Thus, the result is a leaching solution containing a desired metal cations (M^{z+}) and some impurities. Leaching is an aqueous solution containing valuable metal ions suitable for extraction, which may require a process step like precipitation, distillation, or pyrometallurgy. Common leading solutions for hydrometallurgical treatment include copper, nickel, zinc, cobalt, bismuth, lead, uranium, and so on. For instance, solvent extraction is a common process for hydrometallurgical treatment of copper since it is carried out at ambient temperatures. In this particular case, copper ions are reduced to copper metal from the solvent extraction aqueous solution.

Electrometallurgy is an electrochemical process used to extract metal ions from aqueous solutions using electricity. In other words, this engineering discipline is used for recovering or winning some metals from leaching solutions exposed to an aqueous electrolysis. Molten-salt electrolysis is used for recovering aluminum because of hydrogen evolution problems. Nonetheless, the essential device in electrometallurgy is an electrolytic cell that converts external electrical energy into chemical energy or chemical work for the proper transfer of electric charge between anodes and cathodes in ionically conducting electrolyte containing the metal ions. As a result, metal ions are electrodeposited or reduced on cathode electrodes as atoms. Furthermore, electrometallurgy is also divided into subfields shown in Fig. 9.2. Each subfield has its own unique cell characteristics, but they operate under similar electrolysis.

Electrowinning is also an electrodeposition technique for recovering a metal M, such as Cu, Ni, Zn, Ti, Pb, etc. from their low-grade ores. The metal is recovered by electrodeposition in the form of cathodes, which are normally rectangular shapes (plates). If rotating cylinders and disks are used, the final electrodeposition product can be in the form of ribbons and powders, respectively.

Electrorefining (ER) is used for refining electrowon metals to their purest form by dissolving the electrowon cathodes into solution and, subsequently, depositing on new cathodes.

Electroplating (EP) is employed for electrodeposition of metal coatings onto another metal or alloy mainly for protection against corrosion.

Molten-salt electrolysis (MSE) is a high-temperature electrowinning operation for producing metals that cannot be electrowon due to water decomposition, which promotes hydrogen evolution at the cathode before metal deposition occurs. Metals produced or recovered using this technique are Al, Mg, Be, Ce, Na, K, Li, U, Pu, and the like. The electrolysis process may yield different results. For instance, the electrolysis of molten sodium chloride ($Na^{+}CL^{-}$) yields sodium metal (Na) and chlorine gas (Cl_2), while the electrolysis of aqueous sodium chloride gives hydrogen (H_2), chlorine (Cl_2), and aqueous sodium hydroxide (($NaOH$), which remains in solution.

9.3 Electrowinning

Conventional electrowinning (EW), as well as electrorefining, utilizes rectangular electrodes (planar starter sheets) having surface areas that occupy the space 1 mx1 m approximately. The anodic and cathodic electrodes are suspended vertically and alternatively at a distance $3\,\text{cm} \leq x \leq 10\,\text{cm}$ in the cells. The electrodes in one cell are connected in parallel, and the cells are connected in series as shown in Fig. 9.3. This is a classical electrode-cell arrangement for reducing the net cell potential drop and ohmic resistance [2, 43].

Electrowinning is an electrochemical process used to reduce (win or deposit) metal cations on the surface of a cathode sheet from an aqueous solution prepared by a leaching chemical process, which is mainly based on metal cations M^{z+}, sulfuric acid (H_2SO_4), and H_2O. This type of solution is widely used as an electrolyte containing dissolved metal ions in low concentrations. Recent advances in research have uncovered the usefulness of organic substances as part of the electrolyte. One of these substances is the diethylenetriamine (*DETA*), which is used to extract or leach *Zn*, *Ni*, and *Cu* from their hydroxide form (sludge) and reacts to reject *Fe* and *Ca*, but it forms specifically *Cu*:*DETA* complexes [17, 27, 46]. Usually, a small amount of salt, such as $NaCl$ or $NaSO_4$, is added to the electrolyte to enhance ionic conductivity, increase pH values, and reduce the electrolyte resistance, which, in turn, reduces the cell operating potential and energy consumption.

In general, electrowinning (EW) is commonly used to produce atomic *Zn*, *Ni*, and *Cu* from their ionic solutions. The common reactions involved in producing metals by electrolytic deposition (reduction), provided that impurity codeposition is absent and that the electrochemical cell walls and anodes do not dissolve, are:

(1) Leaching steps for producing metal ions M^{z+}:

$$MO + H_2SO_4 \rightarrow MSO_4 + H_2O \tag{a}$$

$$MSO_4 \rightarrow M^{z+} + SO_4^{-2} \tag{b}$$

Here, MO stands for metal oxide, and the most common metal sulfates are $CuSO_4$, $ZnSO_4$, and $NiSO_4$.

(2) Redox reaction: Since electrowinning (EW) is the electrochemical process used to especially recover pure metals from solutions by electrolysis, the main reactions at the electrolyte/electrode interface for a metal cation M^{2+} with an oxidation state of $z = 2$ are

$$2H_2O \rightarrow 4H^+ + O_2 + 4e^- \qquad \text{(anode)} \tag{c}$$

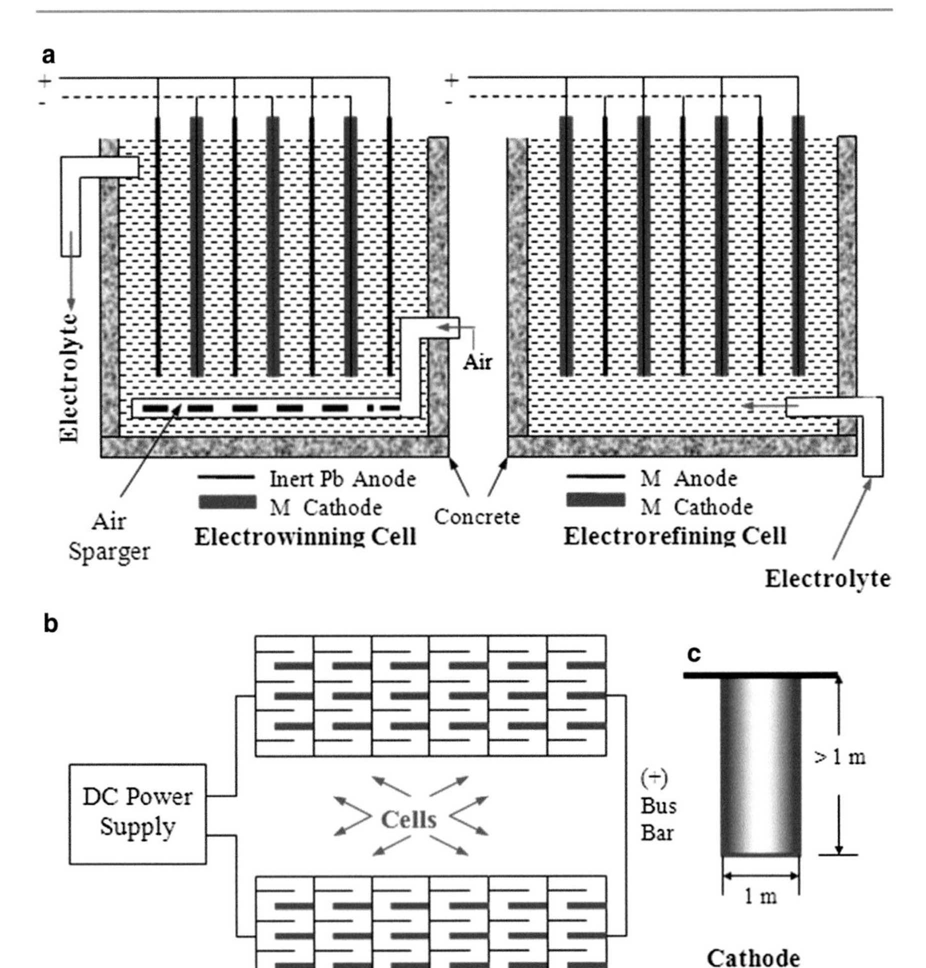

Fig. 9.3 Schematic electrowinning tank cell layout. (**a**) Cross-sectional view, (**b**) top view

$$2M^{2+} + 4e^{-} \rightarrow 2M \qquad \text{(cathode)} \tag{d}$$

$$2M^{2+} + 2H_2O \rightarrow 2M + 4H^{+} + O_2 \qquad \text{(redox)} \tag{e}$$

Note that water molecules decompose to promote oxygen evolution as O_2 gas. Thus far it has been assumed that the electrolytically deposit metal M has suitable characteristics for easy removal from the starter cathode sheets and that hydrogen evolution does not occur; otherwise, metal deposition is obstructed.

(3) Recycling step for producing sulfuric acid:

$$SO_4^{-2} + 2H^+ \rightarrow H_2SO_4 \tag{f}$$

The electrochemistry of electrodeposition is influenced by several factors that must be controlled for producing a metal very efficiently. These factors are:

- Electrolyte composition. Impurities may codeposit.
- Electrolyte conductivity, which can be enhanced by adding a small amount of a salt (*NaCl*, *KCl*).
- The degree of acidity is controlled by pH which determines. In fact, control of pH is desired since the rate of hydrogen evolution may be faster than the rate of metal deposition, and therefore, metal deposition is impaired, and current density for deposition may be reduced.
- Temperature T which is usually less than 70 °C. Apparently, at higher temperatures the cell potential E decreases. This temperature effect can be determined using the Nernst potential equation, Eq. (3.51).
- Applied current and potential. These variables determine the cell power, which, in turn, is linearly proportional to energy consumption.
- Mass transfer by diffusion, migration, and convection. The mass flow rate has to be controlled for an optimum electrodeposition.
- Electrode surface roughness which must be reduced for better electrolytic deposition.
- Plant-cell layout as schematically shown in Fig. 9.3 for conventional electrowinning and electrorefining.
- The electrolyte which consists of at least three components: metal ion source, acid solution, and salt (for adjusting the electrolyte conductivity). The electrolyte is

$$Electrolyte = MSO_4 \cdot nH_2O + H_2O + H_2SO_4 + \ldots\ldots \tag{g}$$

- Starter electrodes. Conventionally, lead (*Pb*) is used as the starter anodic electrode sheets connected to the power supply positive terminal. The cathode sheets can be made out of metal M, stainless steel, or titanium electrode sheets.

Furthermore, anodic reactions occur on the anode-electrolyte interface according to Eq. (c) having a standard potential of $E_a^o = -1.23$ V (Table 2.2) for the oxidation reaction to proceed. The electrons produced by Eq. (9.3) are driven through the external circuit by the DC power and are consumed by the cathodic reactions at the cathode-electrolyte interfaces. Hence, electrodeposition proceeds according to the reaction defined by Eq. (e).

For an optimum production, the electrolyte direct-flow or air-sparging agitation system must be determined because it enhances the current density. However, increasing the solution flow rate may induce a decrease in electrodeposition due to lack of sufficient time for the metal ions to proceed according to the nature of the reduction reactions. At this point, it is uncertain if the current density profile is linear or nonlinear. This is an important aspect in modeling an electrochemical cell for simulation purposes.

In addition, the electrolyte must be kept replenished as electrolysis proceeds; otherwise, the concentration of the metal cations M^{z+} drops very significantly, and the potential vanishes, $E \rightarrow 0$ [49]. In fact, the potential (voltage) E and current I that must be applied to an electrowinning cell depend on conductivity, pH, temperature, and the concentration of all the different species in solution. For metal deposition to occur, in practice, the applied potential must be greater than the sum of the standard potential, $E > E^o$. This requires that the activation and polarization overpotentials (η_a and η_c), ohmic resistance (E_{ohm}) due ionic resistance (E_{IR}), and junction potentials (ϕ_j) must be included in order to determine the total potential across the electrodes. Thus,

$$E = E^o + \eta_a + |\eta_c| + E_{ohm} \tag{9.1}$$

The ohmic effect is defined as $E_{ohm} = E_{IR} + \phi_{j.}$ For example, in copper electrowinning (EW) the applied potential must be

$$E > \left(E^o_{a,H_2O} + E^o_{Cu}\right) = -1.229\text{ V} + 0.337\text{ V} \tag{a}$$

$$E > E^o = -0.892\text{ V} \tag{b}$$

Usually, the potential is in the range of $1.5\text{ V} < E < 4.00\text{ V}$. This is a practical potential range used in conventional Pb anodes and gas diffusion anodes in electrochemical operations [5, 6, 19, 22, 49].

Using gas diffusion anodes [27] and dimensionally stable anodes (DSA®) [5, 6] in electrowinning (EW) of zinc yields high current efficiency in the range of $90\,\% \leq \epsilon \leq 100\,\%$, but the current densities are higher than most conventional planar Pb-anode practices. The common current efficiency range for conventional practices is $80\,\% \leq \epsilon \leq 90\,\%$.

During the electrolytic process, current flows at a suitable rate, but overpotential drop emerges due to connectors and wiring system and electrolyte electric resistance at the cathode-electrolyte interface. Also, overpotential may occur when concentration polarization is due to mass transfer by diffusion [9]. This implies that, in electrowinning, the current density has a significant effect on the rate of metal deposition. The goal is to operate an electrochemical cell at sufficiently high current density so that the rate of metal ionic electrodeposition overcomes the rate of diffusion of other ions through the electrolyte.

In EW and ER, an aggregation of acid drops suspended above the electrolytic cells is a phenomenon known as "acid mist," which comes from bursting bubbles of hydrogen and even oxygen in acid solutions. Therefore, spraying acid drops into the atmosphere is detrimental to health and equipment. One can alleviate the hazardous acid mist by implementing a ventilation system [9, 31] or using close-ended cylindrical cells [14].

As current flow heat generation occurs due to electrolyte resistance and electrical connection resistance. Therefore, a heat exchanger is required to overcome overheating the electrolyte and avoid cathode deterioration and oxidation of anodes. In order to alleviate overheating the electrolyte, a temperature range of $30\,°C \leq T \leq 45\,°C$ is adequate for most EW operations [9]. Overheating can also deteriorate electrical connections and may even cause fire, which is rare but possible.

Once enough metal deposition is obtained, the cathodes are mechanically removed from the cells and peeled off (stripped off), provided that a weak atomic bonding exists at the sheet-deposit interface. Then, the cathode starter sheets are recycled [2, 9, 18, 49].

Figure 9.4 shows a conventional laboratory-scale electrowinning copper-plated cathode exhibiting the common red color of pure copper. Commonly, surface defect sites for preferred initial electrodeposition include vacancies, kinks, ledges, terraces, and individual adatoms (atoms adsorbed on the surface). Yet the electrodeposition of copper from a $CuSO_4$-based electrolyte was done on a clean and shiny stainless steel starter sheet, and it exhibited a uniform surface, indicating that the cell operated near optimum conditions at room temperature since the reduced copper surface is grainy. Despite that the resultant electrodeposited copper surface is smooth, it lacks shininess. Nonetheless, this finished product is not brilliant and shiny red enough to be considered as a high-quality electrodeposited copper for electronic circuit applications. This electrodeposited copper is considered as a uniform medium-quality finished product.

Commonly, an electrowinning cell having several stationary planar electrodes requires that the electrolyte mass flow rate be under the laminar flow theory, as predicted by the Reynolds number. The optimum mass flow rate depends on the cell design shape and related cell dimensions.

Fig. 9.4 Laboratory electrowinning cell for producing copper

Fig. 9.5 ASARCO®. (**a**) Copper electrowinning tank cells and (**b**) cathodes. Ref. http://www.asarco.com/products/copper/resources/

Figure 9.5 illustrates EMEW® industrial cells for producing uniform high-quality pure copper using conventional planar electrodes. Notice the industrial size of cathode electrodes and the shininess and smoothness of electrodeposited copper, which is an excellent final product that exhibits the brightness and shininess of red copper.

In addition, the anode and cathode plates shown in Fig. 9.5 are arranged in parallel so that the electrolyte motion is steady along with slow axial flow rate within the laminar flow limits. This tank cell arrangement induces uniform mass transfer of copper ions, smooth electrodeposition, and high cell efficiency.

From an engineering standpoint, conventional electrowinning technology in the metallurgical industry consists of electrolytic cells arranged in series, having planar electrodes (plates) immersed in electrolytes containing high concentrations of metallic ions, such as copper, nickel, and so on. However, electrowinning of a particular element from low-concentration electrolytes is achievable by modifying the electrolytic cell design. Below is a brief description of a particular electrowinning technology based on stationary cylindrical electrodes.

Figure 9.6 shows an image along with its schematic overview of a vertical electrowinning stationary cell, which consists of a hollow cylindrical cathode having an inner solid cylindrical anode. This cell was designed by Electrometals Technologies Limited (EMEW®) for reducing metal ions from a relative fast-circulating electrolyte at a high upward-spiral flow rate [14]. This unique cell design enhances the ionic mass transfer over axial flow and electrodeposition process, leading to high cell efficiency. Several cells can be connected in series for a large mass production of *Cu*, *Zn*, *Ni*, *Ag*, *Au*, and *Pt* from a variety of concentrations at current densities in the range of $400\,\mathrm{A/m^2} < i < 700\,\mathrm{A/m^2}$, potential at $E < 4.00\,\mathrm{V}$, and current efficiency at $\epsilon > 85\,\%$.

The EMEW® cell consists of concentric cylindrical electrodes enclosed by a *PVC* pipe, and it has several advantages over the conventional planar-electrode cells. For instance, the cell is inexpensive, the "acid mist" problem encountered

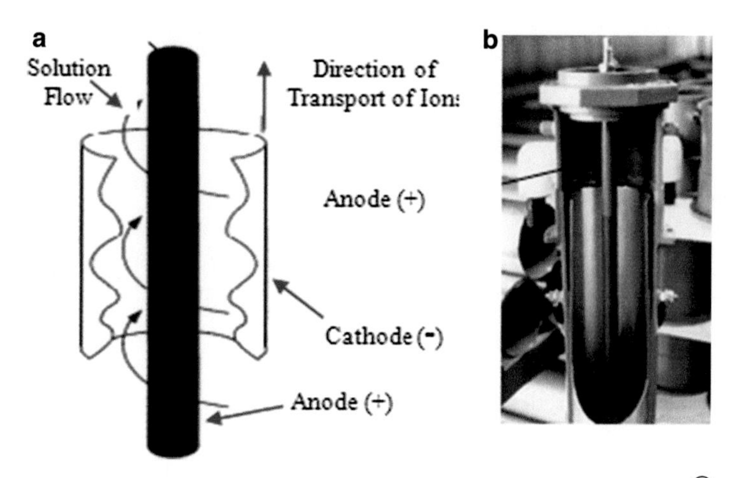

Fig. 9.6 Nonconventional electrowinning cells. (**a**) Schematic and (**b**) actual EMEW® cylindrical cells for a variety of metal recovery [14]

in traditional open-tank cells is eliminated since the cell has closed ends, and metal recovery from alkaline and chloride-containing solutions is suitable at low energy consumption.

Fig. 9.6 Nonconventional electrowinning cells. (a) Schematic and (b) actual EMEWR cylindrical cells for a variety of metal recovery [14]

9.4 Mathematics of Electrowinning

This section is mainly concerned with simple formulae based on Faraday's law of electrolysis and energy consumption for producing high-purity metals from electrolytes through an aqueous electrolysis. Mathematics is simple to understand and easy to use in basic engineering calculations. However, complications may arise when mass transfer by diffusion and migration of solutes are coupled. Diffusion itself is rather a complex subject that depends on the type of diffusion process to be analyzed. On the other hand, the mathematics of mass transfer by convection is included in a later section as a separate topic.

9.4.1 Mass Transfer Effects

Consider again the general half-cell reaction defined by Eq. (5.52). For a reversible reaction,

$$O + 2e^- \rightleftharpoons R \tag{9.2}$$

Henceforth, "O" represents the oxidized species and "R" the reduced species. The rate constants for this reaction at the electrode surface are [4]

$$k_f = k_f^{'} C_O(0,t) = \frac{i_c}{zF} \quad \text{(cathodic)} \tag{9.3a}$$

$$k_r = k_r^{'} C_R(0,t) = \frac{i_a}{zF} \quad \text{(anodic)} \tag{9.3b}$$

Then the net rate constant becomes

$$k_{net} = k_f - k_r = k_f = k_f^{'} C_O(0,t) - k_r^{'} C_R(0,t) \tag{9.3c}$$

$$k_{net} = \frac{i}{zF} \tag{9.3d}$$

from which the net current density takes the form

$$i = i_f - i_r = zF\left[k_f^{'} C_O(0,t) - k_r^{'} C_R(0,t)\right] \tag{9.4a}$$

$$i = i_c - i_a = zF\left[k_f^{'} C_O(0,t) - k_r^{'} C_R(0,t)\right] \tag{9.4b}$$

Assume that $k_f^{'}$ and $k_r^{'}$ have an Arrhenius form so that

$$k_f^{'} = k_o \exp\left[\frac{\alpha zFn}{RT}\right] \tag{9.5a}$$

$$k_r^{'} = k_o \exp\left[-\frac{(1-\alpha) zFn}{RT}\right] \tag{9.5b}$$

where $n = E - E^o$ is the overpotential, E is the applied potential, and E^o is the standard potential for a metal M.

Combining Eqs. (9.5) and (9.4b) yields

$$i = zFk_o\{C_O(0,t) \exp\left[\frac{\alpha zFn}{RT}\right] - C_R(0,t) \exp\left[-\frac{(1-\alpha) zFn}{RT}\right] \tag{9.6a}$$

Similarly, in the bulk

$$i = zFk_o\{C_O(\infty,t) \exp\left[\frac{\alpha zFn}{RT}\right] - C_R(\infty,t) \exp\left[-\frac{(1-\alpha) zFn}{RT}\right] \tag{9.6b}$$

At equilibrium, $i = i_c - i_a = 0$, $i_c = i_a$, and Eq. (9.6) gives

$$C_O(0,t) \exp\left[\frac{\alpha zFn}{RT}\right] = C_R(0,t) \exp\left[-\frac{(1-\alpha) zFn}{RT}\right] \tag{9.7a}$$

$$C_O(\infty,t) \exp\left[\frac{\alpha zFn}{RT}\right] = C_R(\infty,t) \exp\left[-\frac{(1-\alpha) zFn}{RT}\right] \tag{9.7b}$$

Then,

$$\frac{C_O(0,t)}{C_R(0,t)} = \exp\left[\frac{zFn}{RT}\right] \tag{9.7c}$$

$$\frac{C_O(\infty,t)}{C_R(\infty,t)} = \exp\left[\frac{zFn}{RT}\right] \tag{9.7d}$$

Substitute $n = E - E^o$ into Eq. (9.7c), and solve E

$$E = E^o + \frac{RT}{zF}\ln\frac{C_O(0,t)}{C_R(0,t)} \tag{9.8}$$

Raising Eq. (9.7d) to the $-\alpha$ power gives

$$\left[\frac{C_O(\infty,t)}{C_R(\infty,t)}\right]^{-\alpha} = \exp\left[-\frac{\alpha zFn}{RT}\right] \tag{9.9}$$

At equilibrium, $i_o = i_c = i_a$, and Eq. (9.6b) yields

$$i_o = zFk_oC_O(\infty,t)\exp\left[\frac{\alpha zFn}{RT}\right] \tag{9.10a}$$

$$i_o = zFk_oC_R(\infty,t)\exp\left[-\frac{(1-\alpha)zFn}{RT}\right] \tag{9.10b}$$

Combining Eqs. (9.9) and (9.10a) along with $n = E - E^o$ defines i_o as

$$i_o = zFk_o\left[C_O(\infty,t)\right]^{1-\alpha}\left[C_R(\infty,t)\right]^{\alpha} \tag{9.11}$$

Dividing Eq. (9.6a) by (9.11) and using Eqs. (9.7d) and (9.79), one gets the current density with mass transfer effects [4]:

$$i = i_o\left\{\frac{C_O(0,t)}{C_O(\infty,t)}\exp\left[\frac{\alpha zFn}{RT}\right] - \frac{C_R(0,t)}{C_R(\infty,t)}\exp\left[-\frac{(1-\alpha)zFn}{RT}\right]\right\} \tag{9.12}$$

In hydrodynamics, mass transfer by forced convection into the electrochemical cell containing static electrodes is an efficient method for achieving a homogeneous electrolyte solution. In this case, the metal ion concentration $C_O(x,t)$ is uniform, and the current-potential equation, defined by Eq. (9.12), becomes independent of mass transfer effects. Therefore, all $C_O(x,t)$ and $C_R(x,t)$ terms in Eq. (9.12) cancel out, and the resultant expression is the Butler–Volmer classical equation used to polarize an electrochemical cell for reducing metal ions from solution without mass transfer effects. Hence,

$$i = i_o \left\{ \exp\left[\frac{\alpha z F n}{RT}\right] - \exp\left[-\frac{(1-\alpha) z F n}{RT}\right] \right\} \tag{9.13}$$

Logically, a high flow rate of electrolyte or a well-stirred solution induces $C_O(x, t)$ to be uniform at any point within the electrolyte.

Application of Eq. (9.12) or (9.13) in practical venues includes electrowinning of metal ions, where $C_O(0, t) = C_{M^{z+}}(0, t)$ and $C_R(\infty, t) = C_M(\infty, t)$. For example,

$$Cu^{2+} + 2e^- \rightleftarrows Cu \tag{a}$$

where $C_O(0, t) = C_{Cu^{z+}}(0, t)$ and $C_R(\infty, t) = C_{Cu}(\infty, t)$.

9.4.2 Faraday's Law of Electrolysis

The law of electrolysis states that the amount of a species j that is gained or liberated on the electrode surface during the electrochemical process is directly proportional to the quantity of electric charge (Q) that passes through the aqueous electrolyte. The electric charge is defined as the electric current per unit time as previously defined by Eq. (5.99). For a constant current flowing through the electrolyte, the electric charge is

$$\int dQ = I \int dt \tag{9.14a}$$

$$Q = It \tag{9.14b}$$

The theoretical amount (W_{th}) of a species j is defined as the mass gained or liberated due to the flowing electric current during a period of time t. Thus,

$$W_{th} = \lambda_e Q \tag{9.14c}$$

$$W_{th} = \lambda_e It \tag{9.14d}$$

where λ_e = electrochemical constant (g/A)

Q = electric charge (C = A s = J/V)

In addition, λ_e is defined by

$$\lambda_e = \frac{A_w}{z q_e N_A} \tag{9.14e}$$

$$\lambda_e = \frac{A_w}{zF} \tag{9.14f}$$

where F = Faraday's constant

$F = q_e N_A = 96{,}500$ (C/mol = A s/mol = J/mol V)
q_e = electron charge = $1.6022x10^{-19}$ (C/electrons)
N_A = Avogadro's number = $6.02213x10^{23}$ ions/mol
A_w = atomic weight (g/mol)

Combining Eqs. (9.14d) and (9.14f) yields the theoretical weight gain

$$W_{th} = \frac{ItA_w}{zF} \tag{9.15a}$$

$$W_{th} = \frac{QA_w}{zF} \tag{9.15b}$$

Furthermore, the electric current that evolved during electrolysis has been derived. Recall that the electrical conduction is a mass transfer phenomenon within which electrons and ions carry the electric charge through their mobilities. Both positively charged ions (cations) and negatively charged ions (anions) flow in the opposite direction. Thus, cations and electrons move to the negatively charged cathodic electrode surface (−), and the anions move toward the positively charged anodic electrode surface (+).

Example 9.1. *Calculate* ***(a)*** *the number of electrons and* ***(b)*** *the weight of electrodeposited nickel on a cathode electrode when a current of* 1 *A flows through a nickel sulfate* ($NiSO_4$) *electrolyte for 1 h at a temperature* $T > 25\,°C$. *Assume an ideal electrolysis process for recovering nickel cations from solution at* 100 % *current efficiency.*

Solution. **(a)** *The reduction reaction for the electrodeposition is* $Ni^{2+} + 2e = Ni$. *Using the given data yields*

$$Q = It = (1\text{ A})\,(3600\text{ s}) = 3600\text{ C}$$

$$N_e = \frac{Q}{q_e} = \frac{3600\text{ C}}{1.6022x10^{-19}\text{ C/electrons}}$$

$$N_e = 2.25x10^{22}\text{ electrons}$$

$$N_{Ni} = \frac{N_e}{z} = \frac{2.25x10^{22}\text{ electrons}}{2\text{ electrons/atoms}}$$

$$N_{Ni} = 1.125x10^{22}\text{ atoms}$$

(b) *The weight gained is*

$$W_{th} = \frac{QA_w}{zF} = \frac{(3600\text{ C})\,(58.71\text{ g/mol})}{(2)\,(96500\text{ C/mol})}$$

$$W_{th} \approx 1.10\text{ g}$$

9.4.3 Production Rate

Actual electrochemical cells do not operate at 100 % efficiency because of imperfections in the wiring system and possible contaminants or impurities in the electrolyte. Therefore, Faraday's law of electrolysis, Eq. (9.14), must include the cathodic current efficiency parameter for determining the amount of electrolytic metal deposit or weight gain on cathodes. Normally, the current efficiency at the cathodes is less than 100 % due to hydrogen evolution or codeposition of impurities. Hence, the actual weight gain together with the current efficiency and total cathode surface area becomes

$$W = \frac{\epsilon I A_w t}{zF} = \frac{\epsilon i A_s A_w t}{zF} \tag{9.16}$$

$$\epsilon = \frac{W}{W_{th}} = \frac{i}{i_{th}} < 1 \tag{9.17}$$

$$A_s = 2NA_c \tag{9.18}$$

where ϵ = current efficiency

I = current (A)
A_w = atomic weight or molar mass (g/mol)
t = time (s)
z = valence
$F = 96{,}500\ \text{C/mol}\ (= \text{A s/mol} = \text{J/mol V})$
A_s = total cathode surface area (cm^2)
N = number of cathodes
A_c = cathode surface area (cm^2)

Conventional electrowinning cells utilize planar *Pb*-base anodes at current densities in the range of $200\ \text{A/m}^2 < i < 500\ \text{A/m}^2$ [5, 49]. On the other hand, advanced electrowinning cells use hydrogen gas diffusion (HGD) anodes at $2\ \text{kA/m}^2 < i < 8\ \text{kA/m}^2$ [1, 6, 19, 22]. Normally, the current density and energy efficiency (ϵ^*) are less than 100 % due to drawbacks mainly caused by ohmic potential drop and oxygen evolution reaction. The energy efficiency can be defined by

$$\epsilon^* = \frac{\epsilon E_{cell}}{E} < 1 \tag{9.19}$$

where E is the applied potential. The potential E is the sum of electrode potentials and overpotentials, including the ohmic effect [17, 49]. Hence,

$$E = E_a + E_c + \eta_a + \eta_c + IR_s \tag{9.20}$$

Also, E can be defined as [36]

$$E = \frac{IL\rho_x}{A_s} = iL\rho_x \tag{9.21}$$

where ρ_x = solution resistivity (Ω cm)

$\rho_x = 5\ \Omega$ cm resistivity for Cu^{2+} in solution [49]
L = anode-to-cathode distance (cm)

According to Ohm's law, the distribution of the solution resistance can be deduced from Eq. (9.21) as

$$R_s = \frac{E}{I} = \frac{L\rho_x}{A_s} \tag{9.22}$$

Another important parameter for assessing the performance of an electrowinning cell is the production rate, which is defined by

$$P_R = \frac{dW}{dt} = \frac{\epsilon i A_s A_w}{zF} \tag{9.23}$$

In order to analyze an electrowinning cells from economical point of view, the power and the energy consumption needed to operate the cell are, respectively,

$$P = EI \tag{9.24}$$

$$\gamma = P/P_R = \frac{zF}{A_w}\frac{E}{\epsilon} \tag{9.25}$$

The power P is converted to kW and γ to kWh/kg. Thus, Eq. (9.25) is a simple and, yet, essential in analyzing electrowinning processes. One can observe that this is one expression that depends on two independent variables. The behavior of this equation is best understood by plotting a surface (mesh) in a three-dimensional space as shown in Fig. 9.7.

The surface indicates that γ decreases with decreasing E and increasing ϵ. The optimum metal recovery can be achieved when $\epsilon = 1$ and $E = 1$ V so that the energy consumption becomes $\gamma = 0.82$ kWh/kg for Zn, $\gamma = 0.84$ kWh/kg for Cu, and $\gamma = 0.91$ kWh/kg for Ni. However, real electrowinning operations are seldom at 100 % efficiency due to many design factors, such as electrical contacts, electrolyte resistance, and the like. Nevertheless, an idealized electrowinning cell must operate at very low energy consumption so that production cost is kept as low as possible.

Since electrometallurgical processes operate on a continuous schedule, a continuous supply of fresh electrolyte from a leaching plant to the cells places a very important role in producing metals from solutions. Therefore, the volume flow rate (F_r) of the electrolyte entering the cells based on Faraday's law of electrolysis must be controlled mechanically. A pump in the order of F_r is needed for circulating the electrolyte through the electrodes to the leaching plant for recycling purposes and for avoiding an increase in the electrolyte acidity.

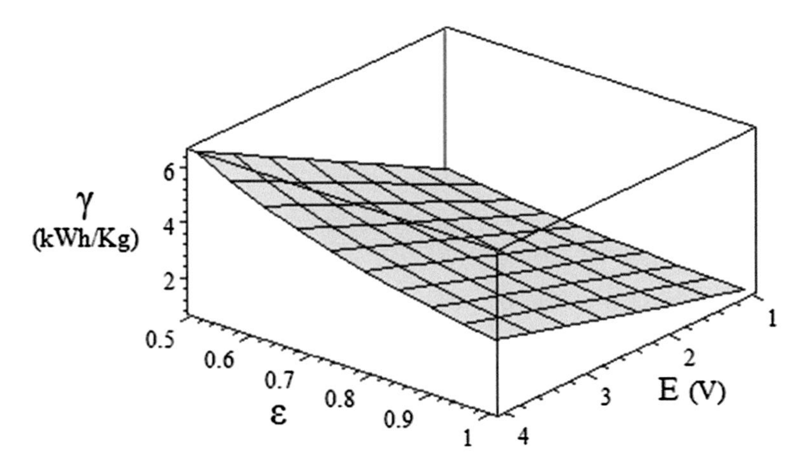

Fig. 9.7 Theoretical energy consumption surface for electrowinning zinc as per Eq. (9.25)

The electrolyte volume flow rate (F_r) is another parameter that must be controlled during the electrolysis of a particular metal under forced laminar flow. Thus,

$$F_r = \frac{P_R}{C_j} = \frac{P_R}{a_j A_{w,j}} \tag{9.26}$$

where C_j = bulk concentration of metal cations M^{z+} (g/l)

a_j = bulk activity of metal cations M^{z+} $\left(\text{mol/cm}^3\right)$
$A_{w,j}$ = atomic mass of metal M (g/mol)
F_r is normally in units of l/ min

Common electrolyte constituents are metal sulfate, water, sulfuric acid, and a small amount of salt for enhancing the electrolyte electric conductivity (K_c). Hence,

$$\textbf{Electrolyte} = MSO_4 \cdot nH_2O + H_2O + H_2SO_4 + NaCl. \tag{9.27}$$

For a fixed total volume of the electrolyte, the weight of hydrated metal sulfate and sodium chloride and volume of sulfuric acid can be determined, respectively, using the following formulae:

$$W_{MSO_4 \cdot nH_2O} = \left(VC_j\right) \frac{A_{w,MSO_4 \cdot nH_2O}}{A_{w,M}} \tag{9.28}$$

$$W_{NaCL} = VC_{NaCl} \tag{9.29}$$

$$V_{H_2SO_4} = \left(VC_{H_2SO_4}\right) / \rho_{H_2SO_4} \tag{9.30}$$

where V = total volume of electrolyte (l)

$\rho_{H_2SO_4}$ = density of sulfuric acid $\left(g/cm^3 \text{ or } g/l\right)$
$C_{H_2SO_4}$ = concentration of sulfuric acid (g/l)
$A_{w,MSO_4 \cdot nH_2O}$ = molecular weight of metal sulfate (g/mol)
$A_{w,M}$ = atomic mass of metal M (g/mol)

Once the electrolyte solution is prepared and circulated through the cell electrodes at a particular flow rate and the electric circuit is completed, the electrochemical deposition of metal M on metal starter sheets involves charge transfer and electron flow toward the cathode surfaces. However, the electrolyte conductivity can be adjusted by adding salt to the electrolyte.

Example 9.2. *Calculate the weight of hydrated copper sulfate* ($CuSO_4 \cdot 5H_2O$) *to prepare a* 10-*l electrolyte containing* 35 g/l *of* Cu^{2+} *and* 180 g/l *of* H_2SO_4. *The density of* H_2SO_4 *is* 1.80 g/cm^3 (1800 g/l). *Ignore additions of NaCl.*

Solution.

$$M = Cu$$

$$MSO_4 \cdot nH_2O = CuSO_4 \cdot 5H_2O$$

$C_{Cu} = 35\,g/l = 5.51x10^{-4}\,mol/cm^3$ $\quad A_{w,M} = 63.55\,g/mol$
$C_{Cu} = 35\,g/l = 5.51x10^{-4}\,mol/cm^3$ $\quad A_{w,M} = 63.55\,g/mol$
$C_{H_2SO_4} = 180\,g/l$ $\quad A_{w,MSO_4 \cdot nH_2O} = 249.55\,g/l$
$\rho_{H_2SO_4} = 1800\,g/l$ $\quad V = 10\,g/l$

Using Eqs. (9.28) and (9.30) yields

$$W_{MSO_4 \cdot nH_2O} = (VC_{o,M})\,(A_{w,MSO_4 \cdot nH_2O})\,/\,(A_{w,Cu}) = 1374.39\,g$$

$$V_{H_2SO_4} = (VC_{H_2SO_4})\,/\,\rho_{H_2SO_4} = 1.00\,l$$

$$V_{H_2O} = 10.00\,l - 1.00\,l = 9.00\,l$$

Mix the components and stir until a homogeneous solution is obtained.

Example 9.3. *A circulating acid system is important in electrowinning copper in order to have a continuous deposition process. However, if the cell is not replenished with fresh solution, the cathode potential vanishes. Based on this information, determine* **(a)** *the minimum concentration of* Cu^{2+} *in* g/l *that can be recovered at* 35 °*C and* 101 *kPa.* **(b)** *When will hydrogen evolution occur?*

Solution.

(a) *Minimum concentration of* Cu^{2+} *in* g/l:

$$Cu^{2+} + 2e = Cu \qquad @\ E^o = 0.337\,V$$

Solving the Nernst equation, Eq. (3.31), for the concentration yields

$$E = E^o - (RT/zF)\ln K_{sp} \ \ \& \ \ K_{sp} = a_{Cu}/a_{Cu^{2+}} = 1/a_{Cu^{2+}}$$
$$0 = 0.337V + (0.0133)\ln a_{Cu^{2+}}$$
$$a_{Cu^{2+}} = 9.31x10^{-12}\ \text{mol/l}$$
$$C_{Cu^{2+}} = \left(9.31x10^{-12}\ \text{mol/l}\right)(63.55\ \text{g/mol}) = 5.92x10^{-10}\ \text{g/l}$$

(b) *Hydrogen evolution will start when* $a_{Cu^{2+}} = 9.31x10^{-12}$ *mol/l.*

9.4.4 Electrowinning of Zinc

A typical industrial production of zinc metal is accomplished by electrolysis of highly purified zinc sulfate ($ZnSO_4$). The ideal cathodic and anodic reactions during electrowinning of zinc (Zn) are

$$2Zn^{2+} + 4e^- \rightarrow 2Zn \qquad \text{(zinc deposition)} \tag{9.31}$$
$$2H_2O \rightarrow O_2 + 4H^+ + 4e^- \qquad \text{(oxidation of water)} \tag{9.32}$$
$$2Zn^{2+} + 2H_2O \rightarrow 2Zn + O_2 + 4H^+ \qquad \text{(redox)} \tag{9.33}$$

However, additional cathodic reactions, such as hydrogen evolution ($2H^+ + 2e^- \rightarrow H_2$) and oxygen reduction ($O_2 + 4H^+ + 4e^- \rightarrow 2H_2O$), may occur at the expense of some amount of current. This, then, conveys to current inefficiency, high consumption of energy, and lower efficiency (ϵ). The energy consumption is in the order of 3.3 kW h/kg for a current efficiency of 90 % at 500 A/m^2 [5]. Since the energy consumption is an economical factor that dictates the feasibility of an electrowinning cell, substantial energy savings can be accomplished by reducing the irreversible energy dissipation due to oxygen evolution on the anodes. In fact, reducing ohmic potential drop induces energy savings. A reduction in energy consumption can be achieved by using catalytic oxygen evolution anodes, such as dimensionally stable anodes (DSA®) [5] and hydrogen gas diffusion (HGD) anodes [10, 13]. The latter type of anode reduces production cost [6, 20, 21, 38]. Thus, HGD anodes require a low potential range, 1.5 V $< E <$ 4 V, and high current densities, 2 kA/m^2 $\leq i \leq$ 8 kA/m^2, as compared with the conventional range of 400 A/m^2 $\leq i \leq$ 800 A/m^2. Moreover, the structure of a gas diffusion anode can be found elsewhere [6, 28]. Some relevant data for electrowinning zinc is displayed in Table 9.1.

According to these data, the main difference between the electrowinning cells containing conventional Pb alloy and nonconventional DSA anodes is the current density. However, E, ϵ, and γ are similar, except for one reported DSA cell [22]. These observations suggest that conventional cells are still promising in metal recovery of zinc from solutions. Despite that catalytic oxygen anodes promote

Table 9.1 Relevant parameters used in electrowinning of *Zn*

i (A/m^2)	E (V)	ϵ (%)	γ (kWh/kg)	Anode	T (°C)	Ref.
500	3.3	90	3.30	Pb alloy	> 25	[5]
5000	4.0	91	3.60	DAS®	> 25	[6]
5000	1.8	90	1.64	DAS®	> 21	[22]
5000	4.0	90	3.10	DAS®	50	[1]
5000	4.0	95	3.50	DAS®	50	[1]

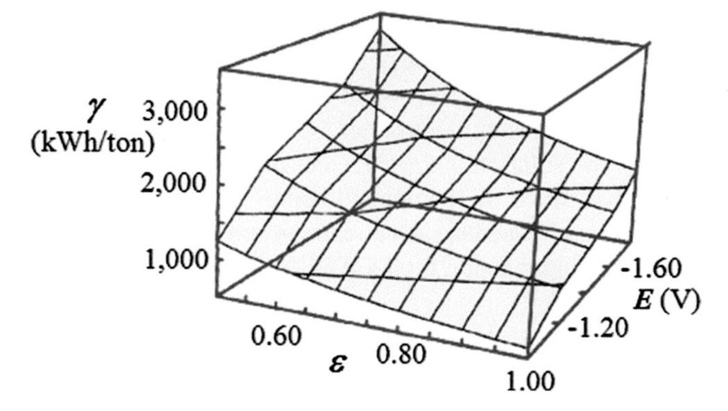

Fig. 9.8 Experimental energy consumption surface for electrowinning zinc using hydrogen gas diffusion (HGD) anodes [38]

substantial savings, impurities in the zinc electrolyte limit their service life, but purification of the electrolyte can reduce this detrimental effect [5]. According to a literature review of HGD anodes conducted by Bestetti et al. [6], these types of anodes offer a feasible technological approach and have the advantage over the conventional *Pb*-alloy anodes to resist higher concentrations of sulfuric acid containing lower concentrations of zinc. Thus, energy consumption is reduced as well as the cell potential.

Figure 9.8 shows a 3D surface plot for electrowinning of zinc [38]. This 3D plot indicates that energy consumption (γ) decreases with increasing efficiency ϵ at fixed values of potential E. On the other hand, keeping ϵ constant, γ increases with increasing E. Therefore, the ideal operating conditions can be achieved by keeping a low E and a high ϵ for low γ. This surface is in accord with Eq. (9.25).

Table 9.2 lists available experimental data from the literature. Observe that the applied cell potential for both conventional anode and hydrogen gas diffusion (HGD) anode is in accord with Eq. (9.20). The HGD anode yields lower cell potential than the conventional electrodes at a constant current density flowing through a strong acid solution. This suggests that HGD is very promising in electrowinning operations since the energy consumption is lower than its counterpart.

These kinetic parameters induce smooth electrodeposition of zinc.

Table 9.2 Components of cell potentials at $i = 450\ \mathrm{A/m^2}$ and $pH = 1$ [10]

Variable	Conventional (V)	*HGD* (V)
$E_a + E_c$	2.04	0.81
η_a	0.86	0.14
η_c	0.06	0.06
IR_s	0.54	0.54
E	3.50	1.55

9.4.5 Degree of Dissociation

Since electrolytes transport charge in the solid, aqueous, or vapor state, it is worthwhile mentioning that the degree of dissociation (α_1) of a compound added to a solvent can be estimated as the fraction of the dissociated substance. Electrolytes hereafter are treated as aqueous electrolytes (solutions) unless otherwise stated. In general, if $\alpha_1 < 1$, the solution is a weak electrolyte, and if $\alpha_1 \geq 1$, the solution is a strong electrolyte since the compound being added to a solvent completely dissociates into ionic components, such as cations and anions.

Consider the following reaction for a salt M_xO_y dissolved in a solvent:

$$M_xO_y = xM^{z+} + yO^{-z} \tag{9.36}$$

where M is a metal and O is an oxide or hydroxide. The ionic components are metal M^{z+} cations and oxide or hydroxide O^{-z} anions.

The standard free energy change for chemical equilibrium and the equilibrium constant are, respectively,

$$\Delta G^o = -RT\ln(K_e) \tag{9.37}$$

$$K_e = \frac{\left[M^{z+}\right]^x \left[O^{-z}\right]^y}{\left[M_xO_y\right]} \tag{9.38}$$

The activity or molar concentration a_i, the degree of dissociation α_1, and the total molar activity $[C_T]$ are defined as follows [12]:

$$\left[M^{z+}\right] = x\alpha_1\left[C_T\right] \tag{9.39}$$

$$\left[O^{-z}\right] = y\alpha_1\left[C_T\right] \tag{9.40}$$

$$\left[M_xO_y\right] = (1-\alpha_1)\left[C_T\right] \tag{9.41}$$

Substituting Eq. (9.39) through (9.41) into (9.38) yields

$$K_e = \left(x^x y^y\right)\left[\frac{\alpha_1^{x+y}\left[C_T\right]^{x+y-1}}{1-\alpha_1}\right] \tag{9.42}$$

If $\alpha_1 << 1$, then Eq. (9.42) becomes

$$K_e = (x^x y^y)\left[\alpha_1^{x+y}\,[C_T]^{x+y-1}\right] \quad (9.43)$$

from which

$$\alpha_1 = \left(K_e x^{-x} y^{-y}\right)^{1/(x+y)}\,[C_T]^{-1+1/(x+y)} \quad (9.44)$$

This expression indicates that the degree of dissociation increases with increasing concentration; that is, $\alpha_1 \uparrow$ as $[C_T] \uparrow$. Further, if $\alpha_1 << 1$, a hydroxyl ion activity $[OH^-]$ is related to a hydrogen ion activity $\left[H^+\right]$ through the dissociation of water. Hence, the dissociation constant for water is

$$K_w = \frac{\left[H^+\right][OH^-]}{[H_2O]} = \left[H^+\right][OH^-] = 10^{-14} \quad (9.45)$$

For a neutral solution, $\left[H^+\right] = [OH^-] = 10^{-7}$. Also, the acidity or alkalinity of a solution is defined by the pH:

$$pH = -\log\left[H^+\right] \quad (9.46)$$

Hence,

$$\left[H^+\right] = 10^{-pH} \quad (9.47)$$

$$[OH^-] = K_w/\left[H^+\right] = 10^{-14}/10^{pH} \quad (9.48)$$

This hydroxyl concentration may be used in the dissociation of a metal hydroxide compound, such as $Ag(OH)$. An example can make this procedure very clear.

The degree of dissociation (α) can also be defined as the ratio of conductivity at any dilution to the conductivity at infinite dilution in an electrolyte:

$$\alpha_1 = \frac{K_{c,x}}{K_{c,\infty}} \quad (9.48a)$$

From Eq. (9.45), the solubility product for water being $K_w = 10^{-14}$ can be used to determine the ionic product of water as [50]

$$k_w = K_w\,[H_2O] = \left[H^+\right][OH^-] \quad (9.48b)$$

This means that $k_w = K_w\,[H_2O] = \left(10^{-14}\right)(1) = 10^{-14}$. If an extra amount of $\left[H^+\right]$ is added to the electrolyte solution, then the ionization of $[H_2O]$ is suppressed so that $k_w = 10^{-14}$ remains constant. Similarly, if excess $[OH^-]$ ions are introduced in the electrolyte, then the ionization of $[H_2O]$ is again suppressed, and $k_w = 10^{-14}$ [50].

Example 9.4. *Determine the amount (W) of silver hydroxide Ag(OH) needed to make 1 l of silver-containing electrolyte having a pH = 11.6 at 25 °C and 1 atm. The dissociation constant of Ag(OH) is 1.10x10^{-4}. The atomic weight of silver is 107.87 g/mol, and the molecular weight of [Ag(OH)] is 124.87 g/mol.*

Solution. *In this problem, the silver hydroxide is assumed to dissolve to an extent in pure water. The reactions of interest are*

$$AgOH = Ag^{+} + OH^{-} \tag{a}$$

$$K_e = \frac{\left[Ag^{+}\right]\left[OH^{-}\right]}{\left[Ag\left(OH\right)\right]} \tag{b}$$

$$K_e = \left[Ag^{+}\right]\left[OH^{-}\right] = 1.10x10^{-4} \tag{c}$$

and

$$H_2O = H^{+} + OH^{-} \tag{d}$$

$$K_w = \frac{\left[H^{+}\right]\left[OH^{-}\right]}{\left[H_2O\right]} \tag{e}$$

$$K_w = \left[H^{+}\right]\left[OH^{-}\right] = 10^{-14} \tag{f}$$

From Eqs. (9.47) and (9.48),

$$\begin{aligned}
\left[H^{+}\right] &= 10^{-pH} = 10^{-11.6} = 2.51x10^{-12}\ \text{mol/l} \\
\left[OH^{-}\right] &= K_w/a\left(H^{+}\right) = 10^{-14}/10^{-11.6} = 3.98x10^{-3}\ \text{mol/l} \\
C_{H+} &= \left(2.51x10^{-12}\ \text{mol/l}\right)\left(A_w\right)_H = \left(2.51x10^{-12}\ \text{mol/l}\right)\left(1\ \text{g/mol}\right) \\
C_{H+} &= 2.51x10^{-12}\ \text{g/l} \\
C_{OH-} &= \left(3.98x10^{-3}\ \text{mol/l}\right)\left(Aw\right)_{OH} = \left(3.98x10^{-3}\ \text{mol/l}\right)\left(17\ \text{g/mol}\right) \\
C_{OH-} &= 0.068\ \text{g/l}
\end{aligned}$$

From Eq. (9.43) with x = y = 1, the total silver concentration in solution is predicted by

$$\left[C_T\right] = K_e/\alpha_1^2 \tag{g}$$

Solving for $\left[Ag^{+}\right]$ in Eq. (c) yields

$$\left[Ag^{+}\right] = \frac{K_e}{\left[OH^{-}\right]} = \frac{K_e\left[H^{+}\right]}{K_w} = 2.76x10^{-2}\ \text{mol/l} \tag{h}$$

From Eq. (9.40),

$$[OH^-] = \alpha_1 [C_T] \tag{i}$$

$$\alpha_1 = \frac{[OH^-]}{[C_T]} \tag{j}$$

Substituting Eq. (j) into (g) gives

$$[C_T] = \frac{[OH^-]^2}{K_e} = \frac{\left(3.98x10^{-3}\ \text{mol/l}\right)^2}{1.10x10^{-4}} = 0.144\ \text{mol/l}$$

$$C_{Ag(OH)} = [C_T]\,A_w = (0.144\ \text{mol/l})\,(124.87\ \text{g/mol})$$

$$C_{Ag(OH)} = 17.98\ \text{g/l}$$

From Eq. (j),

$$\alpha_1 = \frac{[OH]^-}{[C_T]} = \frac{3.98x10^{-3}\ \text{mol/l}}{0.144\ \text{mol/l}} = 0.03$$

which implies that the solution is a weak electrolyte since $\alpha_1 < 1$. The total weight of $Ag\,(OH)$ added to water solvent is

$$W = VC = (1\ \text{l})\,(17.98\ \text{g/l}) = 17.98\ \text{g}$$

Therefore, add 17.98g of solid $Ag\,(OH)$ to 1 l of pure water; stir until a homogeneous electrolyte is attained. This electrolyte can be used for silver plating purposes.

9.5 Molten-Salt Electrolysis

Conventionally, molten-salt electrolysis (MSE) is used for producing molten aluminum and magnesium. Metals less electropositive in the *emf* series than *Al* and *Mg* cannot be electrowon [9, 18, 40, 43, 49]. Conveniently, the Soderberg [56] and Hall–Heroult (HH) [55] cells are the classical cells design for producing molten aluminum from bauxite mineral in a molten salt at high temperatures. A schematic Soderberg cell is shown in Fig. 9.9. The electrodes are made out of carbon (graphite). The carbon anodes are immersed in the electrolyte, but above the molten *Al* layer, while the carbon cathodes are placed below the molten *Al*.

The current is applied through the anode busbar, and it passes through the electrolyte and molten *Al* and drained through the carbon cathodes. Consequently, an electromagnetic field (EMF) is generated due to the applied current, which interacts with magnetic forces, leading to a possible reduction in current efficiency of the cell.

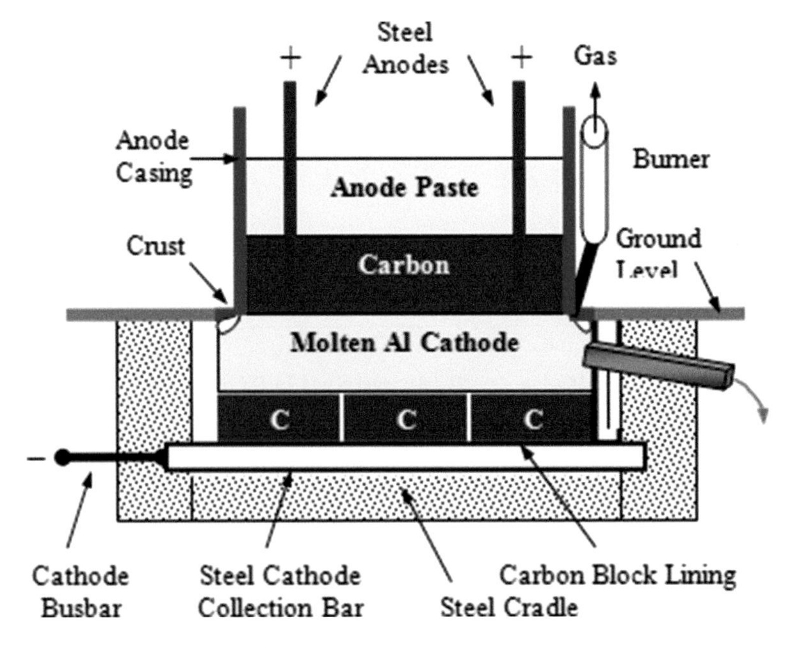

Fig. 9.9 Schematic Soderberg cell

The conventional HH cell [$\approx$ (10 m) x (4 m) x (1 m)] operates at [18]

$$4\ \text{V} \leq E \leq 4.5\ \text{V} \tag{9.49}$$

$$5\ \text{kA/m}^2 \leq i \leq 10\ \text{kA/m}^2 \tag{9.50}$$

$$960\ ^\circ\text{C} \leq T \leq 1000\ ^\circ\text{C} \tag{9.51}$$

$$90\,\% \leq \epsilon < 100\,\% \tag{9.52}$$

These operating conditions offer a range of possibilities for optimizing aluminum production at the highest cell efficiency with minimum energy consumption.

The electrolyte is a molten cryolite salt (N_aAlF_6) which melts at 940 °C and decomposes into sodium and hexafluoroaluminate ions: $\left(AlF_6^{-3}\right)$ [9, 12]

$$N_aAlF_6 \rightarrow 3Na^+ + AlF_6^{-3} + 3e^- \tag{9.53}$$

This electrolyte has a high electrical conductivity, which is a requisite in producing molten aluminum from alumina mineral (Al_2O_3) mixed with N_aAlF_6 at high temperatures and high current densities, which, in turn, cause high energy consumption. In fact, this high energy consumption may be reduced by using recently developed titanium diboride and carbon (TiB_2-C) cathodes containing an adhered thin film of aluminum (Al) [26, 39, 56]. High purity of molten Al is produced and, subsequently, cast into molds of characteristic shapes.

The production of molten Al can be summarized by the following mineral processing step [15] and main reactions [40]:

$$\left|\frac{\text{Bauxite}}{\text{Power}}\right| (Al_2O_3 \cdot nH_2o) \rightarrow \left|\frac{\text{Bayer}}{\text{Process}}\right| \rightarrow \left|\frac{\text{Alumina}}{\text{Power}}\right| (Al_2O_3) \tag{9.54}$$

When Al_2O_3 is added to the molten N_aAlF_6, the following simplified reactions take place:

$$\text{Reduction:} \quad Al_2O_3 = 2Al + \frac{3}{2}O_2 \qquad @ \ \Delta G_{Al} \tag{9.55}$$

$$\text{Oxidation:} \quad 3\overline{C} + \frac{3}{2}O_2 = 3CO \qquad @ \ \Delta G_{CO} \tag{9.56}$$

$$\text{Redox:} \quad Al_2O_3 + 3\overline{C} = 2Al + 3CO \quad @ \ \Delta G \tag{9.57}$$

where $\overline{C}$ is graphite (crystalline carbon).

The Gibbs free energies for these reactions in the written directions are [40]

$$\Delta G_{Al} = -1.68x10^6 + 322.17T \tag{9.58}$$

$$\Delta G_{CO} = 0.354x10^6 + 253.05T \tag{9.59}$$

$$\Delta G = \Delta G_{Al} + \Delta G_{CO} = -1.33x10^6 + 575.22T \tag{9.60}$$

and from Eq. (3.14),

$$\Delta G = -zFE \tag{9.61}$$

where $\Delta G =$ Gibbs free energy (J/mol)

$T =$ absolute temperature (K)

Combining Eqs. (9.60) and (9.61) with $z = 3$ and $F = 96{,}500$ A s/mol yields the theoretical cell potential as

$$E = 4.59 - 2x10^{-3}T \tag{9.62}$$

During electrolysis, the current flows through the molten N_aAlF_6, breaking down the dissolved Al_2O_3 according to Eq. (9.55). Consequently, molten Al and carbon oxide (CO) gas are produced at the cathode and anode surfaces, respectively. Molten Al ($\rho = 2.3\ \text{g/cm}^3$, $K_c = 2.20 \pm 0.2\ \Omega^{-1}\ \text{cm}^{-1}$) settles on the bottom of the cell since it is more denser than molten N_aAlF_6 $\left(\rho < 2\ \text{g/cm}^3\right)$. Then CO gas arises due to carbon consumption of the anode. For a continuous electrolysis, the amount of alumina (Al_2O_3) consumed is compensated by additions of fresh alumina powder. Figure 9.10 shows an oversimplified version of the HH cell shown in Fig. 9.9 for illustrating the above details [55].

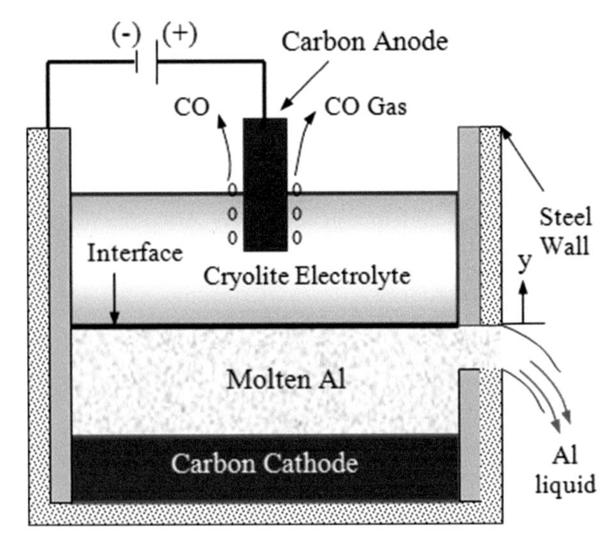

Fig. 9.10 Oversimplified HH cell for producing molten aluminum

According to Faraday's law of electrolysis, the amount of Al during the reduction process can be calculated as per Eqs. (5.29) and (5.30):

$$m = \frac{ItA_w}{zF} \tag{9.62a}$$

$$m = 0.3355It \quad \text{for Al} \tag{9.62b}$$

where A_w/zF = electrochemical coefficient for Al (g/A h)

$A_w/zF = 0.3355\,\text{g/A h}$ for $Al^{3+} + 3e^- \to Al$

z = oxidation state (valence), $z = 3$ for Al

A_w = atomic weight (g/mol)

$F = 96{,}500\,\text{A s/mol} = 26.806\,\text{A h/mol}$ (Faraday's constant)

I = cell current (A)

t = time during electrolysis process (h).

Refractory Corrosion Refractories are porous, brittle, and high-temperature-resistant ceramics used in the furnace or cell as lining materials. However, they are susceptible to corrode at high temperatures in gaseous and liquid environments. In the aluminum production industries, refractory corrosion is a serious and costly problem. The principles of refractory corrosion is described in Yurkov's book [57].

9.5.1 Current Efficiency Model

The current efficiency of a Hall–Heroult (HH) cell depends on internal interactions in the magnetohydrodynamic (MHD) flow, reactions in different zones of the cell,

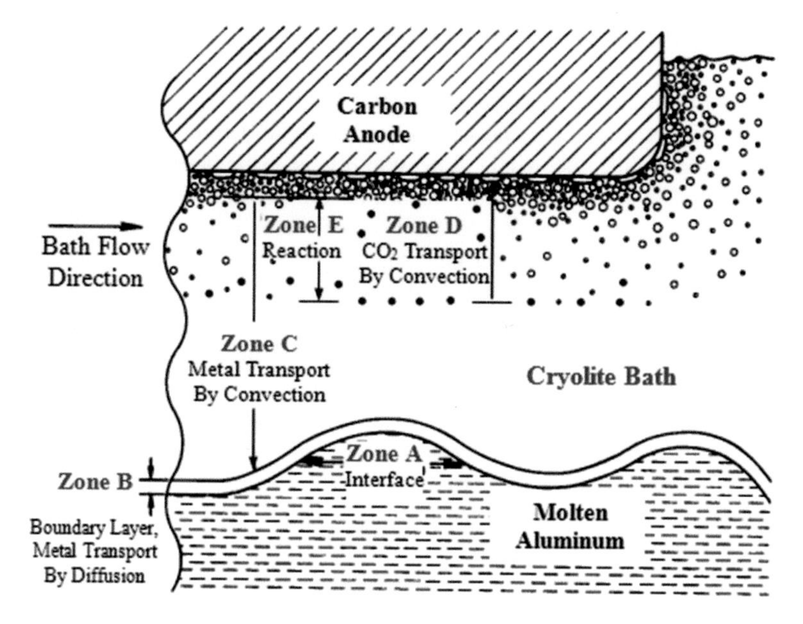

Fig. 9.11 Mechanism of current efficiency [12]

electrical contact system, operating parameters, and anode-to-cathode distance. According to Haupin and Frank [29] model for defining relevant zones (Fig. 9.11) within the HH cell, the possible interactions that are related to current efficiency and energy consumption may be attributable to diverse potentials in these zones and gas evolution during the high-temperature electrolysis. These interactions lead to high energy consumption.

The zones in the model are briefly identified as:

- **Zone A**: It is very sensitive to MHD instability due to oscillations of the molten metal. Therefore, the MHD instability decreases the current efficiency and increases the energy consumption.
- **Zone B**: It is the diffusion layer containing reactants and reduced ions. The metal mass transfer is mainly by ionic diffusion.
- **Zone C**: It is the convective mass transfer of the reduced ions toward the anode. Thus, the motion direction of this mass transfer phenomenon is perpendicular to the bath flow direction as shown in Fig. 9.11.
- **Zone D**: It is the convective mass transfer of carbon dioxide (CO_2) gas away from the anode and that of carbon oxide (CO) gas toward the anode.
- **Zone E**: It is a region of mixing ions.

9.5.2 Magnetohydrodynamic Flow

The theory of magnetohydrodynamic (MHD) deals with the interactions induced by a magnetic field in the electrically conducting fluid (molten aluminum). For incompressible fluids, the MHD flow stability is principally governed by the Maxwell and Navier–Stokes equations which are coupled through the Lorentz force and Ohm's law. In particular, the Lorentz forces may affect or disturb the flow stabilization of the fluid, but the principal sources of MHD instabilities are current flow perturbations that cause distortion of the molten-metal-electrolyte interface and discontinuities of the flow velocity at this interface [47]. In general, the electrometallurgy of molten aluminum depends on the MHD fluid flow, which, in turn, is affected by the inner and outer magnetic fields. The former is generated in the fluid, and the latter is applied through the electrically connected busbars of the HH cells. In general, among many excellent scientific and theoretical sources on MHD, the theory of MHD is clearly covered in Hughes and Young's book [32] and the electrometallurgy of aluminum is adequately written by Haupin and Frank in the *Treatise of Electrochemistry* edited by J. O'M Bobricks et al. [29].

The thermoelectric design of the HH cell requires a careful analysis of the electromagnetic field because of a complicated interaction in the molten N_aAlF_6-Al interface caused by a magnetohydrodynamic (MHD) instability, which leads to high energy consumption and low current efficiency [37, 41]. The MHD instability arises by the presence of electromagnetic and hydrodynamic waves at the interface, where a thermomagnetic turbulence (fluctuations of the molten aluminum height) may generate due to magnetic forces and the magnetic energy is converted to kinetic energy [39, 41]. Therefore, this interface becomes unstable, leading to short circuiting the cell, and consequently the cell potential and energy consumption increase. The reader should be aware of the fact that designing an external busbar system as part of the HH cell requires an analysis of the current fluctuations induced by a magnetic field within the cell. Therefore, a mathematical analysis, if required, can be very complicated due to the effect of coupled magnetic and electric fields [41].

Figure 9.12 shows a volume element as the geometry model of the HH cell (Fig. 9.10). The vectors illustrated in Fig. 9.12 are the current density (**J**), fluid velocity (**v**), magnetic flux (**B**), Lorentz force (**F**), and the vector product (**v**x**B**), which represents the magnitude of the electric field.

This model can be used to predict both MHD and magnetic field characteristics, provided that magnetization of the HH steel shell and the current-carrying steel busbar (positive terminal) are included in the numerical simulation. Application of a *d.c.* magnetic field generates a Lorentz force distribution in the opposite direction of the fluid velocity (v_z). Thus, an electromagnetic force field generates in the HH cell. If this force field is strong enough, it can cause unfavorable physical conditions leading to MHD instabilities since the force field drives the fluid flow to oscillate harmonically. This unfavorable condition is referred to as perturbation at

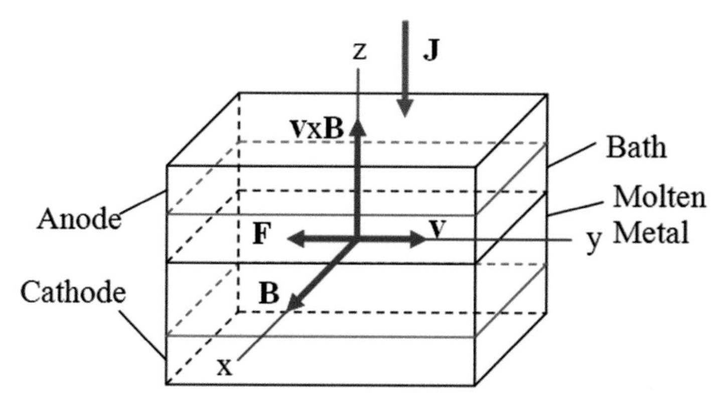

Fig. 9.12 Volume element of an HH cell under d.c. field [41]

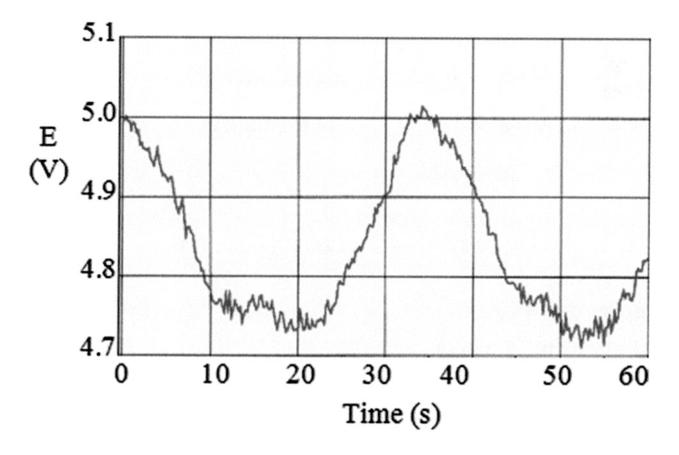

Fig. 9.13 Potential fluctuation in an HH unstable cell [41]

the molten-metal upper face (metal pad), and as a result, the potential fluctuates and energy consumption increases. An example of potential fluctuation in an unstable HH cell is shown in Fig. 9.13 [41].

Furthermore, the key factor in successful HH cells (pot operations) is to predict and maintain MHD stability. This leads to the characterization of the internal MHD flow using Navier–Stokes equation of motion and Maxwell's equations, including the charge distribution, Faraday's law of induction, Ohm's law, Lorentz force law, Poisson's equation, and even LaPlace's equation. The set of these laws and equations constitutes the MHD expressions.

For isotropic and homogeneous fluids, the Navier–Stokes equation of motion in terms of fluid velocity is

$$\rho \frac{\partial \mathbf{v}}{\partial t} + \boldsymbol{\rho} \left(\mathbf{v} \cdot \nabla \right) \mathbf{v} + \nabla P - \eta_{v} \nabla \mathbf{v} - \mathbf{J} x \mathbf{B} = 0 \tag{9.63}$$

and the MHD expressions are defined below in general vector notation (bold-faced letters):

$$\nabla \cdot \mathbf{D} = q \qquad \text{(Coulomb's law)} \tag{9.64}$$

$$\nabla x \mathbf{E} = -\frac{\partial \mathbf{B}}{\partial t} \qquad \text{(Faraday's law)} \tag{9.65}$$

$$\mathbf{F} = \mathbf{J} x \mathbf{B} \qquad \text{(Lorentz law)} \tag{9.66}$$

$$\mathbf{J} = -\sigma \nabla \cdot \boldsymbol{\phi} \qquad \text{(Ohm's law)} \tag{9.67}$$

$$\Delta P = \nabla \cdot \mathbf{F} \qquad \text{(Poisson's law)} \tag{9.68}$$

$$\nabla \cdot \boldsymbol{\phi} = 0 \qquad \text{(LaPlace's equation)} \tag{9.69}$$

$$\nabla \cdot \mathbf{v} = 0 \qquad \text{(continuity equation)} \tag{9.70}$$

$$\nabla \cdot \mathbf{J} = 0 \qquad \text{(continuity equation)} \tag{9.71}$$

$$\nabla \cdot \mathbf{B} = 0 \qquad \text{(in the exterior)} \tag{9.72}$$

$$\nabla \cdot \mathbf{B} = \mu \mathbf{J} \qquad \text{(in the fluid)} \tag{9.73}$$

$$\mathbf{B} = \mu \mathbf{H} \tag{9.74}$$

where $\mathbf{D} = \epsilon \mathbf{E}$ = electric current displacement $\left(\mathrm{C/cm^2}\right)$

ρ = charge density $\left(\mathrm{C/cm^3}\right)$
η_v = fluid viscosity (g/cm s)
$\mathbf{E}$ = electric intensity (gradient) (V/cm)
$\mathbf{B}$ = magnetic flux density $\left(\mathrm{V\,s/cm^2} = \mathrm{weber/cm^2}\right)$
$\mathbf{H}$ = magnetic field intensity (A/cm = C/cm s)
$\mathbf{J}$ = current density $\left(\mathrm{A/cm^2}\right)$
$\mathbf{F}$ = Lorentz force (N)
$\boldsymbol{\phi}$ = electric potential (V)
$\mathbf{v}$ = fluid velocity (cm/s)
P = scalar hydrodynamic pressure $\left(\mathrm{N/cm^2}\right)$
σ = electric conductivity
ϵ = dielectric constant
μ = permeability of the medium (V s/A cm = henry/cm)

Details on mathematically modeling an electromagnetic problem in molten-salt electrolysis can be found elsewhere [37, 41]. In fact, $\nabla \cdot \mathbf{B} = 0$ implies that the magnetic lines are path independent since they close on themselves. Also, $\nabla \cdot \mathbf{B} = \mathit{div}\ \mathbf{B} = 0$ is the divergence of the vector $\mathbf{B}$, $\nabla x \mathbf{E} = \mathit{curl}\ \mathbf{E}$ is the curl of the vector $\mathbf{E}$, and $\mathbf{J} x \mathbf{B}$ is a vector product.

According to Hughes and Young [32], the most general form of the electromagnetic body forces ($\mathbf{F}_e$) in a control volume filled with a fluid that includes the effects of charge and current interactions can be determined by

$$\mathbf{F}_e = \rho\mathbf{E} + \mathbf{J}x\mathbf{B} \tag{9.75}$$

where $\rho\mathbf{E}$ is the electric force and $\mathbf{F} = \mathbf{J}x\mathbf{B}$ is the interaction force between the current and the magnetic field known as Lorentz force. In metallic conductors $\mathbf{J}x\mathbf{B}$ is the predominant force in MHD flow.

9.6 Mass Transfer by Convection

9.6.1 Stationary Planar Electrodes

Consider the electrolyte hydrodynamic flow condition shown in Fig. 9.14. This is a mass transfer found in electrowinning and electrorefining cells, in which the electrolyte motion is upward on the anode surface due to the generation of oxygen bubbles, which enhances the mass transfer. The reader can observe in this model that the electrolyte motion is more intense at the top than at the bottom of the electrode as indicated by the arrows. If the current flows, then the electrolyte motion increases and descends to the bottom of the cathode. This is the case in which a significant convective molar flux is superimposed on the Fick's diffusion molar flux. The concentration gradient in the fluid adjacent to the vertical electrode (plate) surface causes a variation in the fluid density, and the boundary layer (δ_u) develops upward from laminar to turbulent conditions [16, 23, 30, 49].

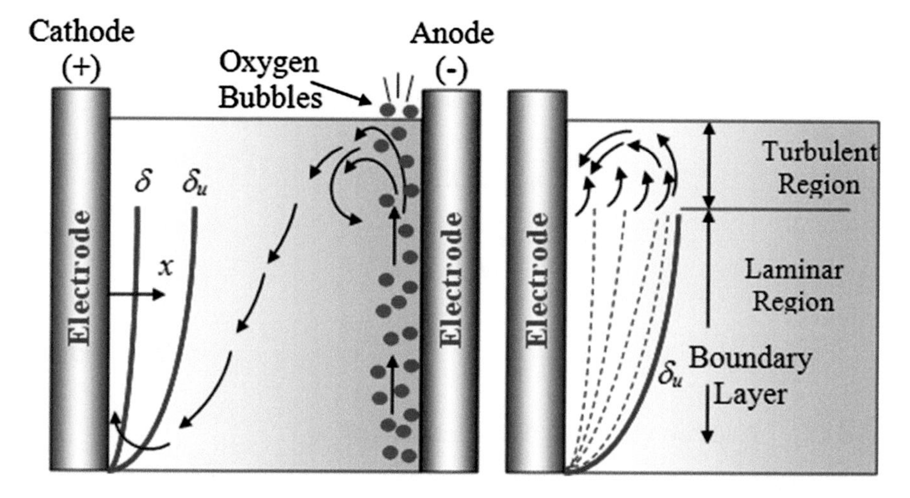

Fig. 9.14 (a) Natural convective mass transfer between two vertical plates and (b) formation of a boundary layer

Table 9.3 Dimensionless numbers and temperature [24]

Nusselt number (N_u)	$\Longrightarrow$	Sherwood number (S_h)
Prandtl number (P_r)	$\Longrightarrow$	Schmidt number (S_c)
Thermal diffusivity (α)	$\Longrightarrow$	Atomic diffusivity (D)
Reynolds number (R_e)	$\Longrightarrow$	Reynolds number (R_e)
Temperature (T)	$\Longrightarrow$	Concentration (C_x)

The forced convection mass transfer emerges provided that $C_s \neq C_b$, where $C_s << C_b$ at $x = 0$ on the cathode surface. According to Geiger and Poirier [24], many forced-convention mass transfer solutions are analogous to heat-transfer cases. This suggests that changes in notations from the latter to the former can be made as given in Table 9.3.

The mass transfer phenomenon is a well-documented engineering field. Thus, books by Evans [18], Geiger and Poirier [24], Holman [30], Gaskell [23], and Incropera and DeWitt [33], among many excellent sources, can be consulted on the current topic for laminar and turbulent conditions. In general, the mass transfer by convection for the case shown in Fig. 9.14 may be further described by the Sherwood Number. Hence,

$$S_h = \lambda_f \left(S_c\right)^{\alpha} \left(R_e\right)^{\beta} \qquad \text{(forced convection)} \tag{9.76}$$

$$S_h = \lambda_n \left(S_c\right)^{\gamma} \left(R_e\right)^{\mu} \qquad \text{(natural convection)} \tag{9.77}$$

where $\lambda_f, \lambda_n =$ constants

$\alpha, \beta, \gamma, \mu =$ exponents

In addition, the Reynolds, Schmidt, Rayleigh (R_a), and Grashof (G_r) numbers are empirically defined by

$$R_e = \frac{Inertia\, force}{Viscous\, force} = \frac{\rho v_x^2/d}{\eta_v v_x/d^2} \tag{9.78}$$

$$R_e = \frac{\rho v_x d}{\eta_v} = \frac{v_x d}{K_v} \tag{9.79}$$

$$S_c = \frac{K_v}{D} = \frac{\eta_v}{\rho D} \tag{9.80}$$

$$R_a = G_r S_c \tag{9.81}$$

$$K_v = \frac{\eta_v}{\rho} \tag{9.82}$$

where $v_x =$ fluid flow velocity in the x-direction (cm/s)

$d =$ characteristic length of a channel (cm)
$K_v =$ kinematic viscosity $\left(\text{cm}^2/\text{s}\right)$
$\eta_v =$ fluid viscosity (g/cm s)

ρ = fluid density $\left(g/cm^3\right)$
D = diffusivity or diffusion coefficient $\left(cm^2/s\right)$

An electrowinning cell can be treated as an open channel (open top) with an equivalent diameter as the characteristic length. This diameter is also known as the hydraulic diameter defined as

$$d = \frac{4A_c}{p} \tag{9.83}$$

Here, A_c is the cross-sectional area of the channel, and p is the wetted perimeter of the channel in contact with the flowing fluid. For square and rectangular cross-sectional areas, d becomes [49]

$$d = \frac{4a^2}{3b} \quad \text{(square)} \tag{9.84}$$

$$d = \frac{4ab}{2a+b} \quad \text{(rectangular)} \tag{9.85}$$

where a = height of the channel (cm)

b = width of the channel

According to Twidwell [49], in most electrowinning cells, $d = 2x$ where x is the distance between electrodes. In addition, the fluid velocity is defined as

$$v_x = \frac{F_r}{A} \tag{9.86}$$

Recall that F_r is the volume flow rate and it is defined in terms of mass flow rate ($\dot{m}$) and density (ρ):

$$F_r = \frac{\dot{m}}{\rho} \tag{9.87}$$

If an electrochemical cell is treated as a pipe, then the conditions for the type of flow may be characterized by the Reynolds and Rayleigh numbers as listed in Table 9.4 [30].

Physically, the Sherwood number (S_h) is defined as the ratio of convection to diffusion mass transfer through the concentration boundary layer at the cathode surface [23, 24, 49]. Thus,

Table 9.4 Characterizing fluid flow by convection

Flow	Forced flow	Natural flow
Laminar	$R_e \leq 2300$	$R_a < 10^9$
Transient	$2300 < R_e < 4000$	$R_a \simeq 10^9$
Turbulent	$R_e \geq 4000$	$R_a > 10^9$

$$S_h = \frac{i}{zFJ_x} \tag{9.88}$$

$$S_h = \frac{i\delta}{zFD\Delta C} = \frac{i}{zFh\Delta C} \tag{9.89}$$

where δ = Nernst diffusion layer (cm),

$\Delta C = C_o - C_b$ = change in concentration (mol/cm^3)
$h = D/\delta$ = mass transfer coefficient (cm/s)

Considering diffusion of a species j through the diffusion layer of thickness δ, the mass flux according to Fick's modified first law of diffusion and that for convection are defined by the following expression, respectively,

$$J_{d,m} = \frac{D(\rho_s - \rho_\infty)}{x} \tag{9.90}$$

$$J_{c,m} = h(\rho_s - \rho_\infty) \tag{9.91}$$

The local Sherwood number for $0 < x \le \delta$ is [23]

$$S_h = \frac{J_{c,m}}{J_{d,m}} = \frac{hx}{D} \tag{9.92}$$

In addition, the forced convection mass transfer coefficient for transfer of the species j to or from the electrode interface can also be defined by [24]

$$h = \frac{J_d}{\Delta C} = \frac{D}{\delta} \tag{9.93}$$

Assuming mass transfer by laminar electrolyte flow in electrodeposition between two relatively large plate electrodes and combining Eqs. (9.92) and (9.93) and letting $x = d$ be the characteristic distance between the electrodes yield a simplified Sherwood number [49]:

$$S_h = \frac{d}{\delta} \tag{9.94}$$

where $d = 2w$ and w is the channel width in the order of 3 cm [49].

In addition, the Sherwood number S_h can be semiempirically defined for natural (free) and forced convection for the case shown in Fig. 9.14.

From Twidwell's Work [49]

(1) Natural convection flow:

$$S_h = \frac{2}{3}(S_c G_r)^{1/4} \qquad \text{(laminar)} \tag{9.95}$$

$$S_h = \frac{13}{42}(S_c G_r)^{7/25} \qquad \text{(turbulent)} \tag{9.96}$$

(2) Forced convection flow:

$$S_h = \frac{9}{5}\left[\frac{d}{L}(S_c R_e)\right]^{1/3} \qquad \text{(laminar)} \tag{9.97}$$

$$S_h = \frac{1}{65}(S_c)^{1/3}(R_e)^{1/2} \qquad \text{(turbulent)} \tag{9.98}$$

where d is defined in Eq. (9.94) and L is the cathode length [49].

From Gaskell [23] and Geiger–Poirier [24] Books

(1) Natural convection flow:

$$S_h = \frac{3}{2\pi}\left(\frac{S_c^2 G_r}{S_c + 5/9}\right)^{1/4} \qquad \text{(laminar)} \tag{9.99}$$

(2) Forced convection flow:

$$S_h = \frac{2}{3}(S_c)^{1/3}(R_e)^{1/2} \tag{9.100}$$

where $10^2 \leq S_c \leq 10^3$ for most fluids.

Particularly, assume that the electrochemical deposition (Fig. 9.14) is under laminar flow. For this practical condition, the Sherwood number can be redefined by inserting Eqs. (9.79) and (9.80) into (9.97). Hence,

$$S_h = \frac{9}{5}\left(\frac{v_x d^2}{LD}\right)^{1/3} \qquad \text{(forced laminar flow)} \tag{9.101}$$

Combining Eqs. (9.88) and (9.101) using $\Delta x = L$ and $J_x = D(C_x - C_s)/\Delta x$ yields the cathodic current density as

$$i = \frac{9}{5}zF(C_x - C_s)\left(\frac{v_x d^2 D^2}{L^4}\right)^{1/3} \qquad \text{(forced laminar flow)} \tag{9.102}$$

If $C_s = 0$ at $x = 0$, then i is the limiting current density [49]. However, Eq. (9.102) is more conveniently used if $C_x = C_b$ at $x = \infty$ from the cathode surface since the bulk concentration is normally known prior to operating an electrolytic cell. Hence, the current density in this case becomes the limiting current density as defined by

$$i_L = \frac{9}{5} zFC_b \left(\frac{v_x d^2 D^2}{L^4} \right)^{1/3} \quad \text{(forced laminar flow)} \tag{9.103}$$

This limiting current density arises due to a high rate of metal reduction at the cathode surface. As a result, ion depletion occurs in the bulk solution unless the electrochemical cell is replenished with fresh solution for a continuous operation.

Nonetheless, the application of Eq. (9.103) is essential for determining the fluid velocity when the limiting current density is known. In fact, this approach may be considered as a general case for the planar electrode geometry. At any rate, the limiting current density in Eq. (9.103) strongly depends on the fluid velocity, and its proportionally level is $i_L \propto v_x^{1/3}$. This implies that Eq. (9.103) can be rearranged in the following form:

$$i_L = \lambda_1 v_x^{1/3} \tag{9.104}$$

$$\lambda_1 = \frac{9}{5} zFC_b \left(\frac{d^2 D^2}{L^4} \right)^{1/3} \tag{9.105}$$

Notice that $i_L = f(v_x)$ gives a cubic parabola as per Eq. (9.104) with constant λ_1, which is treated, for example, as the current density coefficient in forced laminar flow. In addition, the strong dependency of i_L on the fluid velocity v_x is significant at low v_x. In the absence of fluid turbulence, the ionic mass transport from solution to an electrode surface by diffusion mechanism depends on the electrolyte velocity, diffusion coefficient, bulk concentration, and the channel dimensions.

Commonly, these variables are constants, except the v_x which has a cubic parabola effect on i_L as predicted by Eq. (9.104).

Example 9.5. *A hypothetical copper electrowinning cell operates at 35 °C and 101 kPa under forced laminar flow. If the electrolyte/cell has the following conditions:*

$$C_b = 7.10x10^{-4}\ \text{mol/cm}^3\ \textit{(concentration of } Cu^{2+} \textit{ ions)}$$

$$x = L = 100\ \text{cm}\ \textit{(cathode length)}$$

$$d = 5\ \text{cm}\ \textit{(cell equivalent diameter)}$$

$$D = 10^{-5}\ \text{cm}^2/\text{s}\ \textit{(diffusivity of } Cu^{2+} \textit{ ions)}$$

$$K_v = 0.80\ \text{cm}^2/\text{s}\ \textit{(kinematic viscosity)}$$

$$v_x = 0.50\ \text{cm/s}\ \textit{(electrolyte flow velocity)}$$

$$\rho_x = 5\ \Omega\ \text{cm}\ \textit{(resistivity for } Cu^{2+} \textit{in solution)}$$

$$\rho = 8.96\ \text{g/cm}^3\ \textit{(density)}$$

Calculate **(a)** *the Sherwood number,* **(b)** *the mass transfer coefficient,* **(c)** *the molar flux,* **(d)** *the current density, and* **(e)** *the rate of electrolytic deposition* (R_F), *and* **(f)** *is it practical to recover* Cu^{2+} *ions if the electrolyte flow velocity is at its transitional flow condition in a conventional electrowinning cell?*

Solution.

(a) *From Eq. (9.101),* $S_h = (9/5)\left[v_x d^2/(LD)\right]^{1/3}$

$$S_h = (9/5)\left\{\left[(0.50\ \text{cm/s})(5\ \text{cm})^2\right]/\left[(100\ \text{cm})\left(10^{-5}\ \text{cm}^2/\text{s}\right)\right]\right\}^{1/3}$$
$$S_h = 41.77$$

(b) *From Eq. (9.92),* $h = DS_h/x = \left(10^{-5}\ \text{cm}^2/\text{s}\right)(41.77)/(100\ \text{cm})$

$$h = 4.18x10^{-6}\ \text{cm/s}$$

(c) *From (Eq. 9.93), the convective molar flux along with* $\Delta C = C_0 - C_b$ *and* $C_b >> C_o$ *is* $J_x = h\Delta C = hC_b = \left(4.18x10^{-6}\ \text{cm/s}\right)\left(7.10x10^{-4}\ \text{mol/cm}^3\right)$

$$J_x = 2.97x10^{-9}\ \text{mol/cm}^2\ \text{s}$$

(d) *From Eq. (9.88),* $i = zFJ_xS_h = (2)(96{,}500\ \text{A s/mol})\left(2.97x10^{-9}\ \text{mol/cm}^2\ \text{s}\right)(41.77)$

$i = 2.39x10^{-2}\ \text{A/cm}^2 = 239\ \text{A/m}^2$ *which is within a practical range in electrowinning cells.*

(e) *From Eq. (5.1),* $R_F = iA_w/zF = \left(2.39x10^{-2}\ \text{A/cm}^2\right)(63.54\ \text{g/mol})/[(2)(96{,}500\ \text{A s/mol})]$

$$R_F = 7.87\ \mu\text{g/cm}^2\ \text{s}$$

(f) *Using* $R_e = 2300$ *from Table 9.4 as the transitional number, then Eq. (9.79) gives the transitional fluid velocity* $v_x = R_eK_v/d$

$$v_x = (2300)\left(0.80\ \text{cm}^2/\text{s}\right)/(5\ \text{cm}) = 368\ \text{cm/s}$$

Using this transitional fluid velocity, calculate the Sherwood number, the mass transfer coefficient, the molar flux, and the current density. Hence,

$$S_h = (9/5)\left[v_x d^2/(LD)\right]^{1/3}$$
$$S_h = (9/5)\left\{\left[(368\ \text{cm/s})(5\ \text{cm})^2\right]/\left[(100\ \text{cm})\left(10^{-5}\ \text{cm}^2/\text{s}\right)\right]\right\}^{1/3}$$
$$S_h = 377.17$$

$$h = DS_h/x = \left(10^{-5}\ \text{cm}^2/\text{s}\right)(377.17)/(100\ \text{cm})$$
$$h = \left(10^{-5}\right)(377.17)/(100) = 3.7717 \times 10^{-5}$$
$$h = 3.77x10^{-5}\ \text{cm/s}$$

$$J_x = h\Delta C = hC_b = \left(3.77x10^{-5}\ \text{cm/s}\right)\left(7.10x10^{-4}\ \text{mol/cm}^3\right)$$
$$J_x = \left(3.77 \times 10^{-5}\right)\left(7.10 \times 10^{-4}\right) = 2.6767 \times 10^{-8}$$
$$J_x = 2.68x10^{-8}\ \text{mol/cm}^2\ \text{s}$$

$$i = zFJ_xS_h = (2)(96{,}500\ \text{A s/mol})\left(2.68x10^{-8}\ \text{mol/cm}^2\ \text{s}\right)(377.17)$$
$$i = 1.95\ \text{A/cm}^2 = 19{,}500\ \text{A/m}^2$$

Therefore, $i = 19{,}500\ A/m^2$ *is too high, and it is not a practical current density in conventional electrowinning cells containing planar electrodes. This implies that an electrolyte motion at the chosen transition forced laminar flow, the Reynolds number,* $R_e = 2300$*, is not practical. Hence,* $R_e << 2300$ *should give more realistic results.*

Example 9.6. *A hypothetical chemical vapor deposition of a gas is carried out at* $600\,°C$ *and* $101\ kPa$ *on a metal substrate. If the gas has the following properties:*

$$
\begin{aligned}
D &= 0.40\ \text{cm}^2/\text{s}\ \textit{(diffusivity)} \\
\rho &= 3x10^{-3}\ \text{g/cm}^3\ \textit{(density)} \\
\eta_v &= 2.5x10^{-4}\ \text{g/cm s}\ \textit{(viscosity)} \\
v_x &= 97\ \text{cm/s}\ \textit{(gas flow velocity)} \\
A_w &= 95\ \text{g/mol}\ \textit{(molecular weight)}
\end{aligned}
$$

at $x = 100\ mm$ *from the origin of the gas stream to the metal substrate surface, determine* **(a)** *the Sherwood number,* **(b)** *the mass transfer coefficient,* **(c)** *the molar flux, and* **(d)** *the rate of vapor deposition* (R_F) *in* cm/s*. Use a change of concentration equal to* $2x10^{-4}\ mol/cm^3$ *of the gas.*

Solution.

(a) *From Eq. (9.79) with* $d = x$ *and Eq. (9.80), respectively,*

$$R_e = \rho x v_x/\eta_v = \left(3x10^{-3}\ \text{g/cm}^3\right)(10\ \text{cm})(97\ \text{cm/s}) / \left(2.5x10^{-4}\ \text{g/cm s}\right)$$
$$R_e = 11{,}640$$

$$S_c = \eta_v/D\rho = \left(2.5x10^{-4}\ \text{g/cm s}\right) / \left[\left(0.40\ \text{cm}^2/\text{s}\right)\left(3x10^{-3}\ \text{g/cm}^3\right)\right]$$
$$S_c = 0.21$$

From Eq. (9.100),

$$S_h = (2/3)\left(S_c\right)^{1/3}\left(R_e\right)^{1/2} = (2/3)(0.21)^{1/3}(11640)^{1/2}$$
$$S_h = 42.75$$

(b) *From Eq. (9.92),*

$$h = DS_h/x = \left(0.40\ \text{cm}^2/\text{s}\right)(42.75) / (10\ \text{cm}) = 1.71\ \text{cm/s}$$

(c) *From Eq. (9.93),*

$$J = h\Delta C = (1.71\ \text{cm/s})\left(2x10^{-4}\ \text{mol/cm}^3\right) = 3.42x10^{-4}\ \text{mol/cm}^2\ \text{s}$$

(d) *Combining Eqs. (5.1) and (6.8) and dividing the resultant expression by the density* ρ *yield*

$$R_F = J_x A_w/\rho = \left(3.42x10^{-4}\ \text{mol/cm}^2\ \text{s}\right)(95\ \text{g/mol}) / \left(3x10^{-3}\ \text{g/cm}^3\right)$$
$$R_F = 10.83\ \text{cm/s}$$

This vapor deposition seems adequate for most practical applications.

Moreover, mass transfer by convection is a well-documented topic in the literature in which a hydrodynamic viscous convective layer develops due to fluid flow. This convective layer is known as Prandtl boundary layer, which depends on the fluid velocity. According to Erdey-Gruz [16] and Probstein [42] among other excellent authors, the convective layer is defined as

$$\delta_u = xR_e^{-1/2} = \sqrt{\frac{xK_v}{v_b}} \tag{9.106}$$

where $x =$ distance from the electrode surface (cm)

$v_b =$ fluid velocity in the bulk of solution (cm/s)

The correlation of the diffusion and convective layers is graphically shown in Fig. 9.14 and mathematically defined as [16, 23, 24, 47]

$$\delta = \delta_u \left(\frac{D}{K_v}\right)^{1/3} \tag{9.107}$$

The profile of the layer ratio and their linear relationship are shown in Fig. 9.15, which is suitable for determining the convective layer as the slope of the straight line.

Substituting Eq. (9.106) into (9.107) yields the diffusion layer dependent on the fluid velocity:

$$\delta = D^{1/3} K_v^{1/6} \sqrt{x/v_b} \tag{9.108}$$

Combining Eqs. (9.108) and (6.8) along with $J_x = -D(C_x - C_b)/\delta$ gives the current density: for $x > 0$

$$i = zF(C_b - C_x) D^{2/3} K_v^{-1/6} \left(\frac{v_b}{x}\right)^{1/2} \tag{9.109}$$

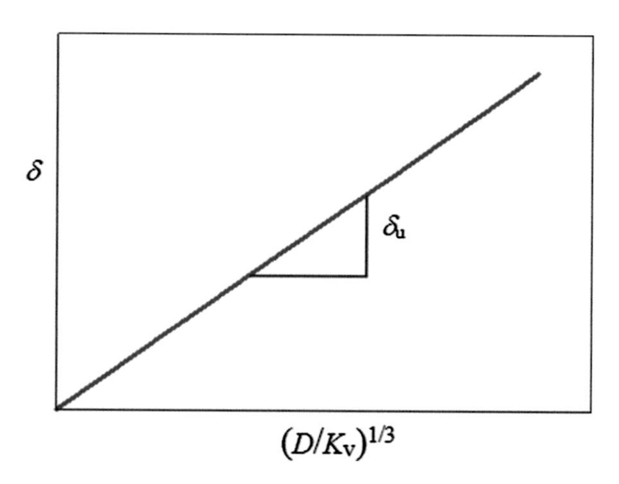

Fig. 9.15 Boundary layer linear relationship as per Eq. (9.107)

If $C_b >> C_x$, then the limiting current density becomes

$$i_L = zFC_bD^{2/3}K_v^{-1/6}\left(\frac{v_b}{x}\right)^{1/2} \tag{9.110}$$

Once again, if all parameters in Eq. (9.110) are constant, except the fluid velocity, then the limiting current density takes the form

$$i_L = \lambda_2 v_b^{1/2} \tag{9.111}$$

where

$$\lambda_2 = zFC_bD^{2/3}K_v^{-1/6}x^{-1/2} \tag{9.111a}$$

This expression, Eq. (9.111a), is the slope of a linear plot of $i_L = f(v_b^{1/2})$. On the analogy of Eq. (9.111), $i_L = f(v_x)$ gives a square root profile and higher values than Eq. (9.104) because of the $1/2$ exponent. Yet λ_2 is also a constant compared to λ_1. Again, i_L defined by Eq. (9.111) strongly depends on the fluid velocity v_x at low values. In this case, the ionic mass transport from solution to an electrode surface by diffusion mechanism depends on the electrolyte velocity, diffusion coefficient, bulk concentration and kinematic viscosity, and distance from the electrode surface.

9.6.2 Rotating-Disk Electrode

In characterizing the effects of solution on corrosion, a rotating-disk electrode is convenient for such an assessment since the angular velocity (ω) must influence the limiting current (I_L). Figure 9.16 shows commercial rotating electrodes, and Fig. 9.17 depicts the hydrodynamic system model for measuring I_L.

The rotating-disk electrode (RDE) is the working electrode and acts as an electrical conductor embedded in an inert polymer (resin). It is connected to an electrical motor for controlling the rate of the solution flow at angular velocities, 100 rpm $\leq \omega <$ 4000 rpm. The side view in Fig. 9.17 indicates that a hydrodynamic boundary layer is dragged by the spinning disk and thrown away from the center of the electrode. This dynamic action is also shown by the top-view drawing.

In this hydrodynamic electrochemical cell, a coupled convection-diffusion mechanism takes place at the WE surface, but it is the solution flow that controls the measurable steady-state current, which is commonly converted to current density by knowing the electrode surface area.

In addition, the role of an auxiliary electrode (AE) in Fig. 9.17 is to balance the current at a hydrodynamic working electrode (WE). The cell potential is measured between the WE and a reference electrode (RE), which does not pass any current to the WE. When the electrode rotates, it induces a flux of electroactive species to the electrode, leading to an increase in steady-state current.

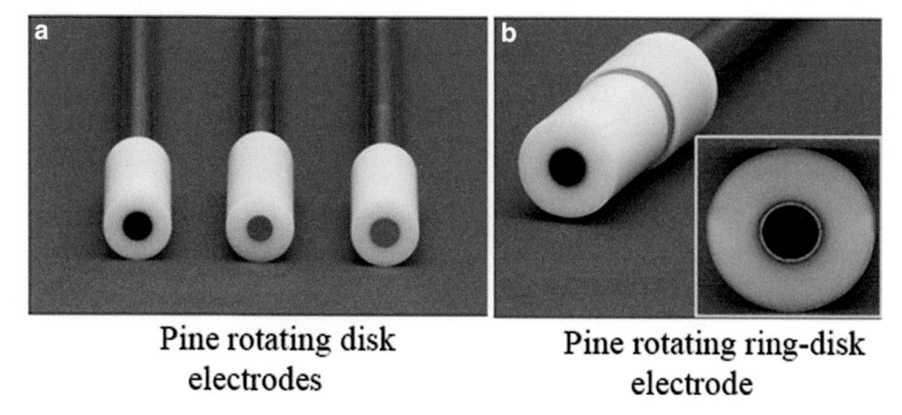

Fig. 9.16 (**a**) Rotating-disk electrodes (RDEs) and (**b**) rotating ring-disk electrode (RRDE). Pictures taken from Pine Research Instrumentation, Technical Note 2005-01, REV 004 (Sep 2007)

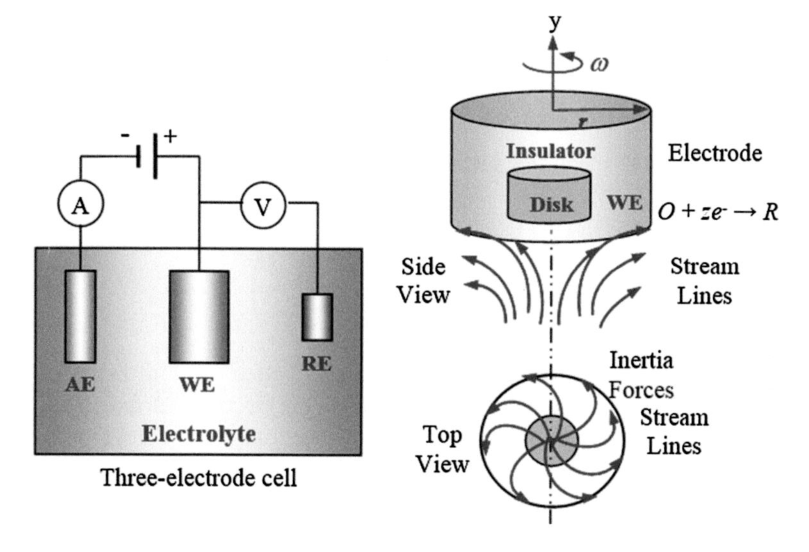

Fig. 9.17 Schematic three-electrode cell and its hydrodynamic cell model having a rotating-disk electrode (RDE) as the working electrode (WE)

On the other hand, the rotating ring-disk electrode (RRDE in Fig. 9.16b) is geometrically more complex than the RDE, and a ring, disk, and shaft share the same y-axis of rotation. The angular velocity range is 1000 rpm $\leq \omega <$ 3000 rpm. The function of the RRDE is to provide a current-potential (I-E) curve on the disk if E_{ring} is constant or vice versa.

For forced convection under laminar flow, the Sherwood number is applicable in the following form [4, 35, 58]:

$$S_h = \lambda_f \left(S_c\right)^{1/3} \left(R_e\right)^{1/2} \tag{9.112}$$

and

$$R_e = \frac{r^2\omega}{K_v} < 10^5 \quad \text{for laminar flow} \tag{9.113}$$

$$\lambda_f = 0.62 \quad \text{(factor)} \tag{9.114}$$

Substituting Eqs. (9.80), (9.113), and (9.114) into (9.112) and inserting the resultant expression into (9.88) along with $\delta = r$ yield the current ($I = iA_s$) for kinetic studies known as Levich equation:

$$I = 0.62\,(zFA_s)\left[\frac{D\,(C_b - C_s)}{\delta}\right]\left(\frac{K_v}{D}\right)^{1/3}\left(\frac{\delta^2\omega}{K_v}\right)^{1/2} \tag{9.115}$$

where $r =$ disk radius

$\omega = v_x/d = 2\pi N =$ angular velocity (rad/s)
$N =$ revolutions per second
$d =$ disk diameter
$A_s =$ electrode surface area (cm^2)

If $C_s = 0$ at the electrode surface, Eq. (9.115) becomes the Levich limiting current, Eq. (6.117) [35]:

$$I_L = 0.62\,(zFA_s)\left(\frac{DC_b}{\delta}\right)\left(\frac{K_v}{D}\right)^{1/3}\left(\frac{\delta^2\omega}{K_v}\right)^{1/2} \tag{9.116}$$

This expression, Eq. (9.116), takes into account the effect of fluid motion on the diffusion layer δ, which is an adherent thin film on the electrode surface. Therefore, diffusion coupled with convection contributes to the total mass transfer process to or from the electrode surface, and it is known as convective diffusion in which δ is considered immobile at the electrode surface [16]. Recall that the limiting thickness of the diffusion layer is illustrated in Fig. 6.7. Also, Levich equation describes the effect of rotation rate, concentration, and kinetic viscosity on the current at a rotating-disk electrode.

In addition, Levich [35] derived an expression for the diffusion layer as a function of angular velocity or rotating speed as

$$\delta = 1.62 D^{1/3} K_v^{1/6} \sqrt{\omega} \tag{9.117}$$

$$\delta = \left(\frac{\pi}{4}\right)^2 D^{1/3} K_v^{1/6} \sqrt{\omega} \tag{9.118}$$

Other hydrodynamic cases for corrosion studies can be found elsewhere [8]. For a continuous metal removal from solution in electrowinning, rotating cylinders and disks are used as cathodes.

The classical rotating-disk cathode is known as **Weber's disk** [11, 44] having a diameter of $2a$. The corresponding differential equation, in cylindrical coordinates, for the relevant concentration of a species j is given by

$$\frac{\partial C}{\partial r^2} + \frac{1}{r}\frac{\partial C}{\partial r} + \frac{\partial^2 C}{\partial r^2} = 0 \tag{9.119}$$

This differential equation must satisfy the following boundary conditions:

$$C = 0 \qquad @\ x = 0, \quad r \leq a \tag{9.120}$$

$$\frac{\partial C}{\partial r} = 0 \qquad @\ x = 0, \quad r > a \tag{9.121}$$

$$C = C_b \qquad @\ x = \infty, \quad r \geq 0 \tag{9.122}$$

$$C = C_b \qquad @\ x \geq 0, \quad r = \infty \tag{9.123}$$

The corresponding closed-form solution of Eq. (9.119) using the modified Bessel function is of the form [11]

$$\frac{C}{C_b} = 1 - \frac{2}{\pi}\tan^{-1}\left(\frac{a}{\mu}\right) \tag{9.124}$$

$$\mu = \sqrt{\left(\frac{1}{2}\left[(r^2 + x^2 - a^2) + \sqrt{\{(r^2 + x^2 - a^2)^2 + 4x^2a^2\}}\right]\right)}$$

where C_b is the concentration in a bulk of solution and μ can take any value based on the disk dimensions.

Using Taylor's series to expand Eq. (9.124) gives

$$\begin{aligned} \tan^{-1}\frac{a}{\mu} &= \sum_{n=0}^{\infty} \frac{(-1)^n}{2n+1}\left(\frac{a}{\mu}\right)^{2n+1} \\ &= \frac{a}{\mu} - \frac{1}{3}\left(\frac{a}{\mu}\right)^3 + \frac{1}{5}\left(\frac{a}{\mu}\right)^5 - \ldots.. \\ &\simeq \frac{a}{\mu} \end{aligned} \tag{a}$$

Thus,

$$\frac{C}{C_b} \simeq 1 - \frac{2}{\pi}\frac{a}{\mu} \tag{9.124a}$$

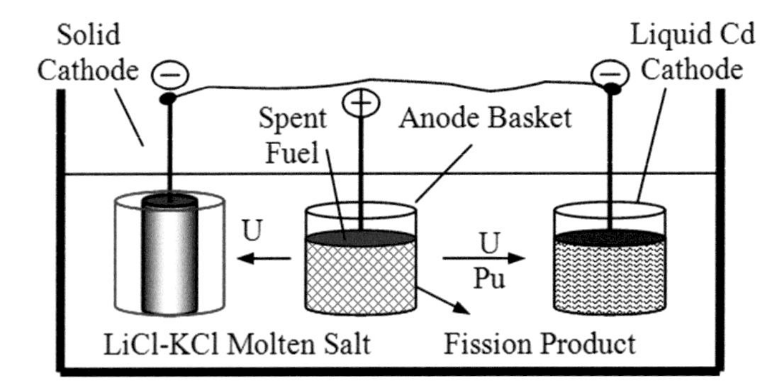

Fig. 9.18 Rotating-basket anode in molten-salt electrorefining [53]

9.6.3 Rotating-Basket Electrode

Furthermore, molten-salt electrolysis of nuclear waste (spent fuel) is an innovative electrometallurgical process essentially developed by CMT Division at Argonne National Laboratory [53]. Figure 9.18 shows a schematic cell for treating nuclear waste, specifically for recovering uranium (U) and plutonium (Pu), and collecting fission products, such as rare earth Cs, Sr, and Ce, at the bottom of the molten $LiCl - KCl$ eutectic electrolyte at 500 °C.

In actual fact, the schematic rotating-anode basket normally contains chopped nuclear waste solids for processing using electrolysis. The system shown in Fig. 9.18 is a high-efficiency electrolytic cell, which consists of a stationary solid cathode where pure U is deposited, a stationary liquid cadmium (Cd) cathode (LCC) for collecting U and Pu due to their chemical stability in liquid Cd. Hence, cadmium reacts with plutonium to form $PuCd_6$ intermetallics. Details on operating parameters and cell conditions and cell design features can be found elsewhere [53].

In essence, all rotating disks or rings are hydrodynamic working electrodes (WEs) used in three-electrode-cell systems. It is typical in a three-electrode system to use a reference electrode (RE) so that the potential difference E between the WE and RE is measured accordingly.

9.7 Codeposition of Impurities

The cathodic overpotential equation derived in Chap. 5 for concentration polarization is very useful in electrowinning operation. In analyzing the electrodeposition of metal M_1^{z+} cations from aqueous solution, the cathodic overpotential and current density for stationary and rotating-disk electrodes are related by

$$\eta_c = \frac{RT}{zF} \ln\left(1 - \frac{i_c}{i_L}\right) \qquad \text{for } i_c < i_L \tag{9.125}$$

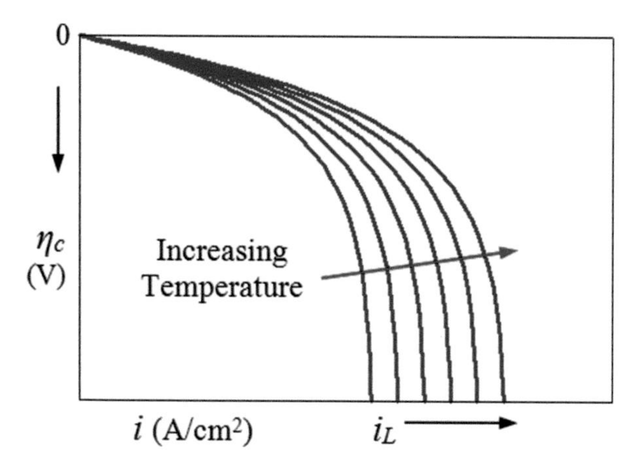

Fig. 9.19 Schematic cathodic overpotential profile showing increasing limiting current density with increasing temperature

In addition, Eq. (9.125) is plotted in Fig. 9.19 for showing the effects of temperature on both overpotential and limiting current density of a hypothetical electrochemical system.

For an electrode containing no metallic impurities under concentration polarization, the applied current density must be a fraction of the limiting current density; that is, $i_c < i_L$ for optimizing the electrolytic deposition of M_1^{z+} cations on the cathode surface electrodes. This current density constraint is the basis for deriving Eq. (9.125), which dictates that $\eta_c \to -\infty$ when $i_c \to i_L$ since $\ln(1 - i_c/i_L) \to -\infty$. Therefore, if $i_c < i_L$ yields $\eta_c >> -\infty$, a regular and uniform (smooth) layer of M_1 atoms is ideally deposited at a relatively high current efficiency ϵ.

According to Eq. (9.125), the optimum overpotential depends on both i_c and T; $\eta_c = f(T, i_c)$. It is known that i_L increases with increasing temperature. This, then, requires that $\eta_c = f(T, i_c, i_L)$, and the optimization scheme is more complex. However, fixing i_L at a temperature T allows Eq. (9.125) be i_c dependent only. On the other hand, if $i_c \geq i_L$, an irregular and porous (powdery) metal deposition and anode passivity result [49].

Since i_L is usually known, a current density factor can be defined as

$$\lambda = \frac{i_c}{i_L} \tag{9.126}$$

Thus, Eq. (9.125) becomes

$$\eta_c = \frac{RT}{zF} \ln(1 - \lambda) < 0 \qquad \text{for } 0 < \lambda < 1 \tag{9.127}$$

which dictates that $\eta_c \to 0$ as $\lambda \to 0$ and electrochemical equilibrium is nearly achieved and polarization is then nearly lost under these conditions. Therefore, the

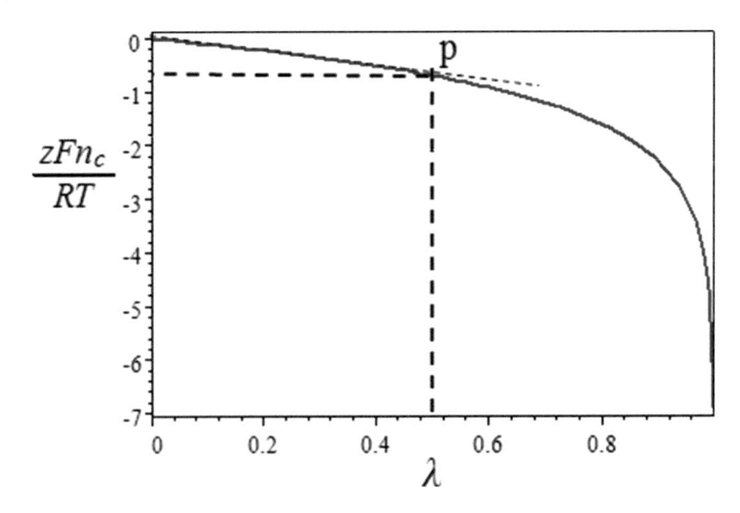

Fig. 9.20 Theoretical trend of the cathodic overpotential showing the idealized point P

cathodic overpotential range is $-\infty < \eta_c < 0$ for electrodeposition of metal M_1^{z+} cations. Thus, η_c is now dependent on two variables, instead of three. Figure 9.20 shows the theoretical trend for η_c as per Eq. (9.127).

Notice that the drawn straight line from 0 to P serves as a crude approximation scheme for predicting $\lambda = 0.50$, which gives $i_c = 0.50 i_L$. This is remarkably in agreement with most practical electrowinning operations [49]. Using $\lambda = 0.50$ in Eq. (9.127) yields

$$\eta_{c,\lambda=0.50} = -\frac{0.69RT}{zF} \tag{9.128}$$

$$\frac{d}{d\lambda}\left(\frac{zF\eta_c}{RT}\right)_{\lambda=0.50} = \frac{1}{\lambda-1} = -0.50 \tag{9.129}$$

Furthermore, if an electrolyte contains M_2^{z+} cations as impurities, the electrodeposition of the principal cation M_1^{z+} can be achieved, provided that the cathodic current density is defined by the following inequality:

$$i_L\left(M_2^{z+}\right) < i_c < i_L\left(M_1^{z+}\right) \tag{9.130}$$

otherwise, M_1^{z+} and traces of other impurities will codeposit [40, 49]. There can be a situation in which codeposition of impurities is required for obtaining an alloyed electrowon layer [25], but this is an unusual case in electrowinning of pure metals from aqueous solutions. If codeposition occurs, electrorefining is used for purifying or refining the electrowon metal M_1. The above current density inequality, Eq. (9.130), is schematically shown in Fig. 9.21 for the principal metal M_1 and impurity metal M_2.

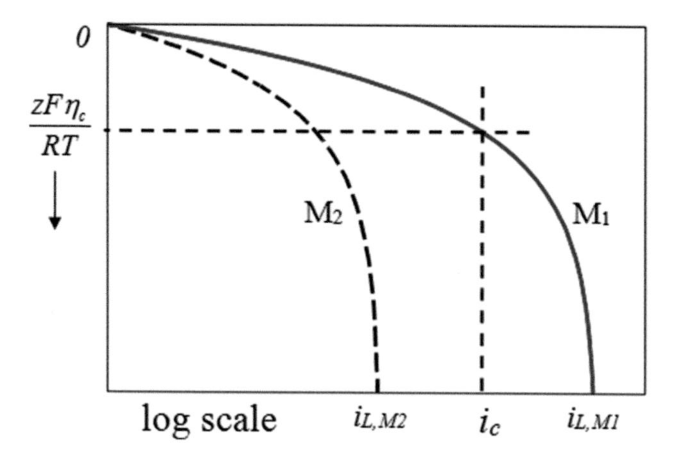

Fig. 9.21 Schematic polarization curves showing limiting current densities

If codeposition is inevitable, a low concentration of impurities must be maintained in solution so that the mass transfer of the impurity M_2^{z+} is minimal and that of M_1^{z+} is maximal since the diffusivity $D_{M_1^{z+}}$ will increase. Also, increasing the electrolyte temperature ($T < 70\,°C$) reduces its viscosity and, therefore, increases the mass transfer of M_1^{z+} cations. For instance, the application of Eq. (9.130) in electrowinning copper provides the constraints of the applied current density for controlling the mass transfer of copper and impurity iron. Hence,

$$i_{L,Fe} < i_c < i_{L,Cu} \tag{9.131}$$

A common applied cathodic current density can be defined by

$$i_c = \frac{i_{L,Cu}}{i_{L,Fe}} \tag{9.132}$$

Apparently, very little $Fe^{3+} + e^- \rightarrow Fe^{2+}$ mass transfer occurs [49]. In addition, the limiting current density due to diffusion mass transfer can be determined using Eq. (6.51):

$$i_L = \frac{zFDC_b}{\delta} \tag{9.133}$$

In the above expression, Eq. (9.133), the diffusivity is treated as a temperature-dependent variable, and it is assumed that it obeys the Arrhenius-type equation:

$$D = D_o \exp\left(-\frac{Q}{RT}\right) \tag{9.134}$$

where D_o = constant $\left(\text{cm}^2/\text{s}\right)$

Q = activation energy (J/mol)

If mass transfer is aided by convection, which is the case for most practical purposes, then i_L can be predicted from Eq. (9.103). However, diffusion, migration, and convection mass transfer influence the electrolytic deposition in electrowinning, electrorefining, and electroplating. In this case, the total molar flux is predicted by Eq. (6.2) and the current density by Eq. (6.9).

In addition, if the limiting current density becomes temperature dependent through the diffusion coefficient, then Eq. (9.133) becomes an Arrhenius-type equation. Hence,

$$i_L = \frac{zFD_oC_b}{\delta} \exp\left(-\frac{Q}{RT}\right) \tag{9.135}$$

This expression clearly shows the temperature dependence of the limiting current density in diffusion control processes [16].

Assume a hypothetical case in which the limiting current density i_L, Eq. (9.135), depends on the temperature (T) and bulk concentration (C_b). Thus, i_L profiles are schematically shown in Fig. 9.22 for a species j into solution.

Clearly, i_L varies nonlinearly according to changes in temperature and bulk concentrations. At relatively low temperatures, i_L is not strongly dependent on temperature. This can be attributed to slow reaction rates.

Moreover, the derivative of Eq. (9.135) with respect to temperature is

$$\frac{di_L}{dT} = \frac{zFD_oC_b}{\delta} \frac{Q}{RT^2} \exp\left(-\frac{Q}{RT}\right) \tag{9.136}$$

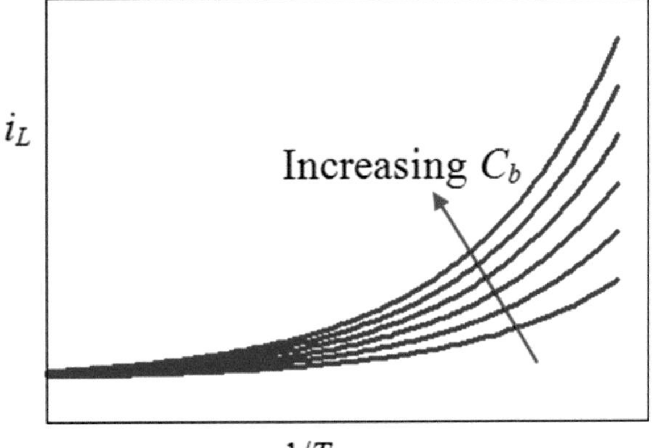

Fig. 9.22 Limiting current density as a function of temperature

This expression provides a maximum di_L/dT when plotted against the absolute temperature T, and it decays very rapidly. This means that di_L/dT takes a near-asymptotic profile

In general, the driving force for electrodeposition is the overpotential induced during cathodic polarization in the range of E_o and E_L, where E_L is the potential at its corresponding i_L response. Hence, the applied cathodic potential range and its corresponding current density response are $E_L < E < E_o$ and $i_o < i < i_L$, respectively. At this time, i_L is defined by Eq. (9.133) or (9.135).

9.8 Electrorefining

Electrorefining (ER) is an electrochemical technique suitable for refining metals produced by electrowinning or smelter. The final metal product from a smelter or an electrowinning (EW) processes contains impurities, but further refinement can be accomplished by electrorefining cells, which deliver high-purity metal M, such as copper, and a secondary recovery of precious metals may follow. A plant layout for electrorefining is shown in Fig. 9.3. The impurities are more noble than the metal being produced. For instance, the nearly pure cathodes are used as anodes in order to force metal dissolution to approximately 85 %, and the remaining is used as scrap for producing starter sheet. Subsequently, metal reduction follows for depositing pure metal at a current density in the range of $200\,\mathrm{A/m^2} \leq i \leq 215\,\mathrm{A/m^2}$ and at a potential $E_{ER} < E_{EW}$. During electrorefining (ER), the impurities fall to the bottom of the cells and are collected as "anode slime," rich in precious metals, such as Au, Ag, and Pt [9, 49]. Furthermore, the reason for using starter sheets made out of a different alloy is to compensate for a weak atomic bonding at the sheet-deposit interface. This would allow easy removal by mechanical means.

Electrorefining has its fundamentals based on the principles of electrochemical kinetics. The anodes are artificially oxidized by losing electrons and freeing cations, which migrate toward the cathode surfaces through the solution under the influence of an applied current. As a result, the cations M^{z+} combine with electrons to form reduction reactions. This is the source of electrolytic metal deposition, which, in turn, becomes adhered as a layer on the cathode surfaces. This layer increases in thickness as electrolysis proceeds up to an extent determined by experimentation or industrial practice. The refining process can be represented by a reversible electrochemical reactions of the form

$$M = M^{z+} + ze^- \tag{9.34}$$

$$M_1 \rightarrow M_1^{z+} + z_1 e^- \tag{9.35}$$

which implies that the metal M looses ze^- electrons at the anode surface and regains them at the cathode surface by a reverse electrochemical reaction. The electrolysis involves exchange of electrons rather than chemical reactions. Some impurities M_1 less noble than metal M can oxidize at the anode surface forming cations dissolved in the electrolyte. Thus far the electrorefining process has been described as an idealized electrochemical process for purifying metals at a macroscopic production. Macroscale operating parameters, such as temperature, electrolyte flow, electrolyte concentration and constituents, applied potential, and current density, have a strong influence on the quality of the electrolytic deposition [51]. Smoothed, adhered, and defect-free (cracks) electrodeposition can be obtained by using adequate amounts of organic additives (inhibitors) in the electrolyte. However, impurities must be eliminated, and a uniform distribution of the current density among the electrodes is a must. Otherwise, the anodes polarize leading to a low rate of dissolution and a reduced rate of metal deposition, and consequently, the plating process yields a rough and nodular surface deposition [13, 49, 51].

9.9 Electroplating

Electroplating (*EP*) is an electrochemical process similar to electrowinning of metal M in which deposition is accomplished as a metal coating having a very thin thickness, and it is based on the principles of electrochemical reactions caused by exchange of electrons, instead of chemical reactions. Also, electroplating can be defined as an electrolytic metal deposition due to reduction reactions on the cathode electrode, where a film (thin layer) of metal is adhered as a surface finishing process. Thus, thin metal films are plated on metal and alloy surfaces to enhance their appearance, corrosion resistance, wear resistance, and aesthetic for jewelry and tableware. Such films must be adherent and uniform on regular or irregular metal surfaces, which must be cleaned prior to electrodeposition; otherwise, poor film adhesion and incomplete deposition will not protect the base metal, and appearance will not be so appealing. In general, cleaning the surface of a base metal (object) is achieved by using organic, alkaline, or acid solutions. The cleaning details depend on the type of solvent, but the main objective is to remove all foreign substances, such as oils, grease, dirt, oxides, and the like.

The most common metal coatings are base on plated Cr, Ni, Cd, Co, Ag, and Au. A metal coating may be electronegative or electropositive to the base metal with respect to the potential difference in a particular environment due to galvanic effects [49]. In general, a metallic or plastic object can be coated with a different plating metal from an aqueous solution, which is the electric conductor, by passing a current from the anode (+) through the solution to the cathode (−). In this electrochemical event, the positively charge ions (cations) are the charge carriers that move in parallel with the current direction toward the cathode.

Many factors are controlled for plating a thin film on a particular solid surface. Thus, the following should be considered [49]:

- Low current density and relatively low temperature promote coherent deposits with fine grains; otherwise a loose and coarse-grained film is produced.

- Small amounts of organic solutes, such as ureas and glues, may enhance the electrodeposition since they hinder metal ion discharge and increase the cathodic polarization
- The electroplating process must be uniform on the surface object. The geometry of the base metal may cause uneven plating distribution.

The most common industrial applications of electroplating can be summarized as follows:

- Tin (*Sn*) coating on steel cans for food storage.
- Chromium (*Cr*) coating of less than 2 μm thick on steel parts, such as automobile bumpers and household towel holders, serves as decorative and protective film. *Cr* coating is hard, wear, and corrosion resistant. In certain decorative applications, a *Cr*coating is plated over a *Ni* undercoating on steel.
- Silver (*Ag*) and gold (*Au*) coatings are individually used as decorative films on jewelry and electric conductive parts. *Ag* coatings are also used on utensils.

It should be mentioned that electroplating is considered as one type of finishing process. Other finishing processes include coating on an object for protecting or decorating purposes, but these processes are not electroplating at all since electricity is not used to accomplish a practical industrial procedure. Some of these nonelectrolytic finishing processes are briefly cited below [48, 54]. Hence:

- Paint coating does not need a detailed explanation at all since it is a familiar technique used to decorate or protect an object or structure.
- Hot dipping is a technique used to coat steel parts with zinc (*Zn*) for protection against corrosion. Thus, the *Zn*-coated part is known as galvanized steel.
- Chemical vapor deposition (CVD), which requires vapor condensation on a solid surface in a vacuum chamber or inside of a tube furnace at relatively high temperatures. Nowadays, this technique is used for coating metal and optical components. Also, CVD is used for producing carbon nanotubes of a few nanometers in diameter on a variety of substrates at relatively low temperatures, but their alignment and direction on a substrate apparently are difficult to achieve. This may be attributable to the graphite sheet size and morphology.
- Aluminum anodizing consists of a chemical reaction between aluminum and oxygen to form a thin aluminum oxide film on an aluminum-base metal. This process occurs as a natural phenomenon in air, and it is an electroless plating process based on chemical reactions.
- Chemical mechanical planarization (CMP) process is used to prepare silicon wafer surfaces for multilayer integrated circuits containing copper interconnections [54].
- Powder coating in which a metal powder is melted on the surface of a base material in a furnace at relatively high temperatures. The molten metal is used to coat sport and playground equipment, washing and drying machines, etc.

- Plasma consists of a cloud of ionized gas for coating a metal substrate at high temperature. This technique is classified as a low-pressure or high-pressure plasma according to a particular application.

9.9.1 Deposition on Nonconductors

Prior to metallization of a plateable nonconductor, nonconductors are pretreated using etching for surface conditioning as a cleaned and contaminant-free surface area. Nonetheless, metallization is the process during which a metal coating is deposited on the nonconductor, such as a plateable polymer. Despite that the electrochemistry of polymers provides a vast number of polymers, it specifically conveys to polyimide polymer and acrylonitrile-butadiene-styrene (ABS) copolymer for metallization of nonconductive components. The plateable nonconductors are used in electronic industry as dielectric coatings for flexible circuitry, photolithography for deposition of copper on a Mylar surface, ion beam techniques such as ion implantation for insulators, electrochemical-autocatalytic method for direct metal deposition with no external electric current required, sputtering method for a well-distributed metal deposition without surface contaminants that leads to poor adhesion, and plasma-induced deposition for thin film of metal [45].

For further treatment, the metal-coated polymers are then single-layer or multilayer electroplated, depending on the design application based on engineering problems. Recall that polymers have a low density compared to most metals and their alloys, and subsequent industrial applications reduce or minimize weight problems.

In addition, a multilayered coating, known as superlattice coating, can be produced during a multilayer deposition from the same electrolyte at different potentials. A modern application of this technique is for computer hard-drive magnetic reading heads using a forced-pulse train-shaped potential. Also, a multilayer coating based on a sequence of metal ion deposition can be applied to either metal or plastic components for corrosion and wear resistance, weight saving, and visual appearance [7, 45].

Electroplating of a metal onto a polymer surface is commonly dominated by ABS due to strong atomic bonding of the final coating. This means that ABS-metal adhesion is good for most applications, such as household appliances and some automotive parts (wheel covers, mirror housings, license plate frames, and the like). The ASTM D3359 standard for measuring adhesion should be consulted. Once a polymer is made a conductor, electroplating of metal is possible using a conventional electrochemical cell. However, the subsequent applications of electroplated polymers depend on their response to mechanical, chemical, electrochemical, and electrical interactions.

In principle, the surface pretreatment of a nonconductor includes improvement of wettability characteristics so that the contact angle is proper for increasing the surface energy of the metal-nonconductor interface. As a result, both nonconductor and metal coating have a high surface energy for a suitable bond strength of the polymer-metal adhesion.

9.10 Problems/Questions

9.1. Copper cations are reduced on a cathode for 8 h at a current of 10 A and 25 °C. Calculate **(a)** the theoretical weight of copper deposited on the cathode and **(b)** the number of Coulombs of electricity in this electrodeposition process. Data: $A_{w,Cu} = 63.55\,\text{g/mol}$ and $F = 96{,}500\,\text{C}$. [Solution: (a) $W = 94.83\,\text{g}$ and (b) $q = 288{,}000\,\text{C}\,(= \text{A s})$.]

9.2. An electrochemical cell operates at 6 A, 25 °C, and 85 % current efficiency in order to electrolytically deposit copper ions (Cu^{2+}) from a leaching solution in an 8-h shift. Calculate **(a)** the amount of electrolytically deposited Cu and **(b)** the thickness of the deposit layer of copper. Data: $A_s = 100\,\text{cm}^2$ (total cathode surface area), $A_{w,Cu} = 63.55\,\text{g/mol}$, $\rho = 8.96\,\text{g/cm}^3$, and $F = 96{,}500\,\text{C/mol}$. [Solution: (a) $W = 48.36\,\text{g}$ and (b) $x = 0.54\,\text{mm}$.]

9.3. It is desired to electroplate chromium (Cr) onto a ferritic-martensitic carbon steel automobile bumper for an attractive appearance and corrosion resistance. An electroplating cell is operated at $I = 6.68\,\text{A}$ and 25 °C. The Cr^{3+} cations are reduced on the bumper (cathode) to form a thin film of 1.5 μm thick. How long will it take to produce a 1.5-μm-thick electroplated chromium film on the bumper if the total surface area is $100\,\text{cm}^2$ and $\epsilon = 60\,\%$? [Solution: $t = 100\,\text{s}$.]

9.4. An electrowinning cell contains $2x10^{-4}\,\text{mol/cm}^3$ of Cu^{2+} ions and operates for 10 min at 40 °C and 1 atm. Assume an ionic copper diffusivity equals to $2.34x10^{-5}\,\text{cm}^2/\text{s}$. Calculate the current density due to diffusion mass transfer. Compare your result with a typical industrial current density value of $200\,\text{A/m}^2$. Explain any discrepancy. [Solution: $i = 43\,\text{A/m}^2$.]

9.5. Calculate the amount of silver that can be electroplated in an electrochemical cell containing Ag^+ ions and operates at 7 A, 25 °C, and 101 kPa for 10 min. Assume a current efficiency of 95 %. [Solution: $W = 4.46\,\text{g}$.]

9.6. What are the differences between galvanic and electrolytic cells? Recall that both are electrochemical cells.

9.7. Consider the bipolar electrode for electrorefining metal M. Thus, the rate of formation (reduction) and dissolution (oxidation) are treated as steady-state quantities. Despite that the initial formation of a thin film at the cathode face and dissolution rates are controlled by reactions at the electrode-electrolyte interfaces, assume a steady-state diffusion mechanism. Derive expressions for the weight gain at the cathode side and the weight loss at the anode side.

9.8. A steel plate is Ni plated for corrosion protection, but its appearance is not so appealing. Therefore, a chromium-electroplating process is carried out for decorative purposes. Eventually, 0.25 g of Cr^{+6} cations are electroplated on the *Ni*-plated surface for a period of 15 min at 8 A. If the density and atomic weight of chromium are $7.19\,g/cm^3$ and $52\,g/mol$, respectively, calculate **(a)** the thickness of the electroplated film of chromium from its highest oxidation state and **(b)** the deposition rate. Data: $A_c = 100\,cm^2$ (cathode area). [Solutions: (a) $x = 3.48\,\mu m$, (b) $P_R = 1\,g/h$, and (c) $\epsilon = 39\,\%$.]

9.9. It is desired to produce a 20-μm-thick film of chromium on a Ni-plated steel part, which has a surface area of $65\,cm^2$. This can be accomplished by setting up an electroplating cell to operate at 7 A and current efficiency of 70 %. The density and atomic weight of chromium are $7.19\,g/cm^3$ and $52\,g/mol$, respectively. Calculate **(a)** the amount of Cr being plated and **(b)** the time it takes to plate the 20-μm-thick film of chromium from a solution containing Cr^{+6} cations at 30 °C. [Solutions: (a) $W = 0.94$ g and (b) $t = 35.40$ min.]

9.10. Predict **(a)** the electric potential E, **(b)** the time in minutes, and **(c)** the film growth rate in μm/min for electroplating a 10-μm-thick Cr film on a Ni-undercoated steel part when the electrolyte contains 10^{-4} mol/l of Cr^{+3} cations at 25 °C. The Ni-plated steel part has a 10-cm^2 surface area. The cell operates at 50 % current efficiency and at a passive current density of $5.12x10^{-3}\,A/cm^2$. [Solutions: (a) $E > E_o = -0.82$ V, (b) $t = 26$ min, (c) $dx/dt = 0.38\,\mu m/min$.]

9.11. The dissociation of silver hydroxide, $AgOH$, is $1.10x10^{-4}$ at 25 °C in an aqueous solution. **(a)** Derive an expression for the degree of dissociation as a function of total activity $[C_T]$, and plot the resultant expression for $0 < [C_T] < 0.50\,mol/l$, **(b)** calculate the degree of dissociation constant α_1 when the molar concentration is $[C_T] = 0.144\,mol/l$, and **(c)** determine the Gibbs free energy change ΔG^o. [Solutions: (b) $\alpha_1 = 0.028$ and (c) $\Delta G^o \approx -22.58$ kJ/mol.]

9.12. Calculate **(a)** the concentration of $AgOH$, Ag^+, and OH^- at 25 °C and **(b)** the pH if $[C_T] = 0.144\,mol/l$ and $\alpha_1 = 0.028$. [Solutions: (a) $C_{AgOH} = 17.48\,g/l$ and (b) $pH = 11.61$.]

9.13. An electrochemical cell contains 75 g/l of Cu^{+2} ions at 35 °C. Calculate **(a)** the Reynolds and Sherwood numbers, **(b)** the diffusion flux (J_x), **(c)** the current density (i), and **(d)** the applied potential (E). Data:

d	Characteristic distance	$= 12\,cm$
$L = x$	Cathode height	$= 100\,cm$
w	Cathode width	$= 80\,cm$
N	Number of cathodes	$= 50$
K_v	Kinematic viscosity	$= 0.80\,cm^2/s$
D	diffusivity	$= 1.24x10^{-5}\,cm^2/s$
v_x	Flow velocity	$= 46.90\,cm/s$
P	Electrical power	$= 32\,kW$

[Solutions: (a) $S_h = 709.19$ and (c) $i = 200\,\mathrm{A/m^2}$.]

9.14. An electrolyte containing Pb^{+2} ions at 35 °C is used in an electrowinning cell. The electrolyte is under steady laminar force convection. Use the data given below to determine **(a)** the Sherwood number, **(b)** the diffusion flux, and **(c)** the current. [Solutions: (a) $S_h = 709.19$, (b) $J_x = 4.49x10^{-11}\,\mathrm{mol/cm^2\,s}$, and (c) $I = 4.91\,\mathrm{kA}$.]

C_b	Bulk concentration	$= 75\,\mathrm{g/l}$
D	Diffusivity	$= 1.24x10^{-5}\,\mathrm{cm^2/s}$
$L = x$	Cathode length	$= 100\,\mathrm{cm}$
S_c	Schmidt number	$= 64{,}516.13$
R_e	Reynolds number	$= 703.50$
A_s	Total cathode area	$= 800{,}000\,\mathrm{cm^2}$
ϵ	Current efficiency	$= 85\,\%$

9.15. A hypothetical rotating-disk cell is shown below for electrowinning copper cations from a solution containing $C_o = 65\,\mathrm{g/l}$ at 40 °C and 101 kPa. Assume that the diffusivity and the electrolyte kinematic viscosity are $10^{-5}\,\mathrm{cm^2/s}$ and $0.60\,\mathrm{cm^2/s}$, respectively. Each disk has a radius of 50 cm and a width of 6 cm, and only a 160° segment is immersed in the electrolyte. Assume a cell current efficiency range of $0.50 \leq \epsilon \leq 1$:

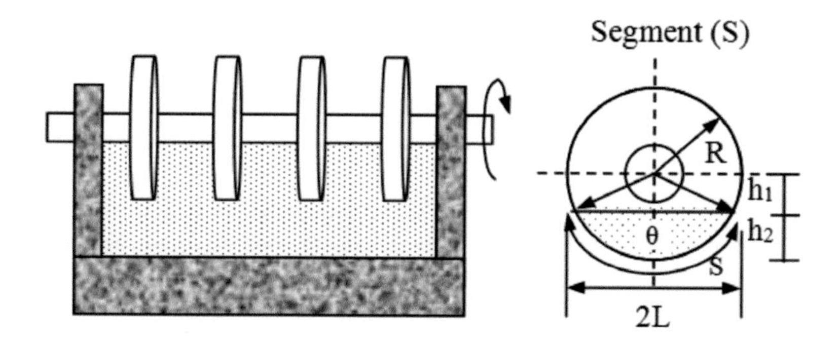

Theoretically, analyze the effect of **(a)** the angular velocity ω on the current density and **(b)** the angular velocity (ω) and the current density (ϵ) on the production rate of metal powder, and **(c)** the potential (E) and the current density (ϵ) on the energy consumption γ.

9.16. Use the hypothetical rotating-disk voltammetry data given in the table below for determining the diffusivity D of a metal cation M^{+2} in solution at room temperature. [Solution: $D = 3.97 \times 10^{-5}\,\mathrm{cm^2/s}$.]

ω (1/s)	100	225	400	625	900	1225
$\omega^{1/2}$ $(s^{1/2})$	10	15	20	25	30	35
i (mA/cm^2)	4.81	7.22	9.58	12.10	14.38	16.80

$C_o = 2.21x10^{-6}$ mol/cm^3
$K_v = 0.056$ cm^2/s
$z = 2$

9.17. It is desired to produce copper according to the electrolytic cell diagram

$$(-)\, Steel|\, Cu^{+2}/Cu,\, H^{+}/H_2|\,|H_2O/O_2\,|PbO_2\,(+)$$

for 24 h at which time the cathodes are mechanically stripped (removed). Copper is reduced from an aqueous copper sulfate ($CuSO_4$) and sulfuric acid (H_2SO_4) solution at 35 °C. The anodes operate at 100 % current efficiency with respect to oxygen evolution, and that for the cathodes is $\epsilon = 90\,\%$ due to hydrogen evolution. These types of electrodes are connected in parallel within 100 tanks that operate at 2.4 V each, and the electrodes have a specific submerged area A_e. Use the following relevant data to perform some calculations given below. Cost of electricity per $kWh, C_e = 0.30$ \$/kW h; market price of copper, $C_{market} = \$0.70$/kg; and

$P = 72$ kW (power)	$A_w = 63.54$ g/mol
$E = 2.4$ V	$D_{Cu^{+2}} = 1.24x10^{-5}$ cm^2/s
$N_T = 100$ (tanks)	$K_v = 0.80$ cm^2/s
$N_a = 30$ (anodes per tank)	$v_x = 0.312$ cm/s (velocity of Cu^{+2})
$N_c = 29$ (cathodes per tank)	$d = 6$ cm (electrode distance)
$A_e = 0.80$ m^2 (electrode area)	$C_e = 0.30$ \$/kW h (electricity)

(a) Write down all the reactions and the standard half-cell potential involved in this electrowinning process. What should the applied potential be? Calculate **(b)** the total current, the anodic and cathodic current densities, and **(c)** S_c, R_e, S_h, and C_b in g/l per tank. Let the anode characteristic length be $L = \sqrt{A_e}$. Use the Sherwood number ($S_h < 100$) and the Reynolds number ($R_e < 2300$) to assure force laminar flow. Determine **(d)** the production rate (P_R) and the total weight produced (W_T) in 24 h and **(e)** the energy consumption (γ). Use your engineering economics skills to estimate **(f)** the cost of energy consumed (C_{te}), the total cost of production (C_{prod}), and the gross income (G_I) in 24 h if the industrial cost of electric energy and the market price of copper are $C_e = \$0.30$ kW h and $C_{market} = \$3.193$/lb $= \$7.041$/kg, respectively. [Solutions: (a) $E^o_{Half-cell} \approx -1.57$ V, (b) $i_c = 0.001293$ A/cm^2, (c) $S_h \approx 39$, and $C_b = 78.80$ g/l.]

9.18. A Hall–Heroult (HH) cell is used to produce molten aluminum-containing hydrogen. The hydrogen content can be removed by allowing a thin film of molten aluminum flow down on an inclined plane during pouring in the vacuum. The inclined plane is shown below, and it is in the L-w-δ space, where δ is the film thickness and w is the width:

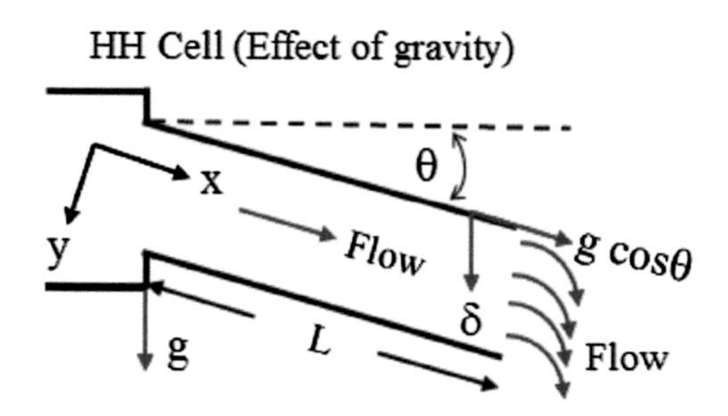

The weight-induced and gravity-induced free fluid flow occurs in the x-direction. Using Newton's law of viscosity yields the force acting on the fluid in the direction of the flow, together with the moment balance. Thus, this phenomenon is described by an ordinary differential equation:

$$\eta_v \frac{d^2 v_x}{dy^2} = -\rho g \cos(\theta) \quad \text{with BC's} \left\{ \begin{array}{l} v_x = 0 \ @ \ y = \delta \\ v_x > 0 \ @ \ y > 0 \end{array} \right\}$$

where η_v = fluid viscosity, v_x = fluid velocity, and ρ = density of the fluid. Solve this differential equation, and calculate **(a)** the average value of the boundary layer, **(b)** the velocity $\overline{v_x}$, **(c)** the molar flux (J_x) of hydrogen due to both diffusion and convection effects, **(d)** the time it takes to remove the dissolved hydrogen along the channel length L and the amount of hydrogen being removed, and **(e)** the Reynolds number R_e. Given data:

$\rho = \rho_{Al} = 2.5\,\text{g/cm}^3$	$w = 50\,\text{cm}$
$\eta_v = 40\,\text{g cm}^{-1}\text{s}^{-1}$	$L = 100\,\text{cm}$
$D_H = 5x10^{-3}\,\text{cm}^2/s$	$\theta = 2^o$
$C_{H,b} = 30\,\text{mol/cm}^3$	$\dot{m} = 100\,\text{g/s}$
$C_H = 0 \ @ \ y = 0$	$g = 981\,\text{cm/s}^2$

[Solutions: (a) $\delta = 0.46\,\text{cm}$, (b) $\overline{v_x} = 4.34\,\text{cm/s}$, (c) $J_x = 130.53\,\text{mol/cm}^2\text{s}$, (d) $m = 2.30\,\text{kg}$, and (e) $R_e = 0.13$.]

9.19. An electrowinning cell produces 10 kg/h of nickel (Ni) at a current efficiency (ϵ) of 80 %. The cell contains 20 cathodes and operates at a current of 215 A/m^2

and a potential of 2 V. Determine (a) the energy consumption, (b) the flow rate of the electrolyte, and (c) the cathode length for a width of 1 m. The concentration of nickel cations is 35 g/l. (d) Plot Eq. (9.25) and explain the result.

9.20. Below is an electrolytic cell diagram used in an electrowinning factory for producing solid copper on cathode electrodes:

$$(-)\, Steel|\, Cu^{+2}/Cu,\, H^{+}/H_2|\, |H_2O/O_2\, |PbO_2\, (+)$$

Recall that the cathodes are mechanically stripped (removed) after the electrowinning run is accomplished. The electrolyte used in this electrowinning operation contains aqueous copper sulfate ($Cu\,SO_4$) and sulfuric acid (H_2SO_4) solution, and it is maintained at 35 °C. Assume that the anodes operate at 100 % current efficiency with respect to oxygen evolution and that for the cathodes is $\epsilon = 88\,\%$ due to hydrogen evolution. These types of electrodes are connected in parallel within 80 tanks that operate at 2.6 V each, and the electrodes have a specific submerged area A_e. Use the following relevant data to perform some calculations given below. The average retail price of electricity (cost of electricity) per kWh,$C_e = 0.14$ \$/kW h; market price of copper, $C_{market} = \$0.17/\text{kg}$; and

$P = 72$ kW (power)	$A_w = 63.54$ g/mol
$E = 2.6$ V	$D_{Cu^{+2}} = 1.24x10^{-5}$ cm^2/s
$N_T = 100$ (tanks)	$K_v = 0.80$ cm^2/s
$N_a = 30$ (anodes per tank)	$v_x = 0.312$ cm/s (velocity of Cu^{+2})
$N_c = 29$ (cathodes per tank)	$d = 6$ cm (electrode distance)
$A_e = 0.80$ m^2 (electrode area)	$C_e = 0.14$ \$/kW h (electricity)

(a) Write down all the reactions and the standard half-cell potential involved in this electrowinning process. What should the applied potential be compared to the half-cell potential? Is the given operating potential (2.6 V) suitable for this electrowinning? Calculate **(b)** the total current, the anodic and cathodic current densities, and **(c)** S_c, R_e, S_h, and C_b in g/l per tank. Let the anode characteristic length be $L = \sqrt{A_e}$. Use the Sherwood number ($S_h < 100$) and the Reynolds number($R_e < 2300$) to assure force laminar flow. Determine **(d)** the production rate (P_R) and the total weight produced (W_T) in 24 h and **(e)** the energy consumption (γ). Use your engineering economics skills to estimate **(f)** the cost of energy consumed (C_{te}), the total cost of production (C_{prod}), and the gross income (G_I) in 24 h if the industrial cost of electric energy and the market price of copper are $C_e = \$0.30$ kW h and $C_{market} = \$3.193/\text{lb} = \$7.041/\text{kg}$, respectively. [Solutions: (a) $E^{o}_{Half-cell} \approx -1.57$ V, (b) $i_c = 0.001194$ A/cm^2, and (c) $S_h \approx 39$ and $C_b = 73.08$ g/l.]

References

1. P. Ardelean, K.J. Cathro, E.J. Frazer, J.F. Kubacki, T. Lwin, R.H. Newnham, L.J. Rogers, High intensity zinc electrowinning at the laboratory and small technical scale, in *Electrometallurgical Plant Practice*, ed. by P.L. Claessens, G.B. Harris (Pergamon Press, New York, 1990), pp. 115–127
2. ASARCO Incorporated, Commercial Department, Denver, CO, USA, www.asarco.com (2001)
3. P. Balaz, *Extractive Metallurgy of Activated Minerals* (Elsevier Science, New York, 2000)
4. A.J. Bard, L.R. Faulkner, *Electrochemical Methods* (Wiley, New York, 1980), pp. 98–100, 283
5. M. Bestetti, U. Ducati, G.H. Kelsall, G. Li, E. Guerra, Use of catalytic anodes for zinc electrowinning at high current densities from purified electrolytes. Can. Metall. Q. **40**(4), 451–458 (2001)
6. M. Bestetti, U. Ducati, G.H. Kelsall, G. Li, E. Guerra, Zinc electrowinning with gas diffusion anodes: state of the art and future developments. Can. Metall. Q. **40**(4), 459–469 (2001)
7. J.J. Bladon, A. Lamola, F.W. Lytle, W. Sonnenberg, J.N. Robinson, G. Philipose, A palladium sulfide catalyst for electrolytic plating. J. Electrochem. Soc. **143**(4), 1206–1213 (1996)
8. P.J. Boden, Effect of concentration, velocity and temperature, in *Corrosion: Metal/Environment Reactions*, 3rd edn. ed by L.L. Shreir, R.A. Jarman, G.T. Burstein (Butterworth-Heinemann, Oxford, 1998), pp. 2:3–2:30
9. H.E. Boyer, T.L. Gall (eds.), Electrolytic and electrothermal reduction and refining, in *Metals Handbook*, Desk edition (ASM Metals, Metals Park, OH, 1985), pp. 21·20–21·22
10. G.M. Cook, *Energy Reduction Techniques in Metal Electrochemical Processes*, ed. by G.R. Bautista, R.J. Wesley (The Metallurgical Society, Warrendale, PA, 1985), pp. 285–296
11. J. Crank, *The Mathematics of Diffusion* (Oxford University Press, Oxford, 1975), pp. 31, 42–43
12. R.T. Dehoff, *Thermodynamics in Materials Science* (McGraw-Hill, New York, 1993), pp. 458–484
13. Ch. Droste, M. Segatz, D. Vogelsang, in *Light Metals: Proc. 127th TMS Ann. Meet.*, San Antonio, TX, USA, 15–19 February 1998, pp. 419–428
14. Electrometals Technologies Limited (2001), www.electrometals.com
15. EPA Office of Compliance Sector Notebook Project, Profile of the Nonferrous Metals Industry, SIC Codes 333–334, EPA/310-R-95-010, September 1995
16. T. Erdey-Gruz, *Kinetics of Electrode Process* (Wiley-Interscience, a Division of Wiley, New York, 1972), pp. 104–106, 115–118
17. N. Eresen, R. Kammel, Nickel extractions from limonite ores by pressure leaching with aqueous solutions of polymines. Hydrometallurgy **7**, 41–60 (1981)
18. J.W. Evans, L.C. De Jonghe, *The Production of Inorganic Materials* (Macmillan, New York, 1991)
19. D. Felsher, J.A. Finch, P.A. Distin, Can. Metall. Q. **39**(3), 291–296 (2000)
20. N. Furuya, N. Mineo, J. Appl. Electrochem. **20**(3), 475–478 (1990)
21. N. Furuya, S. Motoo, J. Electroanal. Chem. **179**, 297–301 (1984)
22. N. Furuya, Y.T. Sakakibara, J. Appl. Electrochem. **2**(1), 58–62 (1996)
23. D.R. Gaskell, *An Introduction to Transport Phenomena in Materials Engineering* (Macmillan, New York, 1992)
24. G.H. Geiger, D.R. Poirier, *Transport Phenomena in Metallurgy* (Addison-Wesley, Readings, MA, 1973), pp. 514–541
25. J.W. Graydon, D.W. Kirks, Can. J. Chem. Eng. **69**(2), 564–570 (1991)
26. Great Lakes Research Corporation (GLRC) (2000), www.es.epa.gov
27. J.P. Gueneau de Mussy, Ph.D. Dissertation, Presses Universitaires de Bruxelles (PUB), Belgium, 2002, Sec. 1.3, Chap. 6
28. C.H. Hamann, A. Hamnett, W. Vielstich, *Electrochemistry* (Wiley-VCH, New York, 1998)
29. W.E. Haupin, W.B. Frank, Electrometallurgy of aluminum, in *Comprehensive Treatise of Electrochemistry Vol. 2: Electrochemical Processing*, ed. by J. O'M. Bobricks, B.E. Conway, E. Yeager, R.E. White (Plenum Press, New York, 1981), pp. 301–324

30. J.P. Holman, *Heat Transfer* (McGraw-Hill, New York, 2002), p. 207
31. C. Hughes, AJ Parker Cooperative Research Center for Hydrometallurgy (2000), www.parkercentre.crc.org
32. W.F. Hughes, F.J. Young, *The Electromagnetodynamics of Fluids* (Wiley, New York, 1966)
33. F.P. Incropera, D.R. DeWitt, *Introduction to Heat Transfer* (Wiley, New York, 1985)
34. K. Kondo, R.N. Akolkar, D.P. Barkey, M. Yokoi (eds.), *Copper Electrodeposition for Nanofabrication of Electronics Devices* (Springer Science+Business Media, New York, 2014)
35. V.G. Levich, *Physicochemistry Hydrodynamics* (Prentice-Hall, Englewood Cliffs, NJ, 1962)
36. J.B. Mohler, *Electroplating and Related Processes* (Chemical Publishing Company, New York, 1969), pp. 119, 225–228
37. A. Moraru, I. Panaitescu, A. Panaitescu, D. Mocanu, A.M. Morega, Rev. Roum. Sci. Techn. Electrotechn. et Energ. **43**(4), 473–484 (1998)
38. S. Motoo, M. Watanabe, N. Furuya, Gas diffusion electrode of high performance. J. Electroanal. Chem. **160**(1–2), 351–357 (1984)
39. New materials improve energy efficiency and reduce electricity use in aluminum production, DOE/CH-10093-140, September 1992, www.es.eps.gov
40. R.D. Pehlke, *Unit Processes of Extractive Metallurgy* (Elsevier, North-Holland, New York, 1973), pp. 201–225
41. H.D. Peters, VAW Aluminium Technologie GmbH, Bonn (2001) www.vaw-atg.de
42. R.F. Probstein, *Physicochemical Hydrodynamics: An Introduction* (Butterworth, A Division of Reed Publishing Inc., Boston, 1989), pp. 59–61
43. T. Rosenquist, *Principles of Extractive Metallurgy* (McGraw-Hill, New York, 1983)
44. Y. Saito, Rev. Polarogr. (Kyoto) **15**, 177 (1968)
45. M. Schlesinger, M. Paunovic, *Modern Electroplating*, 5th edn. (Wiley, Hoboken, NJ, 2010)
46. M. Sheedy, in *Proc. 1997 TMS Ann. Meet.*, Orlando, FL, USA, 9–13 February 1997, pp. 433–441
47. A.D. Sneyd, A. Wang, Magnitnaya Gidrodinamika **32**(4), 487–493 (1996), www..mhd.sal.lv/contents/1996/4/mg.32.4.r.html
48. The American Electroplaters and Surface Finishers Society, Inc. (2001) www.aesf.org
49. L.G. Twidwell, Electrometallurgy, Unit Process in Extractive Metallurgy, NSF Project SED 75-04821, 1978
50. P. Tyagi, *Electrochemistry* (Discovery Publishing House PVT LTD, Grand Rapids, MI, 2006), pp. 73, 81
51. B. Veileux, A.M. Lafront, E. Ghali, Can. Metall. Q. **40**(3), 343–354 (2001)
52. A. Vignes, *Extractive Metallurgy 1: Basic Thermodynamics and Kinetics (ISTE)* (Wiley-ISTE, New York, 2010)
53. J.L. Willit, R.J. Blaskovitz, G.A. Fletcher, *Electrorefining Throughput Studies*, Report ANL-01/06, Argonne National Lab, Chemical Technology Division, Argonne (2001), pp. 1–110. http://www.ipd.anl.gov/anlpubs/2001/04/39219.pdf
54. www.mse.berkeley.edu (2001)
55. www.elmhurst.edu (2001)
56. X.J. Xue, in *Proc. 121st TMS Annu. Meet., Light Metals 1992*, San Diego, CA, 1–5 March 1992, pp. 773–778
57. A. Yurkov, *Refractories for Aluminium: Electrolysis and the Cast House* (Springer International Publishing, Cham, 2015)
58. Z. Zembura, Corros. Sci. **8**, 703 (1968)

High-Temperature Oxidation 10

High-temperature oxidation (HTO) of metals and alloys is a scale-forming oxidation process in gaseous environments. HTO is influenced by metal temperature, gas composition, exposure time, and pressure, and it may be characterized by weight gain, thickness reduction (penetration), and rate of oxide thickness growth, which is a measure of the rate of oxidation in oxidizing (O_2), sulfidizing (H_2S), carburizing (CH_4 or CO), and nitriding (N_2) conditions. HTO is also referred to as high-temperature oxidation, tarnishing, and scaling, and the rate of attack is significantly increased with increasing temperature.

In characterizing HTO, X-ray diffraction is a very useful and reliable technique for indexing patterns of phases leading to phase identification in the corrosion product known as surface scale. In fact, this high-temperature phenomenon is different from the rust layer formation (in the presence of moisture) since it virtually occurs in dry gaseous environments or molten salts. In certain cases, HTO may be an internal oxidation process when oxygen diffusion is faster than the surface oxidation rate. Yet the temperature gradient in semiconductors is the driving force for diffusion of charge carriers (electrons or holes).

Most experimental data available in the literature is based on weight gain per unit surface area. However, weight gain and thickness reduction or penetration are adequate parameters for assessing HTO. For instance, measurements of metal thickness reduction and oxide layer thickness increments are very important because the former is related to structure strength.

The protectiveness of an oxide layer is significantly dependent on the temperature and corrosive environments. In high-temperature oxygen media, the Pilling–Bedworth law can be used for assessing quantitatively the protectiveness of oxide layers. These layers are also known as scales, but they can be treated as oxide coatings as well. These coatings may protect the surface of equipment components from thermal deterioration.

N. Perez, *Electrochemistry and Corrosion Science*,
DOI 10.1007/978-3-319-24847-9_10

10.1 Thermodynamics of Oxides

Thermodynamically, the driving force for the oxidation of a metal in gaseous environments is the Gibbs free energy of formation (ΔG), and consequently, the occurrence of an oxidation chemical reaction depends on the magnitude of ΔG.

Figure 10.1 shows the standard Gibbs energy of formation as a function of temperature, known as the Ellingham diagram, for oxides under oxygen partial pressure (P_{O_2}), carbon oxide/dioxide pressure ratio (P_{CO}/P_{CO_2}), and hydrogen/water pressure ratio (P_{H_2}/P_{H_2O}).

The Ellingham diagram (Fig. 10.1) is a graphical form based on the principles of thermodynamics for metal oxidation reactions at relatively high temperatures. Such reactions depend on ΔG^o, pressure, and temperature. This diagram does not include any information on reaction kinetics, but one can assume that chemical reactions occur at slow rate.

Mathematically, the standard ΔG^o at a constant pressure is defined by

$$\Delta G^o = -RT \ln (K) \tag{10.1}$$

$$\Delta G^o = \Delta H^o - T\Delta S^o \tag{10.2}$$

where K is the equilibrium constant and ΔH^o is the enthalpy change, which is a measure of the heat absorbed or released at constant gas pressure. The term ΔS^o is the entropy change and it is a measure of atomic disorder. The variations of both ΔH^o and ΔS^o with temperature are given by

$$\Delta H^o (T) = \Delta H^o (T_o) + \int_{T_o}^{T} C_p(T) dT \tag{10.3}$$

$$\Delta S^o (T) = \Delta S^o (T_o) + \int_{T_o}^{T} \frac{C_p(T)}{T} dT \tag{10.4}$$

where $\Delta H^o (T_o)$ = standard enthalpy change (kJ/mol) at $T_o = 298\,\text{K}$

$\Delta S^o (T_o)$ = standard entropy change (kJ/mol K) at $T_o = 298\,\text{K}$
C_p = heat capacity (kJ/mol K) at constant pressure

In addition, a plot of the function given by Eq. (10.2) with constant ΔH^o and ΔS^o gives a straight line, for which ΔH^o is the intercept at $T = 0$ and $\Delta S = (\partial \Delta G/\partial T)_p < 0$ is the slope. Thus, a thermodynamic linear behavior of the metal-gas oxidizing system is obtained (Fig. 10.1).

Oxidation of a metal caused by oxygen (O_2), carbon dioxide (CO_2), and/or water vapor (H_2O) at relatively high temperatures is briefly described in this section. According to the generalized reactions given below, the standard Gibbs free energies (ΔG^o) of oxide formation for varying temperature and gas partial pressure at equilibrium are

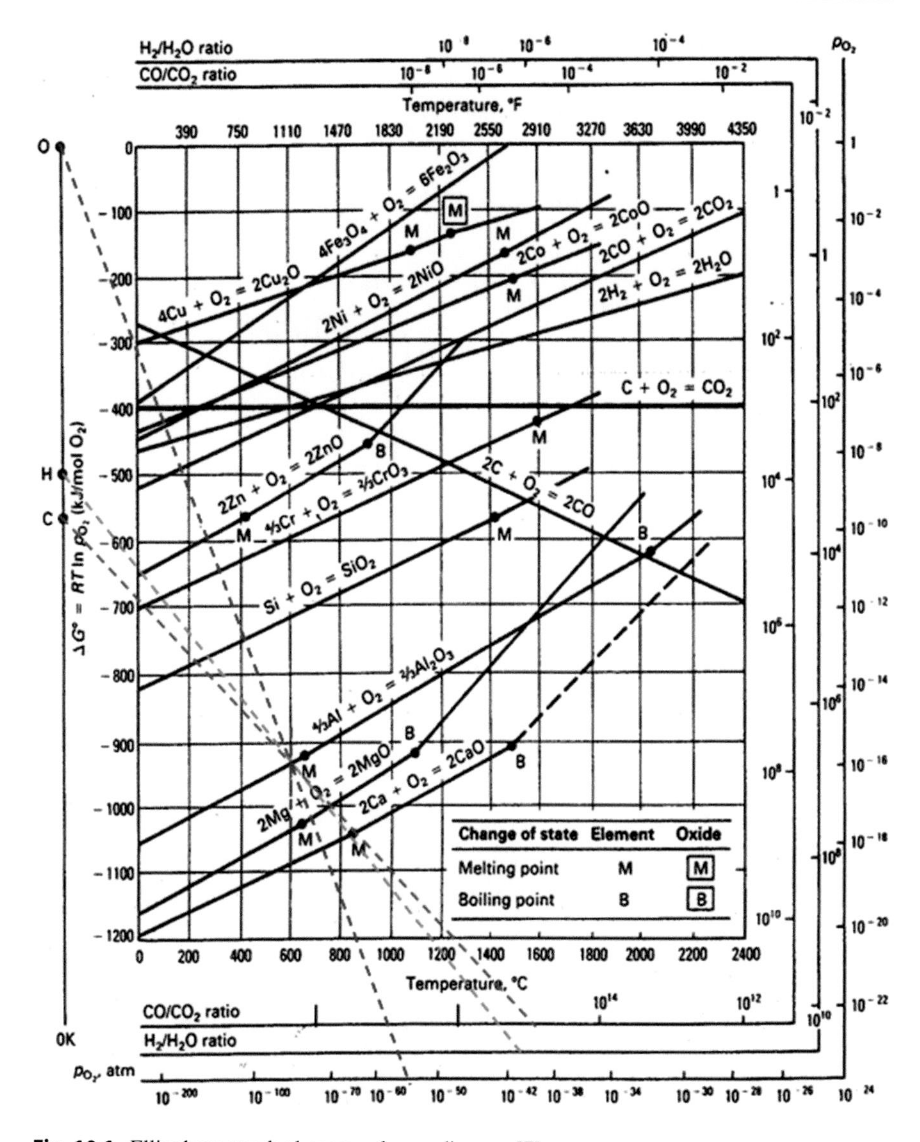

Fig. 10.1 Ellingham standard energy change diagram [7]

$$(1)\ \frac{2x}{y}M + O_2 = \frac{2}{y}M_xO_y \tag{10.5a}$$

$$\Delta G^o = -RT\ln\left(\frac{\left[M_xO_y\right]^{2/y}}{[M]^{2x/y}\left[O_2\right]}\right) = RT\ln\left(P_{O_2}\right) \tag{10.5b}$$

$$(2)\ xM + yCO_2 = M_xO_y + yCO \tag{10.6a}$$

$$\Delta G^o = -RT \ln\left(\frac{[M_xO_y][CO]^y}{[M]^x[CO_2]^y}\right) \tag{10.6b}$$

$$\Delta G^o = -yRT \ln\left(\frac{P_{CO}}{P_{CO_2}}\right) \tag{10.6c}$$

$$(3)\ xM + yH_2O = M_xO_y + yH_2 \tag{10.7a}$$

$$\Delta G^o = -RT \ln\left(\frac{[M_xO_y][H_2]^y}{[M]^x[H_2O]^y}\right) \tag{10.7b}$$

$$\Delta G^o = -yRT \ln\left(\frac{P_{H_2}}{P_{H_2O}}\right) \tag{10.7c}$$

where $[j] = a_j =$ activity of species j $\left(\text{mol/l or mol/cm}^3\right)$

$P =$ pressure (kPa)

Notice that M_xO_y is the common generalized chemical formula for a metal-oxide compound that forms under the influence of P_{O_2}, P_{CO}, P_{CO_2}, P_{H_2}, and P_{H_2O} partial pressures. However, if $\Delta G^o < 0$, then it is a measure of negative deviation from equilibrium, and the reaction proceeds from left to right as written. Conversely, if $\Delta G^o > 0$, then a positive deviation from equilibrium implies that a reaction occurs in the reverse direction from right to left.

Example 10.1. *This example can make it clear on how to use Fig. 10.1. Suppose that one is interested in forming copper oxide (Cu_2O) scale at 887 °C. Then, using the left-hand side vertical pressure line with points "O" for oxygen, "H" for $P_{H_2}/_{H_2O}$ ratio, and "C" for $P_{CO}/_{CO_2}$ ratio, draw straight lines from these points through the copper and oxygen reaction line:*

$$4Cu + O_2 = 2Cu_2O$$

at 887 °C until they intercept the pertinent outer lines for the oxygen and pressure ratios. ***(a)*** *Determine ΔG^o. (b) Which metals are oxidized and reduced based on ΔG^o.* ***(c)*** *What is the usefulness of Fig. 10.1?*

Solution.

(a) *Follow the color dashed lines shown in Fig. 10.1 to read off the pressures and corresponding standard Gibbs free energy change. The results are*

$$P_{O_2} = 2.78x10^{-9} \text{ atm}$$

$$P_{CO}/P_{CO_2} = 10^{-4}$$

$$P_{H_2}/P_{H_2O} = 10^{-4}$$

$$\Delta G^o = -190 \text{ kJ/mol } of\ O_2$$

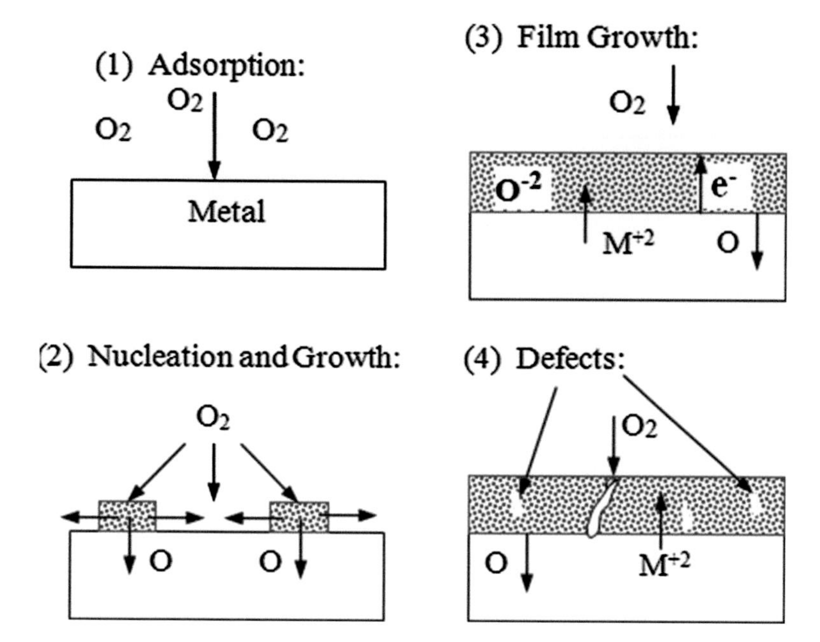

Fig. 10.2 Model for scale formation [13]

(b) *Metals below* $\Delta G^o = -190\ kJ/mol$ *at* $887\,°C$ *can be oxidized, and those above can be reduced. However, Cu can be reduced by Ni because its line is below the aluminum free energy line.*

(c) *It provides a basis for evaluating the possibility of chemical separation by oxidation.*

Furthermore, Fig. 10.2 shows a model for oxide scale formation in gaseous environments at high temperatures.

The schematic sequence of events for scale formation indicates that the mechanism involves adsorption of atomic oxygen on the metal surface at stage 1. Then, nucleation and growth occur at favorable sites until a thin oxide film forms covering the metal surface at stage 2. If the film lacks of macro- and micro-defects, it protects the metal from oxidizing any further. When scale growth occurs as shown in stage 3, metal oxidation occurs, $M \rightarrow M^{2+} + 2e^-$, releasing electrons which migrate through the oxide film to react with atomic oxygen. At stage 4, the oxide scale thickens, and the growth stresses cause defects such as porosity, cavities, and microcracks. Consequently, the oxide scale is not protective, and the metal is oxidized by a different mechanism [4].

Figure 10.3 depicts images of oxide layers on iron and a titanium alloys and aluminum exposed to oxygen-rich environments at high temperatures. Observe how multiphase scales form at the expense of material thickness reduction, which affects the structural integrity to an extent. For example, Fig. 10.3a shows that FeO oxide has a substantial amount of defects next to the metal substrate, but it may be sound at the extreme end. Also, the Fe_3O_4 oxide exhibits defects, including an inclined crack.

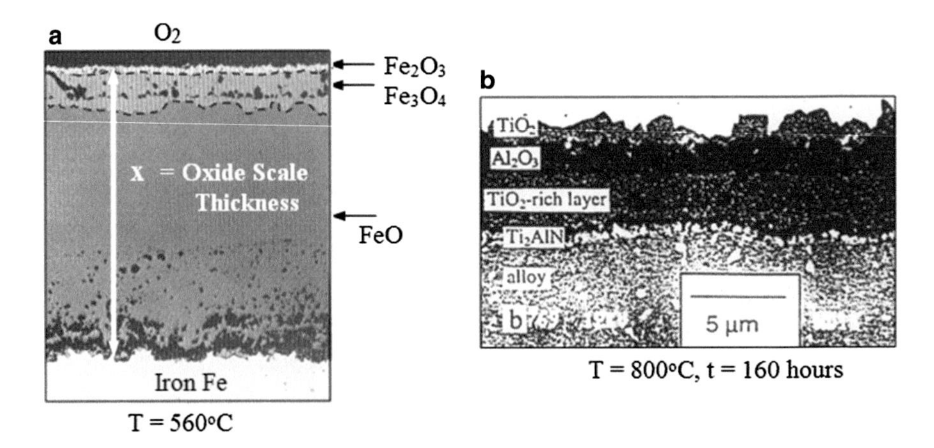

Fig. 10.3 SEM photomicrographs of protective oxide scales on (**a**) iron at high temperatures [6] and (**b**) extruded Ti-46.8Al-1Mo-0.19Si alloy heated in an oxygen environment at 800 °C for 160 h [17]

The scales in Fig. 10.3 indicate that those metals and alloys which oxidize to more than one valence can form a series of oxide layers [6]. The inner layer is the most metal-rich compound and has the lowest valance, such as Fe^{2+} in Fig. 10.3a. This implies that cation vacancies and electron holes are the dominant defects as in p-type FeO. On the other had, the outer layer, such as Fe_2O_3, is oxygen rich and may be n-type oxide because of anion vacancies diffusing inward. Moreover, Fig. 10.3b shows the series of oxide layers that form on a Ti alloy. The characterization of these oxide layers can be found elsewhere [17].

Iron and Aluminum Figure 10.4a shows an oxide layer of less than 1 μm in thickness on a cast Fe-$7.5Al$-$2Cr$ ferrous alloy exposed to a complex gas mixture containing sulfur.

Despite that the oxide has a complex structure, mostly Al_2O_3 phase, the layer breaks down due to the formation of cracks, which act as fast diffusion sites for sulfur-containing nodule growth as indicated in Fig. 10.4b, c. Eventually, the layer becomes nonprotective since these nodules collapse after a prolonged exposure time. The final oxide surface morphology is shown in Fig. 10.4d which represents a nonprotective single-phase oxide scale mostly composed of aluminum oxide (Al_2O_3) [19].

Silicon Thermal oxidation in a water-vapor environment is commonly used to grow silicon oxides in fabrication of wafers or modern silicon integrated circuits (ICs). For instance, the oxidation of silicon may occur in two different environments as indicated by the following chemical reactions:

$$\text{Si}\,(s) + O_2\,(g) \rightarrow \text{Si}\,O_2\,(s) \qquad \text{(dry oxidation)} \qquad \text{(a)}$$

$$\text{Si}\,(s) + 2H_2O\,(g) \rightarrow \text{Si}\,O_2\,(s) + 2H_2\,(g) \quad \text{(wet oxidation)} \qquad \text{(b)}$$

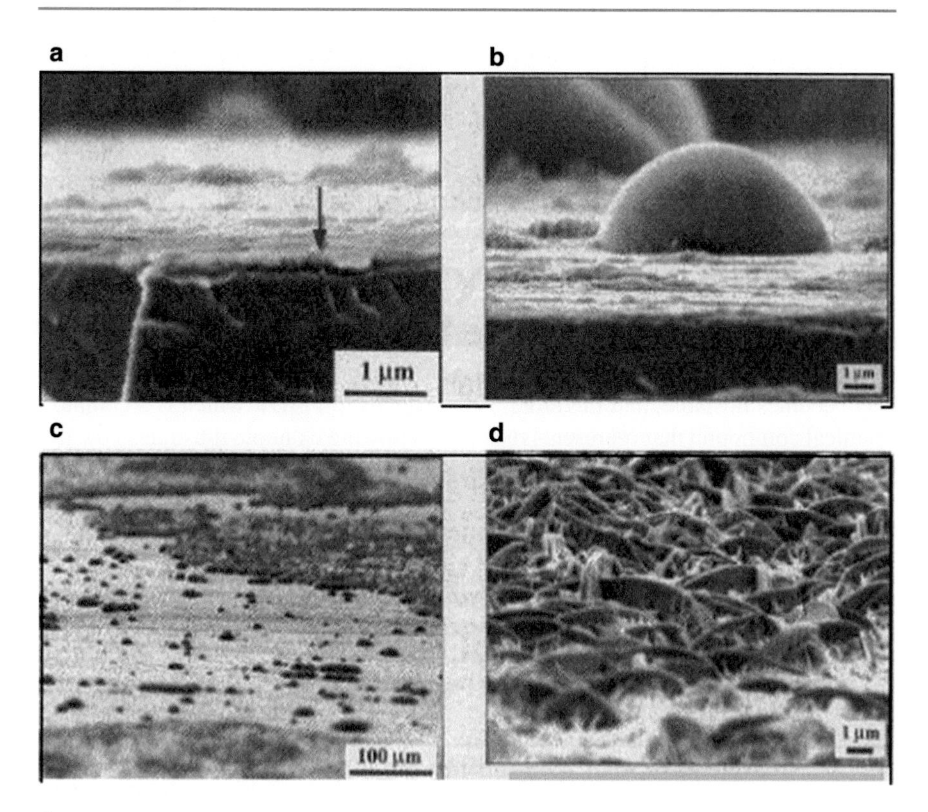

Fig. 10.4 SEM photomicrographs showing the sequence of mostly passive alumina Al_2O_3 scale breakdown. The *arrow* in part (**a**) indicates the initial oxide scale of Fe-7.5Al-2Cr < 1 μm thick on a cast alloy in a complex gas mixture composed of oxygen, carbon dioxide, water, nitrogen, and sulfur at 500 °C. The prolonged exposure time was 100 h [19]. (**a**) Thin protective Al_2O_3 (20,000X), (**b**) nonprotective Al_2O_3 nodule (8000X), (**c**) individual overgrown Al_2O_3 nodule (200X), (**d**) nonprotective Al_2O_3 layer (6500X)

Specifically, the oxidation is a chemical reaction between the oxidants and the silicon atoms, and as a result, a layer of oxide is formed on the silicon surface of the wafer. A typical operating temperature range is $850\,^{\circ}\mathrm{C} < T < 1300\,^{\circ}\mathrm{C}$. This leads to kinetic studies which normally reveal an increase, to an extent, in oxide growth rate with increasing temperature.

Iron and Titanium The oxidation process can be repeated for increasing the thickness of the oxide layer to an extent under control environment. This is evident in Fig. 10.3 for iron (*Fe*) and titanium (*Ti*) exposed to oxygen-based environments at temperature T.

The general steps for oxidizing a metal surface is gas diffusion, solid-state diffusion, and formation of oxides. This sequence of diffusion applies to all high-temperature oxidation processes to form oxide layers, and if this occurs under steady-state conditions, then Fick's first law of diffusion becomes very useful.

The variations in gas pressure at constant temperatures limit the stability of oxide compounds and the solubility of gases. Thus, the higher the pressure, the higher the solubility of gases. This implies that there must exist a chemical stability domain in terms of pressure [2].

Of particular interest is the effect of gas pressure on the structure of carbon nanotubes that is an interesting scientific topic in the chemical vapor deposition process and on the well-known iron oxide phases in high-temperature corrosion. The reader is encouraged to search details on this particular topic at a nanoscale.

Nonetheless, the oxide phases in Fig. 10.3a are a particular case in which wustite (FeO) and magnetite (Fe_3O_4 or $FeO.Fe_2O_3$) compound phases are stable layers at a specific gaseous partial pressure range. Beyond the maximum partial pressure limit, FeO becomes unstable, and therefore, FeO changes to Fe_3O_4, which is a common chemical compound that represents rust. The following example illustrates how the oxygen partial pressure stability domain is determined.

Example 10.2. *The application of thermodynamics in metal oxidation is succinctly presented in this example. One of the simplest methods for characterizing metal-oxide formation is to use a thin plate exposed to an oxidizing environment such as oxygen. The experimental output may be oxide thickness and weight measurements for a given oxygen partial pressure and time at temperatures. However, thermodynamic parameters are significant in this type of work. Suppose that a thin nickel plate is exposed to oxygen for a fixed time at relatively high temperatures and the Gibbs free energy is measured in such an environment. Only the standard free energy* (ΔG^o) *data is illustrated in the table given below along with the overall reaction:*

T (°C)	ΔG^o (kJ/mol)
100	−431
300	−390
500	−348
1000	−252
1400	−170
$2Ni + O_2 = 2NiO$	
$\Delta G^o = -RT \ln K_{sp}$	

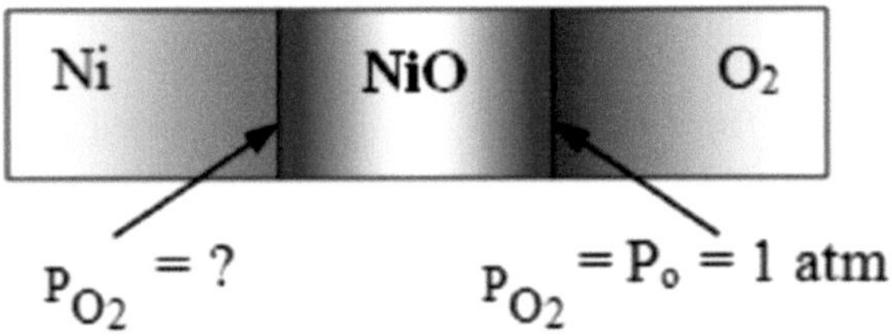

Ni NiO O_2

$P_{O_2} = ?$ $P_{O_2} = P_o = 1$ atm

$$a_{Ni,Ni} = a_{Ni,O_2} = 1\ mol/l$$

Use 100 °C and 1400 °C to calculate **(a)** *the equilibrium constant K_{NiO}.* **(b)** *Plot ΔG^o vs. T (°C) and determine the standard enthalpy change (ΔH^o) and entropy change (ΔS^oin kJ/°C) by regression analysis. Determine* **(c)** *the oxygen partial pressure P_{O_2} and* **(d)** *the activity $a_{O_2\text{-}face}$.*

Solution. **(a)** *The equilibrium constant at the given temperatures:*

$$\Delta G^o = -RT \ln (K_{NiO}) \tag{a}$$

$$K_{NiO} = \exp\left(-\frac{\Delta G^o}{RT}\right) \tag{b}$$

$$K_{NiO} = \exp\left(-\left(\frac{-450 + 0.20T}{8.314\,(T + 273)}\right) x10^3\right) \tag{c}$$

$$K_{NiO} = 1.66x10^{60} \quad @\ T = 100\,°\mathrm{C}$$

$$K_{NiO} = 4.01x10^{3} \quad @\ T = 1400\,°\mathrm{C}$$

Therefore, K_{NiO} is strongly dependent on temperature. The higher the temperature, the lower K_{NiO}.

(b) *Linear regression equation (curve fitting):*

$$\Delta G^o = \Delta H^o - T\Delta S^o = -450\ \mathrm{kJ/mol} + \left(0.20\ \mathrm{kJ/mol\,°C}\right) T \tag{d}$$

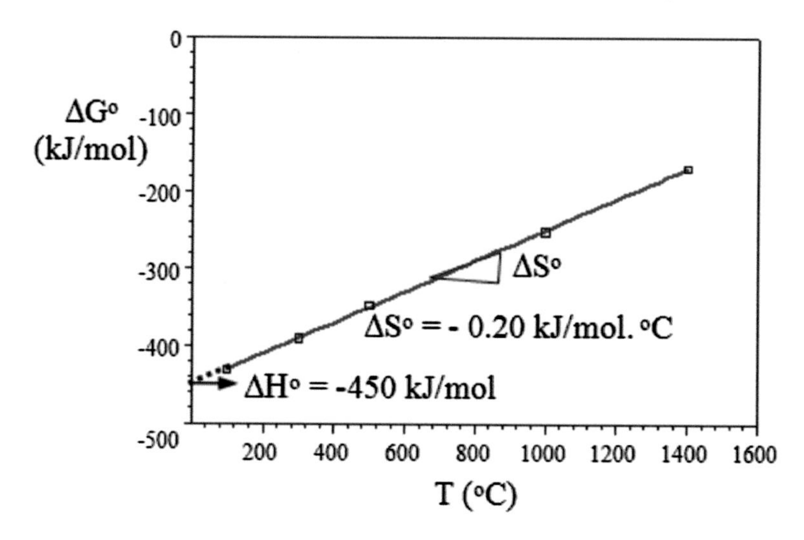

(c) *Oxygen partial pressure at the Ni-NiO$_2$ interface:*

$$K_{Nio} = \frac{[NiO]^2}{[Ni]^2\,[O_2]} = \frac{1}{[O_2]} \tag{e}$$

$$[O_2] = \frac{P_{O_2}}{P_o} \tag{f}$$

Combining Eqs. (c), (e), and (f) yields the oxygen partial pressure in the Ni-NiO$_2$ interface:

$$P_{O_2} = \frac{P_o}{K_{Nio}} = P_o \exp\left[\left(\frac{-450 + 0.20T}{8.314\,(T + 273)}\right)x10^3\right] \tag{g}$$

$$P_{O_2} = 6.04x10^{-61} \qquad @\ T = 100\,°\text{C}$$

$$P_{O_2} = 2.49x10^{-4} \qquad @\ T = 1400\,°\text{C}$$

(d) *The activity of nickel cations at Ni-O$_2$ interface:*

$$\Delta G^o_{Ni\text{-}face} = \Delta G^o_{O_2\text{-}face} \tag{h}$$

$$-RT\ln\left(K_{NiO}\right)_{Ni\text{-}face} = -RT\ln\left(K_{NiO}\right)_{O_2\text{-}face}$$

$$\ln\left(\frac{[NiO]^2}{[Ni]^2\,[O_2]}\right)_{Ni\text{-}face} = \ln\left(\frac{[NiO]^2}{[Ni]^2\,[O_2]}\right)_{O_2\text{-}face}$$

$$\ln\left(\frac{1}{[O_2]}\right)_{Ni\text{-}face} = \ln\left(\frac{1}{[Ni]^2\,[O_2]}\right)_{O_2\text{-}face}$$

Then,

$$\left(\frac{P_o}{P_{O_2}}\right)_{Ni\text{-}face} = \left(\frac{P_o}{[Ni]^2\,(P_{O_2} = P_o)}\right)_{O_2\text{-}face}$$

$$a_{O_2\text{-}face} = [Ni]_{O_2\text{-}face} = \left(\sqrt{\frac{P_{O_2}}{P_o}}\right)_{Ni\text{-}face} \tag{i}$$

Substituting Eq. (g) into (i) gives

$$[Ni]_{O_2\text{-}face} = \sqrt{\exp\left(\left(\frac{-450 + 0.20T}{8.314\,(T + 273)}\right)x10^3\right)} \tag{j}$$

$$[Ni]_{O_2\text{-}face} = a_{O_2\text{-}face} = 7.77x10^{-31} \qquad @\ T = 100\,°\text{C}$$

$$[Ni]_{O_2\text{-}face} = a_{O_2\text{-}face} = 2.22x10^{-3} \qquad @\ T = 1400\,°\text{C}$$

Therefore, the activity of nickel, $[Ni]_{O_2-face}$, is strongly dependent on temperature. The higher the temperature, the higher $[Ni]_{O_2-face}$.

Example 10.3. *This example is adapted from Barsoum [2] in order to determine the oxygen partial pressure stability domain for the iron oxide phases shown in Fig. 10.3a. Use the data given below at* 1000 °*C:*

$$2Fe + O_2 \rightarrow 2FeO \qquad \Delta G^o_{FeO} = -366.10\ \text{kJ/mol} \tag{a}$$

$$6FeO + O_2 \rightarrow 2Fe_3O_4 \qquad \Delta G^o_{Fe_3O_4} = -1263.93\ \text{kJ/mol} \tag{b}$$

$$4Fe_3O_4 + O_2 \rightarrow 6Fe_2O_3 \qquad \Delta G^o_{Fe_2O_3} = -862.96\ \text{kJ/mol} \tag{c}$$

Solution. *Applying Eqs. (10.5a) and (10.5b) to reactions (a) and (b) yields the equilibrium constants and activities as*

$$K_{FeO} = \frac{[FeO]}{[Fe]\,[O_2]} = \frac{1}{[O_2]}$$

$$a_{FeO} = [O_2]_{FeO} = \left(\frac{P_{O_2}}{P_o}\right)_{FeO} \qquad for\ P_o = 1\ \text{atm}$$

Thus, the free energy and oxygen partial pressure equations for the FeO phase (wustite) are

$$\Delta G^o_{FeO} = -RT\ln\left(K_{FeO}\right) = RT\ln\left(P_{O_2}\right)_{FeO}$$

$$\left(P_{O_2}\right)_{FeO} = \exp\left(\frac{\Delta G^o}{RT}\right)_{FeO}$$

For the FeO-Fe_3O_4 and Fe_3O_4-FeO systems, the free energies of formation for reactions (b) and (c) are determined by the oxide energy balance. According to these reactions, the free energy of formation is given by

$$\Delta G^o_{f-Fe_3O_4} = 2\Delta G^o_{Fe_3O_4} - 6\Delta G^o_{FeO}$$

$$\Delta G^o_{f-Fe_2O_3} = 6\Delta G^o_{Fe_2O_3} - 4\Delta G^o_{Fe_3O_4}$$

Thus, the reaction oxygen partial pressure equations become dependent on the free energy of formation of the FeO-Fe_3O_4 and Fe_3O_4-FeO systems. Hence,

$$\left(P_{O_2}\right)_{Fe_3O_4} = \exp\left(\frac{\Delta G^o_{f-Fe_3O_4}}{RT}\right)_{Fe_3O_4}$$

$$\left(P_{O_2}\right)_{Fe_2O_3} = \exp\left(\frac{\Delta G^o_{f-Fe_2O_3}}{RT}\right)_{Fe_2O_3}$$

Now, the calculated free energy change of formation and oxygen partial pressure for each reaction are given below:

Wüstite (FeO):

$$(1)\ 2Fe + O_2 \rightarrow 2FeO$$
$$\Delta G^o_{FeO} = -366.10\ \text{kJ/mol}$$
$$(P_{O_2})_{FeO} = 9.49x10^{-16}\ \text{atm}$$

Magnetite (Fe_3O_4):

$$(2)\ 6FeO + O_2 \rightarrow 2Fe_3O_4$$
$$\Delta G^o_{f-Fe_3O_4} = -331.25\ \text{kJ/mol}$$
$$(P_{O_2})_{Fe_3O_4} = 2.56x10^{-14}\ \text{atm}$$

Iron oxide (Fe_2O_3): *It can be hematite* (α-Fe_2O_3) *or maghemite* (γ-Fe_2O_3):

$$(3)\ 4Fe_3O_4 + O_2 \rightarrow 6Fe_2O_3$$
$$\Delta G^o_{f-Fe_2O_3} = -122.05\ \text{kJ/mol}$$
$$(P_{O_2})_{Fe_2O_3} = 9.81x10^{-6}\ \text{atm}$$

Therefore, the stability pressure domain for each phase is

Fe	is stable at $P_{O_2} < (P_{O_2})_{FeO}$
FeO	is stable at $(P_{O_2})_{FeO} \le P_{O_2} < (P_{O_2})_{Fe_3O_4}$
Fe_3O_4	is stable at $(P_{O_2})_{Fe_3O_4} \le P_{O_2} < (P_{O_2})_{Fe_2O_3}$
Fe_2O_3	is stable at $(P_{O_2})_{Fe_2O_3} \le P_{O_2} \le P_o = 1\ atm$

For the sake of clarity, the reader should verify the calculated values of $\Delta G^o_{f\text{-}Fe_3O_4}$ *and* $\Delta G^o_{f\text{-}Fe_2O_3}$ *from the free energy chart at* 1000 °C. *A similar procedure can be carried out for the oxide formation in Fe-*CO/CO_2 *and Fe-*H_2/H_2O *systems.*

10.2 Point Defects in Oxides

In this section, the phenomenon of point defects, such as vacancies and interstitial atoms, in crystals is briefly introduced. The oxide entropy change (ΔS) increases when more point defects, also known as imperfections, generate within a crystal. Metal oxides at equilibrium may contain nearly equal numbers of cation and anion vacancies.

The number of point defects (n) producing a minimum free energy change, $\Delta G = \Delta H_f - T\Delta S$, can be modeled by the Arrhenius-type equation [1, 5, 11]:

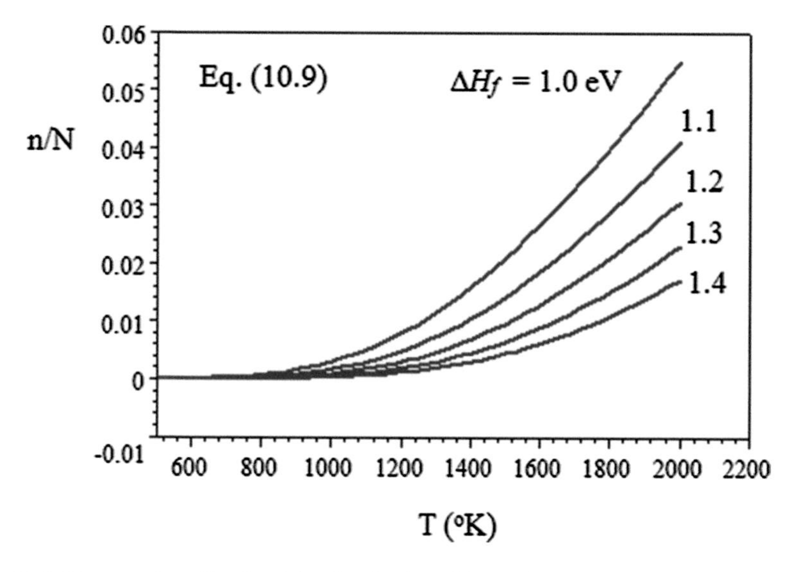

Fig. 10.5 Theoretical profiles for defect fraction in solid crystals

$$n = N \exp\left(-\frac{\Delta G}{2\kappa T}\right) = N \exp\left(\frac{\Delta S}{2k} - \frac{\Delta H_f}{2\kappa T}\right) \tag{10.8}$$

$$\frac{n}{N} \simeq \exp\left(-\frac{\Delta H_f}{2kT}\right) \quad \text{For } \Delta S \to 0 \tag{10.9}$$

where N = number of atoms

ΔG = Gibbs free energy (J/mol)
ΔS = entropy change (J/mol K)
ΔH_f = enthalpy change of formation $\left(eV = 1.602x10^{-19}\ \text{J/mol}\right)$
κ = Boltzmann constant = $1.38x10^{-23}$ J/ (mol K)
$\kappa = 8.62x10^{-5}\ \text{eV/K} = 8.82x10^{-5}\ \text{eV/°C}$
T = absolute temperature (K)

This expression, Eq. (10.9), predicts that the equilibrium defect fraction (n/N) increases exponentially with increasing temperature. The defect fraction profile given by Eq. (10.9) is shown in Fig. 10.5 for several values of the enthalpy change. Observe that n/N is strongly dependent on the temperature above 800 °C for selected values of the enthalpy change $\left(\Delta H_f\right)$.

Furthermore, point defects are mobile imperfections at high temperatures, and eventually, their rate of diffusion may obey the Arrhenius law [11]:

$$R_x = R_o \exp\left(-\frac{\Delta H_d}{\kappa T}\right) \tag{10.10}$$

where $\Delta H_d < \Delta H_f$ and R_o = rate constant.

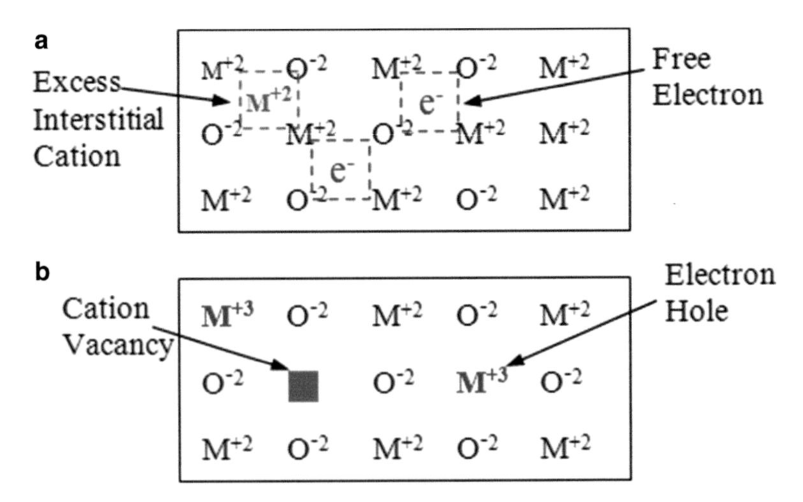

Fig. 10.6 Schematic diagrams for metal-oxide semiconductors. (**a**) n-type oxides, (**b**) p-type oxides

Schottky Defects It is known that ionic compounds have appreciable electric conductivity, which is inseparable from diffusion, due to atomic defects, such as Schottky and Frenkel defects to an extent, and ionic migration and ionic diffusion. For instance, Schottky defects are combinations of cation and anion vacancies necessary to maintain ionic electrical neutrality and stoichiometric ionic structure. In general, ions (cations and anions) diffuse into adjacent sites.

Frenkel Defects These defects are combinations of interstitial cations and cation vacancies. Electrical neutrality and stoichiometry are also maintained. Thus, combinations of the type of defects provide the ionic diffusion mechanism for oxide growth, but stoichiometry may not be maintained due to the electrical nature of oxides having either metal-excess or metal-deficit conditions.

Figure 10.6 schematically shows idealized lattice structures for continuous n-type and p-type oxide semiconductors containing point defects, such as interstitial cations, cation vacancies, and electron holes.

For convenience, the metal is treated as a divalent element (an ion that has a valence of two, 2+). The electron hole is a positive mobile electronic carrier in the valence band within an oxide crystal structure.

The importance of metal oxides in nowadays modern semiconductor technology is the use of a common metal-oxide-semiconductor field-effect transistor (MOSFET), which is a lightly doped (mixed with an impurity) p-type substrate. In fact, the MOSFET has a very high electric resistance to impede easy passage of current. A transistor is like a switch in electronic components.

Analysis of Fig. 10.6 indicates that **n-type oxides** have metal-excess and oxygen-deficit conditions. Also, interstitial cations $\left(M^{2+}\right)$ and electrons are free to move within the oxide lattice toward the oxide-gas interface, where the following reactions occur:

$$\frac{y}{2}O_2 + 2e^- \rightarrow yO^{-2} \tag{10.10a}$$

$$yO^{-2} + xM^{2+} \rightarrow M_xO_y \tag{10.10b}$$

Thus, the nonstoichiometric M_xO_y oxide has its own lattice defects such as vacancies. In this process, both M^{2+} and electrons e^- diffuse through the oxide lattice until the oxide is thick enough. Consequently, the formation of n-type oxides is due to the diffusion mechanism of interstitial M^{2+} cations and oxygen O^{-2} anions in the opposite direction. Eventually, M_xO_y grows at the metal-oxide inner interface.

Oxides such as ZnO, ZrO_2, MgO, BeO, Fe_2O_3, Al_2O_3, TiO_2, and others fall into this category.

p-type oxides are metal-deficit and oxygen-excess compounds in which M^{2+} cations are provided at the metal-oxide interface and diffuse toward the oxide-gas interface through cation vacancies. Consequently, M_xO_y forms at the oxide-gas interface. The electron holes have positive charge and migrate through their conduction band. Therefore, ionic mass transfer occurs by metal vacancy diffusion [6].

Oxides such as NiO, CoO, Cr_2O_3, FeO, Cu_2O, Cu_2S, and others fall into this category. Also, the presence of ionic impurities in the oxide lattice is referred to as the doping effect, which affects the concentration of ionic defects.

In addition, there are highly n-type stoichiometric oxides such as Al_2O_3, MgO, and ZrO_2, since the Gibbs free energy change of formation is large and the cation valence is fixed [11].

On the other hand, for common multivalent cations, nonstoichiometry is pronounced in TiO_{2-x}, $Fe_{1-x}O$ and $Ni_{1-x}O$. For instance, FeO oxide is a face-centered cubic (FCC) structure that may contain a high concentration of cations, and its nonstoichiometry arises due to the unequal number of cations and anions, resulting in $n_{cation}(Fe_xO) > n_{anion}(FeO)$.

Consequently, Fe_xO forms since this oxide departures from its ideal stoichiometric composition FeO.

In this case, the cation vacancy concentration can be determined as $(1 - x)$ since $0.80 < x < 1$ [4,11]. This implies that a fraction of ferrous iron Fe^{2+} cations oxidize to Fe^{3+} cations, resulting in a cation deficiency [5].

In general, metal oxides are considered as compounds having high-temperature chemical stability, but low thermal and electrical conductivity. In fact, low thermal conductivity can be attributed point defects to an extent.

10.3 Kinetics of Oxidation in Gases

Application of metals and alloys in high-temperature environments is of technological importance since the rate of oxide layer formation, mechanism, and degree of protectiveness are the subject matter to be evaluated. Consider the model of oxide formation shown in Fig. 10.7, in which a metal is exposed to an oxygen-rich environment at high temperatures [2, 8, 14, 16].

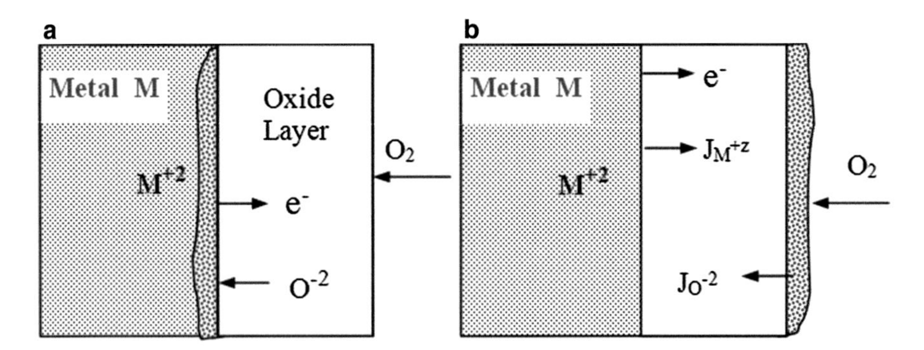

Fig. 10.7 Model for oxidation by diffusion-control mechanism showing the dominant diffusion molar flux. (**a**) Inner layer formation $J_{O^{-2}} > J_{M^{+z}}$, (**b**) outer layer formation $J_{M^{+z}} > J_{O^{-2}}$

The model in Fig. 10.7 predicts that:

- The initial rate of oxide formation is determined by the reactions at the metal/oxygen interface. Thus, M_xO_y forms as a thin layer.
- Once the thin layer is formed, it serves as a barrier separating or insulating the metal substrate. The subsequent oxidation steps are controlled by the diffusion of the species M^{2+} and O^{-2} through the layer.
- If O^{-2} anions diffuse faster than M^{2+} cations, the diffusion molar flux is $J_{O^{-2}} > J_{M^{2+}}$, then an oxide inner layer forms as shown in Fig. 10.6a. In order for O^{-2} anions to diffuse into the metal, there must be metal vacancies in the lattice. As oxide growth proceeds, the volume of the oxide is $V_{M_xO_y} > V_M$ causing a high stress concentration, and consequently, the oxide layer may rupture, exposing the metal substrate to oxygen for further oxidation. Titanium falls into this category due to $J_{O^{-2}} > J_{Ti^{2+}}$.
- If M^{2+} cations diffuse faster than O^{-2} anions, then $J_{M^{2+}} > J_{O^{-2}}$, and an outer oxide layer forms and proceeds through the concentration gradient as shown in Fig. 10.6b. In this situation, the stress concentration is reduced or relieved. Thus, the oxide layer adheres to the metal surface and protects the metal from further oxidation. Nickel falls into the category since $J_{Ni^{2+}} > J_{O^{-2}}$.

Consider the oxidation case in Fig. 10.6b for divalent metal cations $\left(M^{2+}\right)$ and oxygen anions $\left(O^{-2}\right)$. If the ionic radius relationship is $R_{O^{-2}} > R_{M^{2+}}$ and O^{-2} anions diffuse at a negligible rate, then M^{2+} cations diffuse through the oxide layer. Consequently, the diffusion-control process dictates that the diffusion molar fluxes become $J_{O^{-2}} \simeq 0$ and $J_{M^{2+}} > 0$ and metal oxidation reactions proceed until the rate of diffusion slows down reaching steady-state conditions.

Also, the mechanism of high-temperature oxidation may be a combination of metal cations flowing outward and oxygen anions flowing inward due to diffusion and migration mass transfer. This dual mass transfer action leads to beneficial condensation of vacancies and prevention of voids for the formation of a uniformly adhered oxide layer for continuous and nonporous scales. Thus, the ionic mass transfer through the scale is the rate-controlling process. Initially, a thin and protective film forms until the reaction rate decreases. Consequently, film growth forms a scale, which may be attributed to the concentration and potential gradients. In a macroscale, the mechanism involved in scale formation is assessed below for different rate laws, which do not consider the nonstoichiometry of n-type or p-type oxides or sulfides or a combination of these, including carbides and hydrogen effects.

10.3.1 Pilling–Bedworth Ratio

The characterization of metal-oxide scales is commonly based on their quality defined as a characteristic volume ratio known as the Pilling–Bedworth ratio (PB ratio).

Metal-oxide scales can be defined as protective and nonprotective using the classical Pilling–Bedworth law as the ratio given by [18]

$$PB = \frac{V_o}{V_m} = \frac{\left[\rho^{-1}A_w\right]_o}{\left[z\rho^{-1}A_w\right]_m} \tag{10.11}$$

where V_o = volume of the oxide scale (cm^3)

V_m = volume of the solid metal (cm^3)
ρ = density (g/cm^3)
A_w = atomic or molecular weight (g/mol)
z = valence

The *PB* ratio is used to characterize several oxidation conditions. Thus:

- If $PB < 1$ or $PB > 2$, the oxide scale is nonprotective (*NP*) and noncontinuous due to insufficient volume to cover the metal surface uniformly. Thus, weight gain is usually linear.
- If $1 \leq PB \leq 2$, the oxide scale is protective (P), adherent and strong due to compressive stress, refractory due to high melting temperature and low electrical conductor, and nonporous. Because of these factors, diffusion proceeds in the solid state at low rates. Some oxides may not develop compressive stresses, invalidating *PB* law [10].
- If $PB = 1$, then the oxide scale is ideally protective.

Table 10.1 lists several metal/oxide *PB* ratios and their respective quality. Observe that most metals have $PB > 1$. However, it is possible that some oxide scales may be plastically deformed at high temperatures, leading to a porosity-healing and crack-filling processes for $PB < 1$. Other oxide scales, such as WO_3 at $T \geq 800\,°C$, may become unstable and volatile for $PB > 1$. This indicates that the protectiveness is lost [10]. Crystallographic structures and properties of many metal-oxide scales can be found elsewhere [4].

Table 10.1 PB ratios for common metals [10]

Metal	Oxide	PB	Protectiveness
Aluminum	Al_2O_3	1.28	P
Calcium	CaO	0.64	NP
Cadmium	CdO	1.42	P
Cobalt	Co_2O_3	2.40	NP
Copper	Cu_2O	1.67	P
Chromium	Cr_2O_3	2.02	NP
Iron	FeO	1.78	P
Magnesium	MgO	0.81	NP
Manganese	MnO_2	2.37	NP
Molybdenum	MoO_3	3.27	NP
Nickel	NiO	1.70	P
Lead	PbO	1.28	P
Silicon	Si O_2	2.15	NP
Tantalum	Ta_2O_3	2.47	NP
Titanium	Ti_2O_3	1.76	P
Tungsten	WO_3	1.87	P
Uranium	UO_2	1.97	P
Zinc	ZnO	1.58	P
Zirconium	ZrO_2	1.57	P

In general, thermal degradation of metals and alloys, and oxide scales, is a major concern in practical applications, such as furnace parts, heaters, tubes, and the like. Degradation at high temperatures may occur isothermally or cyclically under operating conditions. Therefore, metallic parts must be protected using lesser aggressive environments if possible; otherwise, applying a protective thin film prior to high-temperature applications can be beneficial. Protective oxide films or coatings have been studied for over two decades using rare-earth elements to protect chromia (Cr_2O_3) and alumina (Al_2O_3) oxide scales from oxidizing beyond suitable thickness. A particular paper by Seal et al. [21] addresses the influence of superficially applied CeO_2 coatings ($\approx 2\ \mu$m) on the isothermal and cyclic-oxidation behavior of austenitic stainless steel, such as AISI 316, 321, and 304, in dry air.

10.4 Mathematics of Oxidation

The mathematics of oxidation kinetics involves diffusion and migration mass transfer. According to Fick's first law of diffusion, the molar flux is related to the rate of oxide thickness growth (dx/dt) given by

$$J = -D\frac{dC}{dx} \tag{10.12}$$

$$J = -C\frac{dx}{dt} \tag{10.13}$$

from which the rate of thickness growth (also known as drift velocity) is

$$\frac{dx}{dt} = \frac{D}{C_x}\frac{\Delta C}{x} \tag{10.14}$$

where $D =$ diffusivity (diffusion coefficient) $\left(\text{cm}^2/\text{s}\right)$

$C_x =$ concentration of the diffusing species $\left(\text{ions}/\text{cm}^3\right)$
$x =$ thickness of the oxide scale (cm or μm)

Integrating Eq. (10.14) yields what is known as the parabolic equation for the oxide thickness:

$$x = \sqrt{K_x t} \tag{10.15}$$

$$K_x = \frac{2D\Delta C}{C_x} \approx 2D \tag{10.16}$$

where $K_x =$ parabolic rate constant $\left(\text{cm}^2/\text{s}\right)$.

In oxidation at high temperatures, most variables are temperature dependent. Thus, it is reasonable to assume that both diffusivity and parabolic rate constant obey the Arrhenius relationship (see Chap. 5):

$$D = D_o \exp\left(-\frac{Q_d}{RT}\right) \tag{10.17}$$

$$K_x = K_o \exp\left(-\frac{Q_x}{RT}\right) \tag{10.18}$$

Hence, the oxide thickness becomes $x = f(T)$. It is known that the oxide kinetics of many metals and alloys show parabolic behavior. However, other types of behavior are possible, but Eq. (10.15) can be generalized and modeled as an empirical relationship given by

$$x = (K_x t)^n \tag{10.19}$$

where n is an exponent and K_x becomes the rate constant for the scale thickness growth in gaseous environments.

In fact, x in Eq. (10.19) represents an external thickness in most cases. Now, the generalized equation for the rate of oxide thickness growth takes the form

$$\frac{dx}{dt} = n\,(K_x)^n\, t^{n-1} \tag{10.20}$$

Furthermore, mass gain is another parameter that can be employed in characterizing oxide kinetic behavior. Using the definition of density ($\rho = M/V = M/xA_s$) and

Eq. (10.19) yields the mass gain (m) and the rate of mass gain (dm/dt) during surface oxidation at high temperatures:

$$M = \rho A_s (K_x t)^n = \lambda t^n \tag{10.21}$$

$$\frac{dM}{dt} = n\lambda t^{n-1} \tag{10.22}$$

$$\lambda = \rho A_s (K_x)^n \tag{10.23}$$

where ρ = density (g/cm^3)

A_s = surface area (cm^2)

λ = constant (g/s^n)

Since it is customary to represent weight gain (W) in units of weight/area [23], Eq. (10.21) is rearranged so that the weight gain and its rate (dW/dt) become as

$$W = \frac{M}{A_s} = (K_w t)^n \tag{10.24}$$

$$\frac{dW}{dt} = nK_w^n t^{n-1} \tag{10.25}$$

$$K_w = \rho^{1/n} K_x \tag{10.26}$$

where K_w = rate constant ($g\,cm^{-2}\,time^{-n}$) for weight gain.

The above equations are analyzed based on the exponent n for most observable kinetic behavior. Thus, the physical interpretation of n is given below:

- If $\boldsymbol{n = 1}$, a linear behavior is achieved for a noncontinuous, porous, and cracked oxide scale with $PB < 1$ or $PB > 2$. In this case, the scale is nonprotective, and diffusion of oxygen occurs through the pores, cracks, and vacancies.
- If $\boldsymbol{n = 1/2}$, a parabolic behavior of nonporous, adherent, and protective scale develops by diffusion mechanism. Thus, $1 \le PB \le 2$, and the mechanism of scale growth is related to metal cations (M^{z+}) diffusing through the oxide scale to react with oxygen at the oxide-gas interface.
- If $\boldsymbol{n = 1/3}$, a cubic behavior develops for a nonporous, adherent, and protective oxide scale. Thus, $1 \le PB \le 2$.

In general, the kinetic behavior of a specific oxide scale may be different from the most observable ones cited previously, and therefore, the exponent n may take a different value, which must be determined experimentally. Also, a logarithmic behavior is possible for thin layers at relatively low temperatures [10]. For a logarithmic behavior, the weight gain may be defined by

$$W = K_a \log (a_2 + a_3 t) \tag{10.27}$$

where K_a = rate constant ($g\,cm^{-2}$)

a_2 = dimensionless constant

a_3 = constant ($time^{-1}$)

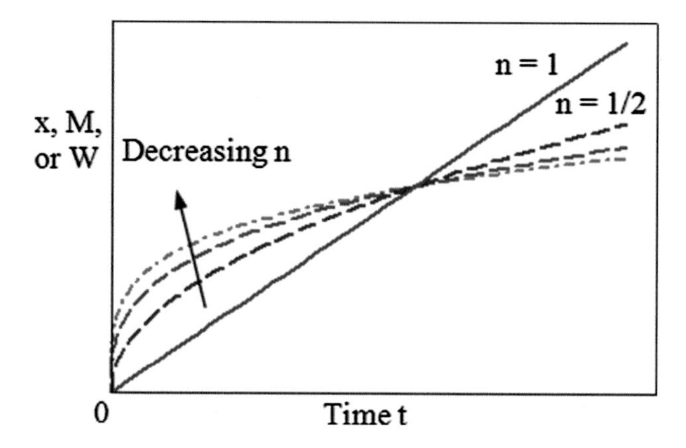

Fig. 10.8 Schematic kinetic behavior cases

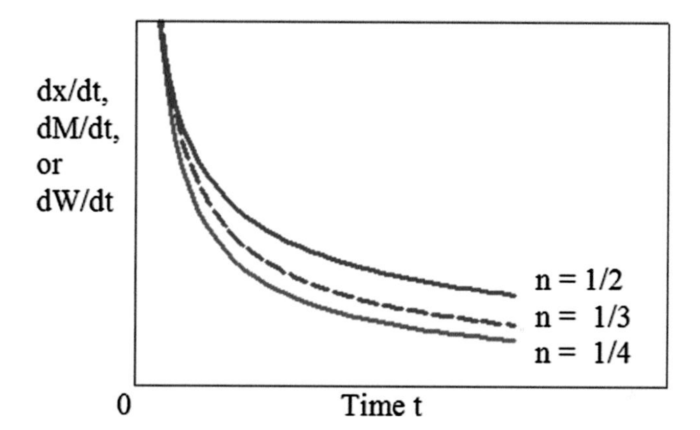

Fig. 10.9 Schematic kinetic rates

Figure 10.8 illustrates general kinetic behavior of metal oxides for some hypothetical cases based on the theoretical background given above, while Fig. 10.9 schematically exhibits the time-dependent rate of metal-oxide formation. The rate of metal-oxide formation is related to the oxide scale thickness. Further analogy of metal oxides is associated with quality, porosity, and crystal structure, including type crystal defects.

Example 10.4. *Calculate the PB ratio for the oxidation of aluminum in hot oxygen. Data:* $\rho_{Al} = 2.70\ g/cm^3$ *and* $\rho_{Al_2O_3} = 3.80\ g/cm^3$. *Will* Al_2O_3 *be protective?*

Solution.
Assume that oxidation of aluminum in oxygen occurs by the reaction

$$2Al + \frac{3}{2}O_2 \rightarrow Al_2O_3$$

and the pertinent atomic and molecular weights are, respectively,

$$A_{w,Al} = 26.98 \text{ g/mol}$$
$$A_{w,Al_2O_3} = 2\,(26.98 \text{ g/mol}) + 3\,(16 \text{ g/mol}) = 101.96 \text{ g/mol}$$

From Eq. (10.11),

$$PB = \frac{\left[\rho^{-1}A_w\right]_o}{\left[z\rho^{-1}A_w\right]_m}$$
$$PB = \frac{[(101.98)\,/\,(3.8)]_o}{[(2)\,(26.98)\,/\,(2.70)]_m}$$
$$PB = 1.34$$

Therefore, Al_2O_3 *is theoretically protective as predicted by the Pilling–Bedworth ratio range* $1 \leq PB \leq 2$.

Example 10.5. *Calculate the thickness of the NiO at a temperature T for 5 h. Assume a 0.10-μm-thick film after* 1*-min treatment. Use the information given in Table 10.1.*

Solution.
From Table 10.1, NiO is protective and parabolic behavior is assumed. Thus, from Eq. (10.15) at $t = 1$ min $= 60$ s,

$$K_x = \frac{x^2}{t} = \frac{(0.10\ \mu\text{m})^2}{60\text{ s}} = 1.67x10^{-4}\ \frac{\mu\text{m}^2}{s}$$

At $t = 5$ min $= 18{,}000$ s, *Eq. (10.15) gives*

$$x = \sqrt{K_x t} = \sqrt{\left(1.67x10^{-4}\ \frac{\mu\text{m}^2}{\text{s}}\right)(18000\text{ s})}$$
$$x = 1.73\ \mu\text{m}$$

10.5 Conductivity and Mobility

The ionic conductivity included in this section is due to ionic motion of a species j within metal-oxide compounds. These oxides may consist of one or more phases making up an oxide scale, which, in turn, is the corrosion product to be analyzed with regard to its protectiveness, conductivity, mobility, diffusivity, and concentration of the species j.

For ionic or atomic transport, the force gradient (F_j) acting on a moving species j along the x-direction can be defined by [5]

$$F_d = -\frac{1}{N_A}\frac{d\mu}{dx} \quad \text{(diffusion)} \tag{10.28a}$$

$$F_m = -\left(z_j q_e\right)\frac{d\phi}{dx} \quad \text{(migration)} \tag{10.28b}$$

where $d\mu/dx$ = negative chemical potential gradient (J/mol cm)

$d\phi/dx$ = negative electric potential gradient (V/cm)
N_A = Avogadro's number = $6.022x10^{23}$ (particles/mol)
q_e = electronic charge = $1.602x10^{-19}$ C (= A s = J/V)
z = valence or oxidation state
zq_e = particle charge (C = Coulomb)

Here, $d\phi/dx$ and $d\mu/dx$ are the driving forces for F. Dividing the ionic velocity (drift velocity v_j) of a species j by the force gradient (F) acting on it defines the **absolute mobility** [5]

$$B_j = \frac{v}{F} \tag{10.29}$$

and

$$B = -\frac{vN_A}{d\mu/dx} \quad \text{(diffusion)} \tag{10.30a}$$

$$B_m = -\frac{v}{(zq_e)(d\phi/dx)} \quad \text{(migration)} \tag{10.30b}$$

where B = absolute ionic mobility (cm^2/J s)

B_m = absolute electrical mobility (cm^2/V s)

The ionic flux due to $d\phi/dx$ and $d\mu/dx$ driving forces is given by the product of C (atomic concentration) and v_j:

$$J_d = Cv = CBF = -\frac{CB}{N_A}\frac{d\mu}{dx} \quad \text{(diffusion)} \tag{10.31a}$$

$$J_m = Cv = CB_mF = -(zq_e)\,CB_m\frac{d\phi}{dx} \quad \text{(migration)} \tag{10.31b}$$

For **diffusion mechanism**, Eq. (3.20) along with $C = a$ yields [5]

$$\mu = \mu^o + RT\ln(C) \tag{10.32}$$

$$d\mu = \frac{RT}{C}dC \tag{10.33}$$

$$\frac{d\mu}{dx} = \frac{RT}{C}\frac{dC}{dx} \tag{10.34}$$

where $\kappa = R/N_A = 1.38x10^{-23}$ J/K = Boltzmann constant

$R =$ gas constant = $8.314\,\mathrm{J/K} = 5.19x10^{19}\,\mathrm{eV}$
$T =$ absolute temperature (K)
$C =$ atomic concentration ($\mathrm{ions/cm^3}$ or $\mathrm{vacancy/cm^3}$)

Inserting Eq. (10.34) into (10.31a) gives

$$J_d = -\frac{RTB}{N_A}\frac{dC}{dx} = -\kappa TB\frac{dC}{dx} \tag{10.35}$$

From Fick's first law of diffusion, Eq. (6.3), the diffusion flux is

$$J_d = -D\frac{dC}{dx} \tag{10.36}$$

Equating Eqs. (10.35) and (10.36) yields the **Nernst-Einstein equation** defined by

$$D = \kappa TB \tag{10.37}$$

For **migration mechanism**, Eqs. (10.31b) and (10.37) along with $B_m = B$ give the migration flux as

$$J_m = -\frac{(zq_e)\,CD}{\kappa T}\frac{d\phi}{dx} \tag{10.38}$$

The derivation of the **electrical mobility** (B_e) follows. Equating Eqs. (10.31b) and (10.38) gives the drift velocity

$$v = \frac{(zq_e)\,D}{\kappa T}\frac{d\phi}{dx} \tag{10.39}$$

Then,

$$B_e = \frac{v}{d\phi/dx} = \frac{(zq_e)\,D}{\kappa T} \tag{10.40}$$

The total flux in the x-direction is

$$J_x = J_d + J_m \tag{10.41a}$$

$$J_x = -D\frac{dC}{dx} - \frac{(zq_e)\,CD}{\kappa T}\frac{d\phi}{dx} \tag{10.41b}$$

Combining Eqs. (6.9) and (10.38) yields the current density due to migration:

$$i = -(zq_e)\,J_m \tag{10.42}$$

$$i = \frac{(zq_e)^2\,CD}{\kappa T}\frac{d\phi}{dx} \tag{10.43}$$

From Ohm's law, the ionic conductivity due to migration is given by [2]

$$\sigma = -\frac{i}{d\phi/dx} \tag{10.44}$$

$$\sigma = \frac{(zq_e)^2\,CD}{\kappa T} > 0 \tag{10.45}$$

$$\sigma = (zq_e)^2\,CB > 0 \tag{10.46}$$

Assume that the ionic conductivity obeys the Arrhenius relationship so that

$$\sigma = \sigma_o \exp\left(-\frac{Q_\sigma}{\kappa T}\right) \tag{10.47}$$

The ionic conductivity arises due to the electrical nature of the diffusing species and inherent defects since the ionic electrical conduction and diffusion are inseparable processes [2, 8]. If the drift velocity (v) is primarily carried by cations, then ionic current density may be given by [2, 5]

$$i = (zq_e)\,Cv \tag{10.48}$$

where the concentration of the gas may be estimated by [2]

$$C = \frac{N_A S_o \rho_{oxide}}{A_{w,M} + A_{w,O_2}} \tag{10.49}$$

where S_o = stoichiometry number of the oxygen in the oxide formula

ρ_{oxide} = density of the oxide compound (g/cm^3)

The total conductivity is simply the sum of the cation conductivity (σ_c), anion conductivity (σ_a), and electron conductivity (σ_e) or electron holes conductivity (σ_h) given by

$$\sigma_t = \sigma_c + \sigma_a + \sigma_e + \sigma_h = \sum \sigma_j > 0 \tag{10.50}$$

The conductivity fraction for each species j is a quantity called the transfer number (t) defined by

$$t = \frac{\sigma_j}{\sigma_t} < 1 \tag{10.51}$$

$$\sum t = 1 \tag{10.52}$$

Example 10.6. ***(a)*** *Derive expressions for the oxygen anion diffusivity and the mobility in*

$$Zr_{0.85}Ca_{0.15}O_{1.85}$$

semiconductor oxide (brittle ceramic material), and **(b)** *calculate σ, D, B_e, and B for oxygen anions. Use the Kingery et al. [12] oxygen conductivity expression given in Barsoum's book [2], page 211, and the oxygen ionic concentration of $5.52x10^{28}$ $ions/m^3$ [2]:*

$$\sigma = \left(1.50x10^5\ \Omega^{-1}\,\text{m}^{-1}\right)\exp\left(-\frac{1.26\ \text{eV}}{kT}\right) \tag{a}$$

This is a temperature-dependent Arrhenius-type equation.

Solution.

(a) $C = 5.52x10^{28}\,\text{m}^{-3}$ $\quad q_e = 1.602x10^{-19}\,\text{C}\ (= \text{A s})$ $\quad z = 2$

$\kappa = 1.38x10^{-23}\,\text{J/K} = 8.61x10^{-5}\,\text{eV}$

From Eq. (10.45), the O^{-2} anion diffusivity is given by

$$D = \frac{\kappa T \sigma}{(zq_e)^2\, C} \tag{b}$$

$$D = \left(2.44x10^{-15}\,\frac{\Omega\,\text{m}^3}{\text{s K}}\right) T\sigma_i \tag{c}$$

Substituting Eq. (a) into (c) gives $D = f(T)$:

$$D = \left(3.65x10^{-10}\,\frac{\text{m}^2}{\text{s K}}\right) T\exp\left(-\frac{1.46x10^4\ \text{K}}{T}\right) \tag{d}$$

Now, combining Eqs. (10.40), (10.37), and (d) yields the sought ionic mobility expressions:

$$B_e = \left(8.46x10^{-6}\,\frac{\text{m}^2}{\text{V s}}\right)\exp\left(-\frac{1.46x10^4\ \text{K}}{T}\right) \tag{e}$$

$$B = \left(2.64x10^{13}\,\frac{\mathrm{m}^2}{\mathrm{J\,s}}\right)\exp\left(-\frac{1.46x10^4\ \mathrm{K}}{T}\right) \tag{f}$$

Both B_e and B profiles for oxygen are exponential.
(b) *At $T = 1200\,°C$, the kinetic parameters are*

$$\sigma = 0.76\ \Omega^{-1}\,\mathrm{m}^{-1} \qquad D = 2.28x10^{-12}\ \mathrm{m^2/s}$$

$$B_e = 4.40x10^{-11}\ \mathrm{m^2/V\,s} \qquad B = 1.37x10^{-11}\ \mathrm{m^2/J\,s}$$

10.6 Wagner Theory of Oxidation

The Wagner theory [23] for a coupled diffusion and migration (also known as ambipolar diffusion) mass transfer during oxidation of metals and alloys is briefly described. This theory treats the parabolic kinetic behavior of high-temperature oxides given by Eq. (10.19) with $n = 1/2$.

First of all, combining Eqs. (10.31) and (10.41) yields the total ionic flux in the x-direction:

$$J_x = -\frac{\sigma}{(zq_e)^2}\left[\frac{d\mu}{dx} + (zq_e)\frac{d\phi}{dx}\right] \tag{10.53}$$

This expression represents the condition of the quantity of particles in motion of a reacting metal-oxygen system through the chemical and electric potential gradients, $d\mu/dx$ and $d\phi/dx$, respectively. Thus, the relationship between these gradients illustrates the connection embodied in the ionic flux for a reacting system in which all moving species are charged particles. Therefore, both $d\mu/dx$ and $d\phi/dx$ are the driving forces of the ionic flux J_x. The oxidizing system under consideration consists of a pure metal exposed to a gaseous environment at relatively high temperature. Particularly, a pure metal and a single gas phase, such as oxygen, define an invariant system since the metal and gas react to form one single phase.

However, if the material is an alloy containing at least two elements, then the more active element oxidizes first, and its cations react with the gas-phase anions to form the initial oxide, sulfide, carbide, nitride, and so on. Other elements in the alloy may react afterward to form other surface compounds. Consequently, the surface scale may be a combination of phases, which may be identified by X-ray diffraction and revealed by scanning electron microscopy (SEM).

Assume that a pure metal is oxidizing and that a single-phase oxide forms by diffusion of cations and anions and migration of electrons. This ideal atomic mass transfer for oxide growth requires that the positively charged cations and electrons flow outward and the negatively charged anions flow inward. Thus, the ionic flux balance for outward cations (z_cJ_c), inward anions (z_aJ_a), and electrons (z_eJ_e) along with $z_e = 1$ is [23]

$$z_cJ_c = z_aJ_a + z_eJ_e \tag{10.54}$$

This relationship derives directly from the conservation of mass that accompanies the mass transfer, and it is succinctly represented in the treatment of ionic flux balance. For ionic electroneutrality, this notation, Eq. (10.54), can be further simplified by letting $J_a \approx 0$, and using Eq. (10.53) for both cations and electrons together with (10.54) yields the electric potential gradient

$$\frac{d\phi}{dx} = \frac{\sigma_e}{q_e(\sigma_c + \sigma_e)}\left[\frac{d\mu_e}{dx} - \frac{\sigma_c}{z_c\sigma_e}\frac{d\mu_c}{dx}\right] \tag{10.55}$$

$$\frac{d\phi}{dx} = \left(\frac{t_e}{q_e}\right)\frac{d\mu_e}{dx} - \left(\frac{t_c}{z_c q_e}\right)\frac{d\mu_c}{dx} \tag{10.56}$$

Inserting Eq. (10.55) into (10.53) and letting $J_x = J_{x,c}$ yield the cation vacancy flux in terms of more easily measurable electric conductivity:

$$J_{x,c} = -\frac{\sigma_c\sigma_e}{(z_c q_e)^2(\sigma_c + \sigma_e)}\frac{d\mu}{dx} \tag{10.57}$$

$$J_{x,c} = -\frac{\sigma_c t_e}{(z_c q_e)^2}\frac{d\mu}{dx} \tag{10.58}$$

and the chemical potential gradient across the scale is described by

$$\frac{d\mu}{dx} = \frac{d\mu_c}{dx} + z_c\frac{d\mu_e}{dx} \tag{10.59}$$

where $d\mu_c/dx =$ chemical potential gradient for M^{z+} cations

$d\mu_e/dx =$ chemical potential gradient for electrons
$d\mu_c/dx < 0$
$d\mu_e/dx < 0$

The expression given by Eq. (10.57) or (10.58) indicates that:

- An oxide scale conducts both ions and electrons
- The chemical potential gradient $d\mu/dx$ of the neutral species is the driving force for growth of the oxide scale
- Cations, anions, and electrons diffuse together to maintain charge neutrality
- The rate limiting is controlled by the slowest diffusion species
- If $\sigma_c = 0$ or $\sigma_e = 0$, then $J_{x,c} = 0$ (vanishes)
- If $\sigma_e >> \sigma_c$ and $t_e \approx 1$, then $J_{x,c}$ may be determined by the rate electrons move through the oxide scale, which is the electronic conductor.
- If $\sigma_c >> \sigma_e$ and $t_c \approx 1$, then $J_{x,c}$ may be determined by the rate cations (M^{z+}) move through the oxide scale (ionic conductor).

For a simple oxidation reaction $(M = M^{z+} + 2e^-)$, $\sigma_e >> \sigma_c$ and $\sigma_e \approx 1$. Consequently, Eq. (10.57) becomes upon integration

$$J_{x,c} = -\frac{1}{x}\frac{\sigma_c}{(z_c q_e)^2}\int_{\mu_i}^{\mu_o} d\mu \tag{10.60}$$

where μ_i = inside metal/oxide chemical potential (J/mol)

μ_o = outside oxide-gas chemical potential (J/mol)

Additionally, the flow of cations in the x-direction occurs through the growing oxide layer; the cation flux $J_{x,c}$ becomes dependent on the thickness growth rate (dx/dt) of the oxide layer. Hence, $J_{x,c}$ and dx/dt, Eq. (10.20), are related as indicated below:

$$J_{x,c} = C_c \frac{dx}{dt} \tag{10.61}$$

$$J_{x,c} = nC_c K_x^n t^{n-1} \tag{10.62}$$

where C_c = concentration of cations $\left(\text{mol cm}^{-3}\right)$

K_x = fundamental parameter

Combining Eqs. (10.60) and (10.61) gives

$$\frac{dx}{dt} = -\frac{1}{x}\left[\frac{\sigma_c}{(z_c q_e)^2 C_c}\int_{\mu_i}^{\mu_o} d\mu\right] \tag{10.63}$$

Using the Gibbs–Duhem chemical potential gradient [2,4]

$$\frac{d\mu}{dx} = -\frac{\kappa T}{2} d\ln(P_{O_2}) \tag{10.64}$$

and substituting it, Eq. (10.64), into (10.63) yield

$$\frac{dx}{dt} = \frac{1}{x}\left[\frac{\kappa T \sigma_c}{2(z_c q_e)^2 C_c}\int_{P_i}^{P_o} d\ln(P_{O_2})\right] \tag{10.65}$$

where P_o = outer oxygen pressure

P_i = inner oxygen pressure

But the term in brackets is defined as

$$K_r = \frac{\kappa T \sigma_c}{2(z_c q_e)^2 C_c}\int_{P_i}^{P_o} d\ln(P_{O_2}) \tag{10.66}$$

Merging Eqs. (10.45) and (10.66) along with $\sigma = \sigma_c$ gives

$$K_r = \frac{D_c}{2}\int_{P_i}^{P_o} d\ln(P_{O_2}) = \frac{D_c}{2}\ln\left(\frac{P_o}{P_i}\right) \tag{10.67}$$

and combining Eqs. (10.65) and (10.67) defines the rate of oxide thickness growth:

$$\frac{dx}{dt} = \frac{K_r}{x} \tag{10.68}$$

Integrating Eq. (10.68) yields the parabolic equation for the oxide thickness:

$$x = \sqrt{2K_r t} = \sqrt{K_x t} \tag{10.69}$$

This expression is simple and yet important in metal-oxide applications. However, increasing film thickness of oxides may change oxide properties and can jeopardize its optimum usage.

Example 10.7. *Let us use some published data for magnesium oxide [5].* ***(a)*** *Determine if MgO containing aluminum* $\left(2x10^{19}\ cm^{-3}\right)$ *is a mixed oxide conductor, provided that the conduction is by electrons and* Mg^{2+} *cation vacancies at* $1600\,°C = 1873\,K$*.* ***(b)*** *Calculate the species diffusivity. Given data:*

Electron	*Cation*
Holes	*Vacancies*
$C_e = 5x10^{13}\ cm^{-3}$	$C_c = 5x10^{13}\ cm^{-3}$
$B_e = 7.00\ cm^2/V\,s$	$B_c = 7.60x10^{-6}\ cm^2/V\,s$

Solution. (a) *Reactions involved in the magnesium oxide formation:*

$$\begin{array}{ll} Mg \rightarrow Mg^{2+} + 2e^- & z_c = 2 \\ \frac{1}{2}O_2 + 2e^- \rightarrow yO^{-2} & e = 1.602x10^{-19}\ C\ (= \mathrm{A\,s}) \\ \hline Mg^{2+} + \frac{1}{2}O^{-2} \rightarrow MgO & \end{array}$$

From Eq. (10.46),

$$\sigma_e = (z_e q_e)^2\, C_e B_e = (1)\left(1.602x10^{-19}\mathrm{A\,s}\right)\left(5x10^{13}\ \mathrm{cm}^3\right)\left(7\ \mathrm{cm}^2/\mathrm{V\,s}\right)$$
$$\sigma_e = 5.61x10^{-5}\,\Omega^{-1}\,\mathrm{cm}^{-1}$$
$$\sigma_c = (z_c q_e)^2\, C_c B_c = (2)\left(1.602x10^{-19}\,\mathrm{A\,s}\right)\left(2x10^{19}\ \mathrm{cm}^3\right)\left(7.6x10^{-6}\ \mathrm{cm}^2/\mathrm{V\,s}\right)$$
$$\sigma_c = 4.85x10^{-5}\,\Omega^{-1}\,\mathrm{cm}^{-1}$$

Thus, the total conductivity and transfer numbers are

$$\sigma_t = \sigma_c + \sigma_e = 10.46x10^{-5}\,\Omega^{-1}\,\mathrm{cm}^{-1}$$
$$t_e = \sigma_e/\sigma_t = 0.54 \quad \textit{for electronic conduction}$$
$$t_c = \sigma_c/\sigma_t = 0.46 \quad \textit{for ionic conduction}$$
$$\sum t = t_e + t_c = 1$$

Therefore, MgO is a mixed conductor since $t_e > 0$ *and* $t_c > 0$ *and it is an n-type oxide since* $0 < t_e < 1$ *and* $0 < t_e < 1$.

(b) *From Eq. (10.40),*

$$D_e = \kappa T B_e / (z_e q_e)$$
$$D_e = \left(1.38x10^{-23}\ \mathrm{A\,s\,V/K}\right)(1873\ \mathrm{K})\left(7\ \mathrm{cm^2/V\,s}\right) / \left[(1)\left(1.602x10^{-19}\ \mathrm{A\,s}\right)\right]$$
$$D_e = 1.13\ \mathrm{cm^2/s}$$
$$D_c = \kappa T B_c / (z_c q_e)$$
$$D_c = \left(1.38x10^{-23}\right)(1873)\left(7.6x10^{-6}\right) / \left[(2)\left(1.602x10^{-19}\right)\right]$$
$$D_c = 6.13x10^{-7}\ \mathrm{cm^2/s}$$

Therefore, the diffusion rate limiting is due to the cations because of the differences in diffusion coefficient, $D_c < D_e$.

10.7 Experimental Data

The fundamental nature of adherence of oxide scales formed on metallic substrates during high-temperature oxidation is significantly dependent on the exposed time and the environment. Basically, temperature and environment are the most important variables to control for gathering reliable experimental data. Isothermal oxidation experiments at a relatively high temperature are conducted on a particular metal in air or any other environment. Despite the degree of porosity in the oxide scale, the oxide thickness also known as penetration is measured and plotted to find a suitable correlation. However, physical and mechanical properties and the interfacial adhesion mechanism of oxide-metal composites must be determined and studied in order to have a fundamental understanding of oxide scale formation phenomenon.

In general, the thermogravimetric technique (*TGT*) can be used to measure the amount and rate of change of sample mass as a function of time or temperature in gaseous environments. The thermogravimetric analyzer (*TGA*) is an instrument that can be used to characterize the thermal kinetic behavior of solid-gas reactions. As a result, a thermogravimetric analysis allows one to continuously measure weight changes, which depends on the instrument sensitivity.

The parabolic weight gain is a typical oxidation analysis of high-temperature corrosion, which is better known as oxidation kinetics. Figure 10.10 shows the influence of time and temperature on the linearized weight gain $\left(W^2\right)$.

This plot is a convenient way for determining the parabolic rate constant K_w as the slope of a straight line. Thus, the kinetic model given by Eq. (10.24) along with $n = 1/2$ is obeyed. Notice the parabolic rate constant in the literature is K_p, but here $K_w = K_p$.

The slopes of the lines in Fig. 10.10 are plotted in Fig. 10.11. Despite that there is a slight data scatter, the parabolic rate constant obeys the Arrhenius relationship as modeled by Eq. (10.10) below. The curve-fitting equation, the slope of the straight line (Q_w/R), and the activation energy (Q_w) are, respectively,

$$\ln(K_w) = \ln(K_o) - \frac{Q_w}{RT} \tag{10.70}$$

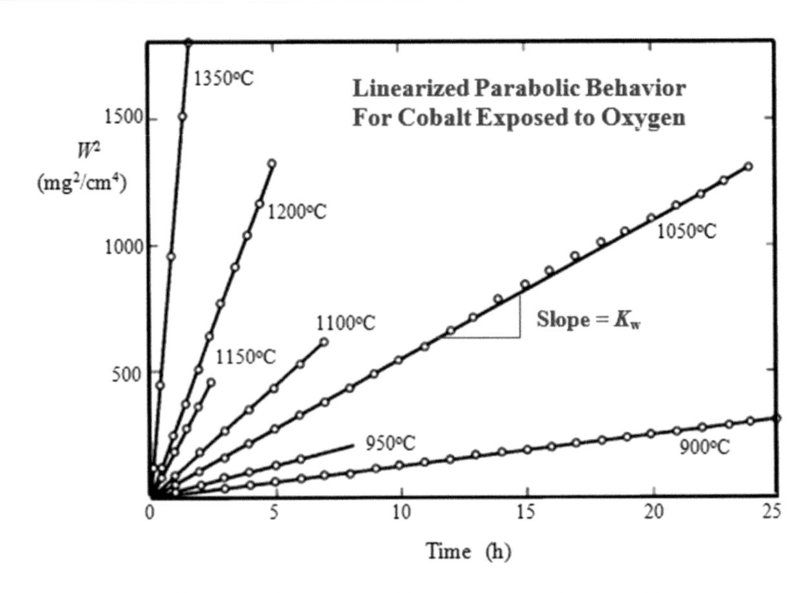

Fig. 10.10 Linearized weight gain for cobalt (Co) in oxygen environment [3]

Table 10.2 Parabolic rate constant values obtained in oxygen [4, 8]

Metal	Oxide	K_w ($g^2/cm^4\,s$)	($\mu g^2/cm^4\,h$)	T (°C)	P_{O_2} (atm)
Co	CoO	$2.10x10^{-10}$	0.76	1000	1.00
Cu	Cu_2O	$6.30x10^{-9}$	22.68	1000	0.08
Ni	NiO	$2.90x10^{-8}$	104.40	1000	1.00
Fe	FeO	$4.80x10^{-7}$	1728	1000	1.00
Fe	Fe_3O_4	$1.40x10^{-6}$	5040	1000	1.00
Cr	Cr_2O_3	$1.00x10^{-10}$	0.36	1000	1.00
Cr	Cr_2O_3	$1.30x10^{-11}$	0.0468	900	0.10
Si	$Si\,O_2$	$1.20x10^{-12}$	0.0043	1000	1.00
Al	Al_2O_3	$8.50x10^{-16}$		600	1.00

$$\frac{Q_w}{R} = 1.176\,°C = 274.176\,K \;\; \& \tag{10.71a}$$

$$Q_w \simeq 2280\,\frac{J}{mol} \quad \text{for cobalt } (Co) \text{ in oxygen gas} \tag{10.71b}$$

Thus, Q_w is high which means that diffusion occurs at a high rate for oxide growth at the high-temperature range $900\,°C \le T \le 1350\,°C$.

Table 10.2 lists some experimental parabolic rate constant values for common oxides. Using the K_w data for selected oxides at $T = 1000\,°C$ and $P_{O_2} = 1.00$ atm, Eq. (10.24) yields the parabolic profiles shown in Fig. 10.12.

Now, combine Eqs. (10.26) and (10.19) or (10.69) to get the oxide thickness growth as

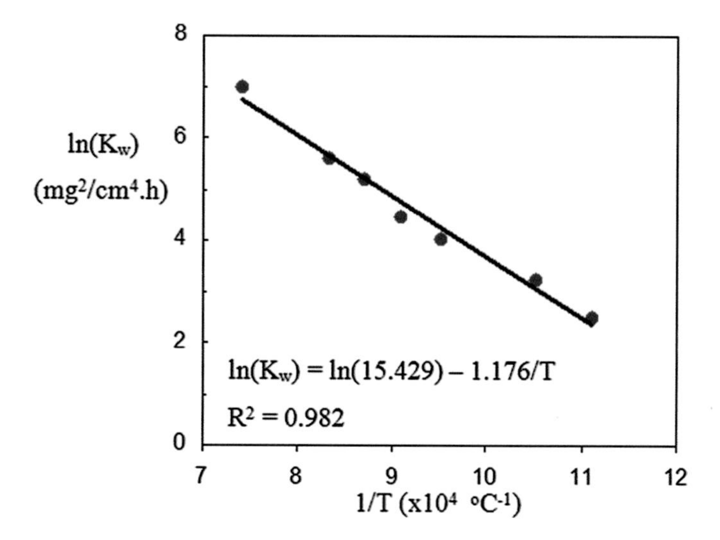

Fig. 10.11 Arrhenius plot for the parabolic rate constant in Fig. 10.10

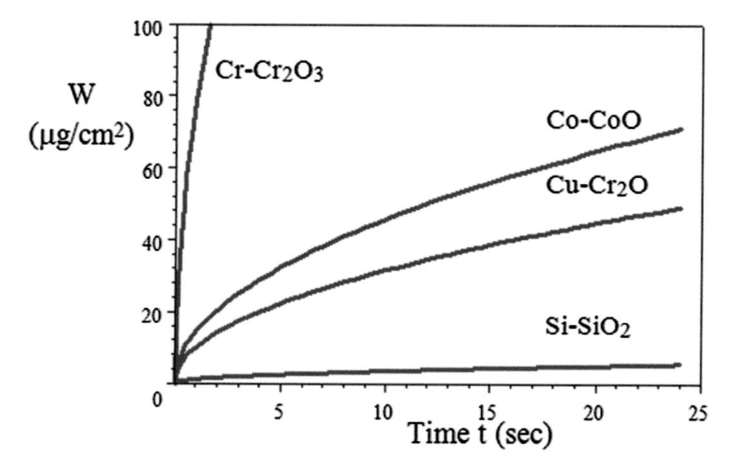

Fig. 10.12 Parabolic behavior due to weight gain of some oxides listed in Table 10.2. Conditions: $T = 1000\,^{\circ}\mathrm{C}$ and $P_{O_2} = 1$ atm

$$x = \frac{1}{\rho}\sqrt{K_w t} \tag{10.72}$$

which is an alternative expression for characterizing oxide thickness growth. Furthermore, the effects of the microstructure on stainless steels are of particular interest despite that chromium enhances oxidation resistance. Grain boundaries can provide suitable sites for accelerated oxidation, but chromium depletion at these boundaries aids the formation of chromium dioxide (chromia) Cr_2O_3 scale. Thus,

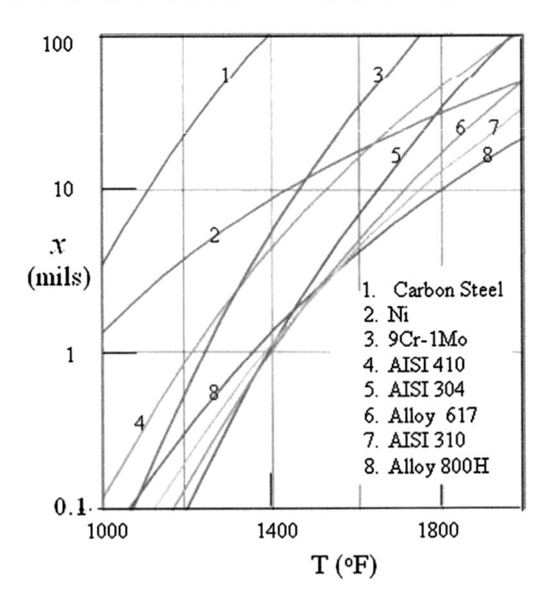

Fig. 10.13 Effect of high-temperature oxidation on the metal-oxide thickness growth of some steel alloys exposed to air (21% O_2) environment for 1 year. 1 mil = 0.25 mm [9]

the oxidation of stainless steels in high-temperature oxygen-rich environments strongly depends on composition and temperature, rather than on microstructure.

The service temperature of a particular stainless steel can be found elsewhere [22]. Recently published stainless steel, carbon steel, and nickel data on oxidation clearly exhibit increasing oxide thickness growth with increasing temperature as summarized in Fig. 10.13 after 1 year of exposure in air [9].

Both carburizing and nitriding are also oxidizing processes, which are important in industrial applications for enhancing the hardness and strength of steels. These methods are out of the scope of this chapter.

In general, it is known that stainless steels exhibit high oxidation resistance in gaseous environments. Using Rohrig et al. data [20], which can be found elsewhere [15], yields plots for oxide thickness formation on some steels exposed to **steam at** $595\,^{\circ}C$. This kinetic behavior is illustrated in Fig. 10.14. These metal oxides are considered as porous scales susceptible to be characterized as protective or nonprotective as per Pilling–Bedworth law being introduced in an earlier section. The data indicates that penetration is significantly reduced as chromium content increases, making the steels corrosion-resistant engineering materials. For instance, the oxide thickness x at 2000 h (vertical line) shows remarkable differences between these alloys.

Example 10.8. *Calculate the thickness of Al_2O_3 oxide when pure aluminum is exposed to oxygen environment at $600\,^{\circ}C$ and 1 atm for 1 day. Use the K_w value from Table 10.2 and $\rho_{Al_2O_3} = 3.80 g/cm^3$.*

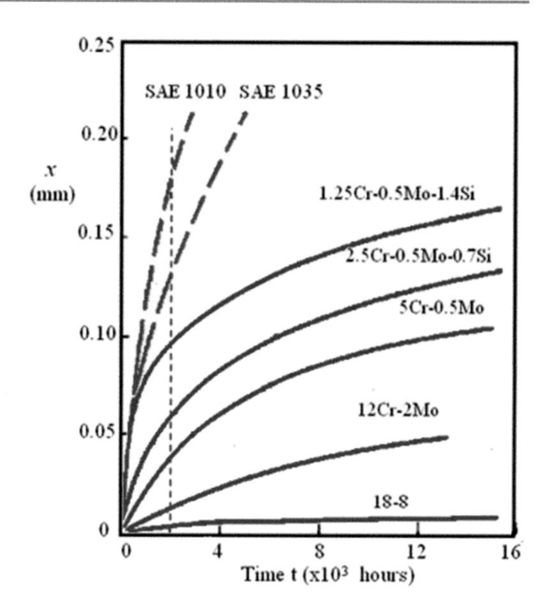

Fig. 10.14 Effect of time on the metal-oxide thickness of carbon and stainless steel bars containing different chromium contents and exposure to steam at 595 °C [20]

Solution.

From Table 10.2 $K_w = 8.50x10^{-16}\ \mathrm{g^2/cm^4\,s}$ *from Eqs. (10.16) and (10.26),*

$$K_x = \frac{K_w}{\rho^2} = \frac{8.50 \times 10^{-16}\ \mathrm{g^2/cm^4\,s}}{(3.80\ \mathrm{g/cm^3})^2}$$

$$K_x = 5.89 \times 10^{-19}\ \mathrm{cm^2/s}$$

From eq. (10.15) or (10.69),

$$x = \sqrt{K_x t} = \sqrt{(5.89 \times 10^{-19}\ \mathrm{cm^2/s})\ (86400\ \mathrm{s})}$$

$$x = 0.23\ \mu\mathrm{m}$$

10.8 Problems/Questions

10.1. A silicon oxide crucible is used to melt pure aluminum in the presence of oxygen at 1300 °C. **(a)** Will the silicon oxide ($Si\,O_2$) corrode? **(b)** What is the oxygen pressure? [Hint: Use Fig. 10.1.] [Solution: (b) $P_{O_2} = 5.71x10^{-7}$ atm $=$ $5.77x10^{-5}$ kPa.]

10.2. An alumina (Al_2O_3) crucible contains molten copper and oxygen (O_2) at 600 °C. Determine **(a)** if the crucible will corrode; if so, calculate **(b)** the oxygen pressure.

10.3. In the case of thick oxide formation at high temperature, the Pilling–Bedworth theory may have a limited applicability. Explain why this may be the case.

10.4. If the protective nature of an oxide film at room temperature is lost at relatively high temperatures, explain the sequence of the oxide thickening process.

10.5. Determine the values of K_w from Fig. 10.10 and plot K_w vs. T. What can you conclude from this plot?

10.6. Use the data given below to plot oxide molecular weight vs. K_p for $T = 1000\,°\mathrm{C}$ and $P_{O_2} = 1$ atm. What can you conclude from the plot?

Element	Atomic weight (g/mol)	Molecular weight (g/mol)	$K_w\ \left[(g^2O_2)/\mathrm{cm}^4\,\mathrm{s}\right]$
Co	58.93	$CoO \rightarrow 74.93$	$2.10x10^{-10}$
Cu	63.54	$Cu_2O \rightarrow 143$	$6.30x10^{-9}$
Fe	55.85	$FeO \rightarrow 71.85$	$4.80x10^{-10}$
Si	28.10	$Si\,O_2 \rightarrow 60.10$	$1.20x10^{-12}$

10.7. Calculate **(a)** the equilibrium constant K and **(b)** the dissociation oxygen pressure P_{O_2} for the oxidation of aluminum at 1100 °C. [Solution: (a) $K = 1.58x10^{31}$ and (b) $P_{O_2} = 6.35x10^{-32}$.]

10.8. If 100 g of pure aluminum (Al) oxidizes according to the reaction $4Al+3O_2 = 2Al_2O_3$, calculate the PB ratio defined by

$$PB = V_{Al_2O_3}/V_{Al}$$

where $V's$ are volumes. Data: $\rho_{Al} = 2.70\,\mathrm{g/cm^3}$ and $\rho_{Al_2O_3} = 2.70\,\mathrm{g/cm^3}$. [Solution: $PB = 1.34$.]

10.9. A chromium bar is exposed to oxygen gas at 900 °C. Calculate **(a)** the oxygen partial pressure and **(b)** the oxygen activity. [Solution: (a) $P_{O_2} = 6.92x10^{-7}$ kPa and (b) $a_{O_2} = 6.85x10^{-12}\,\mathrm{mol/cm^3}$.]

References

1. S.M. Allen, E.L. Thomas, *The Structure of Materials* (Wiley, New York, 1999), pp. 251–260
2. M. Barsoum, M.W. Barsoum, *Fundamentals of Ceramics* (Institute of Physics Publishing, Bristol, 2003)
3. F.R. Billman, J. Electrochem. Soc. **119**, 1198 (1972)
4. S.A. Bradford, Fundamentals of corrosion in gases, in *Corrosion*. ASM Handbook, vol. 13 (ASM International, Metal Park, 1992), pp. 64–115

5. Y.-M. Chiang, D.P. Bernie III, W.D. Kingery, *Physical Ceramics: Principles for Ceramic Science and Engineering* (Wiley, New York, 1997), pp. 101–184, 212–217
6. M.G. Fontana, *Corrosion Engineering*, 3rd edn. (McGraw-Hill, New York, 1986)
7. D.R. Gaskell, *An Introduction to Transport Phenomena in Materials Engineering* (Macmillan, New York, 1992)
8. G.H. Geiger, D.R. Poirier, *Transport Phenomena in Metallurgy* (Addison-Wesley, Readings, MA, 1973)
9. R.C. John, A.D. Pelton, A.L. Young, W.T. Thompson, I.G. Wright, T.M. Nesmann, Adv. Mater. Process. **160**(3), 27–31 (2002)
10. D.A. Jones, *Principles and Prevention of Corrosion* (Macmillan, New York, 1992)
11. A. Kelly, G.W. Groves, P. Kidd, *Crystallography and Crystal Defects* (Wiley & Sons, New York, 2000), pp. 290–298
12. W.D. Kingery, J. Pappis, M.E. Doty, D.C. Hill, J. Am. Ceram. Soc. **42**(8), 393–398 (1959)
13. P. Kofstad, in *High Temperature Corrosion*, ed. by R.A. Rapp (NACE, Houston, 1983)
14. S.Y. Liu, C.L. Lee, C.H. Kao, T.P. Perng, High-temperature oxidation behavior of two-phase iron-manganese-aluminum alloys. Corros. Sci. **56**(4), 339–349 (2000)
15. H.E. McGannon (ed.), *The Making, Shaping, and Treating of Steel*, 9th edn. (United States Steel Corporation, Pittsburgh, PA, 1971), p. 1197
16. C. Newey, G. Weaver (eds.), *Materials Principles and Practice* (Butterworth, The Open University, Boston, 1990), p. 356
17. P. Perez, J.A. Jimenez, G. Frommeyer, P. Adeva, Mater. Sci. Eng. **A284**, 138 (2000)
18. N.B. Pilling, R.E. Bedworth, J. Inst. Met. **29**, 529 (1923)
19. J. Regina, Adv. Mater. Process. **160**(3), 47 (2002)
20. I.A. Rohrig, R.M. Van Duzer, C.H. Fellows, Trans. ASME **66**, 277–290 (1944)
21. S. Seal, S.K. Bose, S.K. Roy, Oxid. Met. **41**(1-2), 139–178 (1994)
22. A.J. Sedriks, *Corrosion of Stainless Steels*, 2nd edn. (Wiley, New York, 1996), pp. 388–400
23. C. Wagner, Z. Physik Chem. **B21** (1933)

Solution of Fick's Second Law

A

The general diffusion equation for one-dimensional analysis under nonsteady state condition is defined by Fick's second law, Eq. (4.19). Hence,

$$\frac{\partial C}{dt} = D\frac{\partial^2 C}{\partial x^2} \tag{A1}$$

Let D be a constant and use the function $y = f(x,t)$ be defined by

$$y = \frac{x}{2\sqrt{Dt}} \tag{A2}$$

Thus, the partial derivatives of Eq. (A2) are

$$\frac{\partial y}{\partial x} = \frac{1}{2\sqrt{Dt}} \text{ and } \frac{\partial y}{\partial t} = -\frac{x}{4\sqrt{Dt^3}} \tag{a}$$

By definition,

$$\frac{\partial C}{\partial t} = \frac{dC}{dy}\frac{\partial y}{\partial t} = -\frac{x}{4\sqrt{Dt^3}}\frac{dC}{dy} \tag{b}$$

$$\frac{\partial^2 C}{\partial x^2} = \frac{\partial}{\partial x}\left[\frac{dC}{dy}\left(\frac{\partial y}{\partial x}\right)\right] = \frac{1}{4Dt}\frac{d^2 C}{dy^2} \tag{c}$$

Substituting Eqs. (b) and (c) into A1 yields

$$\frac{dC}{dy} = -\frac{\sqrt{Dt}}{x}\frac{d^2 C}{dy^2} \tag{A3}$$

N. Perez, *Electrochemistry and Corrosion Science*,
DOI 10.1007/978-3-319-24847-9

Combining Eq. (A2) and (A3) gives

$$\frac{dC}{dy} = -\frac{1}{2y}\frac{d^2C}{dy^2} \tag{A4}$$

Now, let $z = dC/dy$ so that Eq. (A4) becomes

$$z = -\frac{1}{2y}\frac{dz}{dy} \tag{a}$$

$$-2\int ydy = \int \frac{dz}{z} \tag{b}$$

Then,

$$-y^2 = \ln z - \ln B \tag{c}$$

where B is an integration constant. Rearranging Eq. (c) yields

$$z = B\exp(-y^2) \tag{A5}$$

and

$$\int dC = B\int \exp(-y^2)dy \tag{A6}$$

The function $f = \exp(-y^2)$ represents the so-called bell-shaped curves. The solution of the integrals are based on a set of boundary conditions.

A.1 First Boundary Conditions

In order to solve integrals given by Eq. (A6) a set of boundary conditions, the concentration and the parameter y, are necessary. These boundary conditions are just the integral limits. Thus,

$$C = \begin{cases} C_x = C_o & \text{for } y = 0 \text{ at } t > 0 \text{ and } x = 0 \\ C_x = C_b & \text{for } y = \infty \text{ at } t = 0 \text{ and } x > 0 \end{cases} \tag{a}$$

$$\int_{C_o}^{C_b} dC = B\int_o^{\infty} \exp(-y^2)dy \tag{c}$$

$$C_b - C_o = B\int_o^{\infty} \exp(-y^2)dy \tag{A7}$$

Use the following integral definitions and properties of the error function erf(y)

$$\int_{o}^{\infty} \exp(-y^2)dy = \frac{\sqrt{\pi}}{2} \tag{a}$$

$$\int_{-\infty}^{\infty} \exp(-y^2)dy = \sqrt{\pi} \tag{b}$$

$$erf(y) = \frac{2}{\sqrt{\pi}} \int_{o}^{y} \exp(-y^2)dy \tag{c}$$

$$\mathrm{erf}\,c(y) = \frac{2}{\sqrt{\pi}} \int_{y}^{\infty} \exp(-y^2)dy \tag{d}$$

$$\mathrm{erf}(y) + \mathrm{erf}\,c(y) = 1 \quad \text{and} \quad \mathrm{erf}(-y) = -\,\mathrm{erf}(y) \tag{e}$$

$$\mathrm{erf}(0) = 0 \quad \text{and} \quad \mathrm{erf}\,(\infty) = 1 \tag{f}$$

By definition, the function erf $c(y)$ is the complement of erf(y). Inserting Eq. (a) into (A7) yields the constant B defined by

$$B = \frac{2}{\sqrt{\pi}} (C_b - C_o) \tag{A8}$$

A.2 Second Boundary Conditions

The second set of boundary conditions is given below:

$$C = \begin{cases} C_x = C_x \text{ at } y < \infty \\ C_x = C_b \text{ at } y = \infty \end{cases} \tag{a}$$

Setting the limits of the integral given by Eq. (A6) and using Eq. (A8) yield the solution of Fick's second law of diffusion when the bulk concentration ($C_b = C_x$ at $x = \infty$) is greater than the surface concentration ($C_o = C_x$ at $x = 0$). Hence, the solution of Eq. (A1) for **concentration polarization** ($C_b > C_o$) becomes

$$\int_{C_b}^{C_x} dC = B \int_{\infty}^{y} \exp\left(-y^2\right) dy = -B \int_{y}^{\infty} \exp\left(-y^2\right) dy \tag{b}$$

$$C_x - C_o = -\frac{2}{\sqrt{\pi}} (C_b - C_o) \frac{\sqrt{\pi}}{2} \,\mathrm{erf}\,c\,(y) \tag{c}$$

$$\frac{C_x - C_b}{C_o - C_b} = 1 - \mathrm{erf}\left(\frac{x}{2\sqrt{Dt}}\right) \qquad \text{for } C_b > C_o \tag{A9}$$

A.3 Third Boundary Conditions

Similarly, the solution of Eq. (A1) for **activation polarization** $(C_o < C_b)$ upon using the boundary conditions given below as well as Eqs. (A6) and (A8) yields the normalized concentration expression:

$$C = \begin{cases} C_x = C_x \text{ at } y > 0 \\ C_x = C_o \text{ at } y = 0 \end{cases} \tag{a}$$

$$\int_{C_o}^{C_x} dC = B \int_0^y \exp\left(-y^2\right) dy \tag{b}$$

$$C_x - C_o = \frac{2}{\sqrt{\pi}} \left(C_b - C_o\right) \frac{\sqrt{\pi}}{2} \operatorname{erf}(y) \tag{c}$$

$$\frac{C_x - C_o}{C_b - C_o} = \operatorname{erf}\left(\frac{x}{2\sqrt{Dt}}\right) \qquad \text{For } C_o < C_b \tag{A10}$$

This concludes the analytical procedure for solving Fick's second law of diffusion for concentration and activation polarization cases.

Corrosion Cells

B

This Appendix includes three electrochemical cells shown in Figs. B.1, B.2 and B.3 for corrosion studies [1, 2]. Using the proper equipment and instrumentation, one can generate $i - E$ or $E - i$ diagrams for determining the polarization behavior of the working electrode (WE) against the counter electrode (CE). A potentiostat-galvanostat power supply along with an electrometer and a computerized system is a common equipment for conducting electrochemical studies.

In addition, plots like $E = f(t)$ and $I = f(t)$ or $i = f(t)$ can aid in assuring the cell instability during an experiment. It is recommended to connect an electrochemical noise device for avoiding electrical background noise. Hence, the main stream of corrosion studies is the electrochemical methods.

(a) **Two-electrode system**. This cell is suitable for galvanic corrosion between two different metals in a particular electrochemical environment. The power supply can be disconnected for natural galvanic electrochemical studies; otherwise, a forced galvanic corrosion is implemented. In either case, metal ions M_1 are electrodeposited on the cathode counter electrode.

(b) **Three-electrode system**. This cell is suitable for corrosion studies of a metal using a reference electrode and an external source electrons.

(c) **Four-electrode system**. This cell is suitable for corrosion studies using a power supply as the source of electric energy and electrons.

References

1. L.L. Shreir, Outline of electrochemistry, in *Corrosion, Vol. 1, Corrosion Control*, ed. by L.L. Shreir, R.A. Jarman, G.T. Burstein (Butterworth-Heinemann, Boston, 1994), pp. 1:76–1:84
2. A.J. Bard, L.R. Faulkner, *Electrochemical Methods: Fundamentals and Applications*, 2nd edn. (Wiley, New York, 2001), pp. 25–26

N. Perez, *Electrochemistry and Corrosion Science*,
DOI 10.1007/978-3-319-24847-9

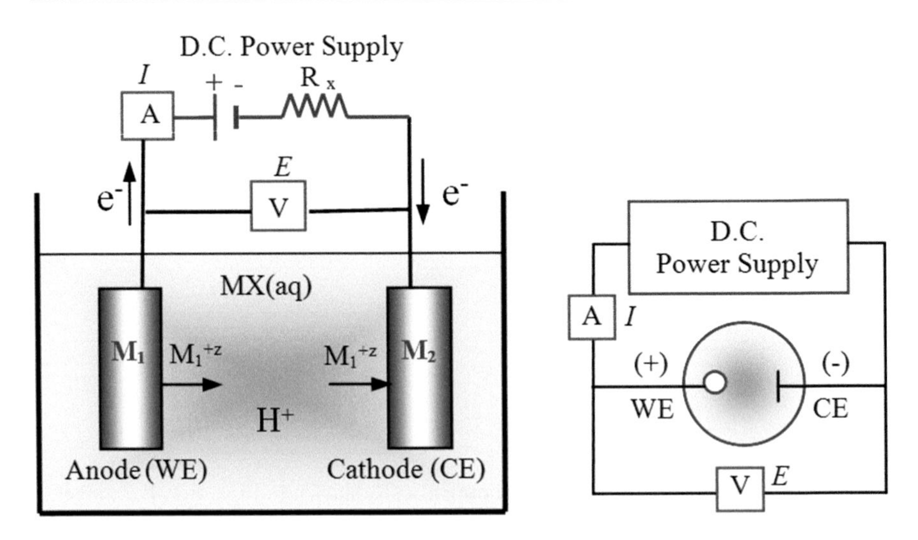

Fig. B.1 Two-electrode system along with the electrical notation

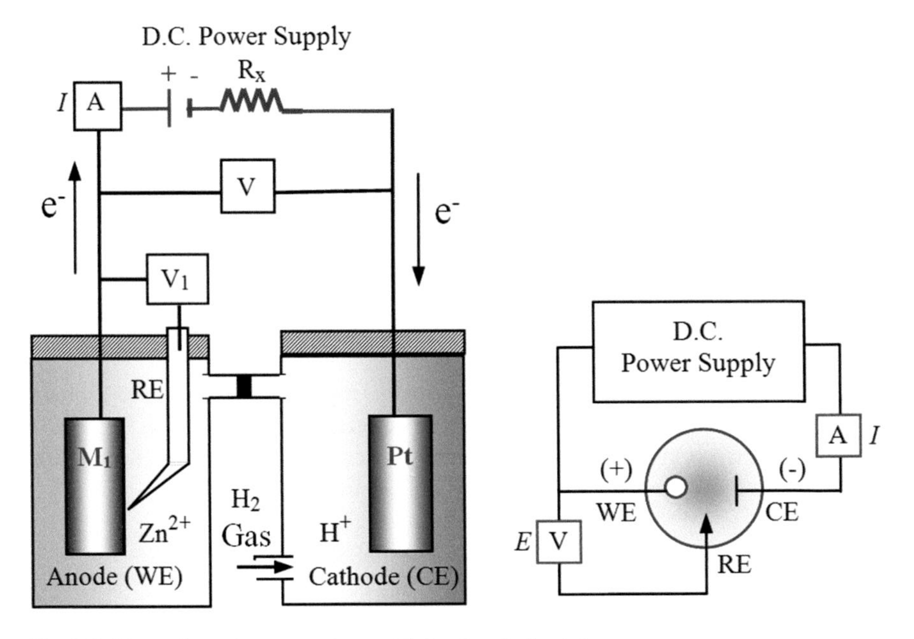

Fig. B.2 Three-electrode system along with the electrical notation

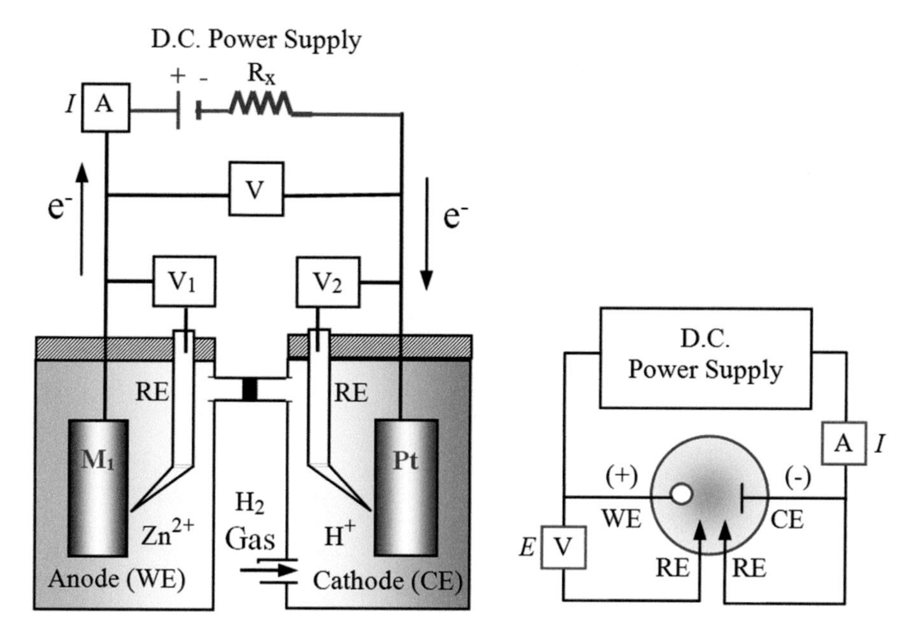

Fig. B.3 Four-electrode system along with the electrical notation

Conversion Tables

C

Prefixes

Prefix	Symbol	Factor	Prefix	Symbol	Factor
Exa	E	10^{18}	Milli	10^{-3}	
Peta	P	10^{15}	Micro	10^{-6}	
Tera	T	10^{12}	Nano	10^{-9}	
Giga	G	10^{9}	Pico	10^{-12}	
Mega	M	10^{6}	Femto	10^{-15}	
Kilo	k	10^{3}	Atto	10^{-18}	

Abbreviations

atm = atmosphere	gal = gallon	MPa = megapascal
A = ampere	hp = horsepower	N = Newton
Å = angstrom	h = hour	nm = nanometer
Btu = British energy	in = inch	Pa = pascal
C = Coulomb	J = joule	psi = pounds/square inch
°C = degree Celsius	K = Kelvin	s = second
cal = calorie	kg = kilogram	T = temperature
cm = centimeter	l = liter	W = watt or weight
eV = electron volt	min = minute	μ = micron
°F = degree Fahrenheit	mm = millimeter	μm = micrometer
g = gram	mol = mole	Ω = ohm = V/A

N. Perez, *Electrochemistry and Corrosion Science*,
DOI 10.1007/978-3-319-24847-9

Greek Alphabet

Alpha	α	A	Iota	ι	I	Rho	ρ	P
Beta	β	B	Kappa	κ	K	Sigma	σ	Σ
Gamma	γ	Γ	Lambda	λ	Λ	Tau	τ	Υ
Delta	δ	Δ	Mu	μ	M	Upsilon	υ	Y
Epsilon	ϵ	E	Nu	ν	N	Phi	ϕ	Φ
Zeta	ζ	Z	Xi	ξ	Ξ	Chi	$\varkappa$	χ
Eta	η	H	Omicron	o	O	Psi	ψ	Ψ
Theta	θ	Θ	Pi	π	Π	Omega	ω	Ω

Area

$1\ m^2 = 10^4\ cm^2$	$1\ cm^2 = 10^{-4}\ m$
$1\ m^2 = 10.76\ ft^2$	$1\ ft^2 = 0.0929\ m^2$
$1\ cm^2 = 0.155\ in^2$	$1\ in^2 = 0.4516\ nm^2$
$1\ cm^2 = 10^2\ mm^2$	$1\ mm^2 = 10^{-2}\ cm^2$

Corrosion Rate

1 mm/year = 39.37 mils/year (mpy)	1 mil/year = $2.54 \text{x} 10^{-2}$ mm/year
1 μm/year = $3.937 \text{x} 10^{-2}$ mils/year	1 μm/year = 25.40 mils/year

Current Rate

$1\ A/m^2 = 9.29 \text{x} 10^{-2}\ A/ft^2$	$1\ A/ft^2 = 10.764\ A/m^2$
$1\ A/cm^2 = 6.452\ A/in^2$	$1\ A/in^2 = 0.155\ A/cm^2$
$1\ A/mm^2 = 6.452 \text{x} 10^2\ A/in^2$	$1\ A/in^2 = 1.55 \text{x} 10^{-3}\ A/mm^2$
$1\ \mu A/cm^2 = 6.452\ \mu A/in^2$	$1\ \mu A/in^2 = 0.155\ \mu A/cm^2$

Density

$1\,kg/m^3 = 10^{-3}\,g/cm^3$	$1\,g/cm^3 = 10^3\,kg/m^3$
$1\,kg/m^3 = 0.0624\,lb_m/ft^3$	$1\,ft^3 = 16.02\,kg/m^3$
$1\,g/cm^3 = 62.40\,lb_m/ft^3$	$1\,lb_m/ft^3 = 0.016\,g/cm^3$
$1\,g/cm^3 = 0.0361\,lb_m/in^3$	$1\,lb_m/in^3 = 27.70\,g/cm^3$

Electricity and Magnetism

$d\phi/dx$ = V/cm	V = volts
1 mho = 1 S	1 S = 1 mho
$1\,S = ohm^{-1} = \Omega^{-1}$	$1\,ohm = 1\,S^{-1}$
$1\,\Omega\,cm = 1.00x10^{-2}\,\Omega\,m$	$1\,\Omega\,m = 100\,\Omega\,cm$
$1\,\Omega\,cm$ = 1 ohm cm	1 ohm = V/A
$1\,maxwell = 10^{-2}\,\mu weber$	$1\,\mu weber = 100\,maxwell$

Energy

$1\,J = 2.778x10^{-7}\,kW\,h$	$1\,kW\,h = 3.60x10^6\,J$
1 J = 0.239 cal	1 cal = 4.184 J
$1\,J = 9.48x10^{-4}\,Btu$	$1\,Btu = 1.05x10^3\,J$
$1\,J = 0.7376\,ft\,lb_f$	$1\,ft\,lb_f = 1.3558\,J$

Flow Rate

1 l/min = 2.1189 ft^3/h	1 ft^3/h = 0.4719 l/min
1 l/min = 3.53×10^{-2} ft^3/min	1 ft^3/min = 28.31
1 l/min = 15.85 gal/h	1 gal/h = 6.31×10^{-2} l/min
1 l/min = 0.2642 gal/min	1 gal/min = 3.7854 l/min
1 m^3/s = 2.1189×10^{3} ft^3/min	1 ft^3/min = 4.7195×10^{-4} m^3/s
1 m^3/s = 35.315 ft^3/s	1 ft^3/s = 2.8317×10^{-2} m^3/s
1 m^3/s = 3.66×10^{6} in^3/min	1 in^3/min = 2.73×10^{-7} m^3/s

Force

1 N = 10^5 dynes	1 dyne = 10^{-5} N
1 N = 0.2248 lb_f	1 lb_f = 4.448 N
1 N = 2.248×10^{-4} kips	1 kip = 4.448×10^{3} N
1 N = 0.1019 kg_f	1 kg_f = 9.81 N

Length

1 m = 10 Å	1 Å = 10^{-10} m	1 mil = 25.40 μm
1 m = 10^9 nm	1 nm = 10^{-9} m	1 yd = 0.9144 m
1 m = 10^6 μ	1 μ = 10^{-6} m	1 mile = 1.61 kg
1 m = 10^3 mm	1 mm = 10^{-3} m	
1 m = 10^2 cm	1 cm = 10 mm	
1 m = 39.36 in	1 in = 25.4 mm	
1 m = 3.28 ft	1 ft = 12 in	

Mass

1 kg = 10^3 g	1 g = 10^{-3} kg
1 kg = 2.205 lb_m	1 lb_m = 0.4536 kg
1 g = 2.205x10^{-3} lb_m	1 lb_m = 45.36 g
1 kg = 10^{-3} metric ton	1 metric ton = 10^3 kg
1 kg = 1.1023x10^{-3} short ton	1 short ton = 2x10^3 lb_m
1 kg = 9.8421x10^{-4} long ton	1 long ton = 1.016x10^3 kg
1 metric ton = 1.1023 short ton	1 lb_m = 5x10^{-4} short ton
1 metric ton = 0.98421 long ton	1 long ton = 1.106 metric ton
1 kg = 6.85x10^{-2} slug	1 slug = 14.59 kg

Physical Constants

Quantity	Symbol	Value
Acceleration of gravity	g	= 9.81 m^2/s
		= 32.2 ft/s^2
Avogadro's number	N_A	= 6.022x10^{23} particle/ mol
Boltzmann's constant	$k = R/N_A$	= 1.38x10^{-23} J/K
		= 8.62x10^{-5} eV/K
Electronic charge	q_e	= 1.602x10^{-19} C
Faraday's constant	$F = q_e N_A$	= 96,500 C/mol
		= 96,500 A s/mol
		= 96,500 J/mol V
Gas constant	$R = kN_A$	= 8.314 J/mol K
		= 1.987 cal/mol K
Plank's constant	h	= 6.626x10^{-34} J s
		= 4.136x10^{-15} eV s
Mass of electron	m_e	= 9.11x10^{-31} kg
Velocity of light	c	= 3x10^8 m/s

Power

$1\,W = 1\,J/s$	$1\,Btu/s = 1\,ft\,lb_f/s$
$1\,kW = 0.9478\,Btu/s$	$1\,Btu/s = 1.0551\,kW$
$1\,kW = 56.869\,Btu/min$	$1\,Btu/min = 1.758x10^{-2}\,kW$
$1\,kW = 3.4121x10^3\,Btu/h$	$1\,Btu/h = 2.9307x10^{-4}\,kW$
$1\,kW = 1.3405x10^{-3}\,hp$	$1\,hp = 746\,kW$

Pressure (Fluid)

$1\,Pa = 1\,N/m^2$	$1\,atm = 760\,mmHg, {}^\circ C$
$1\,Pa = 9.87x10^{-6}\,atm$	$1\,atm = 1.0133x10^5\,Pa$
$1\,MPa = 9.87x10^{-3}\,atm$	$1\,atm = 101.33\,MPa$
$1\,Pa = 1.00x10^{-5}\,bar$	$1\,bar = 1.00x10^5\,Pa$
$1\,Pa = 1.45x10^{-4}\,psi$	$1\,psi = 6.895x10^3\,Pa$
$1\,MPa = 0.145\,ksi$	$1\,ksi = 6.895\,MPa$
$1\,Pa = 7.501x10^{-3}\,torr\,(mmHg, {}^\circ C)$	$1\,torr = 1.333x10^2\,Pa$
$1\,Pa = 2.953x10^{-4}\,in\,Hg, 32\,{}^\circ F$	$1\,in\,Hg, 32\,{}^\circ F = 3.386x10^3\,Pa$
$1\,Pa = 2.0886x10^{-2}\,lb_g/ft^2$	$1\,atm = 14.7\,psi$

Temperature

$T(K) = T({}^\circ C) + 273.15$	$T(R) = T({}^\circ F) + 459.67$
$\Delta T(K) = \Delta T({}^\circ C)$	$T(R) = (9/5)T(K) + 459.67$
	$T({}^\circ F) = (9/5)T({}^\circ C) + 32$
	$T({}^\circ C) = (5/9)[T({}^\circ F) - 32]$

Velocity

1 m/s = 1.1811×10^4 ft/h	1 ft/h = 8.4667×10^{-5} m/s
1 m/s = 1.9685×10^2 ft/min	1 ft/min = 5.08×10^{-3} m/s
1 m/s = 3.281 ft/s	1 ft/s = 0.305 m/s
1 m/s = 39.37 in/s	1 in/s = 2.54×10^{-2} in/s

Volume

$1\ m^3 = 10^6\ cm^3$	$1\ cm^3 = 10^{-6}\ m^3$
$1\ cm^3 = 10^3\ mm^3$	$1\ mm^3 = 10^{-3}\ cm^3$
$1\ m^3 = 35.32\ ft^3$	$1\ ft^2 = 0.0283\ m^3$
$1\ cm^3 = 0.061\ in^3$	$1\ in^3 = 16.39\ in^3$
$1\ l = 10^3\ cm^3$ (cc)	$1\ cm^3 = 10^{-3}\ l$
1 US gal = 3.785 l	1 l = 0.264 gal

Glossary

Activation Energy The energy required for to initiate a reaction or process.

Activation Polarization An electrochemical condition controlled by the slowest step in a series of reaction steps.

Active Material A metal or alloy susceptible to corrode.

Active-Passive Material A metal or an alloy that corrodes to an extent and then passivates due to an oxide film formation on its surface.

Anion A negatively charged ion such as SO_4^{-2}.

Anode An electrode that oxidizes by liberating electrons or an electrode at which oxidation proceeds on its surface.

Anodic Dissolution A corrosion process caused by an anodic overpotential, which depends on the exchange current density.

Auxiliary Electrode An electrode used to provide a uniform current flow through the working electrode.

Battery A galvanic cell that converts chemical energy into electric energy.

Calomel Reference Electrode An electrode that has a potential dependent on chlorine anions.

Cathode An electrode at which reduction proceeds by gaining electrons.

Cation A positively charged ion such as Cu^{+2}.

Concentration Polarization Polarization due to changes in the electrolyte concentration at the electrode-solution interface.

Corrosion Surface deterioration or destruction due to metal loss caused by chemical or electrochemical reactions.

Corrosion Current Density The rate of electron exchange between cathodic and anodic electrodes at equilibrium.

Corrosion Potential The potential difference between the cathodic and the anodic electrodes at their corrosion current density.

Crevice A no-visible narrow area, such as a gap, crack, and fissure.

Crevice Corrosion A form of corrosion that occurs in crevices due to oxygen depletion and dirt.

Crystal Structure A three-dimensional array of unit cells.

Cu/CuSO$_4$ Reference Electrode An electrode that has a potential dependent on copper cations.

N. Perez, *Electrochemistry and Corrosion Science*,
DOI 10.1007/978-3-319-24847-9

Daniel Cell A galvanic cell used as battery to convert chemical energy into electric energy.

Degradation A deteriorative process that may represent permanent damage, such as metal dissolution in metals and swelling in polymers.

Diffusion Mass transport by atomic motion.

Dislocation A linear defect in crystalline materials that represents atomic misalignment.

Diffusion Flux The quantity of mass diffusing through a perpendicular unit cross-sectional area per time.

Driving Force A term used to represent the parameter that causes a process to occur.

Electrochemical Cell A system consisting of a cathode and an anode immersed in the electrolyte.

Electrolysis A reduction process caused by an external current flow.

Electrolyte A solution through which an electric current is carried by the motion of ions.

Electrolytic Cell An electrochemical cell with forced reactions to promote electrolysis.

Electromotive Force (emf) Series The ranking of the standard electrochemical cell potential of metallic elements.

Electrometallurgy The field of engineering that uses science and technology of electrolytic processes for recovering metals from solutions.

Electroplating A metallurgical process used for reducing metal cations on an electrode surface by electrolysis. The resultant product is a metallic coating.

Electrorefining A metallurgical process used for purifying metals from solutions by electrolysis.

Electrowinning A metallurgical process used for extracting or recovering metals from solutions by electrolysis.

Equivalent Weight A quantity determined by dividing the atomic weight by the valence of a metal.

Erosion Corrosion Surface deterioration caused by the combined action of chemical attack and mechanical wear.

Exchange Current Density The rate of electron exchange between cathodic and anodic reactions on an electrode at equilibrium.

Faraday A quantity of electric charge equals to $96{,}500\,C = 26.81$ Ah required to oxidize or reduce one equivalent weight.

Free Energy A thermodynamic quantity that is a function of internal energy and entropy of a system.

Galvanic Cell An electrochemical cell with spontaneous reactions.

Galvanic Corrosion Surface deterioration due to localized galvanic cells imparted by dissimilar microstructural phases or materials.

Galvanic Series The ranking of metallic materials to form galvanic couplings.

Imperfection A deviation from atomic order in crystalline materials.

Inhibitors Organic compounds or inorganic anions that form a protective coating in situ by reactions of the electrolyte and the corroding surface.

Intergranular Corrosion Localized metal dissolution along grain boundaries in polycrystalline materials.

Ion An electrically charged species *j* in an electrolyte.

Limiting Current Density The maximum rate of reduction possible for a given electrochemical system.

Liquid Junction Potential A potential difference between two ionic solutions.

Metallic Coating An electroplated or melted coating that acts as a corrosion-resistant coating or sacrificial coating for cathodic protection.

Mixed Potential An potential caused by anodic and cathodic electrochemical and simultaneous reactions on an electrode surface.

Molality (M) Concentration in moles per kilogram of a species *j* in solution.

Molarity Concentration in moles per liter (or moles per cubic centimeters) of a species *j* in solution.

Mole The quantity of a substance related to Avogadro's number.

Normality (N) Concentration in gram equivalent weight per liter of a species *j* in solution. For example, $1\ M\ H_2SO_4 = 2\ N\ H_2SO_4$ due to $2H^+$ ions in solution.

Open-Circuit Potential The reversible equilibrium potential of an electrode at its exchange current density value.

Oxidation The removal of electrons from an atom, ions or molecule.

Overpotential A potential difference between a working electrode and its open-circuit potential.

Passivity Loss of chemical reactivity of an electrode in an environment due to the formation of an oxide film its surface.

Penetration Rate The corrosion rate for thickness reduction per time in a corrosive environment.

pH A quantity used as a measure of acidity or alkalinity of an aqueous electrolyte.

Pipeline Coatings Thermoplastic coatings based on bituminous coal tar enamels or asphalt mastics.

Pitting Corrosion A very localized metal dissolution forming pits or holes on the surface.

Polarization Deviation of the electrode potential from its equilibrium caused by current flow.

Porcelain Enamel A fused or melted glass powder and, subsequently, cooled on a metal surface that acts as an inert vitreous-glass coating to water.

Potential A parameter used as a measure of voltage.

Reduction Addition of electrons to cations for recovering a metal in its atomic state.

Reference Electrode An electrode of known electrochemical characteristics used for measuring the potential difference between the of the working electrode against it.

Sacrificial Anode A metallic material that corrodes and cathodically protects a structure by liberating electrons.

Stress Corrosion Cracking A form of failure due to the combined action of a tensile stress and a corrosive environment on a metallic material.

Unit Cell A three-dimensional atomic arrangement forming a specific geometry.

Working Electrode An electrode exposed to an electrolyte at a finite current.

Index

N. Perez, *Electrochemistry and Corrosion Science*,
DOI 10.1007/978-3-319-24847-9